Cryocoolers

Part 2: Applications

THE INTERNATIONAL CRYOGENICS MONOGRAPH SERIES

General Editors

K. D. Timmerhaus, *Engineering Research Center
University of Colorado, Boulder, Colorado*

Alan F. Clark
*National Bureau of Standards, U.S. Department of Commerce
Boulder, Colorado*

J. L. Olsen, *Laboratorium für Festkörperphysik
Eidgenössische Technische Hochschule, Zurich, Switzerland*

Founding Editor

K. Mendelssohn, F.R.S. *(deceased)*

H. J. Goldsmid
   Thermoelectric Refrigeration
G. T. Meaden
   Electrical Resistance of Metals
E. S. R. Gopal
   Specific Heats at Low Temperatures
M. G. Zabetakis
   Safety with Cryogenic Fluids
D. H. Parkinson and B. E. Mulhall
   The Generation of High Magnetic Fields
W. E. Keller
   Helium-3 and Helium-4
A. J. Croft
   Cryogenic Laboratory Equipment
A. U. Smith
   Current Trends in Cryobiology
C. A. Bailey
   Advanced Cryogenics
D. A. Wigley
   *Mechanical Properties of Materials
   at Low Temperatures*
C. M. Hurd
   The Hall Effect in Metals and Alloys
E. M. Savitskii, V. V. Baron, Yu. V. Efimov,
M. I. Bychkova, and L. F. Myzenkova
   Superconducting Materials
W. Frost
   Heat Transfer at Low Temperature
I. Dietrich
   Superconducting Electron-Optic Devices
V. A. Al'tov, V. B. Zenkevich, M. G. Kremlev, and
V. V. Sychev
   Stabilization of Superconducting Magnetic Systems
G. Walker
   Cryocoolers, Part 1: Fundamentals
   Cryocoolers, Part 2: Applications

Cryocoolers

Part 2: Applications

Graham Walker

The University of Calgary
Calgary, Alberta, Canada

PLENUM PRESS • NEW YORK AND LONDON

Library of Congress Cataloging in Publication Data

Walker, G. (Graham), 1930–
  Cryocoolers.

  (The International cryogenics monograph series)
  Includes bibliographical references and indexes.
  Contents: pt. 1. Fundamentals — pt. 2. Applications. 1. Low temperature engi-
neering. I. Title. II. Series.
  TP482.W34 1983   621.5'9   83-2166

ISBN-13: 978-1-4684-4432-2   e-ISBN-13: 978-1-4684-4430-8
              DOI: 10.1007/978-1-4684-4430-8
              © 1983 Plenum Press, New York
          Softcover reprint of the hardcover 1st edition 1983

          A Division of Plenum Publishing Corporation
              233 Spring Street, New York, N.Y. 10013

High-capacity, three-stage Vuilleumier cycle cryocooler.
(Courtesy of Hughes Aircraft Co., Los Angeles.)

Foreword

The rapidly expanding use of very low temperatures in research and high technology during the last several decades and the concurrent high degree of activity in cryogenic engineering have mutually supported each other, each improvement in refrigeration technique making possible wider opportunities for research and each new scientific discovery creating a need for a refrigerator with special features. In this book, Professor Walker has provided us with an excellent exposition of the achievements of this period, the fundamental principles involved, and a critical examination of the many different cryogenic systems which have led to a new era of low-level refrigeration.

I feel fortunate to have had a part in the developments discussed in this book. During the early 1930s I constructed several rotary engines using leather vanes. Their performance was not good, but I was able to liquefy air. I had been impressed by the usefulness of leather cups in tire pumps and in Claude-type engines for air liquefaction. I was trying to find a way to avoid that part of the friction generated by a leather cup as a result of the radial force of the working gas on the cylindrical part of the cup. During the 1950s I built two efficient helium liquefiers in which essentially leather pistons were used. A steel core was encased in a stack of leather rings separated by thin steel rings of slightly smaller diameter, the stack of alternate leather and steel rings being compressed to provide a rigid piston which could be machined to fit the cylinder. The wearing quality was excellent. One of these liquefiers provided thousands of liters of liquid helium during about two years of use. Instead of becoming smaller in diameter because of wear, the diameter of the pistons actually increased by absorption of water. The necessity of controlling the water content was a disadvantage.

I was also intrigued by Heylandt's crowned piston concept during the 1930s but I ruled it out because of the sealing problem at the warm end of the piston. I had seen some of them in action in oxygen plants. A slight air leakage was of no importance but a helium leak of such magnitude could not be tolerated. Little did I realize then that by 1960 I would find an adequate solution of the leakage problem by the use of rubber O-rings. During the 1930s I experimented with free piston expanders and compressors and with diaphragm engines. When the war began in 1941 I was able to secure substantial funds for the development of oxygen generators and decided to see if Kapitza's engine could be useful. At least two attempts had been made to reproduce Kapitza's liquefier in the U.S. One was a complete failure and the other almost so.

I had tried several combinations of metals for the piston and cylinder. The problem was to produce a fit between piston and cylinder close enough to reduce leakage to a reasonable value but not close enough to encourage seizure. Of the four combinations I tried, Kapitza's choice of bronze and stainless steel was the least desirable. Kapitza found it necessary to employ a relatively wide annular gap and contrived a way for the piston to travel very rapidly during the power stroke so as to minimize leakage. By making both piston and cylinder of nitrided alloy steel (hardness = 90 RC) I was able to reduce the radial clearance by an order of magnitude and thus reduce leakage to a negligibly small value. It became possible, therefore, to use conventional gear for controlling the motion of piston and valves.

Several dozen engines with nitrided pistons and cylinders were manufactured for oxygen production during the war. After the war this type of engine was incorporated in helium liquefiers and was manufactured by Arthur D. Little and its successor, CTI-Cryogenics. Between 1947 and 1970, 365 units were marketed.

I started using oil-lubricated rubber O-rings to seal piston rods and valve stems in nitrogen and oxygen plants in 1950 and in helium liquefiers a few years later. Finally, I returned to the Heylandt type of piston, first in nitrogen plants, and in 1960 in a helium liquefier. The piston was made from a rod of laminated phenolic plastic (Micarta) about 2 ft long machined to fit loosely in a stainless steel cylinder sealed at the room-temperature end by a single O-ring. The advantages over the earlier engine proved to be enormous. The pistons and valves could be removed easily for inspection without disturbing the insulating vacuum or breaking any piping. There was no friction between piston and cylinder except when massive amounts of air or water gained access to the working fluid. The reliability was very high.

The loss of efficiency from the flow of gas to and fro in the annular space between the hot and cold ends was very slight and mostly from the mismatch of temperature between cylinder wall and piston surface. The thin (0.002–0.006-in.) annular passageway became an effective regenerator.

The inefficiency, as in other reciprocating engines, was mostly the result of irreversible heat transfer between the working gas and the walls of the expansion chamber in response to the relatively great temperature change the gas undergoes. Engines working at lower temperature levels tended to be more efficient because the drop in temperature of adiabatic expansion was smaller. In fact, the efficiency of the two-phase engine was apparently above 90%. Not only was the change in temperature very small, but also the vanishingly small heat capacity of cylinder and piston made the expansion truly adiabatic.

When I retired from MIT in 1964, I joined the engineering staff of A. D. Little and soon thereafter put together an experimental model of their standard liquefier equipped with Micarta pistons sealed by O-rings. A considerable number were produced over the next few years. Concurrently, I developed a slightly larger machine, again with Micarta pistons, which led to the 1400 series, completely replacing the earlier series. There are 200 of these units now in use.

A two-phase engine has replaced the Joule–Thomson valve in all of the recent machines I have built and in some that CTI has manufactured. The size of plant on which this change has been made ranges from 5 to 1000 l/hr. The gain in liquid production has been 25% to 33%.

One of the Naval Research Laboratory liquefiers has an inverted engine, that is the cold end of the cylinders is at the top. It was astonishing that the two-phase engine behaved normally and gave the expected improvement over the Joule-Thomson. It seemed strange to have liquid helium formed on top of the piston without appreciable loss by flowing down the annular gap toward the hot end of the piston. Apparently convection is very ineffective in thin layers of the order of 0.004 in. when so little time (a period of one stroke) is available.

There is a wealth of information in this book. The comprehensive Bibliography and Guide to Cryogenic Engineering Literature (Appendix II of Part 2) is indicative of the prodigious effort which Professor Walker has expended in preparation for writing this book.

Professor Walker and those who contributed certain chapters are to be congratulated on this scholarly and timely treatise on a very interesting subject.

S. C. Collins

Preface

The kindest thing to be said about this book is that it is like the curate's egg—good in parts. Many times in its composition I have felt I would be better engaged in *reading* rather than *writing* a book on cryocoolers. Perhaps my modest effort will stimulate others better qualified than I to do the job.

Completion is due entirely to Klaus Timmerhaus, Associate Dean of Engineering at the University of Colorado, and to Dr. Alan Clark of the National Bureau of Standards, Boulder, Colorado, joint editors of the International Cryogenics Monograph Series published by Plenum Press. They saw a draft of my first book on Stirling-cycle machines and flattered me with an invitation to write another on cryocoolers. With the starlight in my eyes I signed the contract that has been their rod to beat me with since. Now at the end I am most grateful for their unquenchable interest and enthusiasm despite the discouragements with which I have confronted them. They have proved the burr in my saddle that only a completed manuscript will remove.

Luster has been added to my effort by the substantial contributions of others. I am indebted first to Samuel Collins for the Foreword. Ray Radebaugh of the Thermophysical Properties Division of the National Bureau of Standards, Boulder, Colorado, contributed Chapter 11, dealing with his interest and expertise in the fundamentals of cryocooling and Chapter 12, dealing with the systems and techniques for achieving very low temperatures.

Fred Chellis, senior Applications Engineer of Cryogenic Technology Inc., Waltham, Massachusetts, has contributed Chapter 10, dealing with the practical problems of cryocooler design and operation. Fred was highly qualified to do this, having been around cryocoolers longer than he cared to remember and really should have been the author of this book. He died unexpectedly during the production of this volume; his death was a great loss to the cryocooler community.

Evgeny Mikulin, a Professor of Cryogenic Engineering at the Baumann Institute in Moscow, contributed Chapter 13. This review of the state of development in the USSR of cryogenic cooling engines was a most valuable contribution to offset my predominantly North American view. Mikulin spent two months visiting the University of Calgary in 1977 and we came to know him well. In a similar way Yoshihiro Ishizaki of the University of Tokyo contributed Chapter 14, a review of cryocooler development in Japan.

In Chapter 4, I have reproduced, with permission, a great deal of an unclassified report on Vuilleumier cooling engines by Ronald White of the U.S. Air Force Flight Dynamics Laboratory of the Wright-Patterson Air Force Base, Ohio.

Professor Kurti of the University of Oxford convinced me to use the descriptor *cryocoolers* rather than the more prosaic *cryogenic cooling engines* that came naturally to me.

I am most grateful to all these men.

The book is dedicated to a trio who in different ways, contributed much to my interest in this field. Aubrey Burstall, Professor Emeritus of the University of Newcastle, perceived with great foresight, the future significance of Stirling engines and initiated research at Newcastle on the subject soon after World War II. With much persistence and charm he secured delivery of the first Philips Stirling engine liquefier in Great Britain. It was my good fortune to install and research this machine under Burstall's guidance.

Jan Köhler led the team from Philips in their brilliant development of a commercial Stirling cryocooler. He is recognized as a principal figure in modern regenerative cryocooler development. I am most grateful to Köhler for the insight and inspiration I gained from his stimulating public presentations and technical writings.

Samuel Collins completes the trio to whom this book is dedicated. While serving as professor at the Massachusetts Institute of Technology, Collins invented a machine permitting the virtual routine production of liquid helium and thus unlocked the way to widespread superconducting research and development. When the ramifications are worked through and superconducting systems are commonplace, Collins' place in technology will surely be recognized as the equal of James Watt and the condenser: the development that prefaced the industrial revolution.

My own debt to Collins lies principally in his book *Expansion Machines for Low Temperatures*. By good fortune this became available simultaneously with the Philips engine and my need of it at Newcastle. I gained much insight and encouragement from his book.

Since those days, over twenty years ago, many have helped me to understand things better. My old friends Ted Finkelstein and William Beale come quickly to mind. More recently I have benefited much from mutual exchanges with Bill Martini, Costa Rallis, and E. H. Cooke-Yarborough.

My colleagues at the University of Calgary withstand, good-humoredly, my preoccupation with regenerative machines and my insistence on discussing with them matters I suspect are not their principal interest. In particular, John Kentfield is a worthy foil off which I have bounced many ideas. The Head of the Department, Dr. Peter Glockner, has helped in his encouragement and advice.

I wrote this text as my principal activity on a six-month sabbatical leave from the University of Calgary. I am most grateful to the University for the opportunity to devote myself to the task unhampered by my normal academic duties.

Many people helped in the production of the book. Bert Unterberger and his assistants worked tirelessly on the diagrams and illustrations contained herein. Karen Undseth, Edie Schulz, and Pamela Appleton did a good job transforming my chicken-scratching into a readable text. Marlene Stewart and Karen Undseth labored indefatigably in their customary fashion on the index, corrections to the original text, and the galley proofs thereby relieving me entirely of this onerous last lap. I am most grateful for all their help.

My children Josephine and Christopher have been deprived of their rightful allocation of my time but seem to have survived in good order.

Finally I owe my greatest thanks to my wife Ann. For over twenty years now she has heard *all* about Stirling and other engines and still manages a credible show of interest. With unfailing good humor she somehow, in the end, makes it all worthwhile.

G. Walker
Calgary, Alberta

Contents of Part 2

Chapter 9   **Some Aspects of Design**

Chapter 10   Practical Problems in Cryocooler Design and Operation

F. F. Chellis

Chapter 11   Fundamentals of Alternate Cooling Systems

Ray Radebaugh

Chapter 12 Very-Low-Temperature Cooling Systems
Ray Radebaugh

Chapter 13 Cryogenic Engineering and Cryocooler Development in the USSR
Evgeny Ivanovich Mikulin

**Chapter 14 Cryogenic Engineering and Cryocooler
Development in Japan**

Yoshihiro Ishizaki

Bibliography

**Appendix I Glossary of Terms for Cryocoolers and List of
Organizations**

**Appendix II Organizations Having Substantial Interest in
Cryocoolers and Cryocooler Manufacturing**

Appendix III Guide to the Cryogenic Engineering Literature

Contents of Part 1

Chapter 8

Heat Exchangers in Cryocoolers

INTRODUCTION

Nomenclature

Heat exchangers are devices which enhance the transfer of heat and are vital components of all cryocoolers. They exist in a wide variety of types, shapes, sizes, and arrangements and are made of all kinds of material. Metals are commonly used because of high thermal conductivity and relative ease of fabrication.

Heat exchangers can be broadly classified as recuperative heat exchangers or regenerative heat exchangers.

Recuperative heat exchangers have separate flow channels for different fluid streams, the hot fluid usually flowing in the opposite direction of the cold. Regenerative heat exchangers have a single set of flow channels through a porous solid matrix of finely divided wires or balls. Flow through the matrix occurs periodically with the hot flow following the cold flow. The continuous flow of both fluids requires use of two identical regenerators and the means to switch fluids at intervals. One regenerator then experiences the hot flow as the other is experiencing the cold flow.

Siemens conceived the counterflow recuperative heat exchanger in the 1850s. Earlier, in 1817, Stirling incorporated a regenerative heat exchanger in the initial Stirling engine. However, recuperative and regenerative heat exchangers of one form or another, commonly found in biological systems, have preceded these discoveries by several million years.

HEAT EXCHANGERS IN CRYOCOOLERS

A cryocooler usually contains several heat exchangers. The Claude cryocooler shown in Fig. 8.1 is typical. During compression the gas is cooled

1

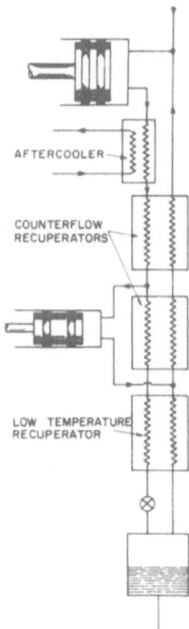

Fig. 8.1. Heat exchangers in Claude cycle system.

between stages in water (or air) cooled recuperative heat exchangers. Another, aftercooler, is used to cool the gas to ambient temperature after compression. The compressed gas then passes through two or three counterflow recuperative heat exchangers and is cooled progressively to low temperatures by heat exchange with the low-pressure effluent streams returning for recompression.

In the low-temperature region the cooled compressed fluid expands. Part of the gas expands in the expansion engine and the refrigeration generated is consumed in the intermediate stage recuperator. The remainder, cooled in the lower stage recuperator, expands through a JT valve and is partly liquefied. The cold vapor returns through the recuperators to the compressor. The liquid remains in a reservoir as the liquid product of the system.

The liquid may itself be the coolant for another recuperative heat exchanger. Liquid air or nitrogen is often used in the precooler for helium, hydrogen, or neon liquefiers. Alternatively the liquid may be removed from the liquid reservoir and used in a cryostat or other form of heat exchanger to provide cryogenic cooling to another fluid or component as required.

In the above system each counterflow recuperative heat exchanger could have been replaced by two regenerators having provision for switching the fluids at regular intervals. This system, first proposed by Frankl and used by the German Linde company in the 1930s, is suited for large-scale low-pressure air liquefaction systems. The contaminants (water vapor, carbon dioxide, etc.) in the air precipitate on the cold surfaces of the regenerator matrix during the hot flow of compressed air *en route* to the cryogenic region. Later, after the flow is switched, the contaminants sublime or reevaporate during the cold flow of low-pressure vapor returning to ambient temperature and eventual discharge to the atmosphere.

Regenerative cycle systems (Stirling, Vuilleumier, Solvay, Postle, and Gifford–McMahon) utilize both recuperative and regenerative heat exchangers. Consider the single-cylinder piston-displacer Stirling cryocooler shown in Fig. 8.2. It consists of two spaces, the expansion space and the compression space, coupled through a regenerative heat exchanger. The reciprocating elements (piston and displacer) operate in the proper phase relationship that the working fluid is mainly in the compression space when the piston is ascending and reducing the total system volume (compression). When the piston is descending (expansion) the displacer descends with the piston and the working fluid is mainly in the expansion space.

To facilitate heat transfer, additional heat exchangers, the cooler and the freezer, are provided, one above and one below the regenerator. These are both recuperative exchangers. The function of the cooler is to enhance the transfer of heat (at ambient temperature) *from* the system (the heat of compression). The cooler may be supplied with either water or air at ambient temperature to remove heat from the system.

The function of the freezer is to facilitate the transfer of heat (at low temperature) *to* the system. The heat thereby transferred is the useful refrigerating effect (the *raison d'etre*) of the cooling system. The heat may

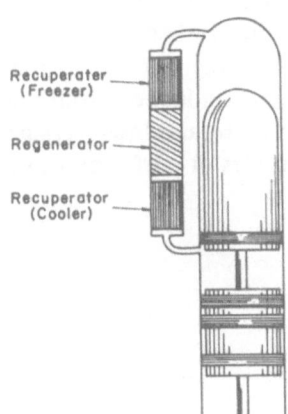

Fig. 8.2. Heat exchanger in Stirling cycle system.

be simply transferred from the surroundings of the cylinder head freezer to produce local cooling. Alternatively, the freezer may be incorporated in a circulatory fluid recuperative heat exchanger to provide continuous cooling to a flow of gas or liquid.

It is important to note in the Stirling cryocooler that the fluid does not move in the same direction continuously as in the Claude cryocooler. Rather, the fluid flows in a complex way sometimes in one direction, sometimes in the other between the expansion space to the compression space. At other times it flows in both directions simultaneously, from the regenerator to both the expansion and compression space. This is called *oscillatory* or *tidal flow* and complicates the design of components such as heat exchangers.

TYPES OF HEAT EXCHANGERS USED IN CRYOCOOLERS

Recuperative Exchangers

Recuperative exchangers may be generally classified as tubular, plate and fin, or perforated plate. Tubular exchangers exist in a range of forms utilizing metal tubes. Tubes are readily available in various metals, are strong, light-weight, and can be easily fabricated into complex assemblies using relatively straightforward techniques (drilling, brazing, etc.).

Plate–fin exchangers are formed from stacks of alternate layers of corrugated, die-formed metal plates closed at the edges with solid metal bars of appropriate shape. The whole assembly is brazed to form a rigid matrix containing fluid flow passages with integral headers welded to the ends.

Perforated plate exchangers consist of stacks of parallel perforated metal plates of high thermal conductivity with gaps between the plates formed by spacers of plastic or other low-conductivity material. The spacers are firmly bonded to the metal plates. They serve not only to minimize axial conduction from one plate to another but also to separate the high-pressure and low-pressure gas streams and confine their flow to particular clusters of perforations in the plates.

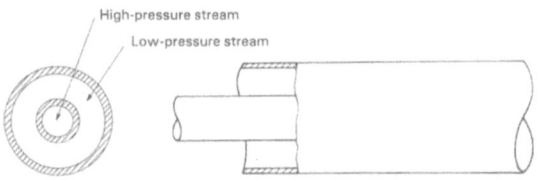

Fig. 8.3. Coaxial tube or double pipe heat exchanger (after Scott, 1966).

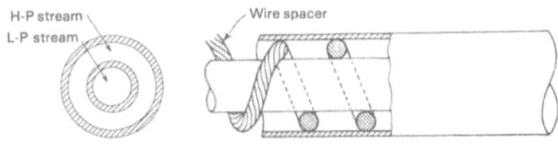

Fig. 8.4. Double pipe heat exchanger with spiral spacer (after Barron, 1966).

Tubular Exchangers. The simplest form of tubular exchanger is the coaxial tube or double pipe heat exchanger shown in Fig. 8.3. Used as a counterflow heat exchanger the high-pressure fluid is contained in the inner tube to minimize use of heavy wall high-pressure tubing. The low-pressure fluid is contained in the annular space between the tubes.

A wire or plastic spacer is sometimes spirally wound on the inner tube as shown in Fig. 8.4. This locates the inner pipe within the outer pipe to assure uniformity of flow and causes low-density fluid in the annulus to follow a long spiral path increasing both the velocity and heat transfer. The gain must be compensated for by increased pressure drop or *pump work.*

Several inner tubes may be contained within a larger diameter outer tube as shown in Fig. 8.5 to form a multitube heat exchanger. A variety of spiral windings, spacers, radial fins, and the like may be used to locate the tubes in the outer tube. Some further development in this direction results in the familiar industrial shell and tube heat exchanger shown in Fig. 8.6. The example shown has only one tube pass. Frequently, however, tube and shell exchangers have two tube passes permitting the use of hair-pin shaped tubes with inlet and outlet of the tube fluid from the same end of the exchanger. Tube and shell exchangers of the type shown in Fig. 8.6 would not be used for cryogenic recuperative counterflow exchangers but as the water-cooled interstage or aftercoolers of the compression system.

Another multitube arrangement is the "cluster" or "bundle" heat exchanger in which a number of tubes containing the low-pressure fluid are clustered around the core tube containing high-pressure fluid. The

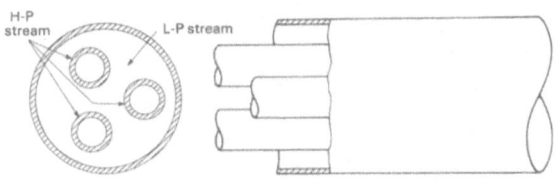

Fig. 8.5. Multitube heat exchanger (after Scott, 1966).

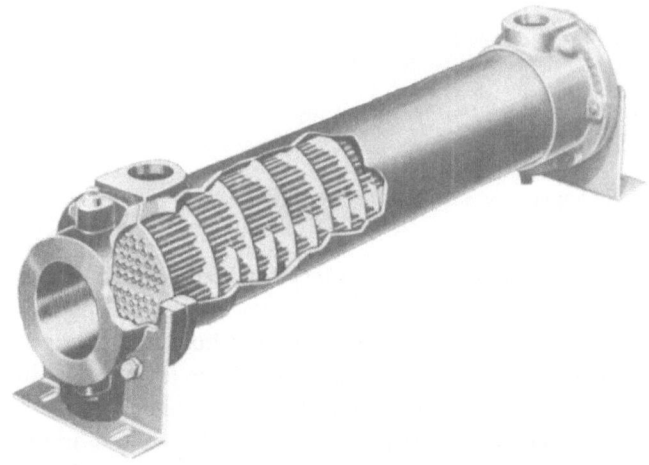

Fig. 8.6. Shell and tube heat exchanger (after Holman, 1976).

tubes are in good thermal contact, being secured by a high conductivity medium, soft solder or conducting Epoxy cement.

For many situations the length of tube required is simply too great for straight tube to be used. Coiled tube exchangers are therefore preferred for cryogenic use. In one form, a long length of coaxial or concentric tube is wound into a spiral coil like a spring and contained within an insulating vessel. In this way a very long length of double pipe heat exchanger can be contained in a relatively short, large-diameter cylinder with comparative freedom in selection of the overall dimensions. This type of coiled exchanger was first used by Linde (1895) in the liquefaction of air and is still occasionally found.

The preferred form of coiled tubular heat exchanger is the Hampson type shown in Fig. 8.7. A coil of thin-wall, small-diameter copper, aluminum, or stainless steel tube is wound onto an inner nonconducting

Fig. 8.7. Hampson coiled tubular heat exchanger (after Scott, 1966).

mandrel or former. A close-fitting sheath of nonconducting material is placed around the outside of the coil and insulation added to the exterior. The high-pressure gas flows *through* the tube and the low-pressure fluid stream flows back over the outside of the helical coil of tube between the inner mandrel and the outer sheath. In miniature systems a single tube may be used (as shown in Fig. 6.6) but for larger systems a number of parallel tubes are used in a single layer or in a number of layers around the mandrel. In their discussion of coiled tube heat exchangers Abadzic and Scholz (1973) show two very large coil tube exchangers with literally hundreds of parallel, small-diameter, coil tubes.

In the construction of Hampson coiled tube heat exchangers it is important to make the tube spacing uniform. Otherwise the low-pressure return gas will flow preferentially through apertures with the largest spacing and thereby decrease the effectiveness of the exchanger. An important advance in the technology of Hampson heat exchangers came with the introduction, by Giauque at the University of California, of the punched brass spacers shown in Fig. 8.8. Giauque used thin strips of cellulose acetate between the tube and the spacer as the tube was wound on. Later the acetate was dissolved away with acetone. Several layers of tubes were used to provide many parallel paths and so reduce the frictional effect. The length of the path was made constant by varying the pitch of the tube helix around the mandrel. Exchangers of this type are frequently called Giauque–Hampson exchangers. Very large exchangers of this type were shown by Kenoldt (1965) in a discussion of European cryogenic practice.

Another variant of the Hampson coiled tube exchanger is shown in Fig. 8.9. This was proposed by Parkinson of the United Kingdom Royal Radar Establishment, Malvern. Lengths of small-diameter tubing are first wound into closely coiled helices and then wound onto a mandrel in much the same way as the classical Hampson construction. The return stream passes through the annular space between the inner mandrel and the outer sheath and is in good contact with the high-pressure coils within that space.

A novel form of heat exchanger element developed by Collins is shown in Fig. 8.10. It consists of a close coiled helix of copper ribbon edge-wound with minimum pitch around a copper tube and soft soldered to it. The tube and ribbon assembly is enclosed in another tube around the outside of which is yet another edge-wound ribbon assembly. This material is manufactured commercially by the Joy Manufacturing Co., Michigan City, Indiana and multiple annular channels are available as shown in Fig. 8.11.

A single tube and ribbon assembly constituting a finned tube may be coiled in classical Hampson fashion around a former and enclosed within a sheath. Collins *et al.* (1958) described this form of construction with the illustration reproduced in Fig. 8.12. Three hundred ft or 0.25-in.-diam

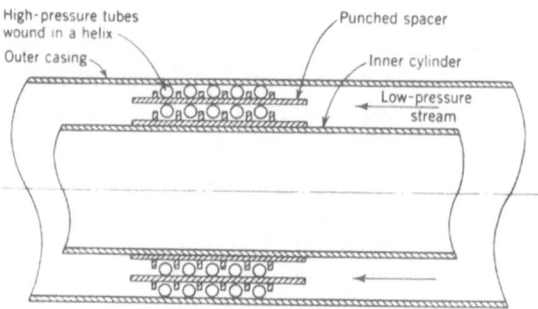

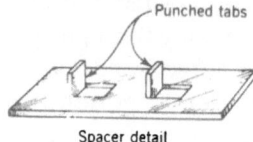

Spacer detail

Fig. 8.8. Giauque spacers for Hampson coiled tube heat exchanger (after Barron, 1966).

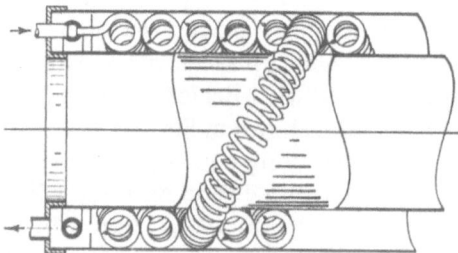

Fig. 8.9. Parkinson coiled tube heat exchanger (after Scott, 1966).

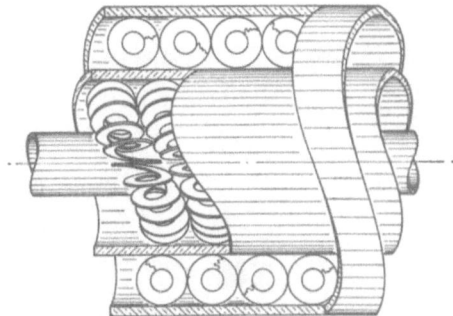

Fig. 8.10. Collins heat exchanger element
(after Scott, 1966).

tubing was formed into a helix approximately 9 in. in diameter and 42 in. long. A cotton cord spiral in the inner and outer interstices of adjacent tubes was used to divert the fluid flow and increase the flow path around the fins to enhance the heat transfer. Later, Collins (1966) described a similar form of construction where the cotton cord was replaced by triangular-shaped tube carrying a third stream of fluid (hydrogen) at low pressure.

Plate–Fin Exchangers. Plate–fin heat exchangers consist of stacks of alternate sandwiches of corrugated die formed metal sheets (fins) between flat metal separation sheets. Typical construction details of a single element are shown in Fig. 8.13. The edges of each element are sealed with bars of appropriate shape brazed to the separator plates to form an integral rigid structure. The flow passages formed within the corrugations are terminated with headers welded or brazed at each end of the stack.

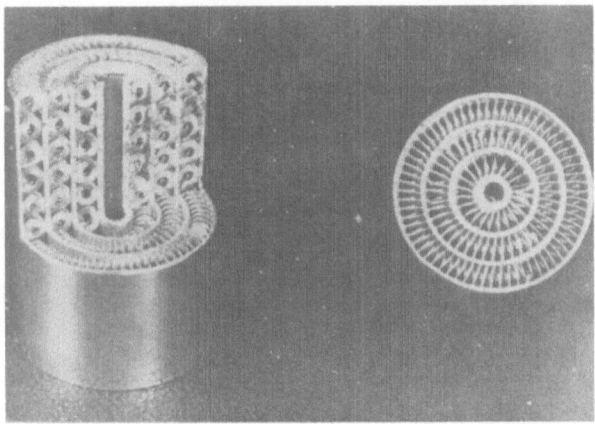

Fig. 8.11. Triple annulus Collins heat exchanger element (after Barron, 1966).

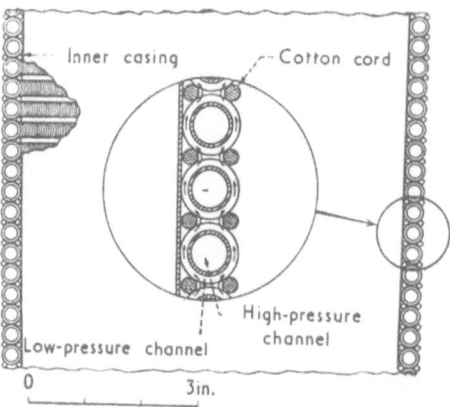

Fig. 8.12. Finned tube counterflow heat exchanger (after Collins, 1958).

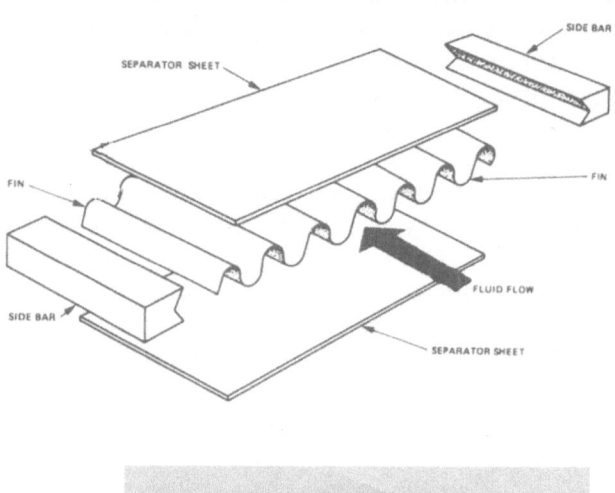

Fig. 8.13. Construction details of a single element of a plate-fin heat exchanger (after Lenfestey, 1961).

Fig. 8.14. Crossflow and counterflow arrangement of plate–fin heat exchanger (after Lenfestey, 1961).

Various stacking arrangements achieve different flow arrangements as illustrated in Fig. 8.14. An endless variety of corrugation geometries may be incorporated. Stacks of plate–fin elements may be assembled in modules of convenient size and then joined and sealed in a dip-brazing process. The modules may be coupled into larger units by welding to produce the large assemblies shown in Fig. 8.15.

Perforated Plate Exchangers. The principle of the perforated plate exchanger is shown in Fig. 8.16. It consists of a large number of parallel perforated plates with gaps between the plates. The plates are made of high-conductivity metal (copper or aluminum) but the spacers between the plates are constructed of low-conductivity plastic. The spacers are bonded to the metal plates and serve to minimize thermal conduction between the plates and confine the high-pressure and low-pressure flows to selected clusters of perforations on the plates. The fluids flow in counterflow and heat transfer occurs laterally from one stream to another *through* the plates.

Very small perforations can be made in the plates so that a large heat transfer surface area per unit volume can be developed. The gaps between plates allow a very uniform flow across the section. Moreover, since the

Fig. 8.15. Two-stream countercurrent block plate–fin heat exchanger in course of manufacture (after Lenfestey, 1961).

length/diameter ratio is very small for each hole in the plate the thermal and hydrodynamic boundary layers are broken up before they have a chance to become fully developed. This results in very high heat transfer surface coefficients (and consequently, high friction factors). Exceptionally compact heat exchangers can be made for a given duty.

Further details of the design and construction of perforated plate exchangers have been given by Wittner (1966) and Fleming (1969). The use of woven wire screens in perforated plate type exchangers was described by Vonk (1969) and by Lins and Elkan (1975).

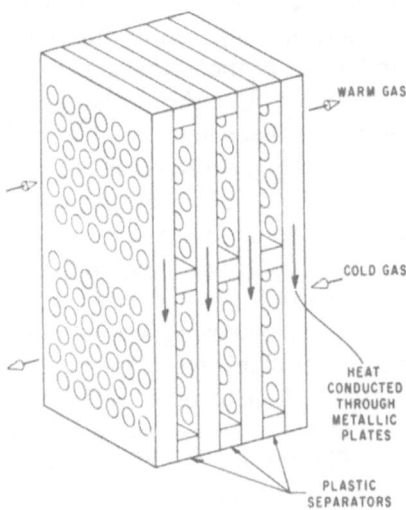

Fig. 8.16. Principle of the perforated plate exchanger.

Fundamentals of Recuperative Theory. The rate of heat transfer in a recuperative heat exchanger in steady state operation may be represented simply by

$$q = UA\Delta T_m \qquad (8.1)$$

where q is the rate of heat transfer, U is the overall heat transfer coefficient, A is the surface area for heat transfer consistent with the definition of U, and ΔT_m is a suitable mean temperature difference across the heat exchanger. This simple but useful equation can be thought of as intuitively based. Clearly the area, A, for heat transfer is important and should be increased to the maximum possible within the bounds of size, weight, cost, and manufacturing limitations. The temperature difference, ΔT, between the fluids has to be a critical parameter. The greater the temperature difference, the greater the amount of heat transferred. The overall heat transfer coefficient U, is a composite factor embracing all the many other parameters affecting the operation of the exchanger. The value is frequently determined experimentally from prototype testing. The values of U and A are interrelated insofar as U is defined in terms of some reference area, A, of the exchanger, usually the internal or external area of the tubes in, say, a tubular-type exchanger.

The temperature difference ΔT_m is usually the logarithmic mean temperature difference (LMTD) defined as

$$\text{LMTD} = \frac{\Delta T_{\max} - \Delta T_{\min}}{\ln (\Delta T_{\max}/\Delta T_{\min})} \qquad (8.2)$$

where ΔT_{max} is the greatest local temperature difference and ΔT_{min} is the least local temperature difference between the two fluid streams at inlet and outlet of the exchanger.

Figure 8.17 shows the possible temperature distribution in a double pipe heat exchanger in various modes of operation. The simplest case shown in Fig. 8.17a is for the special case where hot fluid is condensing at T_h and cold fluid is evaporating at T_c. The temperatures are constant in passage through the exchanger and the temperature difference $\Delta T = T_h - T_c$ and ΔT_{max} is equal to ΔT_{min}.

Figure 8.17b is drawn for the case where one fluid is changing phase, hot fluid condensing or cold fluid boiling, as a result of heat transfer with the other fluid which increases or decreases in temperature in passage through the expander. In both these cases the temperature difference at one end is very different from that at the other.

Figures 8.17c and 8.17d show the temperature distribution when neither fluid changes phase but heat is transferred from one to the other,

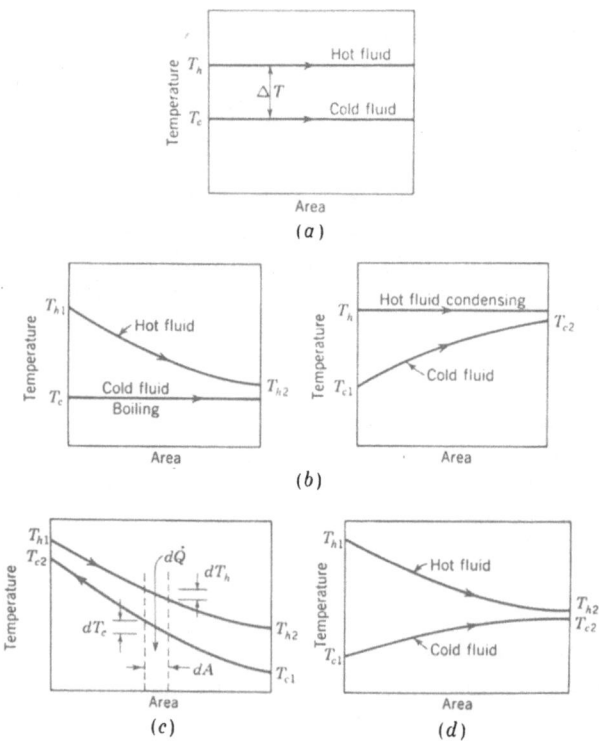

Fig. 8.17. Temperature distribution in a double-pipe heat exchanger in various modes of operation (after Barron, 1966).

cooling one and heating the other. Two cases are possible. In Fig. 8.17c the fluids flow in opposite directions (termed counterflow). In Fig. 8.17d the fluids both flow in the same direction (termed parallel flow). In counterflow the temperature difference is approximately the same throughout the exchanger. It is not exactly the same because of differences in *capacity rate* (the product of the mass rate of flow and the specific heat at constant pressure C_p) of the two fluids. The fluid with *minimum* capacity rate experiences the *greatest* temperature change so as to satisfy the balance of the heat gained by the cold fluid = heat lost by the hot fluid, or

$$(\dot{m}C_p)_c(T_{c_2} - T_{c_1}) = (\dot{m}C_p)_h(T_{h_1} - T_{h_2}) \tag{8.3}$$

In parallel flow the temperature difference between the fluids diminishes continuously in passage through the heat exchanger from a maximum at inlet $(T_{h_1} - T_{c_1})$ to a minimum at exit $(T_{h_2} - T_{c_2})$.

An important point is that in counterflow exchangers the exit (maximum) temperature of the cold fluid can be well in excess of the exit (minimum) temperature of the hot fluid. This cannot be achieved in parallel flow without contravention of the second law of thermodynamics. (Heat will not flow from a cold body to a hot body.) In parallel flow with fluids of equal capacity rate the best that can be achieved is a temperature change of the hot and cold fluids equal to half the maximum temperature difference $(T_{h_1} - T_{c_1})$. In the counterflow exchanger, the greatest temperature change that could be achieved theoretically is equal to the difference between the hot inlet and cold inlet temperature $(T_{h_1} - T_{c_1})$. For the same quantity of heat transfer the counterflow exchanger need be only half the size of a parallel flow exchanger. It is thermodynamically impossible for parallel flow exchangers to induce the large changes in temperature that are characteristic of cryogenic systems.

The Linde coaxial tube exchanger is ideally suited for counterflow operation. The Hampson coil tube type is not a counterflow type of exchanger. The pressurized fluid *in* the tube moves in a direction along the axis of the tube, but the low-pressure fluid passes over the outside of the tube in a direction perpendicular to the tube axis. This is called cross-flow. Plate–fin heat exchangers can be arranged as cross-flow, parallel flow, or counterflow, as required. For cross-flow heat exchangers the same basic principles apply but clearly the situation is more complicated and involves rather more complex theoretical considerations than we have the space for here.

For design purposes the overall heat transfer coefficient U is most reliably determined from experience with similar types of exchanger or experimentally in prototype testing. If this is not possible an approximate value of U may be estimated. For the simplest example consider a double pipe heat exchanger. The equation for estimating a value for U for this case is

$$U_i = \left[\frac{1}{h_i} + A_i \frac{\ln (r_o/r_i)}{2\pi kL} + \frac{1}{h_o}\frac{A_i}{A_o}\right]^{-1} \qquad (8.4)$$

(after Holman, 1976, Chapter 10), where A_i is the inside area of the inner tube, A_o is the outside area of the outer tube, r_i is the internal radius of the inner tube, r_o is the external radius of the inner tube, k is the thermal conductivity of the material of the inner tube, L is the length of the inner tube, h_i is the heat transfer coefficient between fluid in the inner tube and the tube wall, and h_o is the heat transfer coefficient between fluid in the annulus between the tubes and the inner tube wall. The term $[\ln (r_o/r_i)]/2\pi kL$ in Eq. (8.4) represents the thermal resistance to heat flow through the inner tube wall. In many cases the tubes are made of copper, aluminum, or thin-wall stainless steel having a relatively high thermal conductivity. The thermal resistance is therefore minimal and can be frequently ignored without serious error. Similarly the use of relatively thin walls means that the ratio (A_i/A_o) approaches unity. Equation (8.4) then reduces to

$$U = \left(\frac{1}{h_i} + \frac{1}{h_o}\right)^{-1} = \frac{h_i h_0}{h_0 + h_i} \qquad (8.5)$$

The heat transfer coefficient h for convective heat transfer between a fluid and the solid wall of a tube when there is a temperature and velocity difference between the fluid and the tube may be determined from the dimensionless relationship

$$\mathrm{Nu} = B \cdot \mathrm{Re}^n \, \mathrm{Pr}^m = hd/k_f \qquad (8.6)$$

where Nu is the Nusselt number, defined as $\mathrm{Nu} = hd/k_f$, B is a constant characteristic of the flow regimen (i.e., laminar, turbulent, tubular, flat plate, annular, etc.), Re is the Reynolds number defined as $\mathrm{Re} = \rho_f U d/\mu_f$, Pr is the Prandtl number defined as $C_p \mu_f/k_f$, n is an index characteristic of the flow regimen, m is an index characteristic of the fluid, h is the heat transfer coefficient (heat rate per unit area per unit temperature difference between the tube and fluid), d is some characteristic dimension (tube diameter, length of plate, etc.), k_f is the thermal conductivity of the fluid, u is the velocity difference between the fluid and the wall, μ_f is the viscosity of the fluid, and C_p is the specific heat at constant pressure of the fluid. Depending on the flow situation the value of the convective heat transfer coefficient, h, can vary widely from minimal values of less than 5 W/m^2 C for natural convection to rates in excess of 100,000 W/m^2 C for some condensing systems. It is exceedingly dangerous to assume values unless the flow situation is closely specified. Even where the flow situation can be specified it is necessary to resort to empirical relations presented in the heat transfer literature. Confidence limits no better than $\pm 20\%$ should be assumed for first time use of empirical relations to a new or unusual heat exchanger situation.

Exchanger Effectiveness. A useful measure of performance of heat exchangers is the *effectiveness* defined as

$$\text{effectiveness } \varepsilon = \frac{\text{heat actually transferred}}{\text{heat available for heat transfer}}$$

By way of example consider the double-pipe heat exchanger in counterflow operation. The inlet and outlet temperature of the hot fluid are T_{h_1} and T_{h_2}, respectively. The inlet and outlet temperatures of the cold fluid are T_{c_1} and T_{c_2}, respectively. The fluid with the minimum capacity rate will experience the maximum temperature change so as to satisfy the equation

$$C_{\min} \Delta T_{\max} = C_{\max} \Delta T_{\min} \tag{8.7}$$

where $C_{\min}$ is the minimum capacity rate $(\dot{m}C_p)_{\min}$, $C_{\max}$ is the maximum capacity rate $(\dot{m}C_p)_{\max}$, $\Delta T_{\max}$ is the maximum change in temperature of the hot or cold fluid, and $\Delta T_{\min}$ is the minimum change in temperature of the hot or cold fluid. Let the cold fluid have the minimum capacity rate. Then the temperature of the cold fluid at exit T_{c_2} could, theoretically, increase to the inlet temperature of the hot fluid T_{h_1}. In that case

$$\varepsilon_c = (\dot{m}C_p)_c (T_{c_2} - T_{c_1}) / (mC_p)_c (T_{h_1} - T_{c_1}) \tag{8.8}$$

or

$$\varepsilon_c = (T_{c_2} - T_{c_1}) / (T_{h_1} - T_{c_1}) \tag{8.9}$$

By further appropriate algebraic manipulation (Holman, 1976, Chapter 10) it can be shown that for the counterflow double-pipe heat exchanger, the effectiveness is expressed by

$$\varepsilon = \frac{1 - \exp\{(-UA/C_{\min})[(1 - C_{\min})/C_{\max}]\}}{1 - (C_{\min}/C_{\max})\exp\{(-UA/C_{\min})[(1 - C_{\min})/C_{\max}]\}} \tag{8.10}$$

where all terms are as defined before. The grouping of terms $(UA/C_{\min})$ is called the number of transfer units (NTU) and is indicative of the size of the heat exchanger.

The effectiveness–NTU approach is one method for thermal design of heat exchangers. Another method is based on the logarithmic mean temperature difference. The LMTD method is more appropriate when all the required temperatures and mass flows are known and the *size* of exchanger is required. The effectiveness-NTU approach is best suited when the outlet temperatures are to be determined in a situation where the inlet flows and temperatures are known for flow in a given exchanger. Both methods can be used for both situations but in the second case the LMTD method will probably require an iterative solution.

Use of the effectiveness-NTU approach is facilitated by reference to charts produced by Kays and London (1966) and are reproduced in all elementary heat transfer texts (e.g., Holman, 1976, Chapter 10). Kays and London do not include charts for Hampson-type coil tube heat exchangers. The author is unaware of any existing derivation of the effectiveness/NTU equation [similar to Eq. (8.10)] above for Hampson-type exchangers or of effectiveness/NTU charts similar to those mentioned above. In the absence of a specific chart for Hampson coil tube exchangers the author has used the one given by Kays and London for crossflow exchangers with one fluid mixed and one fluid unmixed (specifically Fig. 2.15 in Kays and London, 1966). In heat exchanger parlance a fluid is mixed if no special provision is made to maintain the fluid in separate streams (i.e., tubes, fins, etc.). In the Hampson exchanger the fluid flowing in the tube coils is unmixed. The low-pressure fluid flowing over the outer tube is mixed. Caution must be exercised about applying the chart to coil-type exchangers, particularly in the small, single-layer versions. This is because the existence of the inner and outer sheaths forming the annulus for low-pressure gas flow must exert a large "wall effect" on the exchanger operation.

Design of Recuperative Heat Exchangers. The above review was intended simply to give those not familiar with the field a brief sketch of the important principles. No specific design data and procedures were included on the grounds that excellent texts exist and are widely available. Inclusion of selected design data will lead to its use in situations where it may not be appropriate on grounds of simple expediency—half a loaf is better than nothing. By not including *any* design data one hopes the reader will be led to consult the reference texts discussed below.

The prime repositories of design data and procedures for cryogenic heat exchangers are the proprietary design manuals of the industrial heat exchanger manufacturers. These are based on extensive experience and test data but are closely guarded and not available in the public domain.

The best compilation of design data and procedures in the open literature is *Compact Heat Exchangers* by Kays and London (1966). It is invaluable, and every heat exchanger designer should have it. The book contains heat transfer and friction data for about 100 compact heat exchanger surfaces and much information on design procedures. Other useful texts include *Heat Exchangers Design and Theory Sourcebook*, Afgan and Schlunder (1974), and *Industrial Heat Exchangers*, Walker (1982).

None of these texts are directed specifically to cryogenic heat exchangers. Much of the data and theory is common but there are special problems which have increased emphasis in cryogenic applications. Kays' and London's work originated after the Second World War with the emergence of the gas turbine and the need for compact heat exchangers for exhaust gas/air preheaters. The book edited by Afgan and Schlünder is a compilation of contributions by experts in the field and embraces every aspect of heat exchanger design and theory. It is basically the proceedings of a conference at the International Center for Heat and Mass Transfer, Dubrovnik, Yugoslavia.

The need for very high exchanger effectiveness is among the special problems of cryogenic heat exchangers. In many industrial heat exchangers an effectiveness of 0.8 is perfectly acceptable, and indeed, many operate with values substantially less. In cryogenic exchanger applications a very high effectiveness, of the order of 0.95 and greater, is required. This was illustrated by Barron (1966) in a numerical example for a simple Linde–Hampson system. He found that a 5% decrease in exchanger effectiveness to 0.95 reduced the liquid yield from 0.0872 to 0.0620 and the figure of merit from 0.1413 to 0.0951. A decrease in exchanger effectiveness of 5% caused a 33% decrease in the figure of merit!

Another topic with increased emphasis for cryogenic exchangers is the effect of temperature on fluid properties, particularly specific heat and viscosity. This was discussed in general terms by Kays and London and by Barron with regard to the effects of variable specific heat on exchanger performance.

Timmerhaus and Schoenhals (1974) have contributed an excellent general review article on the design and selection of cryogenic heat exchangers. An important review article by Lenfestey (1961) contains useful design data and discussions of construction methods of cryogenic exchangers. Further details and data were provided in a subsequent paper (Lenfestey, 1968). Abadzic and Scholz (1973) in a review of coil tube heat exchangers included design data for heat transfer in the crossflow of air outside tubes. They also included design data for the increased heat transfer and pressure drop resulting from turbulent flow in a helically wound tube relative to that in a straight tube. This paper is highly recommended for those interested in coil tube exchangers.

Maldistribution of Flow. Most recuperative heat exchangers contain short multiple flow passages for one or both fluids instead of a single long flow passage for each fluid. Except for the very smallest sizes, where a single tube will suffice, most coil tube exchangers have at least three and sometimes many more tubes in parallel, coiled in the helical form. The tubes are fed from toroidal headers located at the ends of the helical coil.

This multitube construction can be clearly seen in all three of the Hampson coil tube exchangers shown in Fig. 8.18. Similarly it is clear from Figs. 8.13 and 8.14 that plate–fin heat exchangers comprise a multiplicity of parallel, identical flow paths for fluid flow between the inlet and outlet headers.

Ideally the fluid flow would be proportioned exactly equal in every parallel channel. In practice it is not possible to ensure equal flow in each channel. The fluid naturally follows the path of least resistance and the major flow is concentrated in channels that are slightly larger in dimension or have smoother walls to offer less frictional drag or which are less contaminated. This phenomenon is called maldistribution of the flow. It is not specific to cryogenic heat exchangers and occurs in all cases of fluid flow in parallel conduits.

A consequence of the maldistribution of flow is to change the local rates of heat transfer and hence the local temperature distribution of the fluid and the material in the exchanger. This was illustrated by Cowans (1974) with the diagrams reproduced in Fig. 8.19 showing the effects of maldistribution on the local temperature distribution. For most heat exchangers in conventional service the effect of maldistribution is barely perceptible unless there is gross unsymmetry of the flow conduits. It becomes more noticeable as the exchanger performance is raised (increasing the number of transfer units, NTU) to gain high exchanger effectiveness. At very high NTU and effectiveness the impact of maldistribution becomes severe.

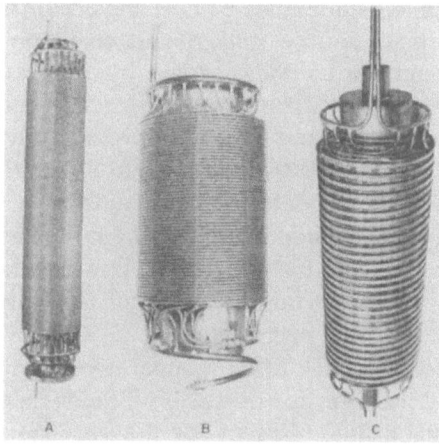

Fig. 8.18. Three Hampson coil tube heat exchangers illustrating the multitube, parallel coupling form of construction (after Scott, 1966).

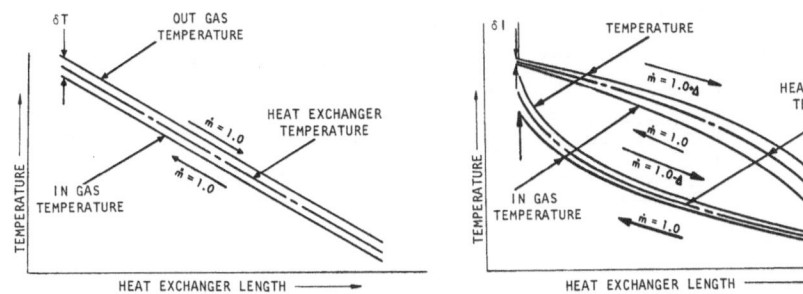

Fig. 8.19. Effect of maldistribution of the flow in the local temperature distribution in a heat exchanger with multiple flow conduits coupled in parallel (after Cowans, 1974).

There is an imperative need for high effectiveness in cryogenic heat exchangers. Therefore awareness of the phenomena of flow maldistribution with multiple parallel conduits is more important to the cryogenics engineer than others concerned with less rigorous applications.

Cowans (1974) attempted to estimate the magnitude of the effect. He noted that it is difficult to fabricate heat transfer surfaces to a dimensional tolerance better than ±5%. Even this requires a normal plate–fin surface 2 mm thick to be held to ±0.10 mm to satisfy the 5% criterion and is typical of the best engineering practice. The close dimensional tolerance would represent a variation in flow impedance of 10% to 20% in a typical exchanger with laminar flow conditions. Cowans concluded the maximum variation in flow would never be less than ±10% and this limit would be difficult to achieve in practice. A more realistic variation would likely be ±20%. At very high NTU and effectiveness the impact of this order of maldistribution is sufficient to reduce effectiveness by four or five percentage points (i.e., 0.98 to 0.94). The dramatic effect on the cycle performance of such a reduction was discussed above.

Cowans outlined a novel approach to solution of the problem of maldistribution. It requires the geometry of the hot and cold flow to be selected so that a shift in temperature in the middle of the exchanger effects only the flow impedance of the hot fluid. Two possible geometries are shown in Fig. 8.20 (after Cowans, 1974). Both operate to increase the impedance to flow of the hot fluid as the temperature increases and thereby reduce the flow rate. A corresponding decline in impedance with temperature will result in an increase in the flow rate. A negative feedback loop has been established that, according to Cowans, can almost completely eliminate the effects of maldistribution. He included a brief discussion of the history of a supercompact high-efficiency recuperative cryogenic heat exchanger incorporated in a life-support system. The original exchanger design achieved an effectiveness of 0.95. This corresponded to an NTU

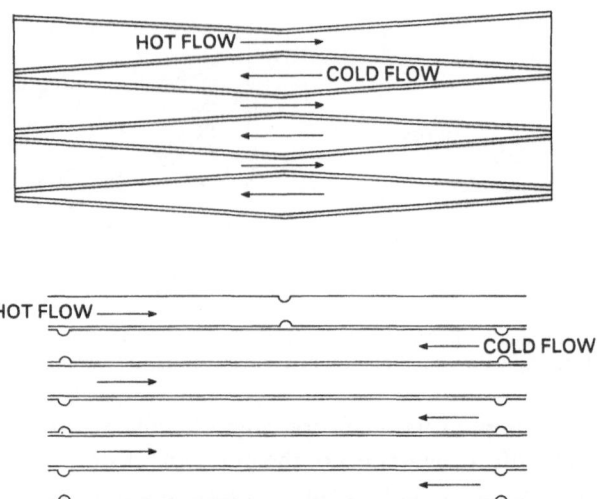

Fig. 8.20. Heat exchanger geometry to counter maladjustment of flow (after Cowans, 1974).

value of 20 compared with the design value of 200. Particular attention to close tolerance fabrication improved the performance to an effectiveness of 0.97, corresponding to an NTU of 33. Then another exchanger, more or less identical to the others, was constructed to incorporate feedback compensation. On test, this unit achieved an effectiveness of 0.994, corresponding to an NTU of 167. This was a fivefold increase in NTU compared with no feedback compensation.

Fleming (1967) produced a milestone paper outlining the effects of flow maldistribution on the effectiveness of high-performance cryogenic heat exchangers. He acknowledged that the problem of poor flow distribution causing a degradation of heat exchanger performance had long been known, with regard to air liquefaction plants, and provided useful references to early studies relating flow distribution to exchanger performance. In the main, these were concerned with exchangers of low NTU where the effects of unequal flow distribution are virtually negligible.

Another important contribution to the study of maldistribution in high-performance cryogenic heat exchangers was made by Weimer and Hartzog (1973). They described a method for simulating both coil wound and plate–fin exchangers and presented detailed results of their computations for selected exchanger configuration.

Axial Heat Conduction. Axial heat conduction along the exchanger passage walls is another factor important in the design of cryogenic heat exchangers but generally neglected in conventional applications. The effect of axial heat conduction was illustrated by Kroeger (1967a) in the figure

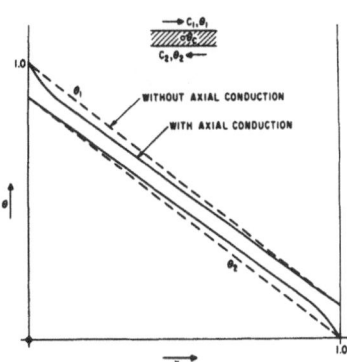

Fig. 8.21. Effect of axial heat conduction on the temperature profile of a recuperative heat exchanger (after Kroeger, 1967).

reproduced in Fig. 8.21 showing idealized temperature distribution of fluids in a recuperative counterflow heat exchanger. The broken lines refer to the case where heat conduction is ignored and the full lines to the situation existing when axial heat conduction is included. The temperature difference between the fluids is generally decreased by the effects of heat conduction. In high-performance heat exchangers, the fluid temperature difference is very small (by the very definition of high-performance exchanger). Decrease of the idealized temperature difference due to axial conduction can only contribute to a deterioration of exchanger effectiveness. The effect is particularly severe in exchangers having large heat exchange surfaces and short flow passages.

An elementary appraisal of the effects of axial conduction is included in the book by Kays and London (1966). A more comprehensive treatment was given by Landau and Hlinka (1960). Kroeger (1967a) extended this further with particular reference to cryogenic heat exchangers. The mathematical relationships involved are relatively complex and Kroeger has provided useful design charts for both balanced and unbalanced flow exchangers.

The effects of axial heat conduction can be mitigated with materials of low thermal conductivity for the passage walls. This is contrary to the requirements for high thermal conductivity of the passage walls in the radial direction to minimize thermal impedance between the fluid streams. Conventional construction materials do not possess asymmetric thermal properties.

An interesting design approach to resolving this paradox was described by Kroeger (1967b). He called it the plated tube heat exchanger, and the basic principle is illustrated in Fig. 8.22. The fluids flow in opposite directions of adjacent small-diameter, thin-wall tubes of relatively low-conductivity material. The adjacent tubes are coupled intermittently by a plating of high thermal conductivity. The plated sections provide good thermal

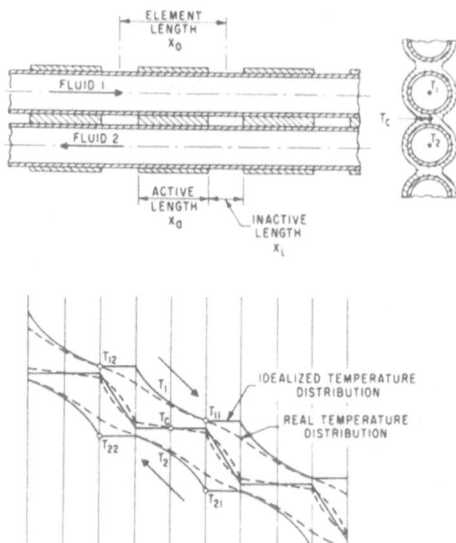

Fig. 8.22. Plated tube heat exchanger concept (after Kroeger, 1967b).

connection to enhance the heat transfer from one fluid to another. The nonplated sections provide a high impedance to thermal flow along the tubes. The equivalent of a structural material with asymmetric thermal conductivity has thereby been created.

Kroeger provided details of a parametric design study of a plated tube exchanger and included a photograph of the elements of a prototype exchanger for a Brayton cycle cryogenic cooling system. No prototype test results were included and none have been published in the open literature. The plated tube exchanger is difficult and expensive to manufacture compared with conventional tubular or plate–fin construction so its use will likely remain confined to small special-purpose high-effectiveness applications.

Friction Effects. When fluid flows in a conduit, there is frictional resistance to flow and work is required to maintain the flow. The effect of friction is manifest in a pressure drop along the conduit. The significant consequence of frictional pressure drop in a cryocooler is to reduce refrigeration generated in the expansion space because the pressure for expansion is reduced. Friction depends on the square of the velocity of the fluid, the area of the flow conduit, and nature of the surface of the wall (i.e., rough, smooth, etc.). This latter effect is accommodated by use of a *friction factor*, f, characteristic of a particular surface geometry and determined experimentally.

Efforts to increase heat transfer invariably increase the friction effects. In a given volume more wall area for heat transfer may be provided. The cross section of the flow passages may be reduced to increase the fluid velocity. Special wall surfaces may be provided to promote turbulence to enhance the heat transfer. All these effects will certainly increase the heat transfer, but the price to be paid is increased pressure drop, and that price may be high. Increasing the fluid velocity will increase the heat transfer, but at a rate somewhat less than the proportional increase in velocity. The pressure drop, on the other hand, increases at a rate proportional to the square of the velocity and the frictional power expended by as much as the cube of the velocity and never less than the square. As Kays and London (1966) point out in their introduction,

> ... for low density fluids, such as gases it is very easy to expend as much mechanical energy (*compression work*) in overcoming friction power as is transferred in heat. And it should be remembered that in most thermal power systems mechanical energy is worth four to ten times as much its equivalent in heat ...

Exactly the same reasoning applies just as strongly in cryogenic cooling systems. Increase in the frictional pressure drop represents aerodynamic heating of the very same fluid we are trying to cool and will directly reduce the pressure available to generate refrigeration in expansion.

Additional pressure losses arise from the abrupt changes in flow section typically found at the inlet and outlet of compact heat exchangers. Further losses arise from sudden changes in the direction of fluid flow, and in passage through control valves. Coefficients to account for these losses have been determined experimentally and in some cases analytically for various shapes and configurations. Design data for computation and assessment of fluid friction effects are summarized very comprehensively by Kays and London (1966).

There is a great difference between heat exchangers for liquids and those for gases. With high-density fluids (liquids) the friction power is generally small relative to the heat transferred and is seldom a critical factor. In exchangers designed for low-density fluids (gases) consideration of friction power is dominant. To achieve a tolerable friction power expenditure low mass velocities must be sought and this, together with the relatively low thermal conductivities of gas, results in low heat transfer rates per unit of surface area. Gas-to-gas heat exchangers incorporate large surfaces areas typically an order of magnitude greater than liquid-to-liquid exchangers of corresponding thermal duty.

Transient Response of Heat Exchangers. For most cryocoolers, it is adequate to analyze and compute the steady state case where flows are assumed to be continuous and independent of time. Calculations are performed for operation at the design point and various off-design conditions. In some cases, it is necessary to investigate time-variant operation of the system such as initial cool-down or load-changing. This requires an analytical or numerical analysis capability much more complicated than the steady flow case. Such capability is becoming increasingly available with the development of large-capacity computers and the more widespread use of general-purpose simulation programs such as the Continuous System Simulation Language CSSL IV.

In most situations where time-variant phenomena are of interest, it is generally necessary to investigate the complete cryogenic system. In some cases, it is sufficient to assess the transient response of single components including the heat exchangers. An excellent compilation of the effects of a sudden step change in input fluid temperature on different heat exchanger configurations, both recuperative and regenerative, was given by Kays and London (1966).

Oscillatory Flow Systems. An important class of recuperative heat exchangers exists where the flow is oscillating. The velocity varies continuously in cyclic quasisteady harmonic fashion that may be approximately sinusoidal, a rounded square wave, or any other relatively smoothly varying cyclic function. Sinusoidal-type oscillatory flows are found in Stirling and Vuilleumier cooling engines. The flow situation is further complicated by the continuously variable pressure and density characteristics of the fluid. The *frequency* of the pressure, density, and velocity variations are identical and correspond to the frequency of operation of the engine but they are seldom in phase. The phase relationship depends on the motion of the reciprocating elements in the engine and the temperatures in the expansion and compression spaces.

By way of example, consider the Philips type-A Stirling cryogenic cooling engine illustrated in Fig. 8.2. This engine contains two recuperative heat exchangers, one above and one below the regenerator. The upper recuperator, the freezer, consists of an annulus of 180 fine radial slots cut into the copper upper cylinder and sealed along the length by a sleeve of thin-wall metal around the displacer. The lower exchanger, the cooler, is a water-cooled tubular exchanger containing a large number of fine-bore, short-length tubes through which the compressed helium gas flows en-route to the compression space from the cooler.

Fluid flow in this machine was investigated by Walker (1961). Using the idealized Schmidt cycle analysis (see Chapter 2) he calculated the mass flows into and out of the expansion and compression spaces to be as illustrated in Fig. 8.23. On this figure, the flows *into* the expansion space and *out* of compression space are plotted above the datum *AA*. These

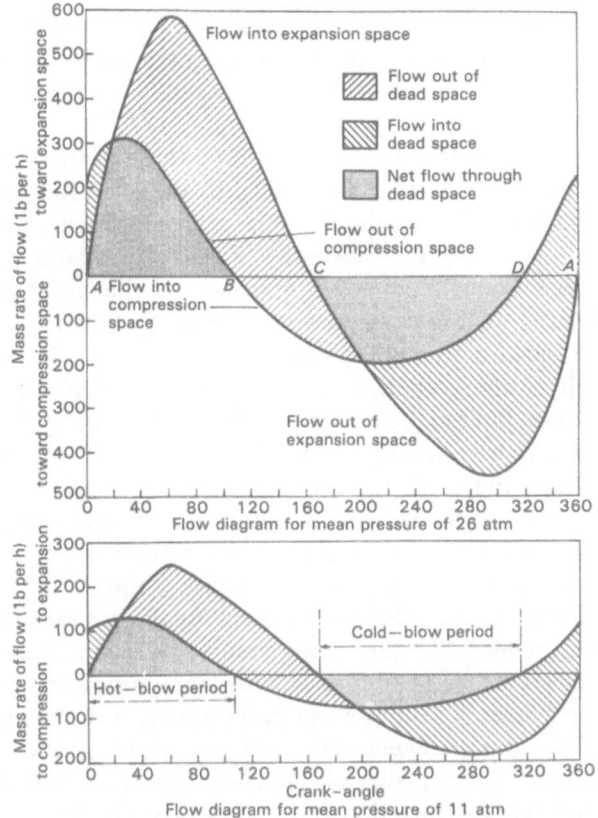

Fig. 8.23. Mass velocity-time characteristic for a Stirling cryogenic cooling engine (after Walker, 1961).

flows are all in the same direction, that is from compression to expansion space. Similarly flows *out* of the expansion space and *into* the compression space are plotted below the datum *AA*. It can be seen that the mass velocities vary continuously in harmonic but nonsinusoidal fashion. The peak mass velocities in the expansion space are approximately double the peak mass velocities in the compression space.

Superimposition of the mass velocity–time characteristic as in Fig. 8.23 revealed some complexities of the fluid flow situation in Stirling engines. For the period *A–B*, the fluid flows through the system from the compression space to the expansion space. For the period *B–C*, the fluid continues to the expansion space and also reverses direction in the cooler to flow *into* the compression space. At *C*, flow in the freezer reverses direction and begins to flow from the expansion space towards the compression space.

Thus, in period *C–D*, both flows are in the same direction towards the compression space. Finally at *D*, the flow into the compression space, through the cooler, now reverses and for the period *D–A*, fluid flows out of the compression space as well as continuing to flow out of the expansion space.

For the period *A–B*, there is a net flow through the regenerator in the direction from the compression space to the expansion space. From *B–C*, there is a general "draining" of the dead space with flow into both the expansion and compression spaces. From *C–D*, there is a net flow through the regenerator from expansion space to the compression space and finally from *D–A*, the dead space is "filled" by flow into it from both the expansion and compression spaces.

The result of the complicated flow process is that the actual fluid movement is quite limited. The fluid movement is probably tidal as calculated by Walker (1961) and shown in Fig. 8.24. This figure indicates the displacement-time diagrams for four fluid particles in the machine. For convenience of display, the reference volumes are all arranged sequentially, with the variable compression volume at the bottom side of the diagram. The upper and lower bounding curves represent the particle trajectories of the fluid particles immediately adjacent to the piston and displacer. These particles never leave the compression space and expansion space. The two intermediate curves represent the trajectories of two arbitrarily selected fluid particles. The important point to note is that no particle ever passes right through the regenerator from the cooler to the freezer. The upper intermediate curve is that of a particle that remains entirely within the regenerator; its residence time in the matrix is infinite.

The consequences of such complex flow in a heat exchanger have not been sufficiently addressed in the literature for meaningful comment to be made at this stage. It is clear the tidal action illustrated in Fig. 8.24 would involve low fluid velocity and hence lead to reduced exchanger effectiveness, partially offset by savings in the fluid friction.

Research attention is presently being focused on the nodal analysis technique for simulation of Stirling engines. Effort is concentrated almost exclusively in prime mover applications (see Walker, 1980) but the technique for cycle simulation could be applied to cooler development. These studies have already led to improved understanding of flow phenomena in Stirling engines, and future work will permit the development of design charts for heat exchangers in Stirling engines. These will be as indispensable as Kays and London's charts have been in conventional applications.

A square-wave oscillating flow is found in regenerative cooling systems where the flow is controlled by valves (Ericsson engines) rather than by volume changes (Stirling engines). Solvay, Postle, and Gifford–McMahon

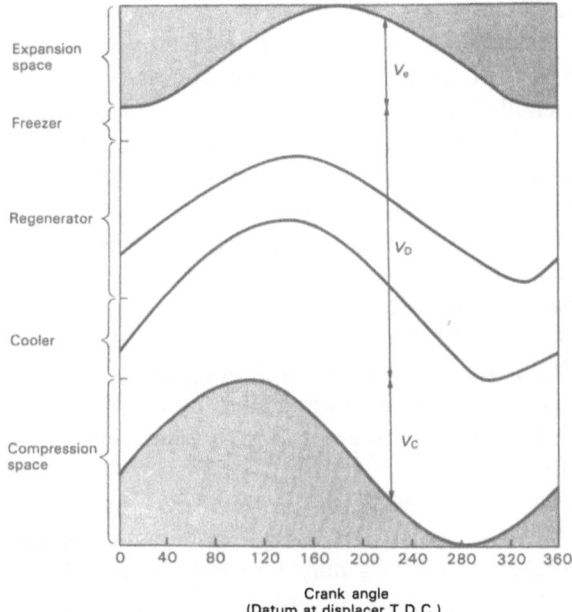

Fig. 8.24. Tidal motion of fluid in a Stirling cryogenic cooling engine (after Walker, 1961).

cryocoolers are typical of systems where the flow is controlled by valves. If the period of the cycle is short, the fluid flow may be tidal and as complicated as Stirling engines.

When the period between flow reversals is long, the transitory effects following a reversal of flow may be neglected. The system is then assumed to operate as a steady flow system with the flow first in one direction and then in the reverse direction. An example of this type of system is the reversing recuperative exchangers used in the Collins–Claude air liquefier. The period between flow reversals is very long (3–4 min).

The assumption of simple steady flow can only be made with low-pressure *recuperative* exchangers of low thermal capacity which rapidly change to the new flow conditions. It *cannot* be applied to *regenerative* exchangers which have a high thermal capacity and therefore a low response rate to changing conditions following switching of the flow.

An interesting study of the recuperative exchanger for the expansion space of a small regenerative cycle engine (Stirling, Vuilleumier, or Gifford–McMahon) was reported by Gifford and Acharya (1968). Using a simplified analysis they found that increasing the NTU (number of transfer units) of the recuperative exchanger was initially beneficial, but on attaining an optimal value, further increase in the NTU resulted in a decline of system performance. This is contrary to normal expectation where an increase in NTU always results in an improvement in heat exchanger performance.

Some aspects of the design of the recuperative heat exchanger at the cold end of a small Gifford–McMahon cryocooler were discussed by Chellis *et al.* (1971). Different designs for the oscillating flow heat exchangers incorporated in Vuilleumier cooling engines were evaluated in the course of the Air Research Manufacturing Inc. program for the NASA Goddard Space Flight Center. Some of the designs are illustrated in Figs. 4.9 through 4.12.

Enhanced Heat Transfer Surfaces. Many cryogenic processes involve heat exchange between fluids experiencing a phase change. This may be condensing from vapor to liquid or boiling from liquid to vapor. Heat transfer with phase change is characterized by very high heat transfer rates several orders of magnitude greater than those attainable in normal gas-to-gas exchangers.

Large-scale gas liquefaction facilities may involve very large heat exchangers. If the cascade cycle for gas liquefaction is used, a variety of fluids condensing and boiling may be found within the system. In such plants, it is vital to operate with minimum temperature differences and compact heat transfer surfaces.

Methods to enhance the heat transfer are, therefore, attractive and have stimulated various research programs over the past 20 years. O'Neill *et al.* (1972) reviewed progress with one type of enhancement technique to promote nucleate boiling involving the attachment of a porous metallic powder lining to the tubes of a recuperative exchanger and its application in liquid natural gas technology.

Regenerative Heat Exchangers

Regenerative heat exchangers may be generally classified as dynamic or static. An illustration of the two types is given in Fig. 8.25 (reproduced from Kays and London, 1966).

Dynamic Regenerative Exchangers. In the dynamic type the hot and cold fluid streams flow in opposite directions in ducts adjacent at the inlet and outlet of the regenerator. The porous matrix of the regenerator is mounted on a shaft and rotates relatively slowly. An element will, therefore, pass alternately through the hot and the cold fluid flow in a regular periodic manner. It will be alternately heated by the hot flow and cooled by the cold flow. The effective result is that heat is transferred from the hot to cold fluid.

Seals are required where the matrix passes from one stream to another. When the pressure of the two streams is significantly different, the fluid

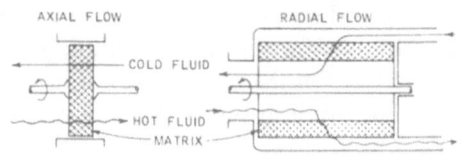

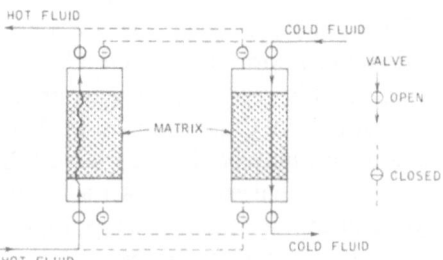

Fig. 8.25. Illustrating the static and dynamic types of regenerative heat exchanger (after Kays and London, 1966).

leakage across the seals may be critically important. Fluid contained in the porous matrix at the time it passes from one stream to another will be lost to the parent stream. With a high pressure difference between streams, this *carryover loss* may become significant particularly where the speed of the matrix is high to compensate for the use of a small matrix.

The axial flow type is favored for automotive gas turbine applications but the material of the matrix remains a serious problem. Continual heating and cooling of the matrix results in alternate small expansions and contractions leading eventually to structural failure of the matrix material by *thermal fatigue*.

The relatively high coefficients of thermal expansion of metals have proved them to be unsuitable for the high temperatures found in gas turbines. This led to research in the use of ceramic materials wound into the disk shape shown in Fig. 8.26 consisting of alternate layers of flat and

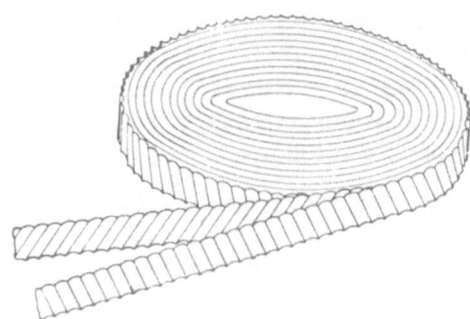

Fig. 8.26. Flame-trap construction of disk-type axial flow dynamic regenerative heat exchanger (after Scott, 1966).

corrugated strips. For less rigorous applications, metal strips may be used in the same form of construction, generally known as *flame-trap*, after the flame traps of the crank cases of large marine diesel engines.

For applications at low temperatures virtually any material may be used provided it can be formed or joined, and retains its integrity for the life required. Axial flow dynamic exchangers made of paper and epoxy cement in honeycomb or alternate corrugations are finding use in heating and ventilation installations for energy conservation in buildings. These devices are known as *thermal wheels*.

Dynamic regenerative exchangers have not been used for cryogenic applications in Claude or Joule–Thomson systems. The reasons are not obvious but may be due to the very high carryover losses arising from the very high pressures used. An assessment of possibilities for application in cryogenic systems (perhaps the upper stage recuperative exchanger) would appear justified. Use of paper, plastic, or light metal matrices could probably result in appreciable savings.

Static Regenerative Exchangers. An illustration of static regenerative exchangers is given in Figure 8.25. Two identical matrices are provided. One experiences the hot flow while the other receives the cold flow. The flows are switched periodically by quick-acting automatic valves.

The porous matrix is contained in a shell designed to withstand the pressure of the high-pressure stream. The matrix can be of any finely divided material, metal wires or strips, metal balls, glass balls, mineral particulates (gravel or brickdust have been used), and can be of any size. Extremely large units (several tonnes) are used in heat recovery units in steel works and in air liquefaction plants. An interesting discussion of large static regenerators in cryogenic service has been given by Ward (1961).

Common Theory for Static and Dynamic Types. Kays and London (1966) point out that the same regenerative heat exchanger design theory applies to both static and dynamic types when reduced to a common basis of matrix mass rate. The mass of the two matrices in the static type, divided by the *period* between valve switching corresponds to the mass of the single matrix multiplied by the speed of rotation.

Advantages and Disadvantages of Regenerative Exchangers. The principal advantages of regenerative exchangers compared with recuperative types are as follows:

  i. A very large area for heat transfer can be readily obtained using inexpensive finely divided material.
  ii. Fabrication and construction is relatively straightforward, and substantial savings can be made for the same heat exchange duty.

iii. Because of the periodic flow reversals the unit tends to be self-cleaning. This propensity for self-cleaning is most advantageous in situations where a contaminated gas is being processed, for example, in air liquefaction plants.

The major disadvantages of regenerative exchangers are the following:

i. Some mixing of the hot and cold streams is inescapable because of the carryover in flow switching and leakage.

ii. With dynamic-type exchangers, leakage at the seals may be appreciable and is important when the fluid streams are at significantly different pressures.

A Low-Temperature Problem; The Regenerator Material Heat Capacity. Regenerative heat exchangers may be thought of as a thermodynamic sponge alternately accepting heat (from the hot flow) and then releasing it (to the cold flow). Another simplistic model is to think of the regenerator as a thermal flywheel accepting energy on the downstroke (hot blow) and releasing it on the upstroke (cold blow). Both these models presume that the thermal capacity (mass times the specific heat) of the matrix is large compared with the thermal capacity of the gas passing through. Normally this is the case because of the high density and relatively high specific heat of the matrix materials. However, the specific heat of solid materials decreases with temperature whereas for gases it increases with decrease in temperature. At very low temperatures the specific heats of gases and solids can become comparable.

Figure 8.27 (after Daniels and du Pre, 1971) shows the specific heat of gaseous helium He (at a pressure of 4 atm), the metals copper, Cu, and lead, Pb, and the rare earth, europium sulfide, EuS, as a function of temperature. The specific heat of helium increases as the temperature decreases whereas those for the metals and the rare earth decrease. At a temperature of 20 K, the specific heat of helium is comparable with that of copper. At low temperatures lead has a high specific heat, substantially greater than any commonly available material. For this reason, lead [in the form of lead shot about 0.1 mm (0.004 in.) diameter] is the preferred material for the low-temperature regenerator in multiple expansion Stirling, Vuilleumier, and Gifford–McMahon cryocoolers. Nevertheless, the specific heat of lead is a strong function of temperature. It declines rapidly as the temperature decreases and is comparable with helium at about 9 K.

Simultaneously with increase in specific heat the density of gases increases as the temperature decreases, whereas the density of most solids is unaffected by temperature change. The result is that at low temperatures the heat capacity of the matrix is no longer that much greater than the

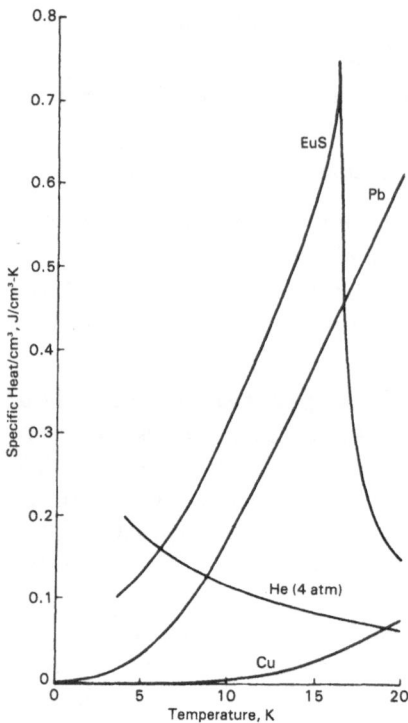

Fig. 8.27. Specific heat as a function of temperature of helium gas, copper, and lead metals and europium sulfide rare earths (after Daniels and du Pre, 1971).

heat capacity of the gas-flow. The matrix is *thermally saturated* and no longer capable of discharging or absorbing heat to and from the gas without appreciable temperature changes. This is the principal limitation of regenerative cryocoolers to attain temperatures less than 6 K.

To combat thermal saturation, proposals have been made to utilize the phenomena of discontinuities in the specific heat-temperature characteristic of various materials. The specific heat-temperature characteristic for the rare earth, europium sulfide, EuS, is given in Fig. 8.27. The characteristic rises steeply from zero at 0 K to a high peak value at 16 K and falls away sharply thereafter. A composite regenerator could perhaps be constructed from several materials having such discontinuities. The regenerator would operate with a temperature distribution corresponding more or less with the peak specific heat temperature of the regenerator matrix component.

The origin of this entertaining concept is unknown. The author recalls a conversation on the topic in 1966 with Dr. D. H. Parkinson of the Royal Radar Establishment, Malvern, U.K. A similar conversation with Professor Gifford also comes to mind. Gifford has also mentioned the concept briefly in a number of publications. Daniels and du Pre (1971) discussed the compound regenerator and reported on prototype tests with a miniature triple-expansion Stirling cycle cooling engine. With lead spheres in the low-temperature regenerator a minimum temperature of 9 K was achieved. With europium sulfide, a temperature of 7.8 K was achieved. They also followed a proposal by Fleming (1968) for the use of activated particulate charcoal as regenerator material and achieved a temperature of 8.5 K. Charcoal itself has a small heat capacity at low temperatures but is attractive as a potential regenerator material because of the ability to adsorb helium. It adsorbs so many times its own volume of gas that the apparent density of the helium approaches that of the liquid. If the adsorbed helium had a heat capacity similar to that of gaseous or liquid helium, the charcoal regenerator would have sufficient heat capacity to be used at very low temperatures.

Pron'ko et al. (1976) investigated the possibility of silica-gel as the matrix material for low temperature regenerators. Silica-gel was attractive because of its high adsorption of helium. Pron'ko found the low conductivity of silica-gel and adsorbed helium limited the usefulness as a regenerative matrix and introduced an unspecified "heat conducting constituent" which improved the situation. Insufficient detail was included to be specific about use of silica-gel as a matrix material.

The work performed by Daniels and du Pre, while of extreme interest, was not sufficient in extent or duration to be convincing one way or the other. The achievement of 8.5 K with activated charcoal and 7.8 K with europium sulfide versus 9 K with lead must be set off against the achievement of 6.5 K by others (Stuart et al., 1970) in a three-stage Gifford–McMahon cooling engine with lead in the coldest regenerator.

One difficulty about the compound regenerator is initial cooling to operating conditions. It is not at all clear that a regenerator could be persuaded to "bootstrap" itself sufficiently to scale, and then remain poised, on the Alpine peaks of specific heat discontinuities. Perhaps over-speed, over-pressure operation would be necessary on start-up with subsequent relaxation to "cruise" conditions for normal operations.

A theoretical study of the effect of variable matrix specific heat on the performance of thermal regenerators was reported by Rios and Smith (1968). Confining their study to high effectiveness regenerators, they found that a temperature-dependent specific heat matrix material resulted in significant deterioration of the regenerator effectiveness and temperature distribution at the moment of flow reversal. The effects were particularly significant at the cold end of the regenerator. Recently, Wheatley et al. (1980) in the course of studies of Stirling engines with dense phase working fluids (the Malone cycle) has suggested the use of

Malone regenerative/recuperative heat exchangers in the low-temperature region of a cryocooler. In this concept, a light, structural contraflow recuperator is equipped with one-way valves in the flow ducts. The ducts remain charged with helium working fluid moving intermittently and relatively slowly and then acting as the principal thermal capacity of the system to effect a measure of regeneration.

No other significant work in this area is known in the literature and there is a real need for fundamental research in an environment relieved of the pressures customary in prototype development.

Ideal Regenerator. The ideal regenerator can be conceived as a thermodynamic "black box" accepting gas at temperature T_c and heating it to T_h. After some time the flow is reversed and gas enters at T_h leaving T_c. The pressure drop across the regenerator would be zero. When used in Stirling engines where dead space is an important parameter, the ideal regenerator has zero void volume.

The ideal regenerator is impossible. Achieving the constant inlet/outlet temperature would require infinitely slow operation or the heat transfer coefficient and heat transfer area to be infinitely great and the heat capacity of the fluid and matrix to be zero and infinite, respectively. The absence of a pressure drop would require the flow to be frictionless. The absence of void volume would preclude the provision of flow passages through the matrix for the fluid to traverse.

Hausen Regenerator. Practical regenerators have characteristics far different from the ideal regenerator. Although they were used as early as 1817, their operation is among the most difficult to analyze mathematically. It was not until the period of 1920 to 1930 that Nusselt and Hausen developed the first rational theories for regenerator operation. Their work has been summarized by Jakob (1957) with extensive references to the source documents of both Nusselt, Hausen, and others.

Various modes of regenerator operation may be postulated, but that of most interest is called the *state of cyclic operation.* This is attained when, after repeated heating and cooling for a fixed time-cycle consisting of one heating and one cooling period, the temperature at any one point in the fluid (or the matrix) is then the same as it was a full cycle earlier.

Figure 8.28 is a representation of a thermal regenerator in counterflow cyclic operation. Hot fluid at a constant inlet temperature, entering from

Fig. 8.28. Thermal regenerator in counterflow operation.

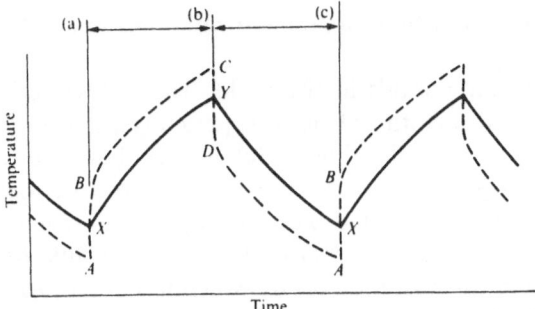

Fig. 8.29. Time–temperature variation of fluid and matrix in a thermal regenerator.

the left-hand end, passes through the matrix, gives up part of its heat, and leaves the right-hand end with a variable temperature, lower than the inlet temperature. The supply of hot fluid is discontinued, and all the fluid is ejected from the matrix through the exit at the right. Cold fluid now enters at a constant inlet temperature from the right, passes through the matrix, is heated by absorbing heat from the matrix, and leaves at the left-hand end with a variable temperature above the inlet temperature. The cold fluid supply is discontinued, and all of the fluid is ejected from the cold end, to complete the cycle of operations.

Figure 8.29 shows the variation with time of the matrix temperature and fluid temperature at any station in the matrix. Figure 8.30 shows the temperature field of the fluid and matrix at the instant of flow-reversal. The upper curves represent the temperature of the fluid and matrix at the end of hot-blow and the start of cold-blow. The lower curves represent temperature conditions at the end of cold-blow and start of hot-blow. At any station in the matrix, the temperatures will fluctuate between the upper

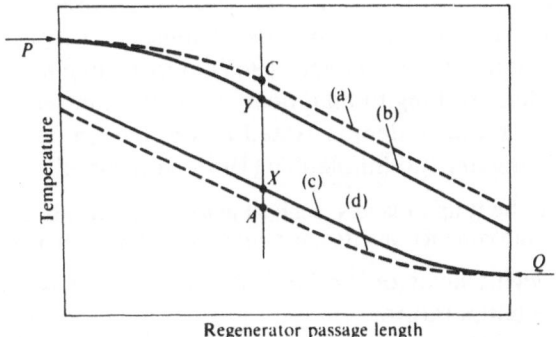

Fig. 8.30. Spatial temperature variation of fluid and matrix in a thermal regenerator.

and lower curves, in a time-dependent relationship similar to that shown in Fig. 8.29.

There are four periods in the cycle. The "blow period" is the time taken for the total quantity of fluid to pass any point in the regenerator; the "reversal period" is the time which elapses between the entry of one fluid and the entry of the other. Blow and reversal periods exist for both fluids. Iliffe (1948) has pointed out that in practical regenerators the blow period is the same as the reversal period. The last fluid to enter is driven out by other fluid through the port by which it entered. In the hypothetical regenerator, the blow period is always less than the reversal period by the time taken for a gas particle to travel from one end of the regenerator to the other. If this effect is ignored, we are assuming that the *time for a particle to pass through the regenerator (the residence time) is small compared with the total blow time.*

Other simplifying assumptions necessary to render the analysis tractable are summarized below:

(a) The thermal conductivity of the matrix. Nusselt considered four cases:

  i. Thermal conductivity of the matrix is infinitely large. There is no temperature difference in the matrix; Nusselt showed this would have poor performance.
  ii. The thermal conductivity of the matrix is infinitely large parallel to the fluid flow, and finite normal to the fluid flow. In practice, this may be approached by a very short regenerator, with a matrix composed of thick walls.
  iii. The thermal conductivity of the matrix is zero parallel to the fluid flow, and infinitely large normal to the fluid flow.
  iv. The thermal conductivity of the matrix is zero parallel to the fluid flow, and finite normal to the fluid flow.

Cases (iii) and (iv) correspond most closely to the practical regenerator, and it is unfortunate that the analyses of these two cases are the most complicated. Schultz (1951), Tipler (1947), and Hahnemann (1948) have examined the effect of longitudinal heat conduction in the walls of regenerator passages, and have demonstrated this to have a negligible effect in certain cases. Saunders and Smoleniec (1948) concluded

> for matrices built up in layers, such as gauzes (screens) or matrices made of refractory, the conduction effect is almost certainly negligible.

(b) The specific heats of the fluids and of the matrix material do not change with the temperature.

(c) The fluids flow in opposite directions, and have *inlet temperatures that are constant both over the flow section and with time.*

(d) The *heat-transfer coefficients and fluid velocities are constant with time and space*, even though they may be different for the two fluids.

(e) The *rate of mass flow of either fluid is constant during the blow period*, even though it may be different for the two fluids, and the blow periods may be different.

Very little theoretical work appears to have been done on regenerators operating under conditions not fulfilling assumptions (b), (c), and (d), and most results are available for operation with equal blow times and equal mass flow. However, Johnson (1952) and Saunders and Smoleniec (1948) have investigated this latter effect. Saunders and Smoleniec also considered the effect of variation in the specific heats of the fluid and matrix. They found the assumption of constant values, made in (b), resulted in less than 1% error in the effectiveness.

An interesting case, considered by Nusselt (see Jakob, 1957), was a regenerator with infinitely small reversal period, in which the fluids had been switched infinitely often. The theory is simple, and corresponds to a recuperator in which the two fluids flow continuously separated by metal walls.

Presentation of Performance Data: Reduced-Length–Reduced-Period Method. The performances of regenerators, assumed to be operating under the conditions discussed above, have been presented in a variety of ways. Among the most useful are the curves given by Hausen reproduced in Fig. 8.31. These have been supplemented by similar curves calculated by Johnson and Saunders and Smoleniec. They show the effect on regenerator effectiveness of two dimensionless parameters called (after Hausen), the reduced length, Λ and the reduced period, Π. The reduced length (in the flow direction) is defined by

$$\Lambda = hAL/VC_p \qquad (8.11)$$

where h is the heat-transfer coefficient between fluid and matrix, per unit surface area, A is the matrix surface area per unit length, V is the fluid-volume flow rate, C_p is the specific heat of the fluid, and L is the matrix length. The reduced period is defined by

$$\Pi = hAZ/MC \qquad (8.12)$$

where h and A are as defined above, and M is the mass of matrix material, C is the specific heat of matrix material, and Z is the blow time. Frequently Λ and Π are combined by the quotient

$$\Pi/\Lambda = U = VC_p/MC \cdot Z/L \qquad (8.13)$$

and called the utilization factor, representing the ratio of the sensible-heat capacity of the fluid per blow to the heat-storage capacity of the matrix.

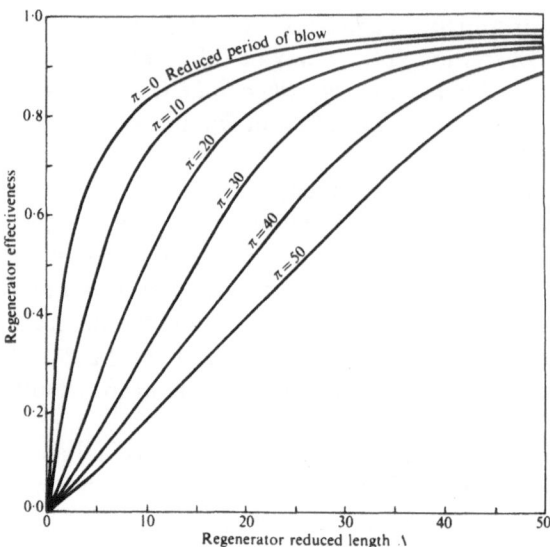

Fig. 8.31. Regenerator effectiveness as a function of the reduced length and the reduced period II.

Regenerators may have different reduced periods and reduced lengths for the hot and cold blows, so that there are four factors to be considered. In these cases, Saunders and Smoleniec recommend that average values be used, suggesting (on the evidence of calculations carried out by Johnson) that the error is small. This is probably because, even when the actual blow times are unequal, the reduced periods are much nearer equality, since a reduction in the actual blow time Z is usually accompanied by an increase in the rate of fluid flow V.

Usefulness of the concept of reduced dimensionless parameters and curves of regenerator effectiveness is limited by the accuracy of the heat-transfer data. This is generally measured experimentally using the "single-blow" transient technique, first described by Furnas (1932) and later by Saunders and Ford (1940), Johnson (1952), Saunders and Smoleniec (1948), Coppage (1952), Rapley (1960), Vasishta (1969), Wan (1971), Walker and Vasishta (1971), and Walker and Wan (1972). In this technique, an isothermal matrix is suddenly subjected to a flow of hot fluid, entering at constant inlet temperature. The change in the fluid exit temperature is measured with time. Very careful measurements are required. The theory for "single-blow" operation was first given by Schumann (1929), and may be used to determine from the measured data the heat-transfer coefficient of test matrix. There is some question whether these data are strictly applicable to regenerators operating cyclically. A reasonable amount of heat-transfer data is contained in the references given above, but comparison is difficult because several slightly different forms of presentation have been used.

An alternative method for determining the heat transfer coefficients for flow in regenerator matrices under steady state conditions was described and used by Kim and Qvale (1971). They called this the *steady enthalpy flux method*. So far as is known, no comparative studies of the single-blow and steady enthalpy flux method have been undertaken.

NTU-Effectiveness Method. Kays and London (1966) included an extensive compilation of experimental data with notes on a rational design procedure and a numerical example for periodic flow regenerator design. Their method was based on the NTU-effectiveness approach discussed above for recuperative heat exchangers. Kays and London (in common with other modern regenerator studies) were principally concerned with heat exchangers for gas turbine applications. Their results are well suited for this and other applications involving relatively slow flow reversals and long blow times compared with residence times for particles in the matrix.

Application of Theory to Regeneration in Stirling-Type Engines with Oscillatory Flow. The theories of regenerator operation discussed above were developed initially for air-liquefaction and gas-separation plants and for air preheaters for boilers. These plants are large and two regenerators per unit are used, one being heated and the other being cooled. The blow times are long, ranging from ten minutes to several hours.

Later, the theory was adapted and extended for application of regenerative heat-exchangers to gas turbines where the blow times are much shorter. Coppage and London (1953) refer to

> a reversal-time of a quarter of a second (two complete cycles per second) which is near the maximum permissible frequency without undue "carry-over loss"

and, again

> the idealization of no flow-mixing is closely met when the flow-passage length is short, and such shortness of length appears to be good design procedure for the most suitable types of surface.

Most regenerators in gas-turbine engines have a relatively large frontal area and a short flow length, so that, although the blow time is short, the residence time of the particle in the matrix is also very short.

The above theory is applicable and realistic to regenerators used in gas-turbine engines and air preheaters. It is less applicable to regenerators used in Stirling and similar engines where the flow pattern is oscillatory (see Figs. 8.24, 8.25 and the accompanying discussion).

Regenerator theory is based on assumptions which do not apply in the Stirling engine. The most important of these is that the time for a particle to pass through the matrix, the residence time, is small compared to the total blow time. In a Stirling engine, the blow times are exceedingly short. For example, at the moderate engine speed of 1200 rpm (20 Hz), the blow time is ten times less than the permissible minimum in a gas turbine. The blow times are so short in fact that no particle ever passes right through the matrix (Fig. 8.21). Figure 8.24 shows that the actual net flow time through the matrix is about half the complete cycle-time, the remaining time being occupied in either filling, or emptying, the dead space. The heat-transfer process that occurs must be very complex, involving a repetitive fluid-to-matrix, matrix-to-fluid, fluid-to-matrix cyclic relationship, rather like the water bucket passed hand to hand in a fire-fighting operation.

Other important assumptions are that the inlet conditions, temperature, rates of mass flow, and fluid velocity remain constant with time. This is not true in Stirling-cycle regenerators for the inlet conditions vary constantly. Figure 8.24 shows extreme variations in the rates of mass flow. The maximum *net* flow through the matrix is only half the maximum flow into and out of the expansion space.

Attempts to analyze regenerators of Stirling engines by any of the above procedures require the estimation of "average" conditions for the flow. Such gross approximation is required to determine these "average" values that the value of the ultimate result is thought to be questionable. No recommendation can be made for the application of any existing theory of regenerator operation to aid regenerator design. In the past, the author has taken flow characteristics such as those contained in Fig. 8.24 and determined the root mean square or some other average flow to permit application of the Kays and London–Hausen theory. The performance determined in this way appears so unrealistic that attempts to develop this procedure have been abandoned.

There is reason to hope for improvements in the future. Smith and co-workers at the Massachusetts Institute of Technology have established an alternative basis for design in which the various thermodynamic, thermal, and fluidic effects are not coupled but assessed separately. They discuss an approximate solution for the thermal performance of a Stirling-cycle regenerator, in which there is provision for non-steady pressure (and mass flow) conditions, including the possibility of sinusoidal variation, with a phase difference in the peak values. By assuming a second-order polynomial form for the temperature field in the regenerator, a closed solution was obtained for the net enthalpy flux. The theory is idealized and assumes the gas and matrix temperature are practically constant with time, and that there are no wall (or fluid-friction) effects. At present, the theory does not appear to be sufficiently well developed to be directly applicable in regenerator design. An excellent summary of the method and its application has been given by Harris *et al.* (1971). Earlier Smith (1965) discussed some aspects of the selection of regenerators for Gifford–McMahon machines.

Little has been published of the Philips repository of knowledge and experience about regenerators. The leader of the cryogenic engineering research team, J. W. L. Kohler, made the study of regenerators in Stirling engines his special interest and is thought to have developed unsurpassed expertise in the field. It is regrettable that several years after retirement so little of his special knowledge has been put in the open literature. The best account in English of the Philips regenerator design method was given by Kohler *et al.* (1975) and another important paper was published by Kohler (1968). Other publications by Kohler listed in the Bibliography also contain references to regenerator research carried out at Philips. Notes of lectures given by Kohler at the Technical University at Delft, about 1970, are interesting and informative but were not translated into English and published in the open literature.

Recently, Finegold and Sterrett (1978), of the Jet Propulsion Laboratory, have reviewed the literature of regenerative heat exchangers from the aspect of application to Stirling engines. Their study embraced a bibliography of some 200 references. The principal topics of interest were (a) regenerator analysis methods, and (b) experimental heat transfer and friction data. The report is highly recommended.

The authors identify four basic methods of analysis:

  i. Reduced-length–reduced-period, Hausen type discussed above.
 ii. Effectiveness–NTU method given in Kays and London (1966).
iii. Enthalpy flux method due to Smith and others at Massachusetts Institute of Technology.
 iv. Numerical analysis methods embodied in the nodal analysis cycle simulation programs for Stirling and other regenerative engines developed by Finkelstein, Urieli, and others.

The report also contains a compilation of test data from a variety of sources and the very extensive bibliography mentioned above.

Regenerator Design for Stirling Engines. The regenerator for a Stirling engine must attempt to satisfy a number of conflicting requirements. To minimize the temperature excursion of the matrix, and thus improve the overall effectiveness, the ratio of heat capacity of the matrix to the gas should be a maximum. This can be achieved by a *large, solid matrix.*

The fluid-friction loss must be limited. The effect of pressure-drop across the matrix is to reduce the range of the pressure excursion in the expansion space adversely affecting the area of the expansion-space $P–V$ diagram and reducing the refrigeration and coefficient of performance. The fluid-friction loss is minimized by a *small, highly porous matrix.*

Dead space is also important because it influences the ratio of maximum to minimum pressure. For maximum specific output, the ratio should be high and the dead space made as small as possible. This can be done with a *small, dense matrix.*

To improve heat-transfer performance and minimize the temperature difference between matrix and fluid, it is necessary to have maximum surface area for heat transfer between the fluid and matrix. The matrix should be *finely divided*, with preferential thermal conduction, maximum normal to the flow, minimum in the direction of the flow.

The regenerator acts as an effective filter of the working fluid and oil, grease, and particulates are retained in the flow-passages. In a cryocooler, impurities in the working fluid condense in the low-temperature region and accumulate in the regenerator. This build-up is cumulative and increases the fluid-friction losses so the cooling performance progressively diminishes. From this aspect a *small open matrix* is required.

We have the following desirable characteristics for a regenerative matrix:

for maximum heat capacity—a large, solid matrix,
for minimum flow losses—a small, highly porous matrix,
for minimum dead space—a small, dense matrix,
for maximum heat transfer—a large, finely divided matrix,
for minimum contamination—a matrix with no obstruction.

It is impossible to satisfy all these conflicting requirements, and with our present understanding of the cycle, it is not possible to quantify the relative significance of the various aspects.

The above is not meant to baffle newcomers to the field but to point out the conflicting factors to be considered. It is tempting to become enmeshed in the complicated mathematics of heat transfer in regenerators and force the continuously variable flow into a quasi steady flow pattern. This enables various sums to be done, some kind of simulation to be completed, and all manner of charts to be prepared. It is an activity beloved by graduate students and academics and really quite meaningless for it neglects much else that is important.

Regenerator designers must, therefore, accept that much of their initial design must be carried out on a "by gosh and by golly" basis. The matrix should be of material as finely divided as possible within permissible economics and contained in a case with a low thermal conductivity and no opportunity for the fluid flow to leak or by-pass the matrix. It will be difficult to achieve a void volume of matrix less than 60% with conventional materials of wires or balls. Wires should be arranged with the strands perpendicular to the flow to minimize conduction effects.

Attempts to increase the heat transfer performance by providing more matrix material will result in increased fluid friction losses and increased thermal conduction along the larger regenerator case. The effect is similar to losing on the swings what one gains in the roundabouts. The best approach is to keep the design flexible for changes to be made in prototype development or following nodal analysis simulation optimization.

A satisfactory compromise design for many Stirling cryocoolers will be found with a regenerator length at least three times the frontal diameter of the flow section, a matrix of wire or spheres having a diameter of 0.025 to 0.05 mm (0.001 to 0.002 in.) and with a regenerator void volume of one to two times the swept volume in the *expansion* space.

For their cooling engine shown in Fig. 3.1, Philips use a regenerator made of randomly packed short lengths of copper wire 0.001 in. in diameter contained in a low-conductivity compressed-paper sleeve mounted in annular form around the displacer.

The author has found woven wire mesh of copper and phosphor bronze to be effective packing for regenerative matrices. These can be obtained in a variety of mesh densities and wire sizes. As the mesh density increases and the wire diameter decreases, the price per unit area increases very steeply and it is doubtful that the material could be used for production machines. An annular regenerator such as the Philips unit is very expensive, because the center section, punched from the screen, is "wasted."

Wire screens can be "sintered" easily, to form a stable semirigid block. One way is to pack the screen in some form that can be loaded with a weight. The wire screen is cleaned by immersion in nitric (or hydrochloric) acid, and the loaded assembly heated for a short period in a furnace with a reducing atmosphere. On removal, it will be found that the screen has "sintered" to a solid assembly and can be lightly machined. It is important to arrange the screen so that the wires are normal to the axis of flow, otherwise axial conduction may be too high. Sintering with light loading does not appear to significantly increase the axial conduction of the screens but does improve the packing ratio because of a considerably reduced porosity.

For very low temperature generators, lead in finely divided form is the preferred matrix material. Lead is not available in wire or woven cloth form but can be obtained as "shot," approximately spherical in shape. Fine lead shot can be obtained from the lead supply companies in relatively small amounts but cannot always be hand sorted in specific sizes. The usual procedure is to purchase 50 or 100 kg of fine shot and then screen it to obtain the hundred grams or so of fine (0.025-mm-diam) shot required for the regenerator. Careful selection of particles of the same diameter will fill the regenerator with maximum mass of material and minimum void volume.

It is important to contain the regenerator so that fluid is caused to pass through the matrix rather than an alternative bypass. Elimination of the alternative flow paths is difficult to achieve unless given careful attention at the design stage. Fluid bypassing the matrix experiences little or no regenerative action and reduces the cryocooler performance.

Experimental Performance. Little has been published about imperfect regeneration on the performance of Stirling-cycle machines, or experimental work with regenerators under conditions approximating those present in Stirling engines. Davies and Singham (1951) reported experiments on a small regenerator of brass and copper wire gauzes subject to the oscillating flow of a constant volume of air at atmospheric pressure and a frequency of five cycles per second. The air was heated on one side of the regenerator and cooled on the other. Continuous records of the temperature of the air were taken on both sides of the matrix. It was concluded from these experiments that (1) for a given gauze matrix, the regenerator efficiency increases with the matrix weight, but the improvement takes place at a progressively diminishing rate; (2) for a given matrix weight, the regenerator efficiency increases with decrease in the diameter of the gauze-wire. Tests, with equal weights of brass and copper gauze, gave approximately the same values for regenerator efficiency. Although the copper had a thermal conductivity three times greater than the brass, it had little effect. It was concluded that with fine wires the conductivity lag is extremely small. In these tests the regenerator efficiency was obtained by analysis of the continuous fluid temperature records measured at each end of the regenerator matrix.

Experiments by Walker (1961) with a series of different regenerators on the Philips gas refrigerating machine confirmed that reduction in wire diameter increases the regenerator effectiveness. The criterion of performance was the quantity of liquid air produced by the machine operating at constant speed and mean pressure of the working fluid. A reduction in the wire diameter, with approximately constant matrix weight and porosity, resulted in an increase in the surface area for heat transfer.

Work by Murray, Martin, Bayley, and Rapley (1961) shed light on the performance of regenerators under sinusoidal flow conditions. Frequency had little effect on the heat-transfer process, but the wave shape had a significant effect. With pulsating flow, the effectiveness of gauze matrices was appreciably below that under steady flow conditions, but with flame-trap matrices an improvement in heat transfer rate with unsteady flow was observed.

Bretherton *et al.* (1971) have presented experimental data for three regenerators of lead shot and copper grains using helium as the working fluid to pressures of 30 atm and with blow times in the range 1 to 190 sec. The temperature difference across the regenerator ranged from 35 K at the cold end to 300 K at the warm end.

Gifford *et al.* (1969) presented results of extensive regenerator tests with dense matrices constructed of wire mesh screens of stainless steel and bronze and small-diameter lead shot. The temperature range across the regenerator was 300 to 80 K. The blow times ranged from 0.5 to 0.2 sec. Their apparatus consisted essentially of twin parallel regenerators and a rotary valve to switch the hot and cold flows back and forth between the test matrices. This series of tests is perhaps the most representative of conditions prevailing in small cryocoolers of the Stirling, Vuilleumier, and Gifford–McMahon type. It is recommended for close study. A feature of particular interest of both this and the previous reference is the very high regenerator "efficiencies" (probably the same as regenerator effectiveness) measured. Values well in excess of 0.95 were consistently recorded for a wide range of conditions. Subsequently, Gifford and Acharya (1970) reported an extension of the work to low-temperature regenerators operating over the temperature range 10 to 80 K, but few test results were included.

Heat-Transfer and Fluid-Friction Characteristics of Dense-Mesh Wire Screens. The heat-transfer and fluid-fraction characteristics of various porous media were given by Coppage and London (1956) and have been supplemented by later data. Few data have been published on flow in *dense* wire screens for sizes of interest in the regenerators of Stirling cryocoolers. Data measured at the University of Calgary by Vasishta (1969) and by Wan (1971) are included here, but no other values are known with which these results may be compared. To validate the experimental apparatus, Vasishta obtained results for stainless-steel mesh in sizes comparable to those studied by Coppage and found results in close agreement.

The heat-transfer and fluid-friction data for two sizes of screen are given in Figs. 8.32 and 8.33, respectively. Both sizes were woven from phosphor bronze wire, having the following composition:

tin 3.5%–3.8%,
phosphorus 0.3%–0.35%,
iron 0.1%,
lead 0.05%,
zinc 0.3%,
copper rem. density 554 lb ft^{-3},
thermal conductivity 47 Btu h^{-1} ft^{-1} F^{-1},
specific heat 0.104 Btu lb^{-1} F^{-1}.

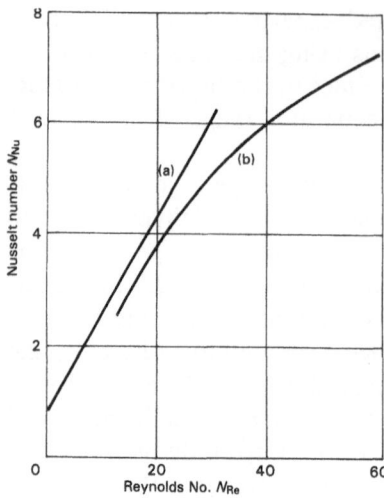

Fig. 8.32. Heat-transfer characteristics of dense-mesh wire screen (after Walker, 1972). (a) 400 × 400 strands per in., 0.001 in. wire diameter. (b) 200 × 200 strands per in., 0.002 in. wire diameter.

The two screen sizes investigated were (a) 200 × 200 strands per in., 0.0021 in. wire diameter, (b) 400 × 400 strands per in., 0.001 in. wire diameter.

The heat-transfer characteristics are presented as the Nusselt number, Nu, as a function of the Reynolds Number, Re, defined as follows:

$$Nu = (4r_h/k_f)(h/f) \tag{8.14}$$

$$Re = \rho_f V d/\mu_f = (4r_h/\mu_f p)(W_f/A_f) \tag{8.15}$$

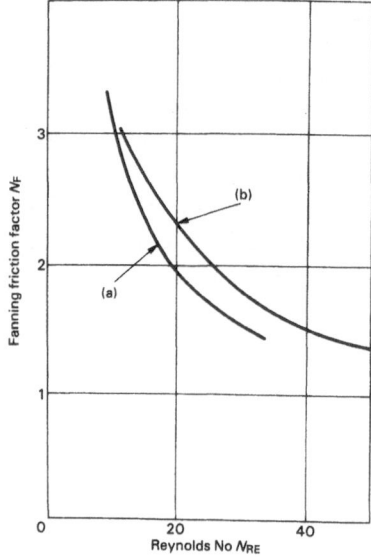

Fig. 8.33. Fluid-friction characteristics of dense-mesh wire screen (after Walker, 1972). (a) 400 × 400 strands per in., 0.001 in. wire diameter. (b) 200 × 200 strands per in., 0.002 in. wire diameter.

where r_h is the calculated hydraulic radius of the screen, h is the heat transfer coefficient, k_f is the thermal conductivity of the fluid, ρ_f is the density of the fluid, V is the volume flow rate of fluid in matrix, W_f is the mass flow rate of fluid in matrix, A_f is the frontal area, p is the calculated porosity, μ_f is the dynamic viscosity of the fluid, p = (volume of matrix − volume of metal)/(volume of matrix), r_h = (total volume of connected void spaces)/(total surface), and, area = (volume of matrix) × (porosity)/(total surface area).

The fluid-friction characteristics are presented as the Fanning friction factor f, as a function of the Reynolds number Re, defined as follows:

$$f = 2\rho_f \,\Delta P r_h p^2 / nLG_A^2 \qquad (8.16)$$

where ΔP is the pressure drop, n is the number of layers of screen, L is the length of the matrix, G_A is the mass flow per unit area, and the remainder are as defined above.

Regenerative Annulus. A simple yet effective regenerator can be created using the clearance space between the displacer and the cylinder wall in Stirling, Vuilleumier, Gifford–McMahon, and Solvay cycle cooling engines. This is called the regenerative annulus or gap regenerator, and is sometimes used in small or miniature cooling engines of low capacity. Conventional and gap regeneration methods are compared in Fig. 8.34 (reproduced from Daniels and du Pre, 1971). The conventional regenerator is shown incorporated in the displacer, a common feature with small regenerative cooling engines. The gas flows through the regenerative-displacer from the compression space (below) to the expansion space (above). There is a slight pressure drop across the regenerator and a dry-rubbing seal is necessary to cause the gas to flow *through* the porous regenerator rather than simply follow the lesser resistance in the annular gap *around* the displacer. The regenerator could be also incorporated as

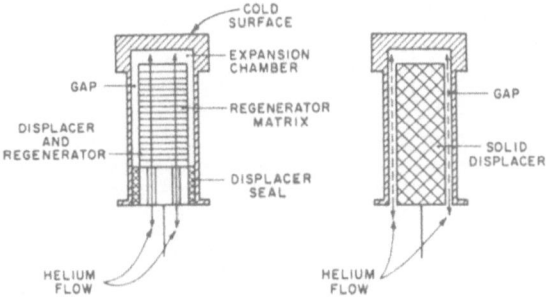

Fig. 8.34. Comparison of conventional regeneration and gap-regeneration methods (after Daniels and du Pre, 1971).

a static device outside the cylinder with the flow of gas through ducts from the compression and expansion spaces to the matrix. In that case the displacer would be solid (or to reduce thermal conduction, a hollow thin-wall solidus). Again a seal on the displacer would be required as before to prevent leakage through the gap.

The alternative gap regenerator is shown in the right-hand diagram of Fig. 8.34. Here the regenerative matrix has been entirely eliminated and the seal removed. The displacer is solid, of low conducting material (or a hollow, thin-wall solidus). As it moves, the displacer causes gas to flow in the annular gap between the displacer and cylinder wall from the expansion and compression space and vice versa. Regeneration is achieved because the temperature of the walls of the gap (the cylinder and the displacer) vary from the cold expansion space (at the upper end) to the ambient temperature compression space (at the lower end).

The regenerative annulus is one limiting case of the compromise in regenerative heat exchangers between effective heat transfer and high pressure drop. The pressure drop in conventional regenerators arises partly from the density of the matrix and partly from the velocity (squared) of the fluid in the voids. In reducing the density of the matrix to a minimum (by eliminating the matrix) the way is clear to increasing the speed of operation of the engine. Since a contribution to the refrigeration effect is generated per cycle, an increase of engine speed increases the rate of refrigeration generated. Conversely, for a given load, a smaller engine may be used.

The possibilities inherent in gap regeneration for cryocoolers were investigated by Daniels and du Pre (1971) using the double expansion engine shown schematically in Fig. 8.35. The displacer for this engine was

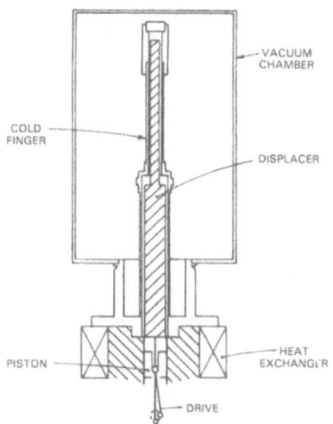

Fig. 8.35. Small high-speed double-expansion Stirling cooling engine with gap regeneration (after Daniels and du Pre, 1971).

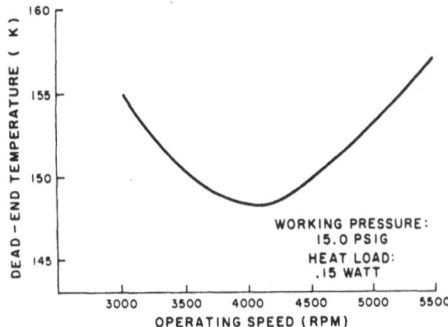

Fig. 8.36. Effect of operating speed on the performance of a cooling engine with gap regeneration (after Daniels and du Pre, 1971).

machined from solid Rulon fluorocarbon. Displacer diameters were 4.3 mm (0.17 in.) and 8.1 mm (0.32 in.) and the length of each section was 100 mm (4 in.) long. In both cylinders the gap width was 0.25 mm (0.010 in.).* The displacer stroke was 5 mm (0.2 in.). The displacer was contained in a stainless steel cylinder having a wall thickness of 0.1 mm (0.004 in.).

The effect of operating speed on performance (expressed in terms of the minimum temperature measured at the extremity of the cold cylinder) is shown in Fig. 8.36. With a mean pressure of only 0.2 MPa (30 psi abs.) the minimum temperature of 130 K was achieved at a speed of 66 Hz (4000 revolutions per minute). It was noted by du Pre and Daniels that "in practical terms, the speed of a conventional Stirling refrigerator is limited to approximately 1750 revolutions per minute." Subsequently with the conical gap regenerator shown in Fig. 8.37 a minimum temperature of 98 K was achieved. The cone-shaped displacer carries to the limit the benefits occurring from increasing the number of stages of expansion. With a cone-shaped displacer, a theoretical infinite number of separate expansions is incorporated. A relatively shallow cone angle and short displacer stroke are necessary to minimize cyclic variation of the regenerator gap width. Increase of the gap decreases the effectiveness of regeneration whereas decrease of the gap increases the fluid friction.

One very difficult problem with machines having gap regenerators is maintaining the concentricity of the displacer in the cylinder. The fluid forces always conspire to push the displacer to one side of the cylinder so that the gap on one side is zero and on the other side is twice the nominal value. The rate of mass flow through a narrow aperture is a function of the *cube* of the width of the aperture. A slight deviation in the concentricity of the displacer in the cylinder will result in a massive concentration of

---

* It is not clear from the reference paper whether the clearance of the displacer in the cylinder was in fact 0.25 or 0.50 mm. Since the gap *width* is described as 0.25 mm a total clearance of 0.50 mm is inferred.

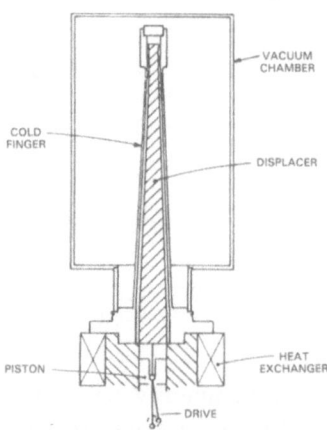

Fig. 8.37. The conical gap regenerator used on a small high-speed Stirling cooling engine (after Daniels and du Pre, 1971).

flow through the large gap where the heat transfer effectiveness is less. Concentricity of the displacer can be maintained using guide rings at both ends of the displacer. It will be recalled that mechanical rubbing friction near the cold end of the cylinder is deleterious to the refrigerating performance of the engine for heat generated by the rubbing friction is immediate charge on the refrigeration generated.

Another technique to maintain the displacer concentric in the cylinder is to mount the displacer on a center guide-post. Although well suited for many applications, this technique is hardly practical for miniature cooling engines where the cold cylinder diameter is perhaps limited to 6.3 mm (0.25 in.).

The problem of maintaining the displacer concentric in the cylinder was not addressed by Daniels and du Pre (1971) or even mentioned. However, it is well known in other applications and would surely be evident in the long narrow structures used by Daniels and du Pre.

We have dwelt at length on this topic because of its importance in small cryocoolers for the future. The experiments reported by Daniels and du Pre are the only ones in the literature specifically directed to a study of the phenomenon. However, gap regeneration has been used with advantage elsewhere. Zimmerman and Radebaugh (1977) utilize gap regeneration in the multiple expansion miniature cooling engine shown in Fig. 1.3. Contrary to the interest of Daniels and du Pre, the Zimmerman engines operate at very low speed (about 1 Hz) but yet are able to achieve low temperatures with no formal regenerators.

Gap regeneration is used in some of the small Stirling engines developed for artificial hearts. These have been described by Walker (1980) and the various contractor reports to the U.S. National Institutes of Health.

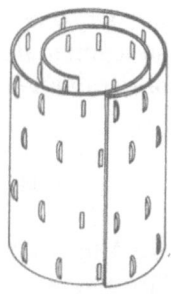

a) DIMPLED   COIL
FOIL REGENERATOR

b) STRAW REGENERATOR

Fig. 8.38. Coiled foil annular gap and straw regenerator.

These engines have had enough development for a variety of ingenious concepts to have been incorporated, including two concerned with annular gap regenerators. The first is the coiled foil regenerator shown in Fig. 8.38. Thin metal foil is wound in a helical coil and inserted in the annular gap between the cylinder wall and the displacer. The foil is corrugated or intermittently dimpled and when coiled the sheets are slightly separated. This provides a multiplicity of narrow annular paths for the gas to pass through in laminar flow with minimal pressure drop and good heat transfer characteristics. The coiled foil regenerator was invented by Martini of Richland, Washington.

The "straw regenerator" is a similar concept providing multiple parallel conduct for laminar flow in the annular gap. It was developed by Hoffman of the Aerojet Liquid Rocket Co., Sacramento, California and incorporated in the Aerojet artificial heart engine. A ring of fine-bore, thin-wall low-conductivity glass straws, arranged as shown in Fig. 8.38, are provided in the annular gap between the cylinder wall and the displacer. This figure shows simple circular cross-section tubes, but more complicated sections are also used, two of which are shown in Fig. 8.39. The "square straw" contains five circular tubes enclosed within a square-section tube. The assembly is first prepared using tubes of a convenient size to handle and then heated and drawn through a die to reduce the cross section to the

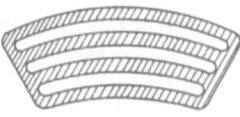

(a) SQUARE STRAW
SECTION

(b) SEGMENTAL SECTION

Fig. 8.39. Flow sections for straw regenerators.

desired size. The square straw is available commercially in a range of sizes from 1 mm width. The segmental section is a later development and is also available in a range of sizes and geometries.

HEAT PIPE

A heat pipe is used to transfer heat (thermal energy) with minimal temperature drop over a substantial distance (up to 2 m). It consists, in principle, of a hollow tube containing a fluid in the vapor and liquid states. One end of the tube is coupled to a heat source and the other to a heat sink. In the tube, the fluid boils at the source end and condenses at the sink end so there is a circulation of saturated vapor from the source to sink ends and saturated liquid from the sink to the source.

Heat transfer in boiling and condensing occurs at very high rates and little energy is consumed in "pumping" the fluids. Therefore, the transfer of heat from one end of the tube to another is accomplished with little temperature difference (a degree or two). The heat pipe corresponds to thermal conductive transfer along a solid bar but with an effective thermal conductivity an order of magnitude greater than any known metal.

In many applications a "wick," a porous liner or wire mesh screen placed around the internal periphery, is helpful in promoting the return of liquid to the source area. It is necessary when heat pipes are used on spacecraft in low gravity environments where the buoyancy forces due to density differences are small. Surface tension forces associated with the internal wick assist the return of the liquid to the source end.

Heat pipes have no moving parts so that reliability is high. Heat can be transferred with minimal temperature drop and there can be appreciable weight savings compared with conductive or forced-convective transfer systems.

Depending on the internal fluid and material of the enclosure, heat pipes may be used at any temperature. Applications in cryogenic systems may include:

i. transfer of heat at high temperatures from an isotope source or solar energy concentrator–absorber to the heater of a Vuilleumier cooler,

ii. coupling the compressor and compression space cooling systems to radiative transfer panels externally mounted on spacecraft,

iii. efficient thermal linkage of central cold sinks and cooled detectors.

The elementary technology of heat pipes is given in standard heat transfer texts. Recent developments of high and ambient temperature systems are reported in the specialist heat transfer literature (*ASME Journal of Heat Transfer*). Relatively recent developments for cryogenic applications of heat pipes have been reported by Joy (1972) and by Foster and Murray (1973). Both contain further references to earlier works, principally the report of Haskin (1967).

HEAT EXCHANGERS FOR VERY LOW TEMPERATURES

The technology for attaining very low temperatures (below 1 K) is well developed and a number of different systems or methods may be used. One of these, the He^3–He^4 dilution refrigerator depends greatly on the operation of contraflow recuperative heat exchangers at very low temperatures. Unusual thermal effects and thermophysical properties of the helium working fluid require the application of special considerations to the design and operation of heat exchangers for these specialist applications.

We shall not develop the topic beyond this cautionary note. Moreover, there is presently no engineering application established or in prospect that appears to require operation at temperatures less than 1 K. Those interested in research in this temperature range are referred to Chapter 12 and to excellent texts by Lounasmaa (1974) and Betts (1974). There are many specialist papers in the principal cryogenic literature.

REFERENCES

Abadzic, E. E., and Scholz, H. W. (1973). "Coiled Tube Heat Exchangers." *Adv. Cryog. Eng.* **18**, 42–51.

Afgan, N., and Schlünder, E. U. (1974). *Heat Exchangers, Design and Theory Sourcebook.* Scripta Book Co./McGraw-Hill Book Co., NY.

Barron, R. (1966). *Cryogenic Systems.* McGraw-Hill Book Co., Toronto.

Betts, D. S. (1974). *Refrigeration and Thermometry below 1 K.* Sussex University Press, Falmer, Sussex (in the U.S., Crane Russak and Co., New York).

Bretherton, A., Granville, W. H., and Harness, J. B. (1971). "Performance of Regenerators at Low Temperatures." *Adv. Cryog. Eng.* **16**, 333–341.

Chellis, F. F., Hosmer, T. P., and Keller, E. (1971). "Closed Cycle Refrigeration for an Airborne Illuminator." *Adv. Cryog. Eng.* **16**, 214–220.

Collins, S. C. (1966). "Helium Refrigeration and Liquefier." *Adv. Cryog. Eng.* **11**, 11–15.

Collins, S. C., and Cannaday, R. L. (1958). *Expansion Machines for Low Temperature Processes.* Oxford University Press, Oxford, England.

Coppage, J. (1952). "Heat-Transfer and Flow-Friction Characteristics of Porous Media." Thesis, Stanford University, Stanford, California.

Coppage, J. E., and London, A. L. (1953). "The Periodic-Flow Regenerator—A Summary of Design Theory." *Trans. ASME* **75**, 779–787.

Coppage, J. E., and London, A. L. (1956). "Heat-Transfer and Flow-Friction Characteristics of Porous Media." *Chem. Eng. Prog.* **52**(2), (Feb) 56–57.

Cowans, K. W. (1974). "A Countercurrent Heat Exchanger that Compensates Automatically for Maldistribution of Flow in Parallel Channels." *Adv. Cryog. Eng.* **19**, 437–444.

Daniels, A., and du Pre, F. K. (1971). "Triple Expansion Stirling Cycle Refrigerator." *Adv. Cryog. Eng.* **16**, 178–184.

Davies, S. J., and Singham, J. R. (1951). "Experiments on a Small Thermal Regenerator. General Discussion on Heat Transfer." *Proc. Inst. Mech. Eng.* 434–435.

Finegold, J. G., and Sterrett, R. H. (1978). "Stirling Engine Regenerators—Literature Review." Report No. 5030-230, Jet Propulsion Laboratory, California Institute of Technology, Pasadena, California (July).

Fleming, R. B. (1967). "The Effect of Flow Distribution in Parallel Channels of Counterflow Heat Exchangers." *Adv. Cryog. Eng.* **12**, 352–362.

Fleming, R. B. (1968). "Regenerators in Cryogenic Refrigerators." Tech. Rept. AFFDL-TR-68-143, Wright–Patterson Air Force Base, Dayton, Ohio (see also U.S. Patent 3,262,277, Fleming—July 26, 1966).

Fleming, R. B. (1969). "A Compact Perforated Plate Heat Exchanger." *Adv. Cryog. Eng.* **14**, 197–204.

Foster, W. G., and Murray, D. O. (1973). "Development Program for a Liquid Methane Heat Pipe." *Adv. Cryog. Eng.* **18**, 96–102.

Furnas, C. (1932). "Heat Transfer from a Gas Stream to a Bed of Broken Solids." *Bull. U.S. Bur. of Mines*, No. 361.

Gifford, W. E., and Acharya, A. (1968). "Optimization of a Cryogenic Refrigerator Heat Exchanger." *Adv. Cryog. Eng.* **13**, 599–606.

Gifford, W. E., Acharya, A., and Ackermann, R. A. (1969). "Compact Cryogenic Thermal Regenerator Performance." *Adv. Cryog. Eng.* **14**, 353–360.

Gifford, W. E., and Acharya, A. (1970). "Low Temperature Regenerator Test Apparatus." *Adv. Cryog. Eng.* **15**, 436–442.

Hahnemann, H. (1948). "Approximate Calculation of Thermal Ratios in Heat Exchangers Including Heat Conduction in the Direction of Flow." N.G.T.E. Mem. 36, National Gas Turbine Establishment, Pyestock, U.K.

Harris, W. S., Rios, P. A., and Smith, J. L. (1971). "The Design of Thermal Regenerators for Stirling Type Refrigerators." *Adv. Cryog. Eng.* **16**, 312–323.

Haskin, W. L. (1967). "Cryogenic Heat Pipe." Report No. AFFDL-TR-66-228, Wright–Patterson Air Force Base, Dayton, Ohio, June.

Holman, J. P. (1976). *Heat Transfer*. 3rd Ed. McGraw-Hill Book Co., NY.

Iliffe, C. E. (1948). "Thermal Analysis of the Contra-flow Regenerative Heat Exchanger." *Proc. Instn. Mech. Engrs.* **159**, 363–372.

Jakob, M. (1957). *Heat Transfer*. Vol. II. Wiley and Sons, New York (see Chapter 35, "Regenerators").

Johnson, J. E. (1952). "Regenerator Heat Exchangers for Gas Turbines." U.K. Aero Res. Council, Tech. Report, R and M, No. 2630, U.K.

Joy, P. (1972). "Optimum Cryogenic Heat Pipe Design." *Adv. Cryog. Eng.* **17**, 438–448.

Kays, W., and London, A. L. (1966). *Compact Heat Exchangers*. 2nd Ed. McGraw-Hill Book Co., New York.

Kenoldt, W. (1965). "Selected Examples of European Cryogenic Practice." *Adv. Cryog. Eng.* **10**, 392–404.

Kim, J. C., and Qvale, E. B. (1971). "Analytical and Experimental Studies of Compact Wire-Screen Heat Exchangers." *Adv. Cryog. Eng.* **16**, 302–311.

Kohler, J. W. L. (1968). "Computation of the Temperature Field of Regenerators with Temperature Dependent Parameters." *Proc. Second Int. Cryog. Eng. Conf.* pp. 44–46, Brighton, U.K., May, Iliffe Sci. and Tech. Pubs. Ltd., Guildford, U.K.

Kohler, J. W. L., Stevens, P. F., de Jonge, A. K., and Beuzekom, D. C. (1975). "Computation of Regenerators Used in Regenerative Refrigerators." *Cryogenics*, **15**, 521–531.

Kroeger, P. G. (1967). "Performance Deterioration in High Effectiveness Heat Exchangers Due to Axial Heat Conduction Effects." *Adv. Cryog. Eng.* **12**, 363–372.

Kroeger, P. G. (1967). "Plated Tube Heat Exchanger: Analytical Investigation of a New Surface Concept." *Adv. Cryog. Eng.* **12**, 340–351.

Lenfestey, A. G. (1961). "Low-Temperature Heat Exchangers." *Prog. Cryog.* **3**, 23–48.

Lenfestey, A. G. (1968). "Compact Heat Exchanger Assemblies for Gas Separation Plants." *Proc. Second Int. Cryog. Eng. Conf.*, pp. 47–49, Iliffe Sci. and Tech. Pubs. Ltd., Guildford, U.K.

Lins, R. C., and Elkan, M. A. (1975). "Design and Fabrication of Compact High-Effectiveness Cryogenic Heat Exchangers Using Wire Mesh Surfaces." *Adv. Cryog. Eng.* **20**, 283–299.

Lounasmaa, O. V. (1974). *Experimental Principles and Methods below 1 K.* Academic Press, New York.

Murray, J. A., Martin, B. W., Bayley, F. J., and Rapley, C. W. (1961). "Performance of Thermal Regenerators under Sinusoidal Flow Conditions." *Int. Heat-Trans. Conf.*, *ASME* 781–796.

O'Neill, P. S., Gottzmann, C. F., and Terbot, J. W. (1972). "Novel Heat Exchanger Increases Cascade Cycle Efficiency for Natural Gas Liquefaction." *Adv. Cryog. Eng.* **17**, 420–437.

Pron'ko, V. G., Amamchyan, R. G., Guilman, I. I., and Raygorodsky, A. I. (1976). "Some Problems of Using Adsorbents as a Matrix Material for Low-Temperature Regenerators of Cryogenic Refrigerators." *Proc. Sixth Cryog. Eng. Conf.* (ed. K. Mendelssohn), pp. 86–88, I.P.C. Sci. and Tech. Press, Guildford, U.K.

Rapley, C. (1960). "Heat Transfer in Thermal Regenerators." M.Sc. Thesis, Durham University.

Rios, P. A., and Smith, J. L. (1968). "The Effect of Variable Specific Heat of the Matrix on the Performance of Thermal Regenerators." *Adv. Cryog. Eng.* **13**, 566–573.

Saunders, O., and Ford, H. (1940). "Heat Transfer in the Flow of Gas through a Bed of Solid Particles." *J. Iron Steel Inst.*, No. 1, 291.

Saunders, O. A., and Smoleniec, S. (1948). "Heat Regenerators." *Proc. Seventh Int. Cong. App. Mech.*, Vol. 3, pp. 91–105.

Schultz, B. H. (1951). "Regenerators with Longitudinal Heat Conduction." *J. Mech. E.— ASME General Discussion in Heat Transfer.*

Schumann, T. E. W. (1929). "Heat Transfer to a Liquid Flowing through a Porous Prism." J. Franklin Inst. **208**, 405–416.

Scott, R. B. (1966). *Cryogenic Engineering.* Van Nostrand Co. Inc., New Jersey.

Smith, J. L. (1965). "Some Aspects of the Selection of Regenerators." *Cryogenics*, **5**, 306–314.

Stuart, R. W., Cohen, B. M., and Hartwig, W. (1970). "Operation and Application of a Three-Stage Closed Cycle Regenerative Refrigerator in the 6.5 K Region." *Adv. Cryog. Eng.* **15**, 428–435.

Timmerhaus, K. D., and Schoenhals, R. J. (1974). "Design and Selection of Cryogenic Heat Exchangers." *Adv. Cryog. Eng.* **19**, 445–462.

Tipler, W. (1947). "A Simple Theory of the Heat Regenerator." Tech. Report No. 1CT/14, Shell Petroleum Co. Ltd., Thornton Research Centre.

Vasishta, V. (1969). "Heat-Transfer and Flow-Friction Characteristics of Compact Matrix Surfaces for Stirling Cycle Regenerators." M.Sc. Thesis, University of Calgary.

Vonk, G. (1969). "A New Type of Compact Heat Exchanger with a High Thermal Efficiency." *Adv. Cryog. Eng.* **13**, 582–589.

Walker, G. (1961). "The Operational Cycle of the Stirling Engine with Particular Reference to the Function of the Regenerator." *J. Mech. Eng., Sci.* **3**, No. 4.

Walker, G. (1980). *Stirling Engines.* Oxford University Press, Oxford.

Walker, G., and Wan, W. K. (1972). "Heat Transfer and Flow Friction Characteristics of Dense Mesh Wire Screen Regenerator Matrices at Cryogenic Temperatures." *Proc. Fourth Int. Cryog. Eng. Conf.*, pp. 93–95, I.P.C. Sci. and Tech. Press, Guildford, U.K.

Walker, G. (1982). *Industrial Heat Exchangers: The User Basic Guide.*, Hemisphere Publishing Corp. Washington, D.C.

Wan, W. K. (1971). "Heat-Transfer and Friction-Flow Characteristics of Dense-Mesh Wire-Screen Regenerator Matrices." M.Sc. Thesis, University of Calgary.

Ward, D. E. (1961). "Some Aspects of the Design and Operation of Low Temperature Regenerators." *Adv. Cryog. Eng.* **6**, 525–536.

Weimer, R. F., and Hartzog, D. G. (1973). "Effects of Maldistribution on the Performance of Multistream, Multipassage Heat Exchangers." *Adv. Cryog. Eng.* **18**, 52–64.

Wheatley, J. C., Allen, P. C., Knight, W. R., and Paulson, D. N. (1980). *Principles of Liquids Working in Heat Engines.* (In press). Dept. of Physics, Univ. of California at San Diego.

Wittner, C. E. (1966). "Design of a Closed Cycle Helium Temperature Refrigerator." *Adv. Cryog. Eng.* **11**, 107–115.

Zimmerman, J. E., and Radebaugh, R. (1977). "Operation of a SQUID in a Very Low-Power Cryocooler." App. of Closed Cycle Cryocoolers to Small Superconducting Devices, Proc. of Conf. NBS, Boulder, Oct., pp. 59–66.

Some Aspects of Design

INTRODUCTION

In the development of cryocoolers, highly competent engineering development teams have exercised great ingenuity and skill. Modern machines exemplify the present limits of manufacturing technology and design technique. It would be presumptuous to suggest that this single chapter could contain an adequate summary of that experience. Rather, we have attempted a cursory introduction to some of the more important and obvious areas of concern.

TARGET DEFINITION

The primary consideration for any new project is to define the application and specify a target performance with any accompanying limitations. For this to be meaningful, it is necessary to allocate priorities. Any practical cooler is essentially the compromise of many conflicting factors.

For practical engines, the critical parameters must be identified at the earliest stage, and then defined in detail as soon as possible. Achievement of these critical targets will, if they front the status of technology, always entail some sacrifice of other desirable, but not vital features.

In some instances, minimum size and weight are of paramount importance. Examples of this are the cryocoolers found in infrared heat-seeking guidance systems of defensive missiles. A reduction in size and weight can only be achieved by working the machine harder. This may entail an increase in the pressure of the working fluid and the speed of operation. Taken to excess, both factors will lead to increased rates of wear and operating losses so that there is a reduction in efficiency and engine life. In the case of a missile system, a short life of a few hundred hours can be tolerated. Quick cooldown and vibration-free operation are also important. Cost is secondary but becomes increasingly significant as development proceeds from one level to another.

For other situations, reliability in terms of the period of maintenance-free operation may be the vital parameter. One example is the cryocooler for a medical office, generating superconducting temperatures for a magnetocardiogram device. Another might be the cryocooler for the superconducting magnets of a magnetohydrodynamic base load power generating system. In both cases, size and weight are not important provided long life can be gained. Thus a relatively large, heavy, slow-running machine with long cooldown times is possible. Noise and vibration are important, more so in one case than in the other. Capital cost is important, but not critical.

RELIABILITY

There are instances where reliability is so important as to be the dominant parameter. One example is the cryocooler used on a spacecraft. In that application, routine maintenance and repair or replacement is not feasible. Sherman *et al.* (1979) have reported on recent developments for long-life spacecraft cryocoolers. In their paper, they discuss interesting approaches by several contractors to achieve the three-year operating life desired by the NASA for future unmanned space explorations. Four principles of design to achieve high reliability are simplicity, redundancy, light loading, and the selection of appropriate materials.

Simplicity of design is important from the aspect of minimizing the chance of something going wrong. William Beale, inventor of the free-piston Stirling engine and engineering designer *par excellence*, is fond of quoting the precept "do not design something complicated until you have failed with something simple."

By way of example, consider a machine consisting of two units each having 90% probability that it will function over the required time of operation. When the two units are combined in a system, the probability that the combination will endure is not 90% but only 90% of 90% or 81%. The chance of success with four units having individual probabilities of success of 90% when combined in a system declines to 65% and so on. Now cryocoolers are complex machines that may contain scores or even hundreds of parts. Successful attainment of a target reliability should depend on each and every part. If it does not, then the design can likely be modified to eliminate the unnecessary parts. An excellent discussion of the technique of component and system failure analysis applied to cryocoolers was given by Pitcher (1975).

Design redundancy is a favored approach to gain reliable operation. This does not mean the provision of unnecessary parts, but rather the incorporation of alternative systems that can assume the necessary duty on failure of the original unit. Sometimes it is adequate to provide supplementary back-up systems for those defined by failure analysis to have a high probability of failure. In other cases, the provision of one or even two complete replacement systems can be justified to ensure the highest reliability. Decisions about the provision of redundant units are made on the basis of the consequences of equipment failure against the limitations of space, weight, and cost.

Development of miniature solid-state electronics with unprecedented reliability has virtually removed these as an area of concern to the cryocooler system designer. Considerations for reliability are now concentrated on the electrical power and mechanical systems, the bearings, seals, rubbing surfaces, springs, and general integrity of structural elements. The life of such components is a function of the work they do—lightly loaded systems last longer and have higher reliability than heavily loaded systems. The temptation can be very high to upgrade a successful system to operate at a higher rate of production. This is almost always accompanied by increased wear and risk of failure unless better materials and improved designs are used.

Selection of appropriate materials is fundamental to good design practice in every area of engineering. Plastics, metals, and surface treatments are being improved all the time and it is a difficult task to keep abreast of all developments. Designers must simply recognize the problem, keep well informed, and be ready to adopt new materials and techniques.

Reliability is important, critical in some cases, and always significant. Yet, the operating life in terms of the period between maintenance or failure is difficult to specify at the design stage. The most rigorous reliability analysis leaves the proof of the pudding to be found in the eating. Increasing certainty comes from extensive component and system life testing. Rarely is it possible to design a new system so exactly that every goal can be achieved in the first prototype. The general experience of engineering development is that, given the will and the means, any new system can be dramatically improved in performance or life by test-bed development. Cryocoolers are no exception. Unfortunately, they are rarely required in numbers that would justify multiple development prototypes and the undivided attention of expert development engineers.

Comprehensive data about reliability is hard to come by. Apart from the Philips Stirling liquefier, there are no substantial numbers in civil applications. A considerable number (in the hundreds) of Collins liquefiers have been fabricated and there is also a sizable community of Gifford–

McMahon coolers. However, the manufacturers are coy about releasing the data they have accumulated. Civil users rarely have enough machines to provide a statistical sample and are not organized to pool their experiences.

Most machines manufactured, particularly in small sizes, have been used for military purposes (heat seeking and thermal imaging). A valuable review of reliability and maintainability of U.S. Air Force flight cryocoolers was given by Clarke (1973). It makes dismal reading and reports failures occurring generally between 100 and 300 hours of operation and with maintenance attention generally required in less than 100 hours. Clarke included operating experience of over 2000 cooler installations. They were principally Joule–Thomson systems but included about 800 Stirling systems and a few Gifford–McMahon machines. As he wryly remarks in the conclusion of his review:

> ...In the mid-Sixties when IR and cryogenics were coming into vogue, manufacturers published advertisements and other literature claiming 2000 to 2500 hours between failures. These claims have not held up, and are indeed, off by an order of magnitude...

Clarke gathered his data from Air Force Logistics Command maintenance data reports for the period 1971–1973. It is possible that in the succeeding interval, substantial improvement in performance has been achieved, but the AFLC reports are not available publicly. There is an urgent need for up-to-date reliability and maintenance reviews along the lines of the Clarke survey by Air Force, Navy, and Army users of cryocoolers.

Harkless (1973) of the Honeywell Radiation Center, Lexington, Massachusetts, has reported on reliability tests carried out for the Air Force on four Vuilleumier cryocoolers manufactured by the Hughes Aircraft Company. Their demonstrated reliability was well below the design predictions due to a succession of minor problems in the noncryogenic areas of the machine. However, as Harkless points out in his introduction, the design was based on cooler technology of the 1969–1970 era. It is to be expected that the subsequent decade has improved the basis for reliability.

In Chapter 14, Ishizaki gives brief details of Japanese experience of cryocooler failure. Pitcher (1973) has written an excellent survey of approaches to extended mechanical life of spacecraft cryocoolers and Sherman *et al.* (1979) have reported recent progress in this area.

COLDFINGER DESIGN

In many electronic and instrument applications of small regenerative cryocoolers, the cold assembly is similar to that shown in Fig. 9.1 for a

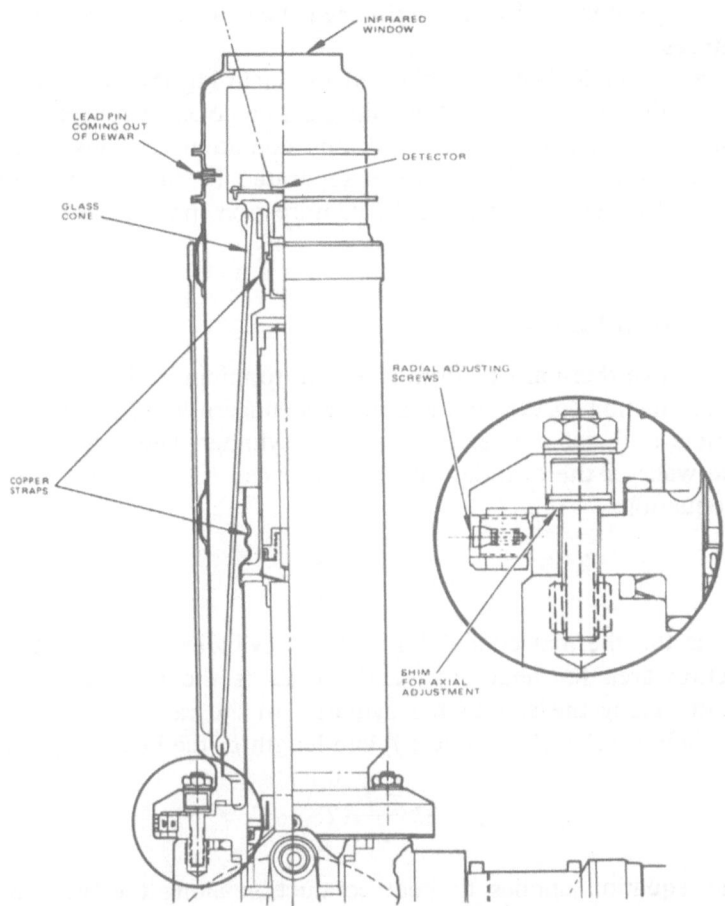

Fig. 9.1. Typical coldfinger design for small Vuilleumier cryocooler (after Russo, 1976).

two-stage expansion Vuilleumier cycle cryocooler (after Russo, 1976). The necessary refrigeration load is a nominal 1 W at 80 K or some lower temperature. This is generated within the cryocooler expansion cylinder, typically, for a single stage of expansion about 50 mm (2 in.) long and 6 mm (0.25 in.) diameter. It is called the "coldfinger" or "sting." The coldfinger is sheathed by a glass or metal vacuum-insulated Dewar flask assembly.

The essential objective of coldfinger design is to maximize the refrigeration generated by expansion of gas in the cold region. To accomplish this objective it is advantageous to site the refrigeration load as close to the expansion space as possible, and thermally isolate the refrigerating space and the refrigerated load to minimize heat leaks from sources at higher temperatures.

Thermal isolation can be achieved by enclosing the coldfinger in a highly insulating environment. Silvered glass or polished reflecting metal Dewar flasks evacuated to a high vacuum condition are most effective. Furthermore, the expansion space is separated by the regenerative or recuperative heat exchanger from the compression space at ambient temperature.

Conduction Heat Leakage

To improve thermal isolation, it is vital to minimize heat leakage by thermal conduction to the cold space. This concern leads to the familiar arrangement of the coldfinger as a long thin cylinder. Thermal conduction along the walls of the cylinder and displacer can be estimated using the Fourier equation:

$$q_{cond} = -kA \frac{dT}{dx} \tag{9.1}$$

where k is the thermal conductivity of the cylinder material, A is the cross-section area for heat flow, and dT/dx is the rate of change of temperature along the axis of the cylinder. In the case of a cylinder of external diameter d, wall thickness t, and length L, the heat conduction is

$$q_{cond} = -k \frac{\pi d t}{L} (T_C - T_E) \tag{9.2}$$

A similar equation applies to heat conduction along the walls of the regenerator.

Clearly, the heat leak due to conduction will be minimized by the use of a long, thin-wall, small-diameter cylinder of low-conductivity material. The wall thickness for the cylinder may be estimated from the thin-wall cylinder equation:

$$\sigma = P_{max} d/2t \tag{9.3}$$

where σ is the design stress (0.8 of 0.1% proof stress) and P_{max} is the maximum cylinder pressure, assumed to be 4/3 of the mean pressure. Stainless steel is favored for the coldfinger cylinder wall because it is a strong material allowing use of a high stress σ, is tough (not subject to the ductile–brittle transformation), and has a relatively low thermal conductivity. It is not subject to cracking, is impermeable to gases, is relatively easily worked, and is readily available in a wide range of qualities and forms.

Precision drawn stainless tube is particularly useful for coldfinger elements and can be found in a wide range of sizes and qualities finished to close dimensional and concentricity limits. The tubing can be used both for the cylinder wall and, in lighter section, for the regenerative displacer wall.

The cold end of the cylinder can be closed by a stainless steel cap of substantial section brazed or soldered to the tube. The warm end of the cylinder can be brazed or welded to a stainless steel flange or other convenient fitting. Flanged or screwed joints penetrating through the cylinder wall at the cold end of the cylinder are to be avoided. It is difficult to maintain a seal subject to the differential thermal contractions arising during cooldown.

The same material should be used throughout the coldfinger cylinder assembly to avoid problems due to differential thermal contraction. Welding of the tube is possible but not recommended. The high temperatures of the welding processes result in distortion and rearrangement of the crystal lattice with deleterious effects on strength unless the part is subsequently heat treated (with further opportunities for distortion). Brazed and soldered joints are satisfactory for most purposes.

Cooldown

The time taken to cool to operating temperature is an important characteristic of most cryocoolers. This cooldown time, P, is simply a function of the refrigeration produced per cycle, R, the number of cycles per minute, N, the heat capacity of the element to be cooled, C, and the temperature difference between the ambient and operating temperature, ΔT.

In simplified terms:

$$P = C\Delta T/RN \qquad (9.4)$$

To reduce the cooldown time, the heat capacity (or the mass of metal to be cooled) should be minimized by using a light section cylinder cap, connecting leads, and refrigerator load. Alternatively, a quick cooldown may be achieved by operating at high speed (increasing N) or greater pressure (increasing R).

The gas in the expansion space experiences cyclic temperature variations that result in moderate surface temperature variations of the cylinder end cap. The detectors applied to the end cap are so incredibly sensitive that sometimes their performance is impared by these minor temperature perturbations. This can be overcome by the use of a relatively massive end cap or alternate layers of thin copper and Teflon sheet between the cylinder head and the detector base. The latter provides a coupling of high thermal inertia which damps out thermal perturbations.

A combination of rapid cooldown time and constant operating temperature, not subject to perturbation, is difficult to achieve. The approaches discussed above to meet these two requirements are mutually opposed.

Regenerator

The displacer reciprocating in the coldfinger may or may not contain the regenerator. In Fig. 9.1, the two regenerators are contained within the body of the displacer and oscillate with it. The regenerative displacer case may be fabricated from thin-wall stainless steel tube which is a clearance fit in the coldfinger cylinder. Sometimes paper–phenolic and glass–epoxy resin plastic tubes are used. The tube is closed at both ends by plugs of stainless steel brazed or soldered in place and drilled in the axial direction to permit gas flow through the regenerator.

The matrix may consist of any very finely divided material. Dense phosphor-bronze screens of $200 \times 200 \times 0.002$ in. mesh to 400×400 (strands per inch) $\times 0.001$ in. (wire diameter) are often used. Lead spheres (0.004 in. diameter) are favored for very low temperature operation because of the high specific heat of lead at low temperatures. Nickel or steel balls of 0.002–0.005 in. diameter have also been used. For temperatures down to 60 K the actual material used is not so important as the surface area for heat transfer. Experimental machines with regenerators made of discarded nylon stockings have operated nearly as well as those with expensive phosphor bronze screens.

The use of a stainless steel tube sliding inside a stainless steel cylinder is attractive to minimize problems due to differential thermal distortion. However, the friction and wear effects of similar metals sliding one upon the other are well known. To prevent contact between them, it is advisable to provide guide rings on the displacer, usually of Rulon or other filled PTFE material containing molybdenum disulfide. It is vital to eliminate friction in the cold region, for this dissipates the refrigeration effect created. It is customary, therefore, to use a guide ring at the warm end of the displacer and another about halfway along the displacer.

If the regenerator is not located within the displacer, a different construction may be used. The displacer may be a hollow tube of stainless steel plugged at both ends or a solid element of low-conductivity plastic, nylon, or Teflon. Solid displacers are particularly suited for machines utilizing the regenerative annulus concept (see Chapter 8). Fluid passing between the compression and expansion spaces flow through the annular space between the displacer and the cylinder wall. The surface layers of the displacer and the cylinder wall serve the function of the regenerative matrix.

Hollow displacers are filled with a low-conductivity insulating material (perlite or Min-K). It is good practice to drill a minute hole in the warm end plug so as to allow the interior to become pressurized to the mean cycle pressure. This minimizes the pressure difference the displacer wall must withstand. The hole in the warm end plug must be so small (<0.001 in. diameter) that little fluid will enter and leave the displacer during cyclic operation. If the hole is too large (>0.001 in. diameter) the effect will be to increase the dead space of the machine. The amplitude of the pressure variation will be substantially decreased and, consequently, the refrigerating performance will be diminished.

Shuttle Heat Transfer

Consider the cylinder–displacer assembly shown in Fig. 9.2 with the displacer at the end of its stroke. It is assumed that the temperature distribution along the cylinder is linear and ranges from a low temperature T_E at the cold end to a high temperature T_C at the warm end. It is further assumed the temperature distribution along the displacer is identical to that of the cylinder when the displacer is at the end of its stroke. The temperature at the cold end of the displacer is T_{DE} and at the warm end is T_{DC}.

Now let the displacer move to the other extreme end point and assume the temperature does not change during the movement. The displacer, everywhere along its length, will be cooler than the adjacent cylinder wall and heat will be transferred to the displacer from the cylinder to equilibrate

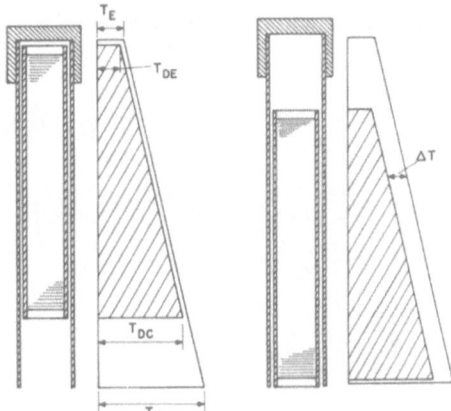

Fig. 9.2. Shuttle heat transfer.

the temperatures. The effect is an additional heat leak to the cold end of the cylinder called shuttle heat transfer. It has been investigated by Zimmerman and Longsworth (1971) and others. The graphic description "bucket-brigade loss," coined by Finkelstein in pioneer studies of the effect, has now been superseded by the more prosaic shuttle heat transfer term. The magnitude of the shuttle heat transfer is minimized by long cylinders having a shallow temperature slope and a short stroke of the reciprocating elements.

BALANCING

In most cryocooler applications it is important to balance the revolving and reciprocating parts as far as possible. Complete balance can rarely be achieved. Lack of proper balancing results in mechanical vibrations, and increases bearing loads and stresses in the structural elements of the system.

We consider here the fundamentals of engine balancing applied to the principal cases likely to be encountered in small cryocoolers, but many different engine configurations are possible and space limitations preclude a comprehensive treatment. For further information, readers are referred to the many excellent texts on the subject (e.g., Bevan, 1946).

Case 1: Single Revolving Mass

Figure 9.3 shows a mass M assumed to be concentrated with the center of gravity a distance r from the axis of rotation and attached to a shaft rotating with angular velocity ω. A centrifugal force, of magnitude $M\omega^2 r$, due to the inertia of the mass will act radially outward.

To counteract the effect of this inertia force, a balance weight B may be attached at a radial distance b to the rotating shaft *in the same plane*

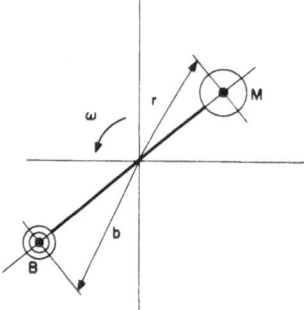

Fig. 9.3. Balance of single revolving mass.

of rotation as the unbalanced mass M. Then for balance

$$B\omega^2 b = M\omega^2 r \qquad (9.5)$$

or

$$Bb = Mr \qquad (9.6)$$

Those unfamiliar with the subject of engine balancing may note that the magnitude of the unbalanced and counter balance force is a function of the square of the shaft angular velocity ω. The angular velocity $\omega = 2\pi N$, where N is the speed of the shaft in revolutions per second. The product Bb of the balancing mass B and its radius b can be apportioned between B and b as desired. It is generally advantageous to select the radius b to be large so as to minimize the extra mass B added to the system. It is rarely possible to add a single B mass which rotates in the same plane as the unbalance mass except in the case of fly-wheels or other rotating disks.

In most other cases it is not sufficient to use a single balancing mass, for although the two inertia forces can be equal in magnitude and opposite in direction, they have different lines of action. This gives rise to a couple which tends to rock the shaft in its bearings as illustrated in Fig. 9.4, a diagrammatic representation of a double throw crankshaft X–X with the throws 180° out of phase. Mass M is attached at radius r from the axis and distance l from the midpoint O of the crankshaft. Mass B is attached at radius b from the axis and distance d from the midpoint O.

Now if $Mr = Bb$, the inertia *forces* due to the two masses will be equal and opposite. However, taking moments about O there is an unbalanced clockwise couple $M \cdot r \cdot l$ due to mass M and another unbalanced couple $B \cdot b \cdot d$ *in the same clockwise direction* due to mass B. When rotating about axis X–X, the shaft would experience a couple tending to rotate the shaft, instantaneously, about axis YY. This unbalanced couple would be

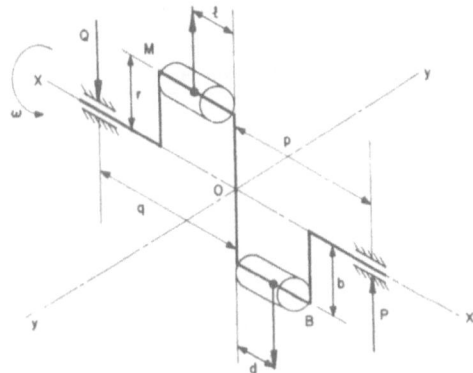

Fig. 9.4. Balance mass not in same plane
as unbalanced mass.

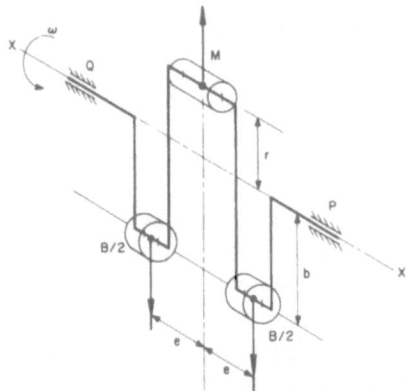

Fig. 9.5. Twin balance masses symmetrically located about the unbalanced masses.

resisted by restoring couples exerted by the bearing load P and Q acting at distances p and q, respectively, from the reference point O.

In cases where the balance weight cannot rotate in the same plane as the disturbing mass, complete dynamic balance of forces and couples can be achieved by dividing the balance mass into two parts. This is shown by the symmetric three throw crankshaft of Fig. 9.5. The balance mass B to achieve the balance force $B \cdot b$ is divided into two equal masses $(B/2)$ located at the same distance e from the reference point O. In this arrangement, there are no unbalanced forces or couples.

Case 2: Several Masses Revolving in the Same Plane

Figure 9.6 shows a system of four masses M_a, M_b, M_c, M_d at different radii r_a, r_b, r_c, and r_d having the angular disposition shown, and rotating about an axis through O perpendicular to the page. All the masses are

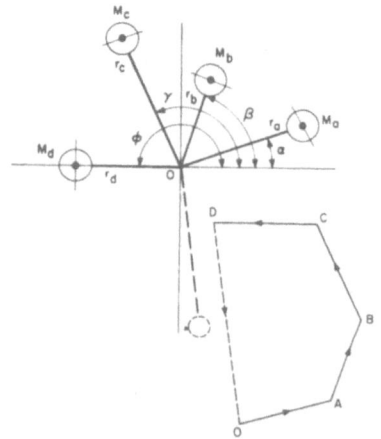

Fig. 9.6. Several masses revolving in the same plane.

assumed to have the same plane of rotation. The system may be balanced by a single balance mass B at radius b from the axis of rotation.

The necessary magnitude and angular disposition of the balance weight may be determined by constructing the vector diagram shown in Fig. 9.6. The vector OA represents the unbalance force $M_a \cdot r_a$ and is drawn along the line of action of the centrifugal force due to M_a. Similarly, the vector AB represents the unbalance force $M_b \cdot r_b$ drawn parallel to the line of action of the centrifugal force due to M_b.

The closing vector DA is the *balance* force $B \cdot b$ necessary to balance the system. The direction of the balance vector indicates the angular disposition of the balance mass.

Case 3: Several Masses Rotating in Several Planes

This case is only encountered in large cryocoolers where there are several cylinders in-line necessitating the use of multiple-throw crankshafts. Balance can be achieved by the use of one or more balance weights attached to the shaft at the correct disposition to compensate for both unbalanced forces and couples.

Reciprocating Masses

Consider the system shown in Fig. 9.7. This is the familiar crank–connecting-rod system used to convert rotary to linear motion. A mass R constrained to move in a straight line X–X', is attached to a connecting rod of length l. The other end of the connecting rod is attached to a crank arm of radius r. It is customary for the line of motion X–X' of the reciprocating mass R to pass through the centre of rotation O of the crank.

It can be shown that the inertia force of the reciprocating mass R is

$$F_R = R\omega^2 r\left(\cos\theta + \frac{\cos 2\theta}{n}\right) \qquad (9.7a)$$

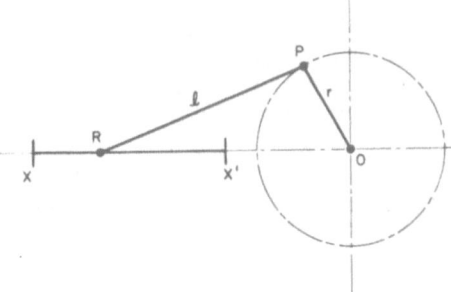

Fig. 9.7. Crank–connecting-rod system.

or

$$F_R = F_p + F_s \qquad (9.7b)$$

where n is the connecting-rod–crank ratio l/r. The component $F_p = R\omega^2 r \cos\theta$ is termed the *primary* inertia force of the reciprocating mass. The component $F_s = R\omega^2 r(\cos 2\theta)/n$ is termed the *secondary* inertia force.

If the piston moves with simple harmonic motion, the primary inertia force only has to be considered. The secondary inertia force arises out of the departure from simple harmonic motion due to the obliquity of the connecting rod. The maximum value of the secondary inertia force is only $1/n$ times the maximum value of the primary force, but occurs *four* times per revolution of the crank compared with twice per revolution of the primary force. Attempts to minimize the secondary inertia force by selecting a high value of the connecting-rod/crank ratio n will result in a long connecting rod and, consequently, an undesirably long and heavy engine.

The unbalanced inertia force due to a reciprocating mass is constant in direction (along the line of reciprocating motion) but varies in magnitude, reaching a maximum value at the ends of the stroke. This contrasts with the inertia force due to a rotating mass which is constant in magnitude but continuously variable in direction. A single revolving mass cannot completely balance a reciprocating mass. The best that can be achieved is partial balancing of the primary inertia force.

Partial Primary Balancing

Consider the balance weight B at radius b in Fig. 9.8 revolving with, and directly opposite, the crank. The horizontal component, F_{bh} of the centrifugal force generated by the balance weight is

$$F_{bh} = -B\omega^2 b \cos\theta \qquad (9.8)$$

The vertical component F_{bv} of the centrifugal force generated by the balance

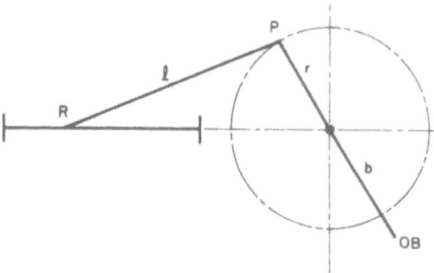

Fig. 9.8. Primary balance of crank–connecting-rod system.

weight is

$$F_{bv} = -B\omega^2 b \sin\theta \qquad (9.9)$$

The primary inertia force F_{Rp} of the reciprocating mass R is horizontal in direction and of magnitude:

$$F_{Rp} = R\omega^2 r \cos\theta \qquad (9.10)$$

The resultant horizontal force may be obtained by summing the primary inertia force due to mass R and the horizontal component due to the balance weight B, i.e., $(Rr - Bb)\omega^2 \cos\theta$. If Bb is selected to be equal to Rr then there would be no out-of-balance force parallel to the line of stroke. Unfortunately, the vertical out-of-balance force due to B remains. If $Bb = Rr$ then the vertical out-of-balance inertia force $F_{bv} = -B\omega^2 b \sin\theta$, will go through the same variation of magnitude as the original horizontal out-of-balance primary force of the reciprocating mass, $F_{Rp} = R\omega^2 r \cos\theta$. The addition of the balance weight has thus simply acted to change the direction of the out-of-balance force by 90°.

Sometimes it is advantageous to select the product $B \cdot b$ to be some fraction z of the product Rr. The horizontal out-of-balance force is reduced to $(1 - z)Rr\omega^2 \cos\theta$ and a vertical out-of-balance force $zRr\omega^2 \sin\theta$ is introduced. By way of illustration, let $z = 0.5$. The forces acting then are shown in Fig. 9.9. The primary inertia force due to the reciprocating mass R is Om. The centrifugal force due to the rotating mass B is Ot which can be resolved in a horizontal component On and a vertical component Ok. Summing the horizontal forces Om and On gives a resultant horizontal force Oy. Combining this with the vertical force component Ok gives a resultant force Oq. The resultant force remains constant in magnitude, $Rr\omega^2/2$ as the crank revolves. Its line of action is inclined at the same angle to the horizontal as the crank *but it revolves in the opposite direction to the crank.*

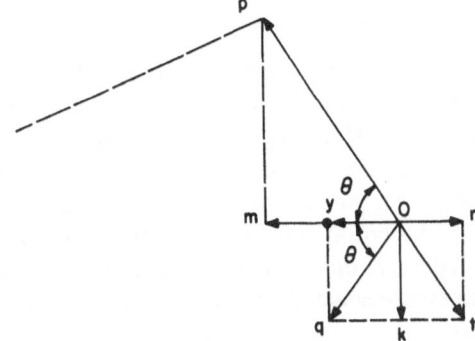

Fig. 9.9. Partial primary balance of crank–connecting-rod system.

If the balance weight B has to balance the rotating parts, say mass M at radius r, as well as partially balance the reciprocating parts, the product $Bb = Mr + zRr = (M + zR)r$. The mass of the connecting rod may be apportioned between the reciprocating mass R and the rotating mass M as appropriate. A division of one third of the connecting rod mass to the reciprocating element and two thirds to the rotating element is frequently assumed.

Secondary Inertia Forces

The primary inertia force of a *reciprocating* mass R is identical to the component, along the line of reciprocating action, of a *rotating* mass of the same magnitude R, attached to the crank. The secondary forces can be considered in the same way. The secondary inertia force due to the reciprocating mass arises as a consequence of the departure from simple harmonic motion because of the obliquity of the connecting rod.

The magnitude of the secondary inertia force is approximately

$$F_s = R\omega^2 r \frac{\cos 2\theta}{n} \tag{9.11}$$

This may be rewritten

$$F_s = R(2\omega)^2 \frac{r}{4n} \cos 2\theta \tag{9.12}$$

which can be recognized as the centrifugal force component along the line of stroke of mass R attached to an imaginary secondary crank of length $r/4n$ revolving at *twice* the speed of the actual crank. The two cranks coincide at the inner dead center position and the imaginary crank always makes an angle 2θ measured from the inner dead center.

Multiple Reciprocating Masses

Many cryocoolers have two or more reciprocating masses rather than the single reciprocating mass considered so far. Figure 9.10 illustrates three mechanical arrangements commonly encountered with integral Stirling or Vuilleumier engines. There are two reciprocating masses, the piston and displacer or two pistons. It is unlikely the magnitude of the two masses will be equal and there is little virtue in striving for this. Pistons are heavy structural elements operating at ambient temperatures and equipped with pressure seals. Displacers are lightweight structural elements with substantial temperature but no pressure gradient along the length. The strokes of

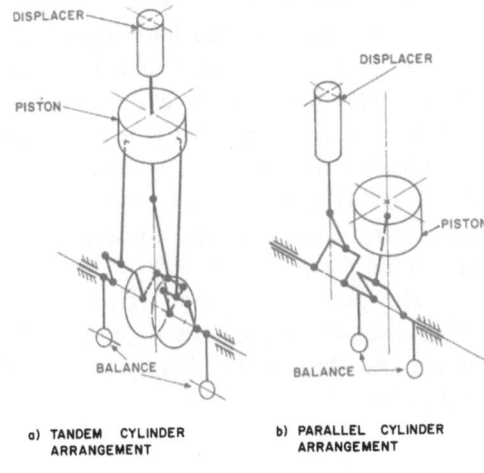

a) TANDEM CYLINDER
ARRANGEMENT

b) PARALLEL CYLINDER
ARRANGEMENT

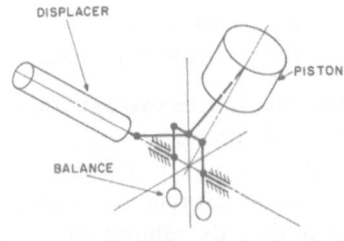

Fig. 9.10. Mechanical arrangements
for Stirling or Vuilleumier cryocoolers.

c) VEE CYLINDERS ARRANGEMENT

the reciprocating elements will be different in most cases. The motion will
always be out of phase with the displacer (or expansion space piston)
leading the piston by an angle of 60 to 120 degrees depending on the
kinematic drive.

Two approaches are possible in the design analysis of linkages with
multiple reciprocating masses. One method is to investigate separately the
unbalance forces and couples arising from individual rotating and
reciprocating masses. The magnitude and location of balance weights to
achieve an acceptable compromise may be determined for each rotating
or reciprocating mass. The spectrum of balance weights determined may
be consolidated into a dynamically equivalent single mass or, for symmetrical designs, twin masses disposed about the engine center-line. This
approach, pedantic in style and tedious in application, is recommended for
those new to the problem of engine balancing and also for machine designers
working creatively at the drawing board. Separate resolution of force–

couple relationships for the various masses involved provides for recognition of the greatest and least significant effects on which to concentrate attention. This extended analysis process seems to create the mental environment for innovative design approaches.

The second approach is to establish equations for the total system of combined multiple reciprocating and rotating masses. This can lead to considerably complex equations but is the best approach for computer parametric analysis of the consequences of change in the geometry of the kinematic drive system.

The preferred design technique is to generate a provisional design using visceral methods by experienced designers working at the drawing board and then refine this by parametric optimization with the aid of a computer.

Design Guidelines for Engine Balancing

The out-of-balance forces due to rotating and reciprocating masses are

i. for the rotating mass $F = M\omega^2 r$,
ii. for the reciprocating mass $F_R = R\omega^2[\cos\theta + (\cos 2\theta)/n]$.

Consideration of these two equations leads to design guidelines to minimize out-of-balance forces.

The most significant parameter is speed. In both the above equations, the out-of-balance forces are a function of the *square* of the angular velocity ω, which is directly related to engine speed, $\omega = 2\pi N$, where N is the engine speed. To minimize out-of-balance forces, the engine should be operated as slowly as possible. A reduction to half speed reduces the out-of-balance forces to one quarter their initial value.

Other parameters affecting the magnitude of the out-of-balance forces are the masses, M (rotating) and R (reciprocating) and the radius r of the crank. Reduction in mass and piston stroke S ($S = 2r$) will decrease the rotating and reciprocating out-of-balance forces. Short-stroke, large-diameter engines with lightweight pistons will have advantages over long-stroke, small-diameter units with heavy pistons. There are further advantages of short stroke in terms of

i. reduced rubbing velocity of the piston seal and guide rings,
ii. reduced shuttle heat transfer losses.

In cryocoolers, it is rare for piston stroke to exceed half the piston diameter, i.e., $S = D/2$ or $r = D/4$.

The reciprocating mass R of the piston may be expressed as the volume of a partially hollow cylinder of diameter D, length L times the density of the material ρ, i.e.,

$$R = \rho \frac{\pi}{4} D^2 L X \tag{9.13}$$

where X is some factor to allow for the fact the piston is not solid (say $X = 0.5$). The length L of the piston is equal to the piston diameter D, approximately. Now writing $r = D/4$, $X = 1/2$, $L = D$, we can see that the reciprocating out-of-balance force is

$$F_R = \frac{\pi}{16} \rho D^4 \left(\cos \theta + \cos \frac{2\theta}{n} \right) \tag{9.14}$$

This indicates the reciprocating out-of-balance forces, both primary and secondary, are strong (fourth-power) functions of piston diameter and to much lesser extent of the density of the piston material—good reasons to select *small* diameter pistons constructed of lightweight material (aluminum, titanium, epoxy–glass).

To minimize out-of-balance forces, therefore, short-stroke, small-diameter pistons running slowly are necessary. Here it may be recalled that the refrigerating capacity Q of a cryocooler is

$$Q = f_n(\bar{p} V_c N) \tag{9.15}$$

where $\bar{p}$ is the mean pressure, V_c is the swept volume of the piston and N is the speed of the machine. Now $V_c = (\pi/4)D^2 S$ and if the stroke is selected to be half the piston diameter, i.e., $S = D/2$, then

$$V_c = \frac{\pi}{8} D^3 \tag{9.16}$$

and so

$$Q = f_n(\bar{p} D^3 N) \tag{9.17}$$

The choice of low-speed N or small-diameter piston, D, would, therefore, require the use of a very high pressure to maintain the refrigeration capacity. With very high pressure, there would be consequent increases in

  i. system weight because of increased wall thickness,
 ii. flow losses because of the increased fluid density,
iii. increased conduction losses because of the increased wall thickness.

Perfect Dynamic Balance

It is not possible to balance the primary and secondary forces of a single reciprocating mass. With two masses of the piston and displacer normally utilized in cryocooler arrangements the problem of dynamic balance will be further compounded. Perfect dynamic balance can be achieved, but the price is an increase in the complexity of the kinematic drive. Figure 9.11 illustrates two machine arrangements which can be completely balanced. Figure 9.11a is a system of two reciprocating masses which are balanced by the addition of a mirror image of the system operating 180° out of phase and incorporating two additional reciprocating masses identical in magnitude to their counterparts.

The second system shown in Fig. 9.11b is commonly known as the rhombic drive. One reciprocating mass is rigidly attached to the upper horizontal bar of a relatively complicated linkage. The other reciprocating mass is attached to the lower horizontal bar. Balance weights of appropriate magnitude are attached to the linkage in the proper location to completely convert the vertical out-of-balance forces to horizontal out-of-balance forces. The two halves of the linkage are symmetric and coupled through contrarotating gears so the horizontal out-of-balance forces are identical but mutually opposed so they cancel each other leaving the system perfectly balanced.

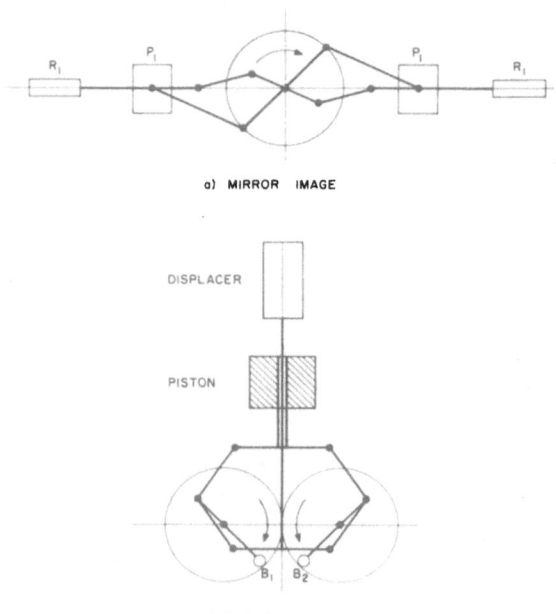

Fig. 9.11. Cryocooler arrangements that can be perfectly balanced.

The rhombic drive was invented in the early 1950s by Meijer (1959) at the Philips Research Laboratories, Eindhoven, Netherlands and was used on Philips Stirling engines of the 1950/1960 period. It was also used on the large Werkspoor cryocooler illustrated in Fig. 3.5b, the small spacecraft cryocooler shown in Fig. 3.9, and the spacecraft Vuilleumier cryocooler shown in Fig. 4.4. Earlier, Lanchester, the celebrated British automotive pioneer used a similar mechanism in his two-cylinder opposed piston automotive engine of 1901 (see Crabtree, 1973).

BEARINGS

Cryocoolers contain three main types of bearings: rotary, linear, and oscillatory. All three types are represented in the crank slider mechanism illustrated in Fig. 9.12. Rotary bearings are used for the main crankshaft bearings. Oscillatory bearings are found at the big and small ends of the connecting rod. Linear bearings are provided on the piston to guide and support the piston motion in the cylinder.

There may be several rings provided on the piston and look superficially to be all the same. Closer inspection will show there are two types. One serves as a seal to prevent the passage of working fluid or lubricant from one side of the piston to the other. The remaining rings serve as linear bearings to guide and support the piston in its reciprocal motion. The principal difference is that the seal ring is usually fluid pressure activated, has a thin flexible section, and is a loose fit in the piston ring groove and the cylinder. The bearing rings, sometimes called the oil scraper or guide rings, are solid rings, may be split for ease of assembly, have a rigid section, and are a good fit in the piston and the cylinder ring groove.

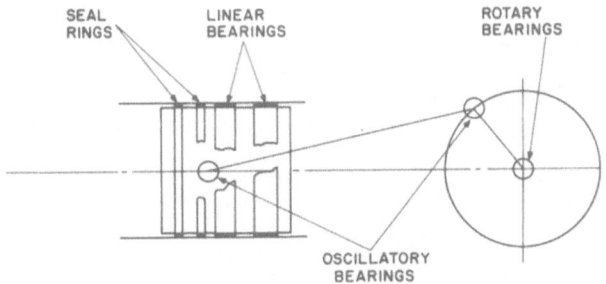

Fig. 9.12. Types of bearings.

Fluid-Lubricated Bearings. Fluid-lubricated bearings utilize hydrodynamic wedge action to maintain a film of lubricant (at high pressure)
between the two surfaces moving relative to one another. This is illustrated
in Fig. 9.13 for both rotating bearings and linear sliding bearings. The
lubricant in the tapered space between the moving surfaces is raised to a
high pressure by the relative motion between the two surfaces and prevents
the surfaces from coming into contact. Forces acting on the two elements
will close the gap between the moving surfaces, but as this occurs the
hydrodynamic fluid pressure in the gap increases to maintain the integrity
of the film. The viscosity of the fluid is an important property affecting the
hydrodynamic wedge. A low viscosity fluid is unable to support the same
load as one of higher viscosity. Of course, the forces acting may be so great
that the film breaks down and the surfaces come into contact, usually with
catastrophic consequences.

Relative motion between the two surfaces is necessary to cause the
flow of lubricant into the wedge and the creation of the fluid pressure that
keeps the surfaces apart. In some bearings the motion reverses periodically
and at the instant of reversal, the relative velocity is zero. Momentarily,
therefore, the hydrodynamic wedge action ceases and contact between the
surfaces may occur if the bearing load is high. Bearings which experience
reversal include reciprocating linear bearings on pistons and crossheads
and oscillatory bearings at the big and small ends of connecting rods.

In addition to providing fluid necessary to establish the hydrodynamic
wedge, the flow of lubricant performs other important functions. It carries
off the frictional heat energy and washes away any detritus or wear debris
resulting from the bearing action. Little fluid is required to maintain the
hydrodynamic wedge. Most of the flow customarily supplied to bearings
performs the supplementary cooling and cleaning functions. Over supply
of lubricant can increase friction of the bearing due to the intense motion
and agitation of the lubricant acting as a fluid brake.

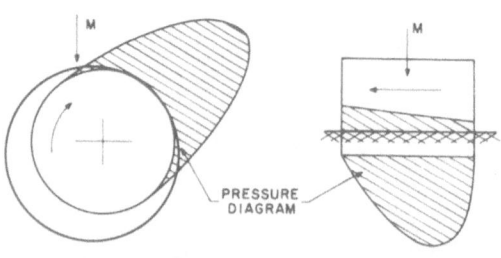

a) HYDRODYNAMIC WEDGE   b) LINEAR BEARING   Fig. 9.13. Hydrodyanamic lubri-
   IN A JOURNAL BEARING   cation.

Lubricant can be supplied in disciplined flows by an oil pump feeding lubricant through ducts or pipes to each bearing. Alternatively, "splash" lubrication can be used in which a variety of scoops and projections dipping into the oil reservoir cause the oil to be agitated so that sufficient will find its way to the bearings.

Oil or Gas Lubrication. Hydrodynamic fluid-lubricated bearings may operate with liquid or gas. The principal difference between the fluids is the high viscosity of liquid compared with gas. This permits relatively high bearing loadings and comparatively relaxed geometrical constraints in the bearing.

Many liquids have good lubricant properties, but oil is generally the preferred choice. Mineral oils specially formulated for machine lubrication are available in a wide variety of grades. These oils range from light distillate oils for delicate mechanisms to heavy-grade oils containing additive suspensions for use in extreme pressure systems, e.g., gears. Oil-lubricated systems have low rates of wear and minimal frictional heat generation. Oils are noncorrosive and have the propensity to "wet" most surfaces to provide an enduring lubricant film.

Grease-Lubricated Bearings. Bearings lubricated with grease are a subgroup of oil-lubricated bearings. Machine grease is a mixture of semi-solid composition containing mineral oil and various additives. In the vicinity of the bearing, the grease liquefies and behaves as a viscous oil lubricant. Grease-lubricated bearings are used in situations where normal oil lubrication is not feasible.

Grease-lubricated bearings require some form of shield to keep the grease in place around the bearings and foreign matter from entering. Manufacturers offer a wide range of pre-lubricated and sealed ball and roller bearings. Plastic seals are provided on both sides of the bearing, and there may be concern at the seal friction loss in complicated drive systems with multiple bearings (e.g., rhombic drive). The seals are not hermetic seals and eventually some of the grease will escape from the bearing. Furthermore, because the frictional heat is not "washed away" from the bearing by a continuous flow of lubricant, there are rather severe limitations on the permissible loadings of presealed grease-lubricated bearings.

Ball and Roller Bearings. Ball and roller bearings are attractive because of their low static friction coefficient ($\mu = 0.04$) compared with journal bearings ($\mu = 0.4$). Once moving, there is little difference in the *dynamic* friction of the various types. Ball and roller bearings are, therefore, most advantageous for bearings subject to repeated stops and starts, such as those found at the large and small ends of connecting rods. Unfortunately, their use tends to complicate construction of the crankshaft or bearing housing. The bearing inner and outer tracks are usually solid, continuous

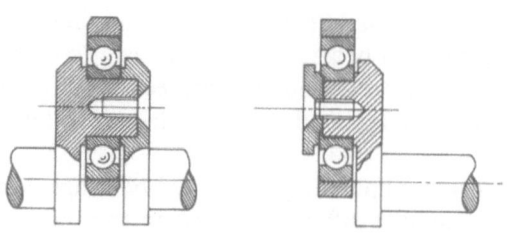

a) COMPOSITE CRANKSHAFT   b) OVERHUNG CRANK

Fig. 9.14. Crankshaft designs for ball or roller bearing.

rings. Therefore, the crankshaft has to be a composite assembly of various parts so that the crank pins can be inserted in the bearing inner track during assembly, as illustrated in Fig. 9.14. There are significant advantages in the use of solid (forged) crankshafts, and so to overcome this difficulty "split" ball and roller bearings are sometimes used.

Superficial inspection suggests that ball and roller bearings utilize a pure rolling motion of the elements. Friction effects might be expected to be zero with no lubrication required. In practice, there is significant rubbing and sliding action taking place between the various elements of the bearing. Lubrication is necessary to establish hydrodynamic wedge action between the balls and rollers at their contact region with the bearing tracks.

Gas Bearings. From the bearing point of view, oil, and to a lesser extent grease, are the best available lubricants. However, the propensity of oil to wet a surface (the very characteristic that makes it so attractive as a lubricant) tends to distribute a film of oil over all the internal surfaces of a machine. In many applications, this does not matter, but it is important in cryocoolers.

One of the functions of the crankcase–working-space seal is to keep oil out of the working space. It proves to be just as difficult to keep the oil out as it is to keep the working fluid in. The principal consequence of contamination by oil of the working space is that the regenerator becomes blocked. Oil coming into the working space will vaporize in the working fluid and eventually be deposited in the regenerator. The finely divided material of the matrix is an excellent filter. These difficulties can be avoided by use of the working fluid itself as the lubricant. Unfortunately, the characteristics of the working fluid that make it attractive from the thermo-dynamic aspect are not congruent with the characteristics best suited for gas-lubricated bearings.

In cryocoolers, the two working fluids customarily used are hydrogen and helium. One reason is that all other fluids liquefy at higher temperatures. The principal reason is that both hydrogen and helium have a low molecular weight and such favorable transport properties (viscosity, thermal conductivity, and specific heat) to make them the heat transfer fluids of choice even in high-temperature applications. It can be easily shown (see Walker, 1980) that the ratio of heat transferred to work done in pumping the gases is, primarily, a function of the molecular weight. This is the primary reason why hydrogen (molecular weight 2) closely followed by helium (molecular weight 4) are preferred as the working fluids for Stirling engine power systems.

For gas bearings, the low viscosity of hydrogen and helium is a disadvantage and a denser fluid of higher viscosity would be more suitable. Nevertheless, gas bearings have been successfully applied to cryocoolers and there is every possibility that further applications will be made for all kinds of shaft oscillatory bearings and reciprocating members.

Gas-Lubricated Pistons. The earliest application of gas bearings to cryocoolers, by Kapitza (1934), was the hydrodynamic gas-lubricated piston of the reciprocating expansion engine of his Claude cycle helium liquefier (see Fig. 7.17). Later Collins (see Chapter 7) used the same principle for the reciprocating expansion engines of his helium liquefiers. It is presently used in the reciprocating expanders of the Doll–Eder helium liquefiers manufactured by the German Linde Company and in a variety of other Claude cycle expansion engines discussed in the literature. The principle of the gas-lubricated piston is shown in Fig. 9.15. It simply involves leakage of working fluid around the piston from a high-pressure region on one side

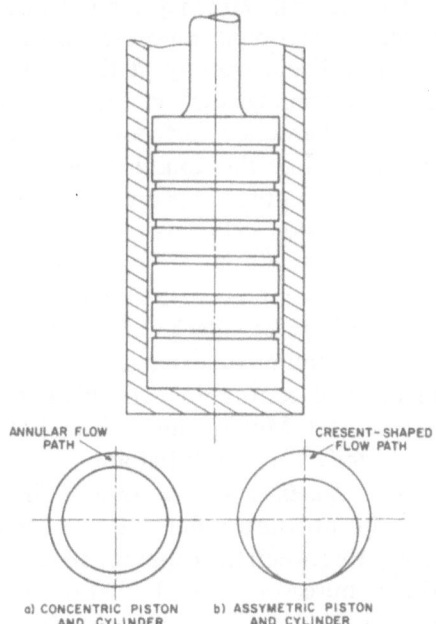

Fig. 9.15. Gas-lubricated piston.

ANNULAR FLOW PATH

CRESENT-SHAPED FLOW PATH

a) CONCENTRIC PISTON AND CYLINDER

b) ASSYMETRIC PISTON AND CYLINDER

of the piston to a low-pressure region on the other side of the piston. Of course, such leakage is highly undesirable from a performance point of view because it represents a direct loss of fluid that could assist in the creation of refrigeration.

The rate of flow of fluid past the piston depends on the pressure difference across the piston, the length of the piston, and, primarily, the cube of the clearance space between the piston and the cylinder. The flow can be minimized by use of a long, small-diameter piston and with a close tolerance between the piston and cylinder diameter. Leakage of up to 10% of fluid involved in the expansion is considered acceptable to gain the advantages of a gas-lubricated piston.

The fluid flow between the piston and the cylinder wall does not act so as to center the piston in the cylinder. The natural tendency is to displace the piston to one side or another as shown in Fig. 9.15. This creates a crescent-shaped flow path with rather less flow resistance than the annular path of a perfectly centered piston. The natural tendency to offset the piston can be countered to some extent by the provision of circumferential grooves along the piston perpendicular to the axis. These serve to distribute fluid uniformly around the circumference of the piston and as labyrinth type sealing to reduce excessive leakage. The grooves are typically of square section of $\frac{1}{16}$ to $\frac{3}{32}$ in. side and about $\frac{1}{2}$ in. pitch on a piston about 2.5 to 3 diameters long.

The natural tendency for the piston to offset makes it necessary to anticipate occasional minor surface contact between the piston and the cylinder wall particularly at the ends of the stroke. Any consequences (scratches, scuffing, etc.) of this occasional contact can be minimized by the use of hardened surfaces (chrome carbide, anodized aluminum) for both surfaces or the combination of a hard surface for one and a soft surface for the other. For cryocoolers, where very low temperatures occur in the expansion cylinders, the combination of two hardened surfaces is preferred to avoid the use of dissimilar materials and the consequence of differential thermal contraction of the piston and cylinder. Hard anodized aluminum is an excellent material from the aspect of wear. However, the use of aluminum for expansion engines poses difficulties because of the high thermal contraction and high thermal conductivity compared with stainless steel. Molybdenum disulfide liberally applied in powder form or as a surface treatment is helpful.

Gas Bearings on Shafts and Flat Surfaces. Gas-lubricated pistons utilize the pressure differential across the piston to establish a lubricant film of gas between the rubbing surfaces. Up to 10% of the fluid compressed in the compressor is simply dissipated to the low pressure region to feed the bearing. The same principle can be applied to gas bearings used for

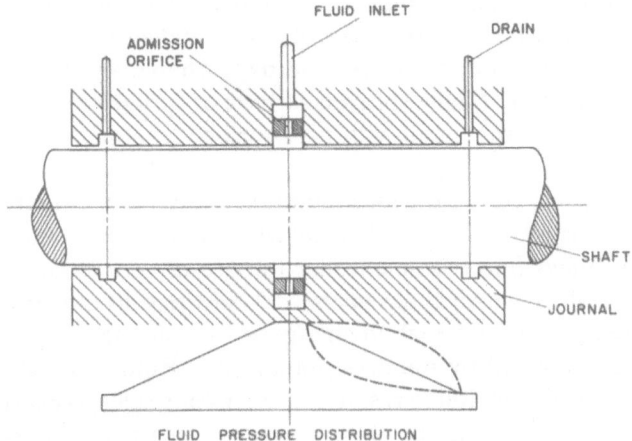

Fig. 9.16. Gas-lubricated bearing.

rotating and reciprocating shafts or flat surfaces. The elements of a gas bearing are shown in Fig. 9.16. A circular shaft is mounted on gas-lubricated journals having an overall length L. At the midpoint one or two rows of fine orifices contained in a shallow groove around the circumference admit pressurized fluid to the bearing. The fluid escapes in an axial direction away from the admission ports to the fluid drains located at the extremities of the bearing. From these, it is conveyed back to the compressor inlet. Alternatively, the fluid may simply drain into low-pressure regions adjacent to the bearing.

The pressure distribution along the bearing is shown in Fig. 9.16. The fluid pressure ranges from a high value in the vicinity of the admission ports to a low value near the drains. The pressure distribution is represented as linear between these two values, but can, in practice, be anywhere between the broken lines shown on the figure. The upper curve is indicative of a bearing capable of carrying a very high load, with minimal clearance between the shaft and journal, or an overly generous gas supply through oversize ports. The lower curve is indicative of a bearing starved of fluid with constricted or blocked admission ports and excessive clearance between the shaft and the journal. This bearing could carry only light loads.

Gases are highly mobile compared with liquids because of their low density and viscosity. This is particularly true for hydrogen and helium. Gas bearings must, therefore, be manufactured with very close clearances between the shaft and the journal, otherwise the fluid flow will be excessive. The clearance will be the smallest achievable at the limits of manufacturing technology, typically less than 0.0005 in. per inch diameter, with the mating surfaces customarily finish ground or ground and lapped. It is good practice to finish both surfaces as hard as possible to avoid surface damage in brief contacts during the initial start-up or perhaps when overloads are suddenly applied to the system. Flame sprayed carbide or furnace carburized surface treatments are recommended with post-surface-treatment finishing to the final fit.

The extremely close clearances necessary between the shaft and journal lead to a very rigid shaft with virtually zero radial movement. As a result of this, it becomes feasible to consider the use of very close tolerance gas seals instead of mechanical gas seals with all their attendant friction and wear problems. The principal advantage is the elimination of all the frictional work generated by the dry-rubbing seal. Equally important is that the wear debris, an inevitable result of dry-rubbing seals, is also eliminated. Fine wear debris entering the extremely close clearance spaces of the gas bearings acts as an abrasive with catastrophic results.

Gas bearings require a different order of technology than the more conventional oil- or grease-lubricated systems. They are not mass-produced as ball and roller bearings are, but must be manufactured to the highest precision and so are expensive. As a consequence, gas-lubricated bearings tend to be confined to applications particularly well suited to their use. Such applications include (a) the very high speed expansion turbines of Joule–Brayton systems, where the low fraction characteristics of gas bearings permit speeds of several hundred thousand revolutions per minute; and (b) long-life (two or three year) cryocoolers with zero maintenance. These include Stirling and Vuilleumier regenerative coolers for spacecraft and the rotating compressors and expanders of large Joule–Brayton or Claude cycle systems for industrial use. Here the principal advantages are low friction and wear without contamination by lubricant of the internal surfaces. Gas bearings are particularly advantageous in free piston engines such as those described in Chapter 3. A conceptual diagram for the hot cylinder free piston of a Vuilleumier cryocooler is shown in Fig. 9.17. It consists of a displacer operating in a cylinder. Motion of the displacer is effected by a linear electric motor located in the ambient temperature region of the machine. The displacer is mounted on a stationary centerpost with two gas bearings, one at the upper end of the displacer and one at the lower end. Movement of the displacer operates a bearing supply compressor which provides pressurized fluid to the bearing reservoir feeding the bearings. Drains for the bearing fluid are provided in the hollow centerpost for the fluid to return to the bearing supply compressor. The working space above the displacer is separated from the "crankcase" below the displacer by a close tolerance seal located at the lower, ambient temperature region of the displacer.

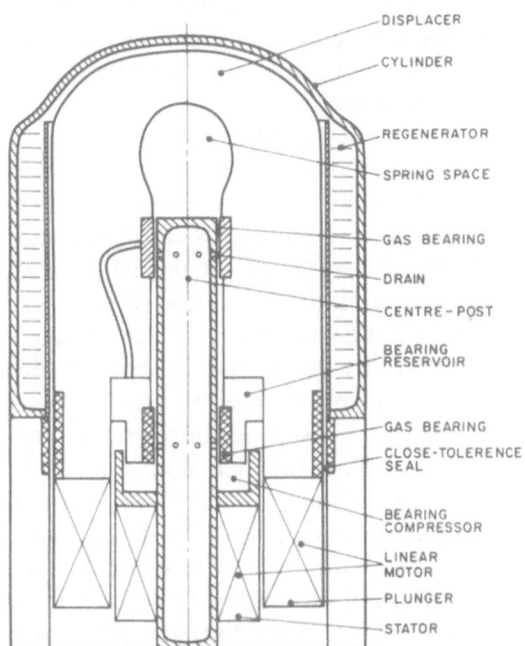

Fig. 9.17. Concept for hot cylinder
of a Vuilleumier cryocooler.

The same principles can be applied to both high- and low-temperature cylinders of a Vuilleumier cooler. Synchronization of the hot and cold displacers can be affected in a variety of ways but most conveniently by positive feedback control of the displacer linear motor drives.

Dry-Rubbing Bearings

The high precision and cost of gas bearings limit their application to a few special machines. At the same time, the need to avoid oil contamination remains imperative in all types of cryocoolers. This has led to much study and development for the use of dry-rubbing surfaces for bearings and seals as an alternative to oil bearings and close-tolerance seals.

The most widely used material for dry-rubbing bearings in cryocoolers is the fluorocarbon plastic material polytetrafluoroethylene, PTFE, known as Teflon or Fluon. PTFE has a number of attractive characteristics. It has a low coefficient of friction, high abrasion resistance, is chemically inert, and remains ductile to very low temperatures.

PTFE is available in pure form or in combination with a wide range of fillers, such as chopped glass fibre, asbestos, lead, silver, bronze, mica, and carbon. One material favored for use in cryocoolers is the glass-filled

PTFE material bearing the trade mark Rulon,* available in different formulations colored distinctively, and a wide variety of sheet, bar, tube or tape form. Design guides and recommended practices for use are available from the manufacturers.

The usual combination for Rulon or other PTFE material is a plastic-faced journal and metal shaft or a plastic guide ring and metal cylinder wall. Best results are gained with hard-surfaced metals, chrome-plated or flame-sprayed metal carbide shafts and hard anodized aluminum cylinder liners. The wear rate of the PTFE plastic rubbing on metal is very high initially, until the interstitial surface of the metal is filled with the accumulated plastic wear debris. Once enough plastic has been transferred to the metal to provide a mutual rubbing surface of PTFE, the wear rate (and friction forces) decline by orders of magnitude. The transfer takes place in the first few seconds of operation and is facilitated by leaving the metal surface slightly roughened or at least unpolished.

The transfer of PTFE to the mating metal surface is essential to satisfactory operation. The deposit can be readily observed on the shaft or cylinder after a few minutes of operation. Those unfamiliar with the material will, attempting to be helpful, clean the surface before reassembly. It is, therefore, important to explain the peculiarity of dry-rubbing bearings and seals to assistants at the earliest stage. Advisory notices on drawings are also helpful.

Results to be expected in the use of Rulon and other PTFE materials are not easily predicted, for the wear rates and slow-speed, light-load friction characteristics are highly variable. This uncertainty is emphasized in closed-cycle cryocoolers where the working fluid is hydrogen or helium. The reason is that any water vapor in the working fluid is precipitated in the cold region of the machine so that the working fluids soon become dry. The presence of slight, even trace, amounts of water contamination in the working fluid results in dramatic reduction of wear and friction. The improvement is apparently continuous but at a progressively diminishing rate as the water content increases. In closed cryocoolers there is no way to prevent the continuous and eventual total "freeze-drying" of the working fluid with consequent elevated rates of friction and wear.

Experience in the use of PTFE materials for nonlubricated compressors was recounted by Muller (1964, 1971). Data on the influence of gas atmosphere and moisture on the sliding wear of PTFE materials were given by Schubert (1971).

---

* Available in North America from the Dixon Corp., Bristol, Rhode Island, and in the United Kingdom from Henry Crossley Ltd., Bolton, Lancashire.

SEALS

All mechanical cryocoolers achieve cryogenic refrigeration by expanding gas from a high pressure to a low pressure. The gas is first compressed, cooled, and then expanded. An elementary rule is that the greater the expansion, the greater the refrigeration produced. Both high and low pressures are normally encountered in cryocoolers, and furthermore, air is not the usual working fluid but rather helium, hydrogen, argon, or nitrogen.

Seals are, therefore, necessary to prevent leakage of working fluid from the system and, to a lesser extent, the ingress of contaminants, air, water, dust, oil, to the system. Internal seals are also necessary to prevent (or at least inhibit) the flow of working fluid at high pressure to a region at lower pressure along the path of least resistance different from the desired flow path. A seal on the displacer of a Stirling engine is required to cause the working fluid to pass through the regenerator rather than simply along the annulus between the displacer and cylinder wall.

Sealing is, therefore, critical to cryocooler engineering and is complicated by the wide selection of seals displayed in manufacturers' literature. All make exaggerated claims about their products that are rarely supported by factual evidence, and the diversity of different types makes seal selection exceedingly difficult.

A useful introduction to seal technology, contributed by Fern and Nau (1976), contains an excellent overview of the subject and a useful bibliography of seal references. This slender booklet is one of 40 Engineering Design Guides prepared by the British Design Center ranging over the entire area of design engineering. A book by Buchter (1979), entitled *Industrial Sealing Technology*, is comprehensive and up-to-date, summarizing a life-time of experience in the sealing field. It is hard to find topics concerned with sealing not addressed somewhere in this book. It is a necessary reference to those professionally engaged in the design of cryocoolers.

Given these texts, each with its own bibliography, we confine our remarks to a few topics specific to cryocoolers. These are divided into the two basic groups, static and dynamic seals.

Static Seals

Static seals are devices to effect a leak-tight joint between two relatively static surfaces. Two methods in general use involve:

    i. gasket materials which require the use of an externally applied compression force to maintain sealing contact, i.e., a bolted flanged joint,

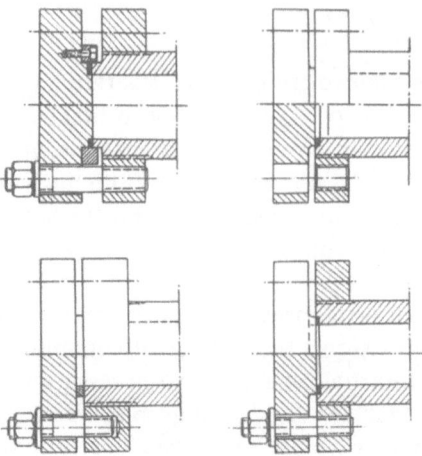

Fig. 9.18. Static seals.

ii. an automatic pressure activated seal such as an O-ring or U-ring in a groove. These units rely on the system pressure to accomplish the sealing but are normally given a small preload to maintain the sealing contact when the system is unpressurized.

Examples of each type are shown in Fig. 9.18. Neither system is completely satisfactory for use when the environmental temperature of the joint experiences substantial variation. At low temperatures, many gasket and seal materials become hard and brittle. Due to differential thermal contraction, the flange compression forces may become relaxed sufficiently to allow a leak.

It is *always* more satisfactory to design the system so as to locate joints at ambient temperatures than in either high- or low-temperature regions of the machine. Joints at cryogenic temperatures which have to be broken periodically for replacement, cleaning, or modification are best accommodated by designing the joint to be sealed with low-melting-point solder. If this is not possible, a stainless steel flange joint with stainless steel bolts may be successful. The preferred gasket material is indium, a soft white metal, and if this is not available, lead may be substituted. Both work best in the spigot and recess type flange joint shown in Fig. 9.19. Both indium and lead can be obtained in flexible "wire" wound onto spools. This can be laid in the recess of the flange and then squeezed by the spigot to seal the joint. (Parenthetically, it may be noted that cryogenic temperature joints should be avoided to also reduce the mass of metal cooled.)

Extensive research on static seals at cryogenic temperatures was carried out at the Cryogenic Engineering Laboratory of the National Bureau of Standards, Boulder, Colorado. The work was sponsored by the U.S. Air

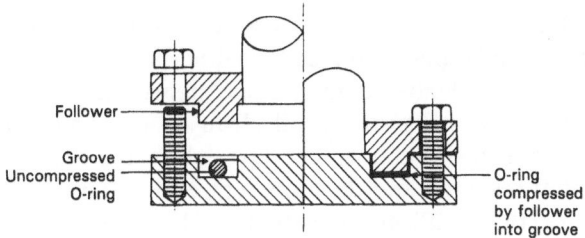

Follower

Groove
Uncompressed
O-ring

O-ring
compressed
by follower
into groove

Fig. 9.19. Spigot and recess static seal.

Force Wright–Patterson Air Force Base. Weitzel *et al.* (1965) reported comprehensively on the program on elastomeric seals and materials. Earlier reports were given in the open literature (Robbins *et al.*, 1962).

Dynamic Seals

Dynamic seals in cryocoolers are found

i. on the reciprocating pistons and displacers of Stirling and Vuilleumier, Gifford–McMahon, and Claude cycle engines, compressors, and expanders,
ii. on rotating shafts emerging from crankcases of compressors and expanders.

In many liquefiers, the compressor is simply a standard Freon refrigerating compressor suitably modified. These are oil-lubricated reciprocating compressors. Much of the modification involves apparatus downstream of the flow to clean oil from the compressed helium or other working fluid. This can be justified since the lubricant improves the performance and extends the maintenance interval of the compressor, one of the weakest links in the reliability chain.

The price to be paid is the additional size, weight, cost, complexity, and additional maintenance of the filtering and high-pressure clean-up equipment. The question facing the designer is whether a larger, dry-lubricated compressor running slower or less heavily loaded would perform better or worse, cost more or less, take up more room, weigh more, and last longer than the corresponding oil-lubricated machine with clean-up equipment downstream.

Concentrated development of the dry-lubricated compressor by the German Linde company and others have improved it by using superior seal materials and design. Muller (1964, 1971) and Schubert (1971) have given interesting accounts of the Linde developments in design of dry-lubricated compressors and the materials used.

Great numbers of small compressors and regenerative displacer cylinders of Stirling, Vuilleumier, and Gifford–McMahon machines have been made for the infrared heat-seeking missile sights and night vision equipment. In these miniature systems, oil clean-up equipment is simply not feasible on grounds of size and weight. Dry-lubricated operation is mandatory. Substantial research and development efforts have been invested in piston seals for this application, with the object of extending life and reliability. Favored candidates which have emerged as a result of this work are shown in Fig. 9.20. They are self-energized, Teflon-based lip-type seals manufactured by the Bal-Seal Engineering Co., Tustin, California, the Omni-seal, and the FCS-Fluorocarbon seal, both of which are manufactured by the Fluorocarbon Co., Anaheim, California. Design guides for application of these seals can be obtained from the companies concerned.

Piston rings are also favored both for sealing and as bearings or guide rings. An excellent reference handbook on the design and application of piston rings is available from the Koppers Co. Inc., Metal Products Division, Baltimore. For dry-lubricated compressors and expanders, the Piston rings are usually made of Teflon-based materials. Rulon A, a glass-filled PTFE (Teflon) material is the popular choice, but other fillers, principally carbon-graphite and molybdenum disulfide, are also widely used.

The principal difficulty with PTFE-plastic based materials is the wide range of variability. A dozen apparently identical piston rings made by the same machinist on the same lathe at the same time and from the same bar of Rulon A will have a wide variation in performance when tested in the laboratory. The wear rates on identical hard anodized or chrome-plated cylinder surfaces will vary by a factor of 3 and the leakage past the seal will vary by an order of magnitude. This experience is extremely frustrating.

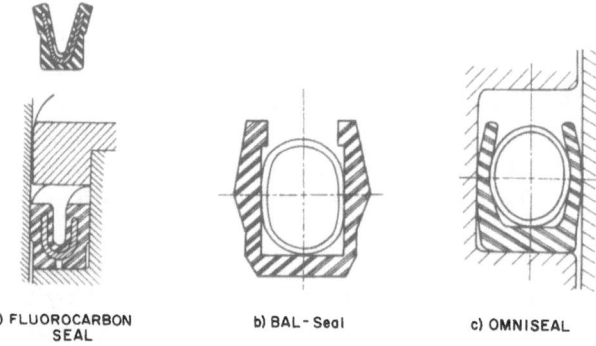

a) FLUOROCARBON   b) BAL - Seal   c) OMNISEAL
   SEAL

Fig. 9.20. Self-energized, Teflon-based, lip-type dynamic seals.

Chellis, an engineer with great experience in the field, has written briefly, in Chapter 10, about the seal problem in cryocoolers. Buchter's (1979) volume on seals contains an extensive discussion of both self-energized lip seals and the design and application of piston rings.

Piston Side Thrust

A difficulty confronting the designer of reciprocating machinery is the nonuniform wear rate of cylinder walls. The problem arises in compressors and expanders where the pistons are activated by some form of kinematic mechanism, such as crank and connecting rod, and is illustrated in Fig. 9.21. A piston connected to a crank–connecting-rod drive reciprocates in a cylinder as shown. The drive force is applied to the piston by the connecting rod along a line of action passing through the big and small end bearings of the connecting rod. The force, represented by the line $A–B$, can be resolved into components parallel to ($C–B$) and perpendicular to ($A–C$) the system center-line. The horizontal component $B–C$ varies from zero when the piston is at the bottom and top dead center to a maximum value near the midstroke position. For more details see the section on balancing in this chapter.

The wear rate of dry-rubbing surfaces is proportional to the contact-pressure between them. Therefore, the wear of cylinder wall and piston seal, due to the horizontal side-force components ($B–C$), will be proportional to the side force applied. The unsymmetrical cylinder wear will not be uniform over the entire stroke but will be least at the ends of the stroke and greatest at midstroke as indicated on Fig. 9.21.

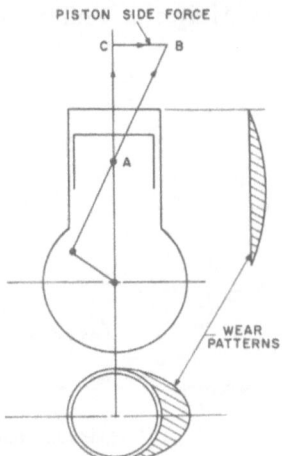

Fig. 9.21. Piston side forces.

The magnitude of the piston side force can be alleviated by increasing the length of the connecting rod. However, this is unacceptable because it increases the size and weight of the compressor crankcase; the designer is always under pressure to produce the smallest, lightest crankcase possible and is driven to use the shortest connecting rod that can be accommodated without interference of the balance weights and piston skirt. The alternative is to use a straight-line mechanism with no side forces. Three possibilities are illustrated in Fig. 9.22.

Fig. 9.22a shows a piston–crosshead system. This is basically the simple crank-slider mechanism with a crosshead added to absorb the piston side forces. The crank and crosshead can be oil lubricated in an unpressurized crankcase while the dry-lubricated piston operates in the cylinder without side forces. It is best suited for large double-acting compressors and expanders or double-acting Siemens-type Stirling engines. The gas seal between the working space and the crankcase is a critical component. One possibility is to use a two-stage cartridge-type seal unit similar to that used in the Swedish United Stirling Siemens-type Stirling engines (see Walker, 1980).

Another possibility for piston drive with no side forces is the "Scotch-yoke" system shown in Fig. 9.22b. This is worth close attention for compact machine design but appears to have eluded the attention of miniature compressor designers. The key component of a Scotch-yoke is the slider bearing operating in the horizontal slot. Satisfactory results can be obtained by mounting a presealed grease-lubricated ball or roller bearing in an overhung crank design. Similar, but larger, sealed bearings can be used also for the main shaft bearings.

Figure 9.22c shows the rhombic drive, another straight-line mechanism invented about the turn of the century by Lanchester and 50 years later reinvented by Meijer at Philips. It has the advantage of no piston and

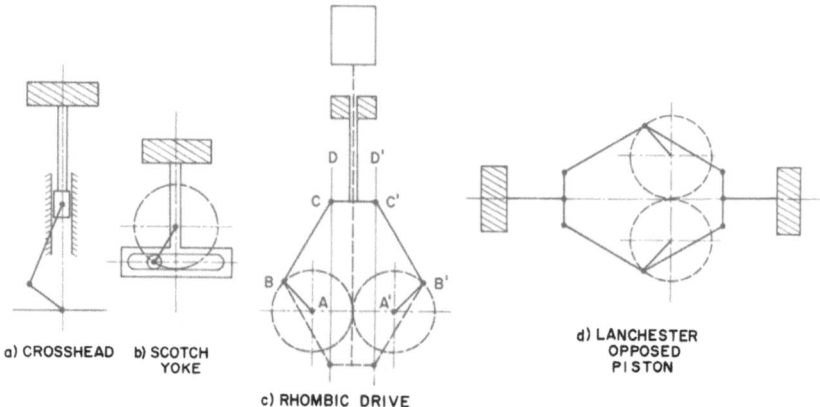

a) CROSSHEAD   b) SCOTCH
                    YOKE

c) RHOMBIC DRIVE

d) LANCHESTER
OPPOSED
PISTON

Fig. 9.22. Straight line piston mechanisms.

displacer side forces and can also be perfectly balanced. With a single reciprocating element (a compressor), the system would include only the full-line elements shown in Fig. 9.22c. Broken line elements would be deleted.

The rhombic drive is complicated, expensive, large, heavy, and hard to justify for a single reciprocating element. Two-stage compressors might be arranged in the original Lanchester form having the opposed pistons shown in Fig. 9.22d. A complete analysis and design manual for rhombic drive systems was given by Meijer (1959).

Hermetic Seals

In closed-cycle cryocoolers, there is no question that oil-lubricated mechanisms are superior in terms of friction and wear to dry-lubricated ones. At the same time, egress of oil in the working space can be disastrous. A reliable hermetic seal between the working space and the crankcase would provide the best solution. Many attempts have been made to achieve such a seal. Basically, they may be divided into three types: diaphragms, both flat and convoluted, rolling diaphragms, and bellows. An example of each type is shown in Fig. 9.23. The basic problem of all three types is reliability. Designs available to date have been unable to endure the repetitive flexing for a sufficient period. Nevertheless, steady improvements in materials and design have dramatically extended lifetimes so they can no longer be summarily dismissed or simply reserved for low-frequency applications.

The Harwell Stirling engine power generator (Cooke-Yarborough and Yeats, 1975) uses a large-diameter convoluted stainless steel diaphragm operating at a frequency of 50 Hz and has accumulated several years of trouble-free operation. Great improvement was made by mounting the diaphragm in an elastomeric seat around the circumference rather than

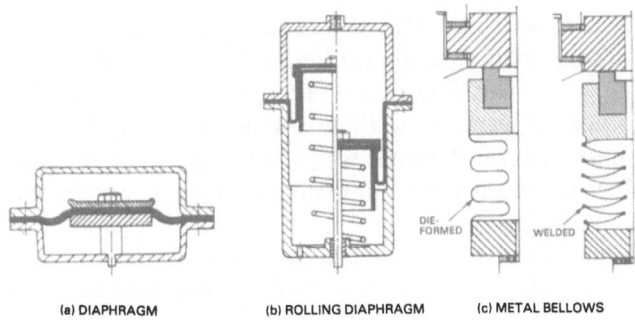

(a) DIAPHRAGM   (b) ROLLING DIAPHRAGM   (c) METAL BELLOWS

Fig. 9.23. Hermetic seals.

the customary rigid flanges. Another factor responsible for the very long life is that deflection is limited to a few thousandths of an inch.

Diaphragms and bellows have been incorporated in the Stirling engine artificial heart developed over the past decade at the University of Washington, Richland, Washington (Johnson *et al.*, 1977) (see also Walker, 1980). Extensive test and development work have been carried out over many years. This team now believes welded metal bellows and convoluted diaphragms can now be used for extended (several years) life provided proper design procedures and operating precautions are employed. Technical design data may be obtained from the Metal Bellows Corp., Sharon, Massachusetts.

Rolling diaphragm seals for Stirling engines have been the subject of intensive research and development at Philips Research Laboratories, Eindhoven, Netherlands. Excellent results have been achieved in the laboratory, but there appear to be difficulties in translating this experience into practice. References to the Philips work on seals are given in Chapter 3.

For many applications of diaphragms or bellows, the moving element must be so thin to have the necessary flexibility that it cannot withstand the full pressure difference between the inside and outside fluids. This difficulty is overcome by supporting the flexing member on a "bed" of oil or other lubricating medium at a pressure near the minimum cycle pressure. This converts the problem from that of a hermetic gas seal to a hydraulic seal from which leakage can be tolerated, provided the oil is the same as the crankcase lubricant and provision is made to replenish the leakage. An ingenious designer will use the pressure of the working fluid to pressurize the oil at a level somewhat below the working space minimum pressure.

Close Tolerance Seals

An alternative to oil or dry lubrication is to use the working fluid itself in gas-lubricated bearings. Use of fine tolerances and clearances provides a rigid mounting with virtually no lateral movement of the reciprocating and rotating parts. This situation is favorable to the use of close tolerance seals. Another imperative to their use is that fine dust generated by dry-lubricated seals can be fatal to gas bearings when it accumulates and impedes flow in the bearing admission orifices or enters the very close clearance bearing lands.

Use of gas bearings carries with it, almost by definition, use of close tolerance seals. The structure of a typical close tolerance seal system is shown in Fig. 9.24. The critical dimensions are length, L, and mean diameter, D, of the seal elements and the clearance, C, between them.

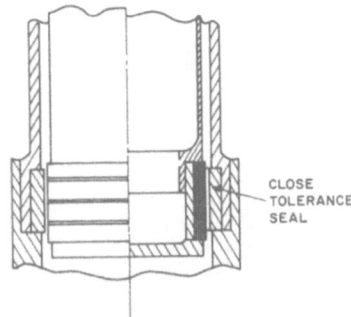

Fig. 9.24. Close tolerance seals.

Analysis of leakage through the seal shows that

$$q = \frac{A\Delta PDC^3}{L} \tag{9.18}$$

where A is the constant characteristic of the seal, ΔP is the difference in pressure across both ends of the seal, and D, C, and L are as defined above. The clearance, C, between the two elements is the most important parameter. The leakage flow varies as the *cube* of the clearance. Increasing the clearance from 0.001 to 0.002 in. will increase the leakage by eight times. The leakage loss may be minimized by increasing seal length and decreasing the diameter.

Further improvement will be gained by machining small grooves perpendicular to the leakage flow in one of the seal elements as shown in Fig. 9.24 to form a labyrinth seal. The principle of the seal is that in passage through the labyrinth of irregular shaped flow passages, the fluid is accelerated and decelerated so that the pressure energy is rapidly dissipated and the leakage flow thereby reduced. The optimum geometry of labyrinth seals is not known. Good results have been obtained with a pitch p of the grooves equal to $(D/4)$, and a groove of square section with $t = D/10$. For large seals, a pitch of 0.75 in. and section of 0.1 in. might be selected.

The clearance, C, between elements is a critical dimension which must be reduced to a minimum, usually controlled by the available manufacturing technology. With the close tolerances demanded for bearings and seals, the concentricity of the machined surfaces becomes important. It is vital for designers to anticipate the manufacturing problem and facilitate the process by concentrating critical concentricities in one or two adjacent components which can be finish ground as an assembled unit in a single setup.

To further facilitate manufacturing, it is sometimes advantageous to separate the seal elements from structural load-carrying as illustrated in

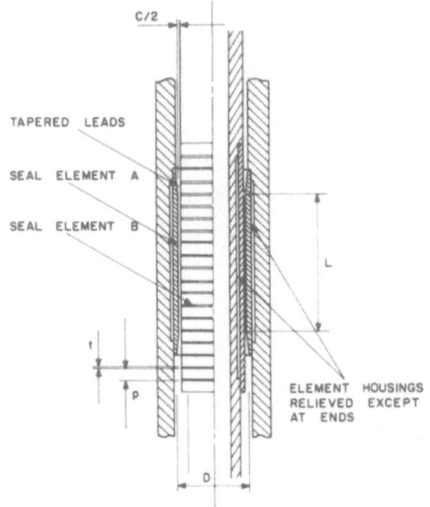

Fig. 9.25. Close tolerance seal elements.

Fig. 9.25. In this design, the two seal elements *A* and *B* are manufactured to very close tolerances and then mounted on the rod and housing. The seal elements are supported only at the ends for both rod and housing. This is to minimize distortion due to dimensional changes or flexing of the load-carrying structural members. At the ends, a larger clearance can be provided to accommodate movements of the structural members.

MATERIALS

The selection and use of materials in closed cryocoolers is so diverse as to warrant a volume in itself. Here we shall refer to the principal literature and briefly discuss a few of the elementary points of general interest. The field can be broadly divided into two sections: materials for low temperature use and materials for use at temperatures near the ambient. Here we concern ourselves only with materials at low temperatures.

An extensive literature on the subject of materials at low temperatures exists. An excellent, brief review of the field has been given by Wigley (1978). The same author earlier published a more extensive work (Wigley, 1971). Both these contain bibliographies and detailed guides to sources of further information

The proceedings of the Cryogenic Engineering Conference held in the United States (now biannually) are published as *Advances in Cryogenic Engineering* (Plenum Press, New York). There are now some 30 volumes in the series and are the prime references to all topics in cryogenic engineering, including materials. Recently (volumes in the 20's) alternate volumes have been devoted entirely to cryogenic materials as the proceedings of the International Cryogenic Materials Conference held simultaneously with the Cryogenic Engineering Conference.

Another good source of information are the proceedings of the International Cryogenic Engineering Conference published by IPC Science and Technology Press. Other articles of interest are published in the scientific and engineering journals, primarily *Cryogenics* and the *Journal of Low-Temperature Physics*.

The Cryogenic Engineering Laboratory of the National Bureau of Standards has contributed extensively to the published information on properties of materials at low temperatures. These contributions take the form of research results in scientific papers and collected results published in handbook or monograph form. Wigley (1978) lists those of principal general interest, and a list of NBS publications may be obtained on application to NBS at Washington, D.C. or Boulder, Colorado.

Significant Properties

Mechanical Properties. Properties of most significance in the selection of materials obviously depend on the use for which they are intended. For cylinder walls, regenerator casings, and other structural applications, strength is important. It is customary to use an appropriate fraction (say 0.75) of the 0.1% proof stress for the limiting design stress. Metals favored for low-temperature applications include copper, aluminum, titanium, and their alloys, high nickel–steel alloys and the austenitic stainless steels. All increase in strength and ductility as the temperature decreases and suffer little change in toughness (measured as the impact energy necessary to cause fracture in a Charpy or Izod test). The degree of change depends on the metal or the composition of the alloy and the heat treatment or mechanical cold working imposed in course of manufacture. Different heat treatments of the same metal can result in radically different strength characteristics. This is important for structures welded during assembly or fabrication where no subsequent heat treatment is possible. Special attention in design and manufacture must be given to the "heat-effected" zones near the weld where changes in crystal structure and strength properties might arise.

Toughness at low temperatures is an important property. The impact energy necessary to fracture the specimen in a Charpy or Izod test is indicative of the resistance to crack propagation and to failure following a sudden accidental overload. At low temperatures, many materials, both

metals and nonmetals, experience a pronounced reduction in their toughness or resistance to fracture on impact. Often the reduction can occur
over a very small temperature change (5–10°C) and result in a reduction
of 90% in the impact strength. The material is said to experience a
ductile–brittle transformation. Above the critical temperature, the material
is ductile and tough. Below the critical temperature it snaps off like a carrot
on sustaining a blow that normally would cause no concern.

For structural applications, toughness is the most important characteristic for use at low temperatures. The crystal structure of metals is the
principal factor affecting toughness at low temperatures. Face-centered
cubic crystal structures are found in copper, nickel, and aluminum based
alloys and the austenitic stainless steels. The strength, ductility, and toughness of these materials all increase as the temperature decreases, and thus
are widely used for low-temperature equipment.

Body-centered cubic materials experience the ductile–brittle transformation. Most carbon steels are in this group and should never be used for
structural applications at low temperatures. Steel alloys with substantial
nickel content (9% or up) are safe to use. The austenitic stainless steels
with 10% to 20% nickel and chromium are safe to use, and in many cases
are the preferred material.

Nonmetals are not used for structural applications at cryogenic temperatures because of their poor toughness characteristics, but fiber-
composite materials are becoming increasingly accepted. Glass and carbon
are the most widely used fibers in a matrix of thermosetting epoxy or
phenolic resin plastic. The fibers and the plastic are inherently brittle at
low temperatures, but in combination, are strong and tough even at the
lowest temperatures. This toughness arises from the heterogeneous structure. A crack started at one fiber does not spread to adjacent fibers as in
homogeneous materials. There is a tendency at low temperatures for the
plastic matrix to "craze" with a multiplicity of fine microcracks with leakage
of gas or liquid under pressure. This can be overcome by electroplating
nickel or copper on the internal surface of finished filament wound pressure
vessels to provide a flexible but impervious membrane.

Physical Properties. To the designer of cryocoolers, the most significant
physical properties are the specific heat, thermal conductivity, and thermal
expansion or contraction coefficient.

Thermal conductivity is an important property of the materials used
for cylinder walls and other structural elements. It controls the heat leak
to the cold region. The ratio of yield strength to thermal conductivity (σ_y/k)
is a good indicator of the *potential* value of a material for applications
involving connections spanning a wide temperature range. A combination
of high strength and low conductivity makes stainless steel the metal of

choice for many cryogenic structural applications. Aluminum alloys with less than half the strength have a conductivity an order of magnitude greater. As a general rule, low thermal conductivities are associated with high alloy materials where the electron thermal conductivity is reduced to a level comparable to the lattice conductivity.

In some instances, high thermal conductivities are required. Pure metals have the highest values. The usual choice is pure copper, (oxygen-free, high conductivity, annealed) or pure aluminum, but in exceptional situations, gold and silver are sometimes used.

Glass fiber reinforced-epoxy composites have remarkably high ratios of strength to thermal conductivity—up to 50 times better than stainless steel in the annealed condition. It is this characteristic which spurs the development of filament wound glass and carbon fiber epoxy composite structures.

Specific heat is an important property for systems where the cool down time is important. Materials with a high specific heat will consume more refrigeration to cool to their operating temperature than those with a low specific heat.

With decrease in temperature, specific heats decline for all solids and liquids but increase for gases. The decline for solids is quite dramatic. Aluminum for example, declines from 900 J/kg K at 300 K to one hundredth of that value at 20 K. A similar decline from 500 to 5 J/kg K is experienced by stainless steel. Lead, much used in finely divided form for very low temperature regenerators, has a more moderate rate of decline ranging from 130 J/kg K at 300 K to 53 J/kg K at 20 K. Ice has the remarkably high values of 2100 J/kg K at 300 K and 114 J/kg K at 20 K. This suggests that powdered ice could be an effective regenerator matrix material, but there seems to be no way to prevent the regenerator melting away when the cryocooler is turned off!

The same unfortunate characteristic is not shared by poly-tetrafluoroethylene (PTFE, Fluon or Teflon). This has a specific heat of 1020 J/kg K at 300 K and the high value of 76 J/kg K at 20 K. PTFE may, therefore, be a good regenerator matrix material, but no studies of PTFE matrices have been reported in the literature.

Dimensional or volumetric contraction experienced on cooling can, in rigid structures, lead to high stresses and cause failure or permanent distortion. During cooldown, the cylinder wall of a cryocooler may contract sufficiently to contact the piston or displacer and cause the engine to "seize." This can be aggravated by the use of dissimilar materials for the cylinder and piston, i.e., a stainless steel piston (with a low contraction coefficient) working in a cylinder with a high contraction coefficient such as PTFE, nylon, or aluminum. Even the use of identical materials does not entirely eliminate problems during cooldown. A thin-wall light-section stainless-steel cylinder will cool very rapidly and contract into contact with a massive solid piston of the same material cooling more slowly because of the higher heat capacity.

Materials of high thermal contraction coefficient will experience greater movements during cool down. Cylinders will become tapered with minimum diameter at the cold end. The use of very thin walls to minimize thermal conduction renders the cylinder prone to thermal distortion. This is particularly so if a thin wall is coupled directly to a massive cylinder head and even more so if the assembly is not symmetrical about the axis.

Problems due to thermal contraction can be minimized by the use of stainless steel (low contraction coefficient), by the use of similar materials of similar cross section in close-fitting situations, or by designing the system as a cantilever so as to allow one end to "float" rather than attempt to secure the system rigidly between "anchor" points. It is always advantageous to locate the principal structural supporting element in the ambient temperature region and to have the cold elements hanging vertically down from the main support.

Fluorocarbons

Polytetrafluoroethylene (PTFE, TFE, Fluon, Teflon), fluorinated ethylene, propylene copolymers (FEP), and polychlorotrifluoroethylene (PCTFE, Kel-F) are all in the same family of fluorocarbon polymer plastics. These materials are good electrical and thermal insulators, inert to most chemicals and solvents. They have low friction characteristics that stimulate their use for dry bearings and seals. Furthermore, they do not become brittle at low temperatures as do the usual elastomers. They can be used as gaskets for static seals and as the sliding elements in dynamic seals (O-rings or self-energized lip seals). The mechanical properties of fluorocarbons can be improved by combination with various fillers, such as glass fiber, and powders of graphite, silver, and bronze.

Closure

It is necessary to emphasize again the inadequacy of this survey. We have said nothing here about comparative cost, ease of fabrication, joining, compatibility, surface treatments, corrosion, wear, failure, fatigue, heat treatments, and many other facets of materials technology. Reference is highly recommended to the more extensive treatments given at the opening of this section.

Designers should never feel that time spent studying materials is ill spent. A knowledge and understanding of materials and their characteristics is the key to successful design in many fields including cryocoolers.

COOLING

Refrigeration can only be produced in a cryocooler by the expenditure of energy, usually electrical or mechanical. The performance of a refrigerator is evaluated by the coefficient of performance. For an ideal Carnot refrigerator this is

$$COP_{(Carnot)} = \frac{T_E}{T_C - T_E} \qquad (9.19)$$

where T_E is the refrigeration temperature and T_C is the cooling temperature. Energy is transferred to the refrigerator at the low temperature, T_E, and from the refrigerator at the high temperature, T_C. Further energy in the form of work is supplied to drive the system. The energy transferred from the refrigerator, Q_C, is the sum of the heat lifted, Q_E, and the work done, WD.

Practical engines are less efficient than the ideal Carnot machines. The best that can be achieved in a practical refrigerator is a coefficient of performance about half that of the Carnot. Many machines, particularly small cryocoolers used in electronic applications, have low coefficients of performance, as low as 1% of the Carnot value. This is because the actual refrigeration load of the detector element is so small as to be virtually negligible. The principal task of the cooler is to cool itself to operating temperature and thereafter produce sufficient refrigeration to compensate for various mechanical, friction, and thermal losses.

A practical cryocooler, therefore, requires far greater work input to produce a unit of refrigeration than the ideal Carnot cycle system. All the energy supplied as work degrades to thermal energy and must eventually be dissipated from the system. Because of the large energies involved, kilowatts per watt of refrigeration, cooling is always difficult and sometimes becomes the critical problem.

Cooling can only be accomplished at temperatures above the ambient. Cooling at subambient temperatures is attractive to enhance the performance of a cryocooler, but can only be achieved if a second refrigerator is used to produce the refrigeration for the subambient cooling. Such multistage operation is frequently used in some form of cryocoolers (see Chapters 6 and 7 on Joule–Thomson and Claude cycle precooled cryocoolers). However, in the end, all the heat absorbed at low temperatures and all the work done in producing the refrigeration has to be rejected from the system at a temperature greater than the ambient. There has to be a temperature differential $\Delta T = (T_C - T_{amb})$ for heat to flow according to the second law of thermodynamics from the hot body (the cooler) to the cold body (the coolant).

The heat transfer from a surface may be calculated as

$$\dot{Q} = hA\Delta T \qquad (9.20)$$

where $\dot{Q}$ is the rate of heat transfer, h is the heat transfer coefficient, A is the area for heat transfer, and ΔT is the temperature differential between the fluid and the surface. The heat transfer coefficient, h for air is small because of the low density and low velocity. The area for heat transfer, A, is small because of the limited fin area to keep the system small and light in weight. The rate of heat transfer, $\dot{Q}$, is fixed by the required refrigeration capacity, Q_E, plus the work done (WD) to drive the system. Now if $\dot{Q}$ is fixed and h and A are small (h by the nature of things and A by design), then according to Eq. (9.20), ΔT must be large.

Now $\Delta T = (T_C - T_{amb})$ so that an increase in ΔT invariably results in an increase in the cooling temperature, T_C. Recall now that the work to be supplied to drive the refrigerator is proportional to the difference between the refrigerating and cooling temperatures, i.e., WD = $f_n(T_C - T_{amb})$. Therefore, an increase in T_C resulting from the increase in ΔT causes an increase in WD. The work done (degraded to heat energy) is actually the prime constituent of the heat to be transferred, $\dot{Q}$, in the cooling process. We can now see that savings in the system weight and volume by reducing fins and eliminating forced convective cooling inevitably results in an increased power requirement to the cooler, which in turn increases the heat to be dissipated and reduces the efficiency.

Further complications ensue when, as in many cases, the miniature cryocoolers are located in confined pods closely packed with electronic and light electric power equipment all generating their own thermal waste. Cooling is rarely adequate, so as a consequence the local ambient temperature, T_{amb}, increases significantly above the atmospheric temperature, T_{atmos}. The elevation of T_{amb} causes an increase in ΔT, resulting in a higher value of T_C, which in turn increases the power input requirement WD and the cooling load, $\dot{Q}$.

Air Cooling

Cooling can be accomplished directly using an air-cooled heat exchanger of the tubular or plate–fin type (see Chapter 8). Heat transfer coefficients on the air side tend to be low because of the low air density.

The heat transfers can be improved by blowing air at high velocities over finned tubes, but this requires substantial work to drive the fan. Direct air coolers tend to be bulky, heavily finned, and consequently expensive devices perhaps equipped with fans to increase their effectiveness but at the expense of increased power input. The temperature differential between the cooled fluid and the cooling air tends to be high.

Water Cooling

In most cases, water cooling is preferred to direct air cooling. The heat transfer coefficients attained with water are orders of magnitude greater than with air. Water coolers are small compact systems with a low temperature differential between the cooled fluid and and the cooling water.

With a continuous supply of cold water at hand, the heated water can simply drain from the system. However, the luxury of an abundant and continuous supply of pure fresh water is hard to find. Many water supplies are polluted and will corrode the system, or mineral deposits will block the flow passages. In other cases, a self-contained mobile system is required. Therefore, heat transfer to the water must eventually be dissipated in an air–water heat exchanger. This can be located remote from the cryocooler, say outside the laboratory or workshop containing the actual cryocooler. The water coolant is simply circulated by a small pump between the cryocooler and the air–water heat exchanger outside. This is an ideal arrangement for permanent installations of large and medium size. The heat rejected can, in cold climes, be used for building heating.

Small machines cannot tolerate the complexities and additional cost of intermediate water cooling. They must be air cooled without provision of cooling fans. Limited finning to enhance heat transfer is usually accepted but without enthusiasm, for it increases the weight and volume requirements.

Spacecraft Radiative Cooling

A particularly difficult cooling problem arises in the case of spacecraft where convective cooling to the atmosphere is simply not available. Heat must be dissipated from the spacecraft entirely by radiative means.

Now the radiant energy flux emitted by a body is given by

$$\dot{Q}_{rad} = \varepsilon A \sigma T^4 \qquad (9.21)$$

where $\dot{Q}_{rad}$ is the radiant energy flux, ε is the emissivity of the surface, A is the area for radiant emission, T is the temperature of the emitting surface, and σ is the Stefan–Boltzman constant. Space radiators are specially treated to provide a surface finish with optimum spectrally selective surfaces. White paints can provide a low-temperature emissivity as high as 0.93 in combination with a solar absorptivity as low as 0.18. Given the maximum possible value of emissivity, ε, the only other variables are the area, A, and the temperature, T. Increase in the area of the radiator increases the heat flux in linear fashion but increase in radiator area is unwelcome, for it increases the size and weight, both severely limited in space systems. Increase in the radiator temperature, T, increases the energy flux as the fourth power. A 10% increase in temperature results in a 46% increase in the heat flux. Doubling the temperature increases the heat flux by 16 times. However, it must be recalled that the temperature, T, is measured on the absolute scale, degrees Kelvin. Increase in temperature from 30 to 60 C is no more than an increase from 303 to 333 K, a mere 10% increase. Nevertheless, as shown above, this increase can result in a near 50% rise in the energy flux.

From the desire to achieve small compact radiators, there are compelling reasons to operate the spacecraft cryocoolers at high cooling temperatures. This results in increased power input and an increased load on the cooling system. A further complication on spacecraft is that the increased power consumption is not so easily come by. Even if the additional power can be had, it increases the cooling load on the radiators of the space power system. Cooling in space requires that the radiator be located at the exterior surface of the spacecraft looking out to space. The cryocooler is located inboard and so must be coupled to the outboard radiator with minimum temperature difference. Heat pipes have proved useful in this application. (See section in Chapter 8 on Heat Pipes.)

ELECTRICAL AND ELECTRONIC SYSTEMS

Those striving to engage in meaningful dialogue with their electrical/electronic colleagues on the basis of barely remembered, half-understood college lectures, will find the book on electromechanical system components by Charkey (1972) enormously useful. In cryocooler design, the mechanical engineer must interface both his activities and interests with those of the electronic/electrical engineers in three areas: electric motor drive to the compressor, refrigeration output control, and refrigeration load requirements.

Drive Motors

Electric motors are not the only way to drive cryocoolers, but are almost invariably used because of their great convenience and economy.

An important design decision is whether the electric motor will be incorporated in the crankcase of the cryocooler. Three possibilities are illustrated in Fig. 9.26. In one case, the motor is mounted external to the crankcase of the Stirling engine or the compressor of a Claude or Joule–Thomson cycle system on a rigid bed plate or frame. The crankshaft of the engine is coupled to the shaft of the motor by a mechanical coupling. A dynamic seal on the engine crankshaft is necessary to contain the crankcase lubricant and working fluid where the crankcase is pressurized.

The principal advantage of an independent motor is that it permits the economy of using standard electric motors. The disadvantage is the need for a crankcase dynamic seal where the crankcase emerges from the crankcase. If the crankcase is not pressurized with working fluid, there is no disadvantage, and the independently mounted motor arrangement would always be chosen.

Pressurizing the crankcase to minimum cycle pressure has such compelling advantages as to be adopted for many systems in miniature, small, and medium class machines. The advantages are that the gas pressure forces acting on the piston, connecting rod, and bearings of the engine are reduced from $(p_{inst} - p_{atmos})$ to $(p_{inst} - p_{min})$, where p_{inst} is the instantaneous cycle pressure, p_{atmos} is the atmospheric pressure in the unpressurized crankcase, and p_{min} is the minimum cycle pressure in the pressurized crankcase. This reduces the structural requirements of the drive mechanism and relaxes the piston sealing requirements. A leak past the piston to the crankcase can be very simply replenished from the crankcase reservoir.

The need for a dynamic seal on the crankshaft can be eliminated by enclosing the drive motor either partially or entirely within the crankcase. Partial enclosure is the term used to describe the system shown in Fig. 9.26b. The permanent magnet rotor of an ac induction motor is mounted directly on an extension of the crankshaft and enclosed by a pressure

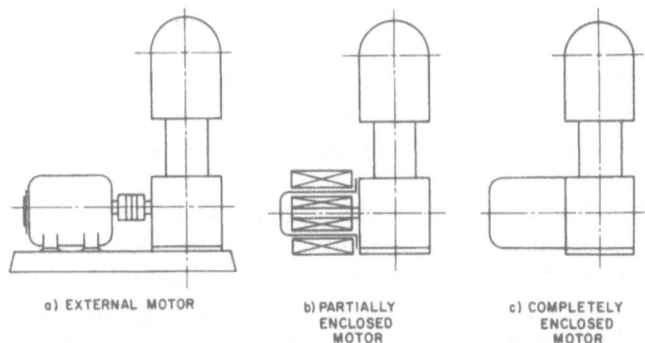

a) EXTERNAL MOTOR   b) PARTIALLY ENCLOSED MOTOR   c) COMPLETELY ENCLOSED MOTOR

Fig. 9.26. Drive motor arrangements for cryocoolers.

membrane (stainless steel shell) in the annulus between the rotor and the external stator carrying the energized windings. It is difficult to design a membrane of adequate strength without degrading the magnetic coupling of the rotor and stator and the motor performance and efficiency.

With such an arrangement, the working fluid is not contaminated by insulation lacquer from the field windings of the stator. The resistance losses of the windings can be more easily dissipated than if they were contained within the crankcase.

The completely enclosed motor shown in Fig. 9.26c permits close magnetic coupling of the stator and rotor, but introduces an additional static seal for the power leads to the motor. The electrical resistance losses of the motor windings now have to be dissipated from the crankcase either directly by external crankcase finning or through the system cooler.

Brushless dc Motors

Many of the applications where miniature cryocoolers are used have only low-voltage (28 V) direct current (dc) power supplies available. Small dc motors with conventional armature, commutator, and field windings or permanent magnet fields may be used. Problems arise because of high levels of electromagnetic interference (EMI) as a result of sparking at the brush–commutator interface. The infrared detectors cooled by such cryocoolers are incredibly sensitive and their signals are badly distorted by brush commutator EMI.

The brush materials, principally graphite, have satisfactory friction characteristics and wear rates when the gas in their environment contains at least traces of water. In a cryocooler with helium working fluid, the water and other contaminants present precipitates in the cold regions, and the helium becomes extremely dry. Once the dew point falls below −40 C, graphite materials experience a dramatic increase in their friction and wear characteristics. In small cryocoolers, the brushes wear rapidly and generate substantial quantities of wear debris.

Another factor to consider is that conventional dc motors with commutators do not have high efficiencies. In small sizes, the usual range of efficiency (defined as the ratio of mechanical work out to electrical power) is from 30% to 50%.

Combination of the above phenomena has forced the investment of substantial effort in the development of small brushless dc motors of high efficiency. This effort has been remarkably successful, and brushless dc motor efficiencies in excess of 80% are being regularly reported. The characteristics shown in Fig. 9.27 were reproduced from the excellent report by Lindale (1978) and refer to a small brushless dc motor manufac-

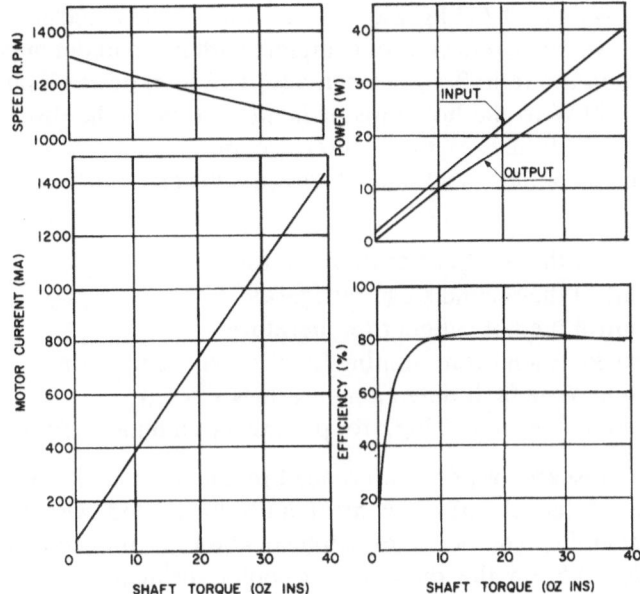

Fig. 9.27. Performance characteristics of a small brushless dc motor (after Lindale, 1978).

tured and tested by H. C. Roters Associates, Inc. A brushless dc motor consists essentially of a rotating armature with permanent magnets and a series of stationary field coils on the stator. The field coils are switched externally by an electronics unit which replaces the commutator switching system of conventional dc motors. The electronics unit is an elaborate system triggered by signals received from a position indicator magnetic disk mounted on the crankshaft.

Lindale (1978) includes a considerable discussion of the motor-controller system of the spacecraft Stirling cryocooler described in the report. A more comprehensive and fundamental discussion of brushless dc motors was given by Radziwill (1969), of the Philips Zentrallaboratorium gmbh, Aachen. Another company known to be producing small brushless dc motors is Aeroflex Laboratories Inc., Plainview, Long Island, New York.

Electronic Controls

Electronic controls are virtually obligatory for cryocoolers subject to stringent performance standards. Russo (1976) has given a good description of the electronic interface unit (EIFU) necessary for the Vuilleumier cooler of a limb scanning infrared radiometer for NASA. The cooler was designed to produce about 0.5 W of cooling at 65 K within a limit of ±2 K. The total system weight of 6.3 kg was divided among the electronic interface unit and the two-stage refrigerator complete with hot cylinder and displacer driver motor in the ratio 3 : 4, respectively. The total power input of 77 W results from 50 W to the hot cylinder heater, 16 W to the displacer drive motor, and 11 W to the electronic interface unit.

The six essential functions of the electronic interface unit were described as

    i.   control the refrigerator drive motor,
    ii.   control the second stage refrigerator,
    iii. control the hot cylinder temperature,
    iv. provide signal conditioning for the instrumentation,
    v.   contain the fault detection–interlock circuits,
    vi. contain the control logic for processing commands for the ground.

Many of the research reports referenced in Chapters 3, 4, and 5 contain descriptions of the electronic control units associated with the coolers described therein. The level of sophistication of the control systems naturally depends on the duty of the cooler, principally the degree of temperature control required and the operating life and reliability.

REFERENCES

Bevan, T. (1946). *The Theory of Machines*. Longmans, Green, and Co. Ltd., London (see Chapter XIV, "Balancing").

Buchter, H. H. (1979). *Industrial Sealing Technology*. Wiley-Interscience, New York.

Charkey, E. S. (1972). *Electro Mechanical System Components*. Wiley-Interscience, New York.

Clarke, W. D. (1973). "Reliability and Maintainability of USAF Cryogenic Coolers." Proc. Closed Cycle Cryogenic Cooler Conf., USAF Academy, Co. (Oct.), pp. 17–26, (AFFDL-TR-73-149, Vol. 1, No. 918234).

Cooke-Yarborough, E. H., and Yeats, F. W. (1975). "Efficient Thermo-Mechanical Generation of Electricity from the Heat of Radioisotopes." Proc. 10th I.E.C.E.C., Paper No. 759150, pp. 1003–1011, Newark, New Jersey.

Crabtree, L. F. (1973). "Engineering—Art and Science." *Aerospace* **3**(7), 22–25.

Fern, A. G., and Nau, B. S. (1976). *Seals. Engineering Design Guides*, No. 15, Oxford University Press, Oxford.

Harkless, L. B. (1973). Reliability Test Results on V–M Coolers." Proc. Closed Cycle Cryogenic Cooler Conf., USAF Academy, Colorado, pp. 93–110 (AFFDL-TR-73-149, Vol. 1, A.D. No. 918234).

Johnson, R. P., Bennett, A., Emigh, S. G., Griffith, W. R., Hoble, J. D., Penrome, R. E., and White, M. A. (1977). "Stirling-Hydraulic Artificial Heat Power Source." Proc. 12th I.E.C.E.C., Paper No. 779016, pp. 104–111, Washington, D. C.

Kapitza, P. (1934). "The Liquefaction of Helium by an Adiabatic Method." *Proc. R. Soc. London Ser. A* **147**, 189.

Lindale, E. (1978). "Stirling Cycle Refrigerators for Gamma-Ray Detector." Report No. PL-42-Cr78-0713, Johns Hopkins Univ., Applied Phys. Lab., Laurel, Maryland.

Meijer, R. J. (1959). "The Philips Hot-Gas Engine with Rhombic Drive Mechanism." *Philips Tech. Rev.* **20**(9), 245–276.

Meijer, R. J. (1959). "The Philips Thermal Engine." *Philips Res. Rep. Suppl.* No. 1, Philips Research Labs, Eindhoven.

Muller, H. J. (1964). "Improvements in Non-Lubricated Compressor Design." *Linde Rep. Sci. Technol.* **6**, 3–8.

Muller, H. J. (1971). "Advances in Non-Lubricated Compressor Design." *Linde Rep. Sci. Technol.* **17**, 3–11.

Pitcher, G. K. (1973). "Mechanical Life of Space Cryocoolers." Proc. Closed Cycle Cryogenic Coolers Conf., USAF Academy, Colorado, pp. 211–224 (AFFDL-TR-73-149, Vol. 1, AD No. 918234).

Pitcher, G. K. (1975). "Spacecraft Vuilleumier Cryogenic Refrigerator—Final Report." AFFDL-TR-75-114, WPAFB, Philips Laboratories.

Radziwill, W. (1969). "A Highly Efficient Small Brushless D.C. Motor." *Philips Tech. Rev.* **30**(1), 7–12.

Robbins, R. F., Weitzel, D. H., and Herring, R. N. (1962). "The Application Behavior of Elastomers at Cryogenic Temperatures." *Adv. Cryog. Eng.* **7**, 343–352.

Russo, S. C. (1976). "Study of a Vuilleumier Cycle Cryogenic Refrigerator for Detector Cooling on the Limb Scanning Infrared Radiometer." NASA CR 145078 (Hughes Aircraft Co., Culver City, California).

Schubert, R. (1971). "The Influence of a Gas Atmosphere and Moisture on Sliding Wear in PTFE Compositions." *Linde Rep. Sci. Technol.* **17**, 12–20.

Sherman, A., Gasser, M., Goldowsky, M., Benson, G., and McCormick, J. (1979). "Progress on the Development of a 3–5 Year Lifetime Stirling Cycle Refrigerator for Space." Cryogenic Engineering Conference, Madison, Wisconsin.

Walker, G. (1980). *Stirling Engines*. Oxford University Press.

Weitzel, D. H., Robbins, R. F., and Ludtke, P. R. (1965). "Elastomeric Seals and Materials at Cryogenic Temperatures." Report No. ML-TDR-64-50, Pt. II, U.S. Air Force, Wright–Patterson Air Force Base, Ohio.

Wigley, D. A. (1971). *Mechanical Properties of Materials at Low Temperatures*. Int. Cryogenics Monograph Series, Plenum Press, New York.

Wigley, D. A. (1978). *Properties of Materials at Low Temperature*. Engineering Design Guides, No. 27, Oxford University Press, Oxford.

Zimmerman, F. F., and Longsworth, R. C. (1971). "Shuttle Heat Transfer." *Adv. Cryog. Eng.* **16**, 342–351.

Chapter 10

Practical Problems in Cryocooler Design and Operation

F. F. Chellis

INTRODUCTION

Much has been written on the thermodynamic design of cryogenic coolers, and professional papers describing successful cryogenic equipment may be found in a variety of publications, but little has been written concerning the practical problems associated with the selection, design, and operation of a cryocooler.

The user of a cooler operating at cryogenic temperatures is often faced with a dilemma when choosing a cryocooler for his specific application. The user must choose between the need for extremely long operating life, and the desire for small size and minimum weight. Like the "universal shoe," the perfect all-purpose closed-cycle cryocooler does not exist. Each application has its own special requirements which will affect the cryocooler configuration: a parametric amplifier cooler for use in a ground-based satellite communications antenna is not limited as to size, weight or power input, but must operate 24 hours a day continuously for years at a time without repair or maintenance; whereas a cooler in an airborne infrared set must be designed for minimum size, weight, and power input.

One of the earliest closed-cycle cryocoolers was developed in 1958 for cooling a three-channel ruby laser in a target tracking radar set. This system included three stages of expansion, plus a Joule–Thomson stage to provide cooling at 3.8 K. Soon thereafter similar systems were installed in

F. F. Chellis • CTI-Cryogenics, Waltham, Massachusetts. (EDITOR'S NOTE: Deceased 1981.)

the Telstar antenna in Andover, Maine for intercontinental telecommunications via satellite, and in the giant antenna at Goldstone, California, for deep-space tracking. The Goldstone installation provided the principal communications link with our astronauts and is still in service today providing communication with the Mariner deep space probes, as they travel the outer reaches of our solar system.

With the development of the parametric amplifier (paramp), which operates at 20 K, a simplified and somewhat smaller two-stage cryocooler was developed with an emphasis on high reliability and long unattended operating life. These machines typically operate continuously for over two years without repair or overhaul. To accomplish this they are conservatively designed with an oil-lubricated compressor and low operating speeds, thereby reducing mechanical loads and the wear effects of friction.

In the early 1960s the closed-cycle cryocooler was adapted to the cooling of small infrared devices for military airborne applications, including reconnaissance and target acquisition. Here the emphasis was upon miniaturization to reduce size and weight of the cryocooler and to minimize cooldown time. In order to get the most cryocooler in the smallest package, it was necessary to reduce the size of mechanical components and to increase operating speed and charge pressure levels. Many of these coolers were built and are presently in service. However, the cost of maintenance and the training of specialists in field repair of these systems has prompted the search for a low life cycle cost system for future applications. Here the emphasis is not really on miniaturization and infinite operating life, but upon an acceptable system size compatible with an operating life of about 3000 hr with no maintenance. Today, with the rapid improvement in electronic design and ever smaller electronic packages, the cryocooler stands out as the largest, most expensive single component in an infrared system. Therefore, there is a continuing effort to develop more reliable cryocoolers in smaller sizes, at lower cost of ownership.

To achieve this balance of reliability, cost, and size, the cryocooler designer must be aware of the mechanical problems associated with the design and operation of cryocoolers. It is beyond the scope of this work to cover in detail the diversity of specialized disciplines and technologies which are required in the construction of a cryocooler, but we will highlight some of the specialized problem areas which must be confronted in the design and operation of a typical miniature cryocooler. Many of these considerations apply to larger cryogenic systems as well.

A COMPARISON OF CRYOCOOLER TYPES

Most closed-cycle cryocoolers in general usage today operate on a cyclic thermodynamic process and employ the following components: a

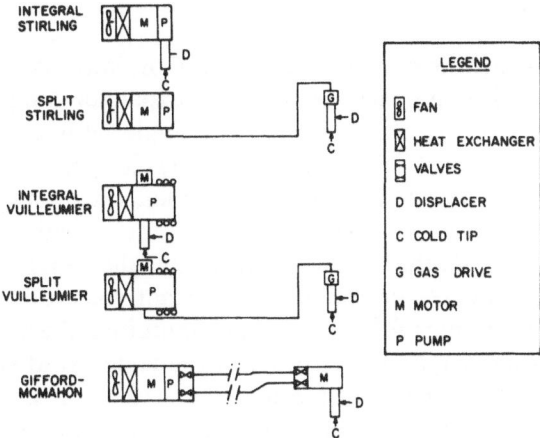

Fig. 10.1. Cryocooler types—a comparison.

compressor with an aftercooler heat exchanger that compresses and subsequently removes the heat of compression from the working gas, and an expander with a reciprocating displacer element and a regenerative heat exchanger that removes heat from a load applied to its cold end. The regenerative heat exchanger alternately removes heat from and restores heat to the working gas flowing through it, thereby maintaining a thermal gradient between the ambient end of the expander and the cold end that is cooled to cryogenic temperature. Closed-cycle cryocoolers of this type may be subdivided into the following classes in accordance with their operating principles as discussed in the earlier chapters (see Fig. 10.1).

Here we review briefly some of their advantages and disadvantages with respect to design.

Integral Stirling

The basic integral Stirling cryocooler consists of a compressor section and an expander section combined in one integrated package. The reciprocating elements of the expander and compressor are mechanically driven from a common crankshaft with the proper phase relationship so that no valves are required in the system. The integral Stirling cycle cryocooler is inherently the most efficient of all cryocoolers and has a particular appeal because of its basic design simplicity, compactness, and lower power input. The principal disadvantage of the integral Stirling is that the compressor, with its heat and vibration output, must be located in close proximity to the cooled device, since the compressor and expander are combined in one assembly.

Split-Stirling

In the split-Stirling system, the compressor, with its motor and heat exchanger, comprise one package while the expander comprises a second package and the two are interconnected by a single gas line. This overcomes the principal objection to the integral Stirling design, since it allows the compressor to be separated from the expander, thus removing the bulk, heat, and vibration of the compressor from the sensitive region of the detector, amplifier, or other cooled device. Since the system uses no valves, the connecting gas line constitutes an important void volume which must be minimized by limiting the line length and, therefore, the distance between the compressor and expander. Because of the added void volumes associated with the interconnecting line and the lack of a direct mechanical linkage between the displacer and compressor, the split-Stirling cooler is somewhat less efficient than the corresponding integral Stirling.

Integral Vuilleumier (VM)

The VM cryocooler may be thought of as a Stirling cycle cooler wherein the mechanical compressor component has been replaced by a thermal compressor. The only mechanical input is that needed to shuttle the displacers back and forth, which can be accomplished with a very small electric motor (compared with the large drive motor in the integral Stirling machine). Since the mechanical forces required to drive the VM are substantially lower than those in other types of cryocoolers, it follows that mechanical wear would be significantly reduced with a consequent increase in operating life and reliability. However, because the compression is provided by transfer of electrical heat to the working gas, this cycle is inherently less efficient than its Stirling counterpart, and an added electronic package is required to control the heater at the hot end. Because of high operating temperatures in the compressor, wear of the hot displacer sealing and riding surfaces is an important design consideration.

Split-Vuilleumier

The split-VM configuration bears the same relationship to the VM as the split-Stirling does to the Stirling configuration. Here we have a thermal compressor (instead of a mechanical compressor) communicating with a remotely located pneumatically actuated expander through a single gas line. This system combines the advantages of the split-Stirling configuration with those of the VM and shares the disadvantages of both.

Gifford–McMahon (GM)

In Gifford–McMahon cryocoolers, the compressor assembly and the expander assembly are typically widely separated from one another, and two interconnecting gas lines join them. The expander requires a small motor to move the displacer up and down and to operate the valves in the expander head. The presence of valves in both the compressor and the expander isolates them from one another and allows the compressor to operate at a different speed from the expander. This arrangement allows the interconnecting lines to be of almost any desired length since their volumes are effectively isolated from the thermodynamically active volumes in the system. The long lines allow separation of the compressor package from the expander, thereby removing the principal mass and source of heat and vibration from the critical region of the device being cooled. These advantages are somewhat offset by the larger size and higher power requirements of this type of system.

DESIGN CONSIDERATIONS AND SYSTEM TRADE-OFFS

Heat Rejection

A prime consideration in the design of any cryocooler is the adequate provision for removal of heat. Very small closed-cycle cryocoolers (in the capacity range of about 0.5 W at 77 K or smaller) have a sufficiently low input power requirement (100 W or less) that they may be conductively cooled by bolting them to a large metal mass or convectively cooled by ambient air. However, cryocoolers with a higher input power require forced air cooling to move air over external fins on the compressor section of the cryocooler for proper heat rejection (see Chapter 9).

With very large systems, where the input power is over 400 W, heat rejection may require special liquid cooling techniques involving liquid coolant, a circulating pump, and a liquid-to-air heat exchanger with a fan or other source of forced air cooling. These additional elements add considerably to the size and weight of the closed-cycle cryocooler system.

Because of mechanical wear and other considerations, it is unreasonable to expect that a cryocooler will maintain the same level of cooling capacity throughout its operating life. The wear of compressor piston seals and displacer seals degrades performance in two ways. Firstly, seal leakage increases, which allows gas blow-by, and, secondly, the wear products accumulate in gas passages and the regenerator matrix to restrict gas flow. Additionally, outgassed organic materials from bearing lubricants, motor windings, etc., migrate to and freeze out in the regenerator matrix and in the clearance spaces at the cold end of the expander. It is, therefore, necessary to provide excess cooler capacity in the basic cooler design, in order to have sufficient capacity at the end of the required operating period. This added cooling capacity requires a larger system, and a larger system requires more power input. Since a cryocooler is a 100% loss device, all the electrical power which goes in to drive the system must be rejected as heat. So, the larger the system, the larger the fan, fins, heat exchanger, etc., needed to get rid of the heat.

Therefore, there is an important trade-off between cooling capacity (system size) and operating life requirements. If the physical size of the closed-cycle cryocooler is of great importance, then the systems designer should concentrate upon minimizing his cooling requirements. This can be done through careful design of the actual cooled device to minimize thermal loads caused by heat conduction and radiation and by the elimination of unnecessary factors of safety in his design. A realistic overall factor of safety is best arrived at through close and early cooperation between the cryocooler design engineer and the designer of the device to be cooled.

Some of the most difficult conditions are created in military airborne applications where a cryocooler must operate in a 55°C environment and often in a totally enclosed chamber close-packed with other heat-rejecting apparatus. A cryocooler's main purpose in life is to reject heat, and this can only be accomplished when it receives proper cooling.

Microphonics

Closed-cycle cryocoolers have a reciprocating displacer/regenerator element in the expander which is driven by either mechanical or pneumatic means. The motion of the displacer is, typically, not balanced and so creates a vibration which is transmitted to the cooled device that is mounted on the cold tip. This vibration can induce false signals due to varying capacities of signal leads or by magnetic fields and currents generated by moving conductors. This is not generally a problem so long as the expander, with its cantilevered cold tip, is rigidly mounted. However, some of the newer electronic devices are so sensitive to microphonics that integrating them with a closed-cycle cryocooler presents a serious problem, which must be addressed either by the cryocooler designer or by the cryocooler user.

Thermophonics

Another problem that is often encountered is that of thermophonics. The problem is inherent in the cyclic operation of the cryocooler wherein the actual temperature at the cold tip fluctuates a few millidegrees with each expansion event of the cooler. This problem is especially noticeable with some of the new supersensitive devices such as laser diodes, infrared detectors, SQUID magnetometers, and Josephson junctions. The thermal "ripple" can often be smoothed out by the interposition of a thermal mass (such as a lead spacer) between the cold tip and the load (cooled device). However, the cooldown time of the system is increased, and the system's designer must choose between signal interference due to thermophonics and minimum system cooldown time.

SPECIAL PROBLEMS RELATED TO CRYOCOOLER OPERATION

Gas Contamination

It is important to realize that the cryogenic cooler has all the necessary elements to make it a very effective still, including a boiler and a condenser. In any totally enclosed, gas-filled system, the coldest point in that system determines the partial pressures of all the condensible gases within the system and, at the low temperatures at which we are operating (80 K and below), all gases (except hydrogen, helium, and neon) will be condensed to a liquid state or frozen into a solid. The compressor end of the cryocooler may be likened to the boiler portion of a still, since most of the electrical power input to the system is converted into heat through electrical losses in the motor windings, heat of compression, and mechanical friction. The compressor typically will operate 30 or 40°C above normal room ambient temperature, which will boil off gaseous constituents from the various organic compounds in the compressor assembly. The outgassed material then mixes with the helium gas and eventually migrates to the cold end of the expander. Here, it freezes out as a frost, or condenses as a liquid in the lower end of the regenerator matrix or in the narrow clearance spaces around the cold end of the displacer, thus impeding gas flow and creating a mechanical interference which degrades the performance of the cryogenic cooler. The degraded performance results in a warming up of the cold end of the refrigerator which will revaporize the frozen contaminants allowing them to again mix with the helium gas. The entire process may occur repeatedly, resulting in temperature instability ("temperature cycling") at the cold end of the refrigerator.

It is impossible to overemphasize the importance of proper choice of engineering materials to be used in the internal parts of a closed-cycle cryocooler. Every single part must be given careful consideration. The principal trouble areas are the drive motor and the lubricants used for bearings and gearing. The mechanical designer commonly thinks of a motor as an off-the-shelf component which can be supplied by any of a number of prominent manufacturers. However, the electric motor should be considered as an assembly, with each component carefully specified as to material. The insulating varnishes, waxes, and resins used in commercial motor manufacture are all undesirable, as they will outgas profusely and cause helium gas contamination problems. Teflon, Nomex, polyimide, and certain low-vapor-pressure epoxy adhesives will meet the low outgassing requirements and are, therefore, suitable for use in closed-cycle cryocooler construction.

The choice of a bearing and gear lubricant for use in high and low ambient temperature extremes (say −40 to 55°C) is difficult enough, but now add the requirement of low vapor pressure and the need for the lubricant to "stay put" inside a bearing or in a gear box, and the candidates narrow down to a handful of very special greases. Krytox-AB* is one of these, being a fluorinated lubricant especially formulated for use in moon probes, space vehicles, and such.

All materials (and especially organic materials such as motor insulation) will retain a certain amount of water attached in a surface layer or absorbed into the material itself. Therefore, it is necessary, just before final assembly of a closed-cycle cryocooler, to vacuum bake all the components at about 60°C. This procedure drives off the water as well as residual high-vapor-pressure fractions from organic materials—all of which, if left in place, would later outgas into the helium gas charge of the cryocooler and then freeze out in the cold end of the expander. It is also important to guarantee the cleanliness of the helium gas which is used to charge the cryocooler system. It is not uncommon to receive an occasional bottle of "dirty" helium gas. Therefore, only helium of certified purity should be used to charge the cryocooler.

Helium Gas Retention

The helium gas molecule is extremely small and can find its way through the apparently solid walls of pressure vessels (which easily contain air, or other gases). Some of the very small cryogenic coolers that are being manufactured today have a total internal volume of 0.2 to 0.4 liters, are charged with helium gas to pressure levels as high as 1000 psi, and are required to have a helium leak rate of less than 1×10^{-8} standard cm^3 per

---

* Krytox is a registered trademark of Dupont, Petroleum Chemicals Division, Wilmington, Delaware 19898.

second. It is therefore reasonable that some of the more common leak paths by which helium can escape from a small cryogenic cooler system are identified.

Rubber O-Ring Seals. Rubber O-ring seals are widely used in pneumatic and hydraulic systems as a static seal to retain the operating fluid. However, helium gas will leak out through the rubber seal by permeation. Some rubbers are worse in this respect than others, and the leak rate is so small that it may be acceptable in those cases where helium make-up gas is readily available. But in the very small cryocooler where the helium gas charge must be retained for months or years, the rubber seals are not acceptable and it is necessary to resort to crushed metal sealing techniques. Soft copper, indium, and even gold have been used for this application.

Casting Leaks. Casting leaks are a particular problem where aluminum castings are used for cryocooler pressure vessel parts. It is especially difficult to produce helium-leak-tight aluminum castings because of voids and entrained microporosity in the metal structure. Some manufacturers have resorted to multiple impregnations of the casting to seal the leaks, but this is not a really satisfactory solution to the problem. It is best to select a highly qualified casting house and work with them to resolve the special problems relative to casting a newly designed part.

Metal Porosity. Metal porosity also occurs in metals other than cast aluminum. It is not uncommon to find helium leaks in parts made of wrought stainless steel, for instance. Defects in the original cast ingot are retained within the metal structure right through to the finished rod and bar stock. We have seen leaks in cylinder end caps (cut from bar stock) that were easily detected by the "bubble test" when immersed in water. The solution to this problem is to specify rolled plate stock for such parts, because the metal defects run parallel to the faces of the plate rather than through it.

Weld Joints. Weld joints, although structurally sound and made by expert technicians, are often a source of helium leakage due to intergranular cracking in the weld zones. It is important to realize that some types of stainless steel are more prone to this problem than others.

Seal Problems

Miniature cryogenic coolers employ reciprocating parts in both the compressor and the expander, and the dynamic seals associated with those reciprocating elements must perform flawlessly over the entire operating life of the cryogenic cooler. Malfunctioning seals are probably responsible for more cryogenic cooler failures than any other single item. Since the split-Stirling cooler is especially dependent upon faultless seal performance,

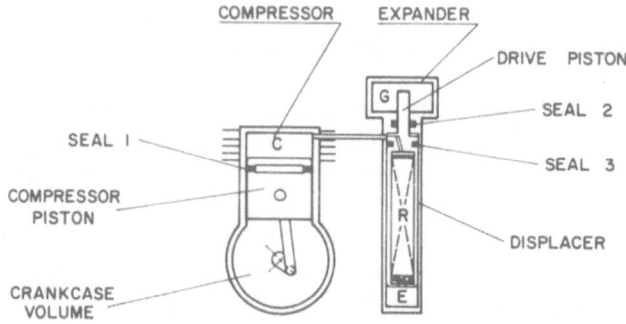

Fig. 10.2. Mechanical configuration details. Split-Stirling cryocooler.

we will use it as an example to illustrate the operating problems that may be associated with seal leakage in closed cycle cryocoolers.

The split-Stirling cryogenic cooler (Fig. 10.2) is comprised of a compressor assembly and an expander assembly joined by a short length of small-bore metal tubing. The system is hermetically sealed to retain the pressurized charge of helium gas, which acts as a refrigerant, for a year or more without important loss of gas pressure. The helium charge is contained in three well-defined spaces within the system which we will designate as the crankcase volume, the working volume, and the gas spring volume. The crankcase volume is the larger of the three and includes the drive motor, motor bearings, flywheel, crankshaft, connecting rod, and piston, all located within the compressor assembly. The working volume includes the swept volume, or compression space C, above the compressor piston, the volume within the gas line which interconnects the compressor and expander, all the clearance spaces above, below, and around the displacer, and the internal void volumes within the regenerator matrix R. The gas spring volume G includes the space in the expander above the drive piston. There are three dynamic gas seals within the split-Stirling configuration, all of which must function faultlessly if the system is to operate properly. The piston seal $S1$ is located on the compressor piston and separates the crankcase volume from the working volume. The drive seal $S2$ is located in the expander and separates the working volume from the gas spring volume. The displacer seal $S3$ is mounted at the warm end of the displacer, and moves back and forth with it. Now, let us examine the functions of these three seals and describe how the malfunction of any one of them has unique and important effects upon the performance of the split-Stirling cryogenic cooler system.

Under normal operating conditions, the pressure level in each of the three volumes rises and falls in a cyclic manner in response to the motions of the compressor piston and the expander displacer. However, the *average pressure* is the same in all three volumes and approximates the charge pressure at initial startup. As the compressor piston moves upward, the working volume decreases and, at the same time, the crankcase volume increases by a like amount. Thus, the motion of the compressor piston creates a nearly sinusoidal pressure variation in the working volume and in the crankcase volume, but 180 degrees out of phase with each other. In a typical operating system, charged to 500 psi, the pressure swing in the working volume will cycle between 350 and 750 psi, while the crankcase volume would cycle between 520 and 480 psi (Fig. 10.3).

The *piston seal* must be designed to be as leak-tight as possible in operation. However, the generally accepted practice is to charge the cooler through a self-sealing valve (much like a tire valve), that is installed in the crankcase wall. The high-pressure charge gas enters the system through the crankcase and must pass by the piston seal to reach the working volume, and thence, it must pass by the drive seal to reach the gas spring volume. Having charged the total system to full pressurization, it is common practice to vent the system to atmosphere, and then to repeat the procedure four or five more times. This will dilute the initial atmospheric air, which was in the system at assembly, to less than one part in a million, which is perfectly acceptable for cooler operation. This charging procedure requires that the piston seal be somewhat leaky in both directions (as also must be the drive piston seal). Thus the designer is faced with a dilemma. The piston seal must, at the same time, be leak-tight for proper operation and yet leaky for the purge/charge operation. Cooler designers have experimented with a variety of ideas to solve this problem, including bypass valves and dual charge fittings (one to the crankcase, the other to the working volume), but in the end have resorted to the leaky seal concept in their final design, since it is practically impossible to obtain a completely

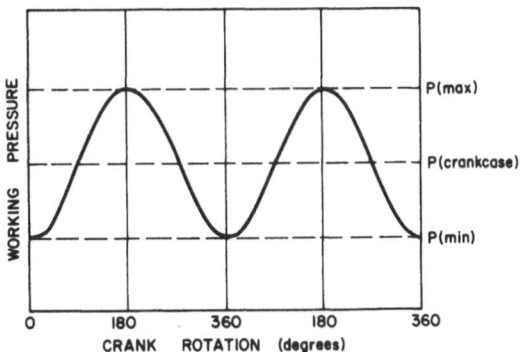

Fig. 10.3. Pressure vs. crank position—normal operation. Split-Stirling cryocooler.

helium-leak-tight dynamic seal. Now, given that the seal does leak in operation, it can leak in either direction and, so long as the leakage is the same in either direction, the cooler will operate in the normal fashion wherein the average working pressure is equal to the average crankcase pressure (as in Fig. 10.3). However, the cooler designer often finds this not to be the case, and the result is a phenomenon commonly referred to as "check valving." In other words, the seal leaks more in one direction than in the other. If the leakage is primarily from the crankcase toward the working volume, then the effect is to supercharge the working volume such that the mean pressure of the working volume is substantially above that of the crankcase volume. In a fully "check-valved up" system, the minimum working pressure value becomes approximately equal to the average crankcase pressure. The result of this is that a system charged at say 500 psi may now operate with a pressure swing of 750 psi on the high side to 490 psi on the low side (Fig. 10.4, curve A). This operational mode results in greatly increased cooling capacity, accompanied by a higher power input to the system. Now if the piston seal leaks primarily from the working volume to the crankcase volume, an opposite effect occurs wherein the working pressure drops below that of the crankcase pressure and, with a 500 psi charge and a fully "check-valved down" condition, we might operate from 300 to 150 psi in the working spaces, with this operating mode resulting in substantially reduced refrigerator performance accompanied by a lower power input to the compressor (Fig. 10.4, curve B). It is not uncommon to have a faulty piston seal that alternately check-valves up and down, thus producing a very unstable refrigerator performance, accompanied by a wide fluctuation of input power. It is therefore incumbent upon the designer of the cryogenic refrigerator to provide a piston seal which will have consistent sealing characteristics throughout a long operating life and under a variety of temperature extremes. The piston seal is generally

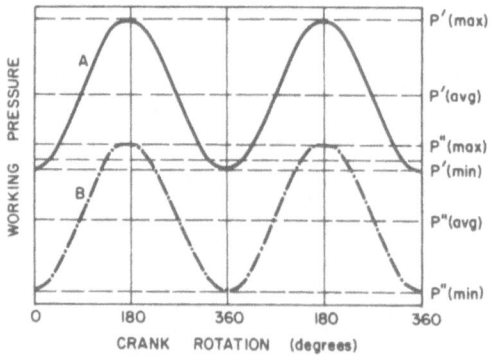

Fig. 10.4. Pressure vs. crank position—check valved operation. Split-Stirling cryocooler (curve A, check-up; curve B, check-down).

a. SPLIT RING SEAL   b. LIP SEAL (FLUOROCARBON)

Fig. 10.5. Dynamic seals for cryocooler applications.   (a) Split-ring
seal, (b) fluorocarbon seal, (c) bal seal.

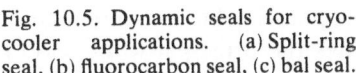

c. LIP SEAL (BAL SEAL)

of a Teflon-based material (the composition of which is proprietary to the manufacturer). Split-piston ring-type seals with C-shaped metal backup springs are commonly used (Fig. 10.5a), as are various varieties of lip seals with built-in expansion springs (Figs. 10.5b, 10.5c).

These same seal types have been used for the *drive piston seal* and for the displacer seal. However, most split-Stirling expanders that are being built today employ no discrete sealing element on the drive pistons but rely upon extremely close fit between the drive piston and its mating cylinder bore (on the order of 0.0001 in. clearance). This seems to be a satisfactory solution in that such a clearance provides adequate sealing to maintain the gas spring volume at the desired pressure level while still allowing sufficient leakage for charging and purging the system. Preferential leakage of the drive piston seal, in one direction or the other, will seriously affect the stroking characteristics of the gas-driven displacer. As mentioned before, the gas pressure in the working volume is principally controlled by the reciprocating compressor piston, and varies sinusoidally between high and low limits. For proper stroking of the displacer, it is mandatory that the pressure in the gas spring volume be maintained as closely as possible to the average working pressure. If the pressure in the gas spring volume is too low, then the displacer will not complete its full stroke but will tend to short stroke while remaining at the warm end of the cold finger cylinder. This operating mode effectively increases the cold end dead volume in the expander and substantially degrades the performance of the cooler. On the other hand, if the pressure in the gas spring volume is substantially above that of the mean pressure in the working volume, then the displacer will tend to shorten stroke and hover around the cold end of the cold finger cylinder, thus increasing the warm end void volumes. This operating mode is much more acceptable than the previous condition but still has a minor effect in performance degradation.

The *displacer seal* is a dual-function device; not only does it serve as a sealing element to prevent gas from bypassing the regenerative heat exchanger, but it also serves as a very important braking element in controlling the reciprocating dynamics of the displacer itself. The refrigerator designer must carefully match the frictional drag (plus dynamic gas effects in the regenerator) against the gas driving forces provided by the pressure differential across the drive piston. Once established, it is extremely important that the value of the frictional drag provided by the displacer seal remain constant throughout the operating life of the machine. This is a nearly impossible task and constitutes the principal weakness of the split-Stirling concept.

Now, with the help of some diagrams, let us explore in more detail the interaction between the gas pressure driving forces and the opposing seal friction and the resultant effect upon displacer motion and refrigerator performance. In Fig. 10.6, curve (a) shows a pressure-versus-volume diagram for the cold end volume of the expander in a split-Stirling cryocooler which is operating normally. Figure 10.7 shows the sinusoidal variation in working pressure as the cycle progresses. (Note that one complete cycle takes only 0.050 sec, if the operating speed is 1200 cycles/min.) At reference point 1 on the diagram, the displacer is all the way toward the cold end and the cold end expansion volume is at a minimum. At this point the compressor piston is located at bottom dead center which corresponds to a maximum volume in the compression space, and the working pressure is at its minimum value, $P(\text{min})$. The compressor piston now moves upward, and at midstroke, the working pressure reaches an intermediate level, $P(\text{avg})$ corresponding to point 2 on the diagram. Since the pressure in the gas spring volume is maintained at the same level, there is no pressure difference across the drive piston. As the compressor piston continues to move upward and the working pressure increases, point 3 is reached where the difference in pressure between point 2 and point

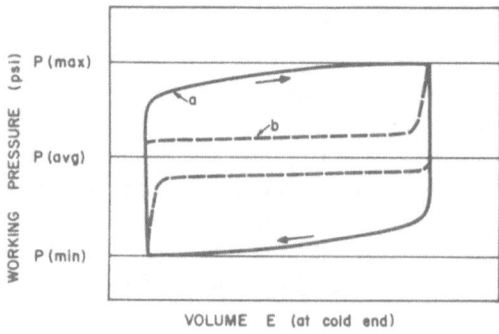

VOLUME   E   (at cold end)

Fig. 10.6. *P–V* diagram—split Stirling.

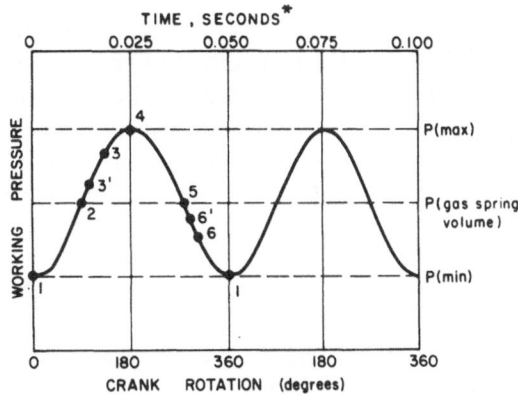

Fig. 10.7. Pressure vs. crank position—split Stirling.

3, multiplied by the area of the drive piston, creates a driving force which is sufficient to overcome the frictional drag of seal $S3$, and the displacer starts to move toward the warm end of the coldfinger cylinder. At point 4 on the diagram, the displacer has reached the warm end of the cylinder, the cold end volume is now maximum, and the compressor piston has reached top dead center, corresponding with maximum pressure P(max). The compressor piston now starts to move downward, and at midstroke, the working pressure is at point 5, again at P(avg). The compressor piston continues to move downward and the working pressure continues to drop until, at point 6, the pressure difference between point 5 and point 6 creates a downward force against the area of the drive piston to overcome frictional drag of $S3$, and the displacer starts to move back toward the cold end of the coldfinger. The compressor piston continues to move downward, accompanied by a decrease in working pressure, and at point 1 we are back at the starting point where the working pressure is at its lowest point, the compressor piston is at bottom dead center, and the displacer is all the way toward the cold end. The cycle now repeats.

But now let us assume that the seal friction on the displacer is substantially lower than that in the previous example. Now referring to Fig. 10.6, curve (b) and starting again at point 1 as the compressor piston moves upwards, accompanied by a rise in the working pressure, we reach point 2 at P(avg) and at a slightly higher pressure we reach point 3'. Since the frictional drag of seal $S3$ is very low in this example, only a small ΔP is required to create enough driving force to move the displacer, and once started, the displacer moves abruptly to the warm end, and reaches the warm end before the working pressure has reached its maximum. The compressor piston continues to move toward the top dead center, raising the working pressure to point 4, and a similar condition exists on the return stroke so that the $P-V$ diagram encloses a very small Z-shaped area. The area inside the $P-V$ diagram is directly proportional to the cooling capacity of the cryocooler. Therefore, we see from this example that the seal friction must be maintained at the proper level to provide a full $P-V$ diagram which corresponds to optimum cryocooler performance.

REFERENCES

Griffith, R. W., and Bertaldo, F. J. (1968). "Aerospace Electric Motors." *Westinghouse Engineer* **July**, 112–17.

Kingsbury, J. E. (1968). "Nonstructural Material for Aerospace." *Mechanical Engineering* **Aug.**, 32–36.

Koppers Company, Inc. *Engineer's Handbook of Piston Rings, Seal Rings, and Mechanical Shaft Seals.* 9th Ed. Metal Products Division, Baltimore, Maryland.

Leven, S. (1964). "A Design Analysis of Piston Rings for Nonlubricated Reciprocating Compressors." ASME Paper No. 64-FE-22.

Chapter 11

Fundamentals of Alternate Cooling Systems

Ray Radebaugh

INTRODUCTION

In spite of great scientific strides over the last century, the refrigeration principles used today are, for the most part, the same as those used for the last century. These principles were refined between the period of the first liquefaction of air in 1877 by Cailletet and Pictet and the first liquefaction of helium by Onnes in 1908. The mechanical work required for these gas systems presents difficult engineering problems. However, considerable engineering progress has been made since that time so that such cryogenic refrigerators or liquefiers have moved from the category of laboratory devices to industrial machines. Still further engineering strides are needed to make such refrigerators sufficiently reliable, inexpensive, and efficient to be used regularly for many other potential application areas.

The previous chapters have dealt with cooling engines, or mechanical refrigeration systems. Such systems dominate the cryogenic cooling field today. The emphasis of this chapter will be to encourage the reader to think of new approaches to refrigeration. If the new refrigeration system is inherently more simple than present-day mechanical systems, then less engineering effort is needed to make the system competitive. A technique of entropy comparison will be discussed in some detail. The comparison

Ray Radebaugh • Thermophysical Properties Division, National Bureau of Standards, Boulder, Colorado 80303.

allows one to quickly assess the refrigeration potential of a new system and to develop some feel for how to look for a new system.

THERMODYNAMIC CONSIDERATIONS

Refrigeration means that some heat, Q, is absorbed at a temperature below ambient. It should be realized that this Q may be due entirely to the heat leak from ambient. Such will be the case in refrigeration of some small superconducting devices which dissipate nanowatts of power. The absorption of Q at some low T is the desirable part of a total refrigeration cycle. Unfortunately the first and second laws of thermodynamics tell us that this process does not come to us free. How much we pay for such refrigeration can be determined only by considering the rest of the refrigeration cycle. At some point in the cycle heat is given off at ambient temperature. Also, the second law of thermodynamics requires that work be done to the system to supply the refrigeration.

Though it is not obvious to the laboratory user of liquid helium, the rest of the cycle still needs to be considered. The liquid helium in a Dewar is part of a discontinuous flow of the working fluid to a separate location from some other part of the cycle. The cost and availability of the helium is determined by the efficiency and reliability of that part of the cycle. When the boil-off gas is vented to the atmosphere instead of returned to the compressor, the cycle is not completed and the process is often called an open cycle. Though often convenient for laboratory use, the discontinuous supply of refrigerant in a Dewar can be troublesome for use in remote locations and for use where the expertise does not exist to transfer liquid from a discontinuous supply. In those cases it is better to provide a continuous supply of refrigeration, which means that the rest of the cycle must be considered as an integral part of the refrigeration process. The refrigerant system, e.g., helium, then undergoes a complete cycle so the term closed-cycle refrigeration is used to describe the process. Any other refrigerant system must also execute a complete cycle to provide refrigeration.

Ideal Refrigeration Cycles

The path used in any refrigeration cycle can be drawn in a temperature–entropy plane, or T–S diagram. The T–S diagram and the paths followed on it by various refrigeration cycles are discussed in detail in Chapter 2. Figure 11.1 shows the three most common cycles—the Carnot cycle, the Stirling cycle, and the Ericsson cycle. We emphasize that these are ideal cycles which can be followed only approximately in practice. In the Stirling cycle, the two isotherms are connected by constant-volume paths whereas constant-pressure paths connect the two isotherms in the Ericsson cycle. At this point we can generalize the Carnot, Stirling, and Ericsson cycles to include systems other than gas systems undergoing pressure–volume variations. Any material whose entropy can be changed can serve as the working substance. Thus instead of just constant-volume process for the Stirling cycle, we can use any constant generalized displacement process. For the Ericsson cycle the paths connecting the isotherms can be any constant generalized force path. Examples of some generalized forces and their associated generalized displacements are shown in Table 11.1. These forces and displacements will be discussed in more detail later.

In any of the cycles discussed above the heat absorbed at the temperature T is related to the entropy change by

$$dQ \leq T\,dS \qquad (11.1)$$

If the process is performed reversibly, the heat absorbed is

$$dQ_{\text{rev}} = T\,dS \qquad (11.2)$$

According to this equation, the heat Q absorbed at the temperature T will be the area under the line at T in Fig. 11.1a, 11.1b, or 11.1c. The heat Q_0 given off at T_0 is the area under that line. It is assumed that all the steps in going around the cycles in Fig. 11.1 are reversible. According to the first law of thermodynamics the total amount of work required to complete the cycle is

$$W = Q_0 - Q \qquad (11.3)$$

This work is represented by the enclosed area shown in Fig. 11.1a, 11.1b, or 11.1c as discussed in Chapter 2. From Eqs. (11.2) and (11.3) we may write

$$W = T_0 \Delta S - T \Delta S = \Delta S(T_0 - T) \qquad (11.4)$$

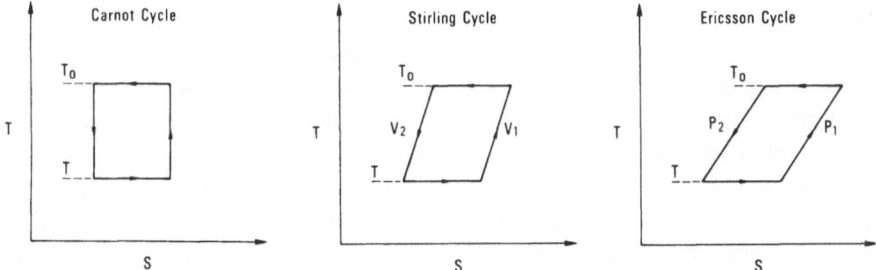

Fig. 11.1. Temperature–entropy diagrams for the (a) Carnot cycle, (b) Stirling cycle, and (c) Ericsson cycle.

Table 11.1. Examples of Generalized Forces and Their Associated Generalized Displacements in Thermodynamic Systems

F (force)	X (displacement)
Pressure, P (N/m^2)	Volume, V (m^3)
Tension, $-\tau$ (N)	Length, L (m)
Surface tension, $-\sigma$ (N/m)	Surface, A (m^2)
Magnetic field, $-H$ (A/m)	Magnetic moment,[a] $\mu_0 m$ (Wb m)
Electric field, $-E$ (V/m)	Dipole moment,[a] p (Cm)
Electric potential, $-\varepsilon$ (V)	Charge, Z (C)
Centrifugal potential, $-r^2\omega^2/2$ (m^2/sec^2)	Mass, M
Chemical potential, $-\mu$ (J/mol)	Moles, n

[a] $m = MV$, where M is the magnetization and V is the volume. $p = PV$, where P is the polarization.

We have assumed ideal behavior in the refrigerant so that ΔS is the same at T and T_0. The efficiency, η, of any refrigeration process is defined by $\eta = Q/W$. From Eqs. (11.2) and (11.4) the efficiency of the Carnot, Stirling, or Ericsson cycle beomes

$$\eta_c = T/(T_0 - T) \tag{11.5}$$

where η_c is called the Carnot efficiency (even though the Stirling and Ericcson cycles have the same efficiency) and represents the highest possible efficiency for a cycle operating between T and T_0. Any real cycle will have irreversibilities which lead to a lower efficiency.

In the Carnot cycle heat crosses the system boundary only during the isothermal processes at T and T_0. However, in the Stirling and Ericsson cycles heat crosses the system boundary during the constant-volume or constant-pressure processes. This heat transfer must also be at a varying temperature. For a reversible heat transfer the total heat transferred is given by

$$Q_e = \int T\,dS$$

where the integral is along one of the lines between T and T_0 or vice versa. Such heat transfer is simply the area under each constant-volume or constant-pressure curve. The system absorbs heat in going from T to T_0 and gives off heat in going from T_0 to T. In an ideal system, the two lines are parallel so that the two heat transfers are equal. In practice this heat transfer is handled by placing a heat exchanger between the two lines. A practical heat exchanger, however, is a source of irreversibility because of the necessary temperature differences. As far as this author knows, the Ericsson cycle always requires a greater heat transfer during the paths between T and T_0 than does the Stirling cycle. The reason is best explained by comparing the entropy change per unit temperature change for the two cases. In terms of the specific heat, C, at a constant force, F, and at a constant displacement, X, this entropy change is

$$(\partial S/\partial T)_F = C_F/T \quad \text{and} \quad (\partial S/\partial T)_X = C_X/T$$

The difference in specific heats (commonly written in terms of P, V, and T) can be expressed in generalized terms as

$$(C_F - C_X)/T = -(\partial X/\partial T)_F^2(\partial F/\partial X)_T \tag{11.6}$$

For all known systems the term $(\partial F/\partial X)_T$ is negative. Examples for F and X are given in Table 11.1. Hence $C_F > C_X$ and $(\partial S/\partial T)_F > (\partial S/\partial T)_X$. Thus the constant F process requires a larger heat transfer than the constant X process. This disadvantage of the Ericsson cycle is often offset by the fact that the maximum generalized force required (upper left corner of the cycle in Fig. 11.1(c)) is less than that of the Stirling or Carnot cycles in single phase systems. For cases where $T_0 - T$ is large the Carnot cycle is often impractical in single phase systems because of the large generalized forces required to complete the cycle, even though no heat exchangers are needed for the variable temperature paths. The Stirling cycle is intermediate in regard to both the necessary forces and the heat exchanger size. In two phase systems with first-order phase transitions, e.g., liquid–gas, the constant pressure lines are also isothermal lines. In that system, the Ericsson cycle cannot be executed and the Carnot cycles requires no higher forces than the Stirling cycle. Finally in regard to the cycles discussed here, it must be emphasized that in practice with high-speed operation the isotherms cannot be followed even closely. Instead nearly adiabatic paths will be followed, which requires even higher generalized forces to complete the cycle.

There are two ways in which a system, or working substance, can be made to go around a cycle in a T–S diagram. The first is the time domain. In this case the system remains physically stationary and is connected to the reservoirs at T and T_0 by heat switches which open and close at the appropriate time. A generalized force does work on the system or receives work from the system at the appropriate time to carry out the cycle. The advantage is that some systems then require no moving parts. Obviously, refrigeration occurs only intermittently in this time-domain operation. The use of rapid cycle times, or two equal systems operating 180° out of phase, eliminates this problem. The other mode of cycle operation is in the physical or space domain. In this case the working substance moves to different locations for the various parts of the cycle in the T–S diagram. This is particularly easy to do with a liquid or gas system and has the advantage of continuous refrigeration. However, solid systems also can go around the cycle in the space domain by placing the material on a rotating wheel. Unfortunately, it is then sometimes difficult for the solid system to interact with the generalized force during a part of the cycle. The space mode of operation does not always imply that there will be moving mechanical parts, as we shall see later for the thermoelectric refrigeration. Besides the two modes of cycle operation discussed above, it is possible to combine the two, as is done in the gas Stirling cycle. Certain combinations can make it difficult to show the cycle on a T–S diagram and to analyze it.

Interaction of Force with System

So far we have discussed complete cycles and the net work required to complete the cycle in terms of the heat Q absorbed at a temperature T. Nothing has been said about how the work is accomplished and what is the maximum Q or W a system is capable of absorbing in one cycle. For a fixed Q in each cycle, the refrigeration power is directly proportional to the molar flow rate around the cycle or to the number of moles of refrigerant times the cycle frequency. Each system will have a characteristic minimum cycle time, below which extraneous heating effects become comparable to the refrigeration power. The extraneous heating effects may come from such things as pressure drops, eddy currents, joule heating, etc. The maximum refrigeration power per mole of working substance will be the product of the maximum cycle frequency and the maximum heat absorption capacity of the substance in one cycle. We shall now consider the maximum heat absorption capacity per cycle in more detail.

Equations (11.1) and (11.2) show that the heat absorption capability of a substance is a maximum when the process is reversible. In addition, these equations show that when a system absorbs heat at a constant temperature, the entropy of the system increases. The system then goes from a relatively ordered state with low entropy to a disordered state with higher entropy. For the case of boiling liquid helium, heat is absorbed reversibly as the system goes from the ordered liquid state to the disordered gaseous state. For an isothermal process the reversible heat absorption capacity of a system with n moles is given by

$$Q_{rev} = nT(S_d - S_0) \tag{11.7}$$

where S_d is the molar entropy in the disordered state and S_0 is the molar entropy in the ordered state. A system with good potential for refrigeration would have a relatively large value of $S_d - S_0$. Such a requirement implies S_d must always be large. The maximum heat absorption capability of various systems can be compared by viewing curves of entropy vs. temperature for the various systems.

Before making such a comparison, we first need to discuss how a generalized force acting on the system can lower its entropy to S_0 or to permit the entropy to change from S_0 to S_d in a reversible manner. It is this generalized force acting on the system which carries out the necessary work to complete the cycle. In a closed cycle the first law of thermodynamics relates, via Eq. (11.3), the *net* work required is equal to the heat transferred. It should be noted that during part of the refrigeration cycle the system may do work on the surroundings. In a closed cycle the state of the system or the energy of the system is never considered. When only part of the cycle is considered, the system undergoes a change in state. The most important part of the cycle we wish to consider is the absorption of heat at some low temperature. For a closed system where there is no mass flow (a cycle operating in the time domain) the heat absorption from the first law of thermodynamics is given by

$$dQ = dU + dW_c \qquad (11.8)$$

where U is the internal energy of the system and W_c is the work done to the surroundings by the closed system. When there is mass flow, as in an open system (a cycle operating in the space domain), the first law of thermodynamics gives

$$dQ = dH + dW_0 \qquad (11.9)$$

where H is the enthalpy and W_0 is the work done to the surroundings by the open system. For a reversible process the entropy change is related to the heat absorption by Eq. (11.2) and thus to the work by Eqs. (11.8) and (11.9). Such a relationship is written as

$$dQ_{\mathrm{rev}} = T\,dS = dU + dW_c \qquad (11.10)$$

$$dQ_{\mathrm{rev}} = T\,dS = dH + dW_0 \qquad (11.11)$$

Also for a reversible process the work terms become

$$dW_c = F\,dX \qquad (11.12)$$

$$dW_0 = X\,dF \qquad (11.13)$$

where F is a generalized force (intensive variable) which can bring about a generalized displacement X (extensive variable). The enthalpy is given by

$$H = U + FX \qquad (11.14)$$

Some examples of generalized forces and their associated generalized displacements are given in Table 11.1. The reader is encouraged to think of other examples not listed here. The negative sign for some of the forces is necessary to give the right sign in Eqs. (11.10)–(11.14).

Equations (11.10) and (11.11) show that the entropy can be changed by either of the following: (i) $dW = 0$, in which case a change in internal energy or enthalpy is responsible for the entropy change or (ii) $dW \neq 0$, in which case the system does reversible work on the surroundings. However, a large work term does not always imply a large entropy change because sometimes the work term may be dominated by a change in internal energy or enthalpy which occurs at the same time work is done. Case (i) occurs when the disordered and ordered states are in equilibrium with each other at the temperature T. The transition from the ordered to the disordered state is then a first-order transition. Boiling liquid helium is a classical example. A magnetic material at the transition from the ferromagnetic to paramagnetic states is another example. This case of phase equilibrium is an especially attractive means of refrigeration because of the simplicity of no work. The refrigerant can be separated a considerable distance from the rest of the cycle. The disadvantage is that the refrigeration can occur at only one temperature, unless the generalized force is varied to shift the temperature of phase equilibrium. There are usually practical and intrinsic limits as to how far the phase-equilibrium temperature can be shifted. The advantage of case (ii) is that phase equilibrium is not required, which means that the system may work over a wide range of temperatures. The disadvantage of case (ii) is that a means must be provided for the system, which is at the low temperature T, to interact and do work on the surroundings. In case (ii) only one phase exists, but its entropy can be changed by the action of the generalized force.

Table 11.2 lists several systems which have ordered and disordered phases in equilibrium. The use of these two phases permits refrigeration in the manner of (i). The temperature of the phase equilibrium can be altered by the force shown in the third column in Table 11.2. It is also this force which can be used to change the entropy of the disordered phase and achieve refrigeration in the manner of (ii). Some of the systems, e.g., liquid–gas, have sharp transitions whereas others are more broad. The question mark in the force column for some systems means that the best generalized force to act on such a system is uncertain. Again the reader is encouraged to extend the list of systems given in this table. In some cases an ordered phase of one system may still have a high entropy relative to the ordered phase of another system because of some other internal disorder. An example is a solid, which is ordered relative to a liquid on a molecular basis, but which can still be magnetically disordered.

There are also some systems where a force acting on one phase changes the entropy, but an appropriate second phase does not seem to exist at any temperature unless a more microscopic view is taken. Consider the case of tension on a wire. The tension tries to disorder the system, hence increase the entropy [not obvious from Eqs. (11.10) and (11.12)], but what

Table 11.2. Ordered and Disordered Phases of Some Systems and the Generalized Force Necessary to Change the Equilibrium Temperature of the Entropy S_d

System		Generalized force
Disordered phase	Ordered phase	
Gas	Liquid	Pressure, P
Liquid	Solid	Pressure, P
Paramagnet	(Anti-)ferromagnet	Magnetic field, H
Paraelectric	(Anti-)ferroelectric	Electric field, E
Normal metal	Superconductor	Magnetic field, H
Fermi gas	Fermi liquid	Fermi pressure or density, n
Surface film	Bulk liquid	Surface tension, σ
Interstitial atoms	Martensite	Tension, τ
Flexible polymer	Crystalline polymer	Tension, τ
Disordered β brass	Ordered β brass	?
Uniform solution	Precipitate	Pressure, P, or concentration, x
Molecular rotation	Rotational order	?
Zn + CuSO$_4$	Cu + ZnSO$_4$	Electric potential, E
Dilute solution	Concentrated solution	Chemical potential, μ, Or concentration, x

is the phase which it is trying to approach—a liquid? Or is it a gas phase? On a microscopic basis the excitations in the solid are phonons, which when excited represent a disorder in the system. A tension on a wire reduces the density of the solid, which causes more phonons to be excited at the same temperature. This will be discussed in more detail later.

For some of the systems shown in Table 11.2, the effect of increasing the generalized force is to decrease the entropy. In that case it is clear that when dealing with one phase it should be the disordered phase. The maximum entropy change the force can produce is just equal to the absolute entropy of the disordered phase. The force can do no better than to completely order the system and give it zero entropy. For some systems shown in Table 11.2, an increasing force can increase the entropy of the ordered phase. Examples are the antiferromagnet, antiferroelectric, and superconductor. The effect of the increasing field is to lower the transition temperature. At any given temperature a field is reached where the ordered phase transforms to the disordered phase. Any further increase in field decreases the entropy. A field always decreases the entropy of paramagnets and paraelectrics but has a negligible effect on normal electrons. Thus, we can say that in these cases where the field increases the entropy of the ordered phase, the entropy can increase to a value no greater than the entropy of the associated disordered phase.

ENTROPIES AND REFRIGERATION
PRINCIPLES OF VARIOUS SYSTEMS

As stated in the previous section, the maximum refrigeration power of any refrigeration system is proportional to the temperature times the maximum possible entropy change at that temperature. Thus, the merit of any system can be evaluated roughly but quickly by a comparison of its entropy curve with those of well-established refrigeration systems. To evaluate the entropy of a given system, we make use of the second TdS equation (derived in any textbook on thermodynamics):

$$T\,dS = C_F\,dT - T\left(\frac{\partial X}{\partial T}\right)_F dF \qquad (11.15)$$

In this equation C_F is the heat capacity of the system at a constant generalized force. The entropy at (T, F) can be evaluated by integrating Eq. (11.15) from $(0, 0)$ to (T, F) by any convenient path. Figure 11.2 shows the two easiest paths. As shown in this figure, the entropy is zero at $T = 0$ for any F. This follows from the third law of thermodynamics for any system which does not have any frozen-in disorder at $T = 0$. The third law then requires that $(\partial X/\partial T)_F = 0$ at $T = 0$. The entropy determinations rely on specific heat measurements from temperatures near absolute zero to the temperature T. Specific heat measurements as a function of T for various F allow the entropy to be evaluated for different F. Otherwise specific heat measurements at $F = 0$ combined with measurements of $(\partial X/\partial T)_F$ vs. F at various T can be used to find $S(T, F)$. A peak in the specific heat of a system is a very desirable feature for refrigeration since it indicates a change in phase and a large change in entropy. Whether any practical force can change the entropy is another question. The value of $(\partial X/\partial T)_F$ gives an indication of the size of F required to change the entropy, as is evident from Eq. (11.15).

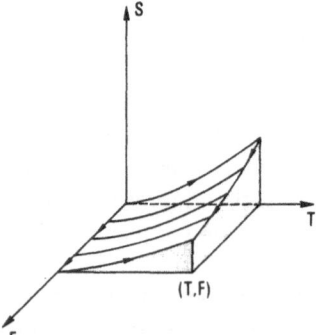

Fig. 11.2. Integration paths for evaluating entropy at specified values of temperature and generalized force.

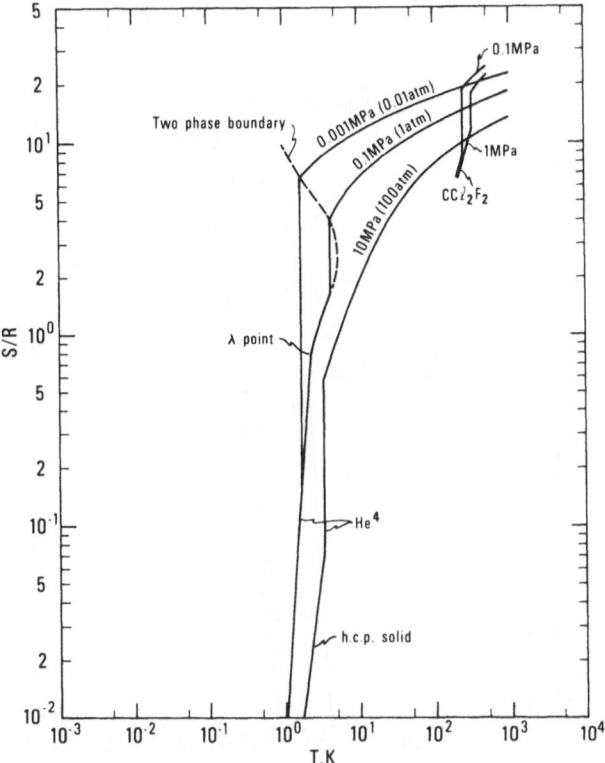

Fig. 11.3. Reduced entropy as a function of temperature for helium and CCl_2F_2 at different pressures.

Figures 11.3 through 11.6 show entropy as a function of temperature for many systems. The entropies are divided by the gas constant $R = 8.3143$ J/mol K to make them dimensionless. Notice that these figures show S as a function of T instead of vice versa, as was done for the thermodynamic cycles shown in Fig. 11.1 and in figures of previous chapters.

Helium-4 and CCl_2F_2

Figure 11.3 shows the entropy of He^4 in the gas, liquid, and solid phases (McCarty, 1972; Wilks, 1967). This system is by far the most common refrigerant for use with superconductors. Its entropy in the gas phase is very high and remains in this disordered phase down to temperatures below superconducting transition temperatures. For a gas system Eq. (11.15) becomes

$$T\, dS = C_p'\, dT - T(\partial V/\partial T)_p\, dP \tag{11.16}$$

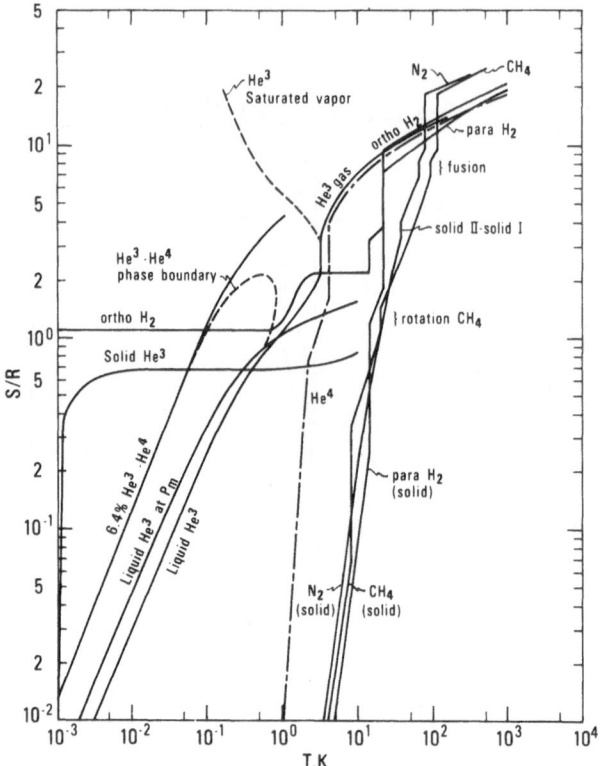

Fig. 11.4. Reduced entropy as a function of temperature for several gas–liquid–solid systems.

The isothermal entropy change in an ideal gas is then given by

$$\Delta S/R = \ln (P_1/P_2) \qquad (11.17)$$

where P_1 and P_2 are the initial and final pressure. Thus, even pressure ratios of 10 can give a $\Delta S/R$ of 2.3. In comparison with other systems at low temperatures this entropy change is high and shows why helium is such an important refrigerant. Figure 11.3 shows how the entropy drops significantly from the gas phase to the liquid phase. As discussed previously, this entropy difference in the phase equilibrium region is very useful. At lower temperatures the entropy of the liquid drops suddenly at the superfluid transition temperature (λ point) due to a quantum ordering process. Another type of order takes place when sufficient pressure is applied to solidify the helium. As seen in the figure, the entropy change associated with melting is small compared with vaporization and thus would not be very useful for refrigeration. Refrigeration at 1.8 K could be accomplished by using a porous plug to separate the superfluid component from

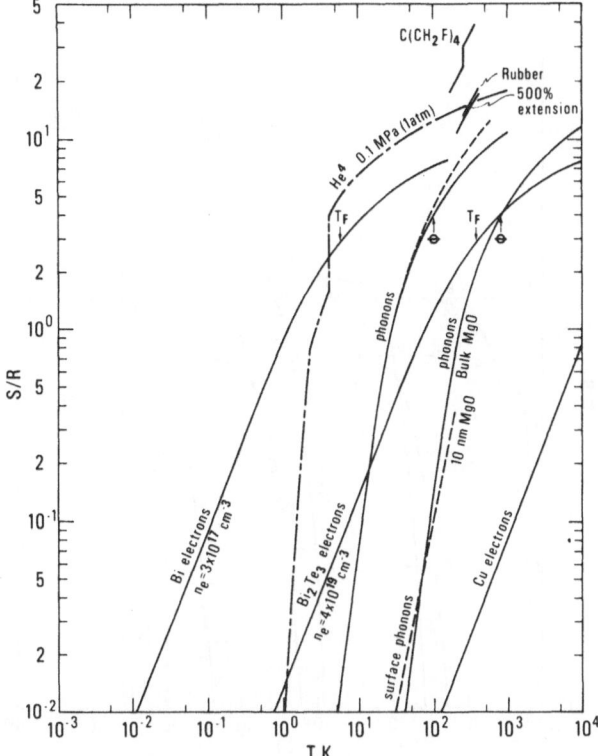

Fig. 11.5. Reduced entropy as a function of temperature for solid systems.

the normal component. Heat is then absorbed by raising the entropy of the superfluid fraction from zero to about $S/R = 0.2$ (the value for liquid He^4 at 1.8 K). This change is small compared with the change in S/R of about 6 for vaporizing the liquid at 1.8 K. This disadvantage could be offset by the advantage of a much smaller molar volume in the liquid state compared with the gas phase at 0.001 MPa (0.01 atm). In fact, the rapidly increasing molar volume in the gas phase as the temperature is reduced places a lower limit of the order of 1 K for refrigeration with He^4. In this case the limit occurs because the molar flow rate is severely restricted by the large molar volume.

Also shown in Fig. 11.3 is the entropy (Van Wylen et al., 1965) of the common refrigerant, dichlorodifluoromethane, CCl_2F_2. The absolute entropy is uncertain, but the values shown are reasonable. The important thing to observe is the entropy change during vaporization. This system is shown in comparison with He^4 to point out its advantage when working over a small temperature span near room temperature. This system is

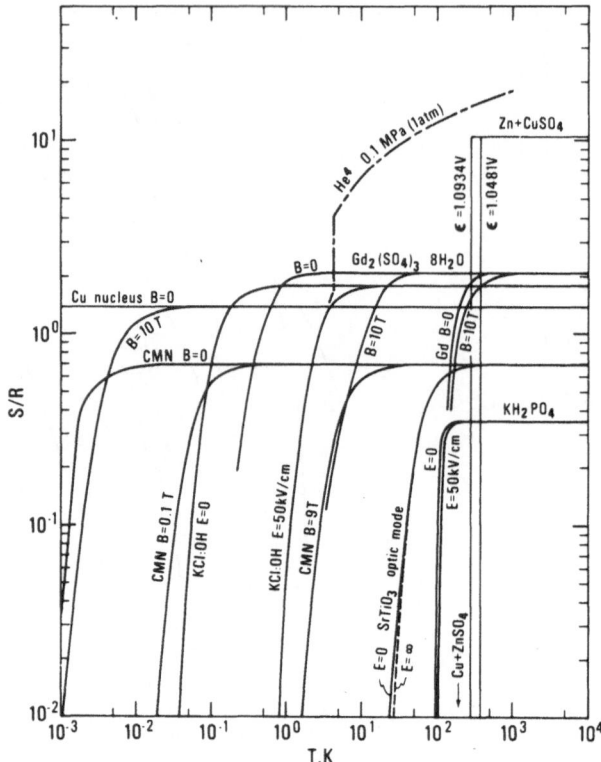

Fig. 11.6. Reduced entropy as a function of temperature for magnetic, dielectric, and electrolytic systems.

commonly used in household refrigerators and air conditioners, and the pressures used are typically those shown in Fig. 11.3. Notice that at 315 K only 1 MPa (10 atm) is required to liquefy the system. After being cooled in a heat exchanger to about 245 K the liquid is expanded to 0.1 MPa (1 atm) and evaporated. During the evaporation the entropy changes by $\Delta S/R = 9.8$. A similar entropy change in He[4] at that temperature would require a pressure ratio of 10^4 compared with 10 for CCl_2F_2. This comparison points out the ease of changing the entropy of a system just above the ordering temperature. Thus it is clear that a He[4] refrigeration system would require a much higher molar flow rate or pressure ratio to achieve the same refrigeration power as a CCl_2F_2 vapor-compression system. If that were done, it is obvious that inefficiencies in the He[4] system would be higher. The curves shown in Fig. 11.3 then tell us immediately that a Stirling-cycle refrigerator using He[4] or some other ideal gas would not be as efficient or useful as a CCl_2F_2 refrigerator for use near room temperature.

The same conclusion is reached by Nesselmann (1957) using more detailed calculations on actual cycles.

Other Gas–Liquid–Solid Systems

Figure 11.4 shows the reduced entropy of several other systems which exist in the gas phase at room temperature. Helium 4 at 0.1 MPa (1 atm) is also shown for comparison purposes. The absolute entropy of solid and liquid N_2 is based on the work of Giauque and Clayton (1933) and the liquid–gas values of Strobridge (1962) are added to this. There is an entropy change associated with a structural change in the solid at 35.6 K, as well as the entropy change with fusion. This transition is one in which a generalized force has little effect, except possibly pressure.

The entropy of methane, CH_4, is taken from several sources (Goodwin 1973; Franck et al. 1937; and Colwell et al., 1962). In the solid phase there is a cooperative rotational ordering at 20.4 K which results in a sudden drop in entropy below that temperature. A similar but less-studied transition occurs at about 8 K. Again, these transitions are ones that we have little control over by the use of some known generalized force. The transition temperatures are determined primarily by the molecular moment of inertia, which cannot be changed much by external forces. Thus these transitions are of little use for refrigeration unless we can find some "handle" to change the moment of inertia. In addition, the entropy changes associated with them are rather small.

The hydrogen molecule can take on two modifications due to the relative orientations of the two nuclear spins. The term orthohydrogen is used for the form with the spins in the same direction and the term parahydrogen is used for the form with the spins in the opposite directions. Since parahydrogen has no net spin, it has the lowest energy state. Thus at 0 K, equilibrium hydrogen will be 100% parahydrogen. The high-temperature equilibrium concentration is 75% orthohydrogen, which is closely approached at room temperature. As shown in Fig. 11.4, there is a difference in entropy between parahydrogen (Johnson, 1960; and McCarty, 1975) and orthohydrogen (Woolley et al., 1948). The drop in entropy of orthohydrogen below about 1.6 K is due to a splitting of the lowest rotational energy state by internal electric field gradients on the nuclear quadrupole moment. Because of the entropy difference between ortho- and parahydrogen, it is tempting to think of some external force which could change the equilibrium concentration at some temperature. Practical levels of electric field gradients or magnetic fields would have only a minor effect on the equilibrium concentration, though further studies may be useful.

Helium 3 is a very useful refrigerant for temperatures below 1 K and will be discussed in more detail in Chapter 12. Because the nucleus has a spin of $1/2$, it follows Fermi–Dirac statistics. The main feature of a fermi fluid is that ordering into lower energy states is a gradual process. For instance, compare the entropy of liquid He^3 with that of liquid He^4. In Fig. 11.4 the liquid and gas entropy for temperatures above 1.5 K are from the work of Gibbons et al. (1967, 1968) and Daunt (1970). The work of Radebaugh (1967) is used for temperatures below 1.5 K and for He^3–He^4 mixtures. Solid entropies are taken from the book of Betts (1976) and the publication of Dugdale and Franck (1964). Because the He^3 entropy falls off so slowly as the temperature is reduced it has enough entropy for refrigeration over a temperature range from 10^{-3} to 1 K, a wider range than any other system. The entropy of the saturated vapor can be used for reaching temperatures of about 0.3 K. The He^3 density becomes too small to reach any lower temperature with the gas phase.

The entropy of solid He^3 remains at the high value of $S/R = \ln 2$ until the nuclear spins order magnetically at about 1.2 mK. At the solid–liquid equilibrium pressures the liquid entropy is changed from that at atmospheric pressure to that shown in Fig. 11.4. Still there is a considerable entropy change between the liquid and solid for temperatures below about 0.1 K. Application of pressure to solidify the He^3 then becomes a useful refrigeration technique for the range 1 mK to 0.1 K. This technique is known as Pomeranchuk refrigeration, based on a proposal in 1950. Details of this refrigeration technique will be discussed in Chapter 12. Practical use of Pomeranchuk refrigeration has occurred only since 1965. This refrigerator operates in the time domain where the refrigerant is stationary and heat switches are used. So far only a heat switch between the high-temperature reservoir at about 30–50 mK has been used, which limits it to single cycle operation rather than continuous operation.

If He^4 is added to He^3, the He^4 acts to separate the He^3 atoms and to make them behave more like an ideal gas. Because the entropy of He^4 itself is so small, below 1 K, it contributes nothing to the entropy of the solution. Because of its gaslike behavior, the entropy of the He^3 in solution with He^4 is higher per mole of He^3 than that of liquid He^3. This is shown in Fig. 11.4 by the curve for 6.4% He^3 in He^4. H. London (1951) first proposed the use of He^3–He^4 mixtures for refrigeration. At that time the existence of phase equilibrium in He^3–He^4 mixtures was not known, so it was thought that some generalized force was needed to interact with the system to bring about a reversible dilution of the He^3. It was discovered about four years later that the He^3–He^4 mixtures phase separated and that at temperatures below about 50 mK the maximum He^3 concentration in the dilute phase was 6.4%. Any excess He^3 floats on top of the mixture in the form of nearly pure He^3. The thermodynamics of the mixtures are described in detail elsewhere (Radebaugh, 1967). This phase separation meant that pure He^3 could be diluted reversibly to 6.4% in a manner very much like evaporation of a liquid. Since the dilution could be done without doing external work, the process would be much simpler than if phase separation did not occur. In 1965 the first successful dilution refrigerator was built and now these have become off-the-shelf items for continuous temperatures down to 10 mK. Continuous temperatures of 2 mK have been reached recently (Frossati *et al.*, 1978). The operating temperature span of 2 mK to 1 K for the dilution refrigerator is considerably larger than any other refrigerator when one considers temperature on a logarithmic scale.

The dilution refrigerator is discussed in detail in Chapter 12, but we describe its operation here in order to compare it with other refrigeration systems. Liquid helium-3, precooled in a recuperative heat exchanger, enters a mixing chamber in which there is a phase boundary between the helium-3 on top and 6.4% He^3 in He^4 on the bottom. Helium-3 continuously passes into the dilute 6.4% solution from the upper concentrated He^3 phase as He^3 is removed from the dilute solution downstream from the mixing chamber. The continuous dilution process absorbs heat and cools the mixing chamber. The cold dilute He^3 diffuses through the He^4 until it reaches the still where He^3 boils off from the dilute solution. Because He^3 has a much higher vapor pressure than He^4, the separation at the still is very good. Before reaching the still at about 0.6 K, the cold dilute He^3 passes through the heat exchanger to precool the pure He^3 before it enters the mixing chamber. The He^3 gas coming from the still is compressed at room temperature and is reliquefied at about 1 K by a separate pumped He^4 bath. The refrigeration rate in the mixing chamber is simply $\dot{Q} = \dot{n} T \, \Delta S$, where $\dot{n}$ is the molar flow rate and ΔS is the entropy difference between the dilute and concentrated He^3 phases as shown in Fig. 11.4.

Electron Systems

Figure 11.5 shows the entropy of electrons in several materials. In the previous figure we saw that the relatively slow ordering in Fermi-fluid He^3 was useful for providing refrigeration over a wide temperature range below 1 K. Helium 3 is a nuclear Fermi fluid, but now we consider a much lighter Fermi fluid—electrons in a metal, semimetal, or semiconductor. The Fermi temperature of a system of spin-$\frac{1}{2}$ particles is given by

$$T_F = \frac{h(3\pi^2 n)^{2/3}}{2km^*} \tag{11.18}$$

where n is the density of particles, k is Boltzmann's constant, and m^* is the effective mass of the particle. The Fermi temperature is the approximate temperature below which ordering begins to take place. The specific heat and entropy of a Fermi gas are linear in T much below T_F whereas the specific heat approaches the constant value $(\frac{3}{2})R$ above T_F. Using the mass of the He^3 nucleus in Eq. (11.18) gives T_F on the order of 1 K for He^3. The much lighter mass of electrons results in a T_F of about 6×10^4 K for electrons in most metals, such as copper. At room temperature or below these electrons are ordered quite well, as suggested by the low entropy. Additional ordering of the electrons can occur when they pair via virtual phonons to become superconducting. The entropy would then drop off very rapidly below the transition temperature. Similar behaviour occurs in liquid He^3 when it becomes a superfluid below 2.7 mK.

According to Eq. (11.18) the Fermi temperature can be lowered by reducing the concentration of electrons. The resultant entropy per mole of electrons is thus increased at a given temperature. Figure 11.5 shows how this reduced density gives higher electron entropies (Goldsmid, 1964) to the semiconductor Bi$_2$Te$_3$ and semimetal Bi. It is clear that heat will be absorbed if electrons go from a normal metal (or superconductor) to Bi$_2$Te$_3$ or Bi. This can be done by joining the two materials and passing an electric current through the junction. The heat absorbed is known as the Peltier heat, and the technique of utilizing this effect is called thermoelectric refrigeration. A voltage source (ε) provides the power necessary (εI) to drive the electrons around the cycle. The electrons return to the metal from Bi or Bi$_2$Te$_3$ at the other junction. This process gives off heat. The similarity to the dilution refrigerator is evident, but in this case it is electron dilution which produces refrigeration. A current of 10 A gives an electron flow rate of 10^{-4} mol/sec, which is a typical He^3 flow rate in dilution refrigerators.

The He^3–He^4 dilution refrigerator is a powerful and successful refrigerator for temperatures down to 2 mK. The thermoelectric refrigerator has been very successful in some cases for temperatures around room temperature. In comparison with the dilution refrigerator, what would one expect for the low-temperature limit of a thermoelectric refrigerator? To answer that question we need to consider several points. First, consider the case of a metal or semimetal with a molar electron density of 6.4% per mole of metal of semimetal, which is the same as the He^3 concentration in the dilute side of a dilution refrigerator. The entropy curve would be increased over that of the Cu electrons by the same amount as the He^3–He^4 curve is above the pure He^3 curve. At a lower practical limit of about 5 mK for a dilution refrigerator the dilute He^3 entropy is about $S/R = 6 \times 10^{-2}$. For a similarly dilute electron system a value of 6×10^{-2} for S/R occurs at 200 K, which at present is somewhat below that possible with single stage thermoelectric refrigerators.

In n-type Bi_2Te_3 the electrons are diluted to the point of $n = 4 \times 10^{19}$ cm^{-3}. The entropy per mole of electrons is then quite high at 100 K but the low density makes it difficult to obtain a reasonable electronflow rate, i.e., the electrical resistance becomes high. A figure of merit for a material for thermoelectric refrigeration is given by Goldsmid (1964) in the form

$$Z = \alpha^2 \sigma / k \tag{11.19}$$

where α is the thermoelectric power, or Seebeck coefficient, σ is the electrical conductivity, and k is the thermal conductivity. The electronic part of the Seebeck coefficient is simply proportional to the electron entropy and is given by

$$\alpha = [8.6 \times 10^{-5} \ V/K]S/R \tag{11.20}$$

For the case where α, σ, and k do not vary much with temperature, the maximum temperature drop is given by (Goldsmid, 1964)

$$(T_h - T_c)/T_c = \tfrac{1}{2}ZT_c \tag{11.21}$$

Larger temperature drops can be achieved by staging.

The presence of phonons in the thermoelectric materials is both a disadvantage and advantage. The disadvantage is that the phonons transport considerable heat from the hot junction to the cold junction and provide resistance to the flow of electrons. The advantage is that as the electrons collide with the phonons, the phonons are dragged along with the electrons in what is called the phonon-drag effect. The reason that this is good is because the phonons then contribute their entropy to the flow process. As seen in Fig. 11.5 the phonon entropy can be quite high, especially for a material with a low Debye temperature, Θ. The phonon-drag effect dominates the Seebeck coefficient of pure metals in the 5–50 K temperature range. Optimization of the phonon-drag effect could be useful to enhance α for temperatures below 200 K. The presence of spins from magnetic impurities can give rise to a similar effect known as spin-drag (MacDonald, 1962). In a particular type of electron-spin scattering, the electrons may drag along the relatively high spin entropy. (See Fig. 11.6.) This effect deserves further study for enhancement of the figure of merit.

A problem with the drag effects is that they also contribute to the electrical resistivity. A more desirable way of increasing the figure of merit is to increase the effective mass of the electrons and thus reduce the Fermi temperature according to Eq. (11.18). A distortion of the Fermi surface causes such an enhanced effective mass. An effective mass of at least 100 electron masses may be necessary to give a high enough figure of merit for refrigeration below 100 K. This high effective mass must occur with high values of electron density and mobility.

In practical thermoelectrical refrigerators one side of the junction is an n-type material like Bi_2Te_3 where the charge carriers are electrons. The other side of the junction is not a normal metal, but a p-type material in which the charge carriers are holes. Thus instead of electrons circulating around the system, electrons and holes are created at the cold junction with the electrons leaving the junction through the n-type material and the holes leaving the junction through the p-type material. The electrons and holes are annihilated when they meet at the hot junction. In the p-type material the entropy is associated with the holes and is generally comparable to the electron entropy in n-type materials. The advantage of using both n- and p-type materials for a junction now becomes clear. There is not only an entropy flow away from the junction in the n-type material but also an entropy flow away from the junction in the p-type material. The net result is that the total entropy flow, and hence the heat input to the junction, is roughly double that for the case where the p-type material is replaced with copper. The overall figure of merit for a refrigerator made with material 1 and 2 for the junction is given by

$$Z_{12} = (\alpha_1 - \alpha_2)^2 / [(k_1/\sigma_1)^{1/2} + (k_2/\sigma_2)^{1/2}]^2 \qquad (11.19a)$$

For $\alpha_2 = -\alpha_1$ and $k_2/\sigma_2 = k_1/\sigma_1$, Z_{12} is the same as Z for either material. For material like copper, $\alpha \approx 0$ but k/σ is approximately equal to that of a good thermoelectric material. Thus Z_{12} would be about $\frac{1}{4}$ that of Z for the good thermoelectric material.

Figure 11.7 shows the highest reported values of Z as a function of temperature for several thermoelectric materials. Shown for comparison is a curve for $ZT = 1$. Temperature drops for each stage, according to Eq. (11.21), become quite small for ZT, much less than 1. For refrigeration just below room temperature the Bi_2Te_3 alloys are used in both the n and p legs of the couple. Values of Z for these two Bi_2Te_3 alloys are from the work of Yim et al. (1966) and Yim and Rosi (1972). The n-type Bi_2Te_3 alloy is $(Bi_2Te_3)_{90}(Sb_2Te_3)_5(Sb_2Se_3)_5$ doped with SbI_3 to give a carrier concentration of 3.6×10^{-19} cm^{-3} at room temperature. The p-type material is $(Sb_2Te_3)_{72}(Bi_2Te_3)_{25}(Sb_2Se_3)3$ doped with excess Te to give a carrier concentration of 2.1×10^{-19} cm^{-3} at room temperature. The Z value for these Bi_2Te_3 alloys is reasonably high at room temperature but falls off rapidly at low temperatures, especially that of the p-type material. By changing the doping levels, Click and Marlow (1970) were able to increase the Z value of the p-type Bi_2Te_3 alloy at low temperatures as shown in Fig. 11.7. This is the highest reported Z value for a p-type material below 220 K. For n-type materials Fig. 11.7 shows that the low-temperature Z values are much higher. The data for undoped $Bi_{85}Sb_{15}$ and $Bi_{92}Sb_8$ are from the work of Yim and Amith (1972) and represent

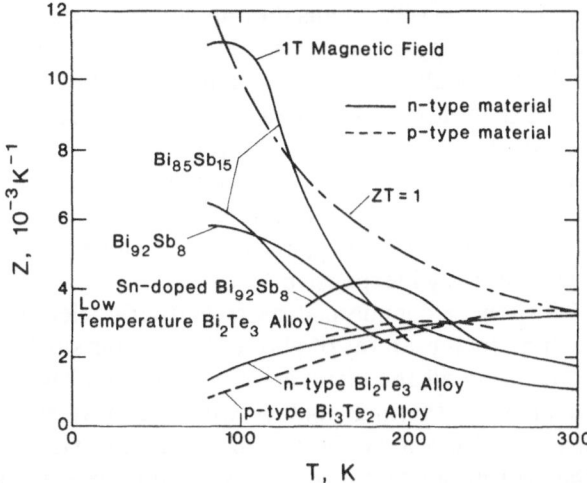

Fig. 11.7. Thermoelectric figure-of-merit, Z, as a function of temperature for several high-Z materials. Exact compositions are discussed in the text.

the highest reported Z value for any material. Wolfe and Smith (1962) first showed that the application of a transverse magnetic field can enhance Z significantly in Bi–Sb alloys. Yim and Amith (1972) applied a transverse field of 1 T to their $Bi_{85}Sb_{15}$ sample and obtained the results shown in Fig. 11.7. Higher fields will not increase the peak value but they can increase the higher-temperature Z values. Figure 11.7 shows that at intermediate temperatures the $Bi_{92}Sb_8$ alloy doped with $1 \times 10^{18} \, cm^{-3}$ of Sn gives the best Z values (Click and Marlow, 1970). Low-temperature thermoelectric refrigeration is severely hampered by the lack of a good p-type material below 200 K.

The lowest temperature ever reached with a thermoelectric device is 128 K with a combined thermoelectric-magnetothermoelectric refrigerator (Yim and Amith, 1972). This device used a transverse field of 3 T on the Bi–Sb lower stages. The lowest temperature ever reported for a device with no field enhancement is about 134 K (Buist *et al.*, 1971, 1976) using eight stages. Click and Marlow (1970) reported a temperature of 145 K for a nine stage refrigerator with the various elements chosen from those shown in Fig. 11.7 to give the best Z value in the operating temperature range for that element. Both of these refrigerators, developed under contract to the U.S. Army Night Vision Laboratory, required less than 50 W of input power.

Thermoelectric refrigerators have rather low efficiencies because of relatively large loss terms such as joule heating and thermal conductivity. The coefficient of performance (COP) at 200 K for a multiple-stage device may be of the order of 1%–2%. Thus thermoelectric refrigerators are used where other criteria are more important than efficiency. Their big advantages are simplicity, small size, and extreme reliability. Thermoelectric refrigerators have demonstrated a mean-time-between-failure (MTBF) of 350,000 hours which is equivalent to 35 years of continuous operation. No mechanical system can ever approach such high reliability. Cooling of small heat loads such as small infrared detectors appears to be the major application of thermoelectric refrigerators. In small size thermoelectric refrigerators are relatively inexpensive.

Figure 11.8 shows a commercial six-stage thermoelectric refrigerator capable of reaching temperatures of about 170 K. This stock item sells for about $500. Generally all the elements are connected together in series. Each stage is in good thermal contact with the next stage but yet is electrically insulated from it by the use of beryllium oxide plates. Figure 11.9 shows the typical performance curves for this six-stage refrigerator. The optimum design of such multistage refrigerators becomes quite complex and is usually not discussed in books on thermoelectric refrigeration. The reports by Buist (1974) and Buist et al. (1976) discuss in detail the design of multistage devices. These reports also point out several fabrication problems such as differential thermal contraction, copper contamination of the elements, and the effect of cutting damage on Z. Problems associated with single-crystal materials are overcome by the use of sintered materials in many commercial refrigerators.

Fig. 11.8. Commercial six-stage refrigerator capable of reaching 170 K. The base dimensions are 28×22 mm and the height is 20 mm. The top is the cold side and the bottom is the hot side. (courtesy of Marlow Industries, Inc.)

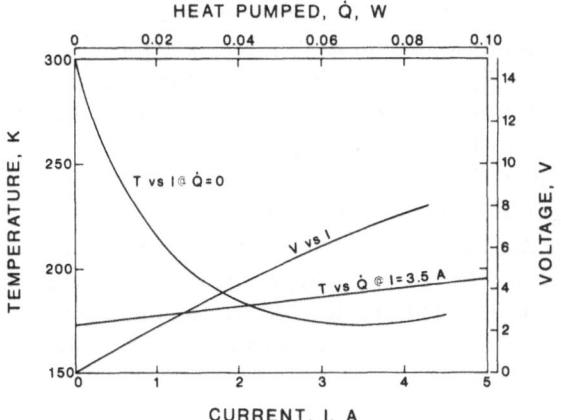

Fig. 11.9. Performance curves for the thermoelectric refrigerator in Fig. 11.8. Data taken in a vacuum but with no radiation shielding and with the hot side at 300 K.

The magnetothermoelectric effect discussed above should not be confused with thermomagnetic effects. In the latter a magnetic field is placed at right angles to the current flow, which then gives rise to a temperature gradient perpendicular to both the current flow and the magnetic field. In essence, the magnetic field deflects the faster, higher-energy electrons more than the slower ones. The lower-energy electrons tend to go to one side of the sample and can absorb heat. The effect is even greater when both electrons and holes are present. In that case an electron–hole pair is created on the cold side (an entropy increase) and driven toward the hot side where the excess pairs are annihilated (an entropy decrease). This thermomagnetic effect is called the Ettingshausen effect after the Austrian botanist and geologist who discovered it in 1887. Thermomagnetic refrigeration is usually better than thermoelectric refrigeration for temperatures below about 150–200 K since a single exponentially shaped thermomagnetic element provides the necessary cascading over a wide temperature drop. Combining the two could give temperatures on the order of 70 K (McCormick et al., 1965). Because efficiencies of such systems are so low, these devices are useful only for cooling small devices with little power dissipation. Cooling of small superconducting devices would be an ideal application if high enough figures of merit could be achieved to reach a temperature on the order of 10 K.

Before leaving the electron systems it should be pointed out that the electron entropies discussed here and shown in Fig. 11.5 are for electrons in equilibrium with the lattice. Certainly higher electron entropies occur when high energy electrons are forced into a material. The resulting high electron entropy decreases as the electrons come into equilibrium with the lattice. If the equilibrium time is long enough, the high electron entropy in nonequilibrium systems may be used for cooling a local region. The return to equilibrium is made to occur downstream at the point where heat is to be rejected to the surroundings. Such a concept is employed in the recent proposal of Melton *et al.* (1980) for a superconducting tunnel-junction refrigerator. In that refrigerator superconducting electron pairs (low entropy) enter a superconducting film but electron quasiparticles (high entropy) are forced to leave the superconductor through the other side. However, the quasiparticles rapidly return to equilibrium a short distance away and the resultant heat can be transferred back to the superconductor being cooled.

Phonon Systems

The two curves shown in Fig. 11.5 for the phonon entropy represent the approximate upper and lower limits for most solids. The upper curve is for a solid with a Debye temperature Θ of 100 K whereas the lower curve is for MgO, which has $\Theta = 800$ K. The dashed line rising above the solid line for the $\Theta = 100$ K curve shows a typical deviation from the ideal Debye behavior. Previously we discussed how phonon-drag effects in the thermoelectric effect utilize some of the phonon entropy for cooling. Other methods of utilizing this entropy may be available. In analogy to thermoelectric refrigeration or the dilution refrigerator, cooling should occur at a junction between two solids as phonons pass from the high-Θ solid to the low-Θ solid. A suitable driving mechanism is not obvious. Electronic or ultrasonic means may exist to drive the phonons across the junction. Considerable problems may exist because of phonon reflections at the junction.

Refrigeration with a force acting on a single solid is possible if Θ can be changed by the force. The Debye temperature of a solid can be given by

$$\Theta = (h/k)c(18\pi^2 n)^{1/3} \tag{11.22}$$

where c is the sound velocity and n is the atomic density. Because of the high short-range forces in a crystal, pressure has only a small effect on n and on c, and thus on Θ. The small effect of pressure on the entropy is seen from Eq. (11.15), which for this system becomes

$$T dS = C_p dT - T(\partial V/\partial T)_p dP \tag{11.23}$$

The term $(\partial V/\partial T)_p$ is just the thermal expansion. Some solid hydrocarbons (low Θ) may have values for $(1/V)(\partial V/\partial T)_p$ of $6 \times 10^{-4}\,\text{K}^{-1}$ at room temperature. Because an ideal gas has the value $3.4 \times 10^{-3}\,\text{K}^{-1}$ for $(1/V)$

$(\partial V/\partial T)_p$ at room temperature, it is clear that the refrigeration power of the solid hydrocarbon will be 18% as much as that for an equal volume of gas. For 10 MPa (100 atm) $\Delta S/R$ for the hydrocarbon may be about 10^{-1} at 300 K. The effect will drop off like T^3 at lower temperatures. Because of the high specific heat of hydrocarbons, a 10-MPa pressure change will change the temperature by only 1–2 K at 300 K.

For a linear system of the same material $(1/L)(\partial L/\partial T)_\tau$ will be about $\frac{1}{3}$ of the volumentric thermal expansion coefficient. The entropy change and cooling effects for the same volume of material for a stress equal to 10 MPa will thus be $\frac{1}{3}$ as large as for the case of pressure on a volume. A 10-MPa stress will cool the material somewhat less than 1 K at 300 K. Thus, useful systems appear impractical.

Other Solid Systems

An interesting exception to the linear system discussed above is the case of vulcanized rubber, such as an ordinary rubber band. This high polymer consists of long chains, which in the relaxed state can be twisted at random in many different ways, i.e., a disordered state. That leads to a high entropy in the relaxed state. As the rubber is stretched, one configuration of the molecular chains becomes preferable and so the entropy of the chain approaches zero according to the relationship

$$S = R \ln m \tag{11.24}$$

where m is the number of microconfigurations. Figure 11.5 shows the entropy of natural rubber for both a relaxed state and for a 500% elongation. These entropies include the normal phonon entropy in addition to the configurational entropy. Unlike the case of a normal solid, which has only the phonon entropy, stretching a rubber band reduces the entropy. For a 500% isothermal elongation the entropy is reduced by an amount $\Delta S/R \simeq 0.8$. The temperature increases during elongation by about 8 K at room temperature if adiabatic conditions are maintained (Treloar, 1958). According to the second TdS equation, Eq. (11.15), the linear thermal expansion coefficient under tension must be large and negative to give the required entropy change. That behavior is indeed the case for rubber.

The entropy change in rubber is sufficiently large to utilize it for a motor (Strong, 1971), although it may be useful only for demonstration purposes. Another interesting demonstration would be to use a separate motor to drive the previously mentioned motor in reverse and obtain refrigeration. Figure 11.10 is a sketch of such an arrangement which could give a maximum temperature reduction of about 8 K. The demonstration is useful in showing how a solid material can be made to go through a

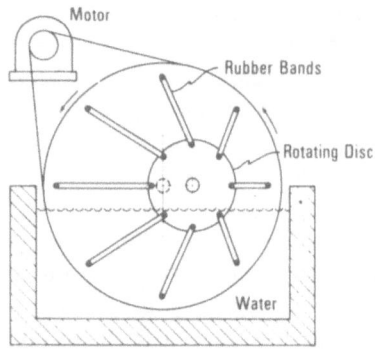

RUBBER BAND REFRIGERATOR

Fig. 11.10. Schematic of a rubber band refrigerator based on a scheme for a rubber band engine. The small disk rotates along with the larger disk but on a different center.

refrigeration cycle in the space domain and provide continuous refrigeration. The type of cycle that this system goes through is actually rather inefficient and is not quite like any of those discussed earlier. The cycle followed is the Stirling cycle (constant length paths) except in the sketch of Fig. 11.10 no provision exists for heat transfer between the two constant-length paths in the cycle. Hence, the refrigeration power is reduced because of the additional heat load of cooling the rubber bands from T_0 to T. If a heat transfer mechanism were provided (a moving oil layer on top of the water bath), the demonstration refrigerator could possibly freeze water.

Another material which can be used in the same way as rubber is 55-NITINOL (Jackson *et al.*, 1972). This alloy (55% Ni–45% Ti) has an unusual martensitic transition at about 50°C, but this temperature can be varied by changing the concentration. This alloy has a memory in the sense that a sample deformed below the transition temperature will revert back to its original shape when heated above the transition temperature. The material can exert considerable force in reverting to the original shape. The entropy of the high-temperature disordered phase is reduced by applying a tensile force. A sufficiently high force can drive the material into the ordered phase. The entropy change between the two phases is on the order of $\Delta S/R = 0.5$, comparable to that of rubber. The material has been considered for use as a motor in the manner described for the rubber band motor previously.

A solid material with a very high entropy change during a solid–solid transition at about 245 K is pentaerythrityl fluoride $C(CH_2F)_4$ (Westrum, 1961). The entropy change of $\Delta S/R = 6.4$ would have considerable practical significance if it could be controlled by a reasonable force. Because it may be a rotational ordering transition, such a reasonable force may not exist.

All of the entropies discussed so far have been for bulk samples. In very small dimensions (nm sizes) the finite energy level spacing of phonons or electrons will bring about an entropy decrease compared with bulk values. However, the surface phonons or electrons will then tend to dominate the total entropy of the material. The entropy of surface phonons or electrons will be higher than for bulk behavior since they are less ordered. Figure 11.5 shows an example of the entropy for surface phonons in 10-nm particles of MgO. This entropy dominates the bulk entropy of MgO only below about 50 K. Other two-dimensional systems such as adsorbed gases can have higher entropies than the ordered bulk phase. An adsorbed gas is also an ordered phase which can exist at temperatures much higher than the critical point of the bulk liquid. For instance, He^4 adsorbed on a substrate at 15 K is in equilibrium with the gas phase at a pressure of about 10 kPa (0.1 atm). The value of S/R in the adsorbed phase is about 1.6 (Goodstein, 1979). Thus the transition from the adsorbed phase to the gas phase requires an increase of S/R by about 8, which is very large. The transition at a constant pressure is quite broad. Generalized forces other than pressure may be applicable to such a system.

Magnetocaloric Systems

Figure 11.6 shows other sources of entropy useful for refrigeration, particularly magnetic systems. The entropy of the disordered-paramagnetic phase can be quite high. In addition, this disordered phase in some cases can exist down to very low temperatures before ordering to a ferromagnetic or antiferromagnetic phase takes place and the entropy is removed. The nucleus has a small magnetic moment and hence a low ordering temperature. The entropy of the copper nucleus is shown in Fig. 11.6 for the case of an applied magnetic field* of 0 and 10 T. Ordering for $B = 0$ takes place at about 10^{-7} K. Thus adiabatic demagnetization from about 5 mK and 10 T will result in a nuclear temperature of about 10^{-6} K or less. Chapter 12 discusses this in more detail.

Magnetic moments of electron spin systems are much higher, which means that ordering temperatures are higher. For a magnetic system the entropy in the disordered phase is given by

$$S/R = \ln(2J + 1) \qquad (11.25)$$

where $(2J + 1)$ is the number of possible orientations of the magnetic dipole of angular momentum J. The three electronic magnetic systems shown in

* Strictly speaking, the applied magnetic field H has SI units of A/m. In this book we are using the magnetic flux density B with SI units of $T(10^4 G)$ as the applied field in free space, i.e., $B = \mu_0 H$, where μ_0 is the permeability of free space.

Fig. 11.6 are $Ce_2Mg_3(NO_3)_{12}\cdot24H_2O$ (CMN), $Gd_2(SO_4)_3\cdot8H_2O$, and pure Gd. Values of J for these systems are $\frac{1}{2}$, $\frac{7}{2}$, and $\frac{7}{2}$, respectively. The ordering temperature for these three systems covers the range from about 10^{-3} K to room temperature. Cerium magnesium nitrate (CMN) has been used for refrigeration down to 1 mK for many years and will be discussed in Chapter 12. The entropy for CMN, shown in Fig. 11.6, is based on the recent work of Giauque et al. (1973). The material $Gd_2(SO_4)_3\cdot8H_2O$ has an entropy of $S/R = 2.1$ which is easily removed by a field of 10 T at 4 K. Because it is a solid with a resulting low molar volume, it may prove to be competitive with helium for refrigeration in the 4 K temperature range. The system could use the Carnot cycle with time domain operation by using heat switches such as magnetoresistive beryllium (Radebaugh, 1977). Such a refrigerator would have no moving mechanical parts. Stirling or Ericsson-cycle operation may also be used by placing the material on a rotating wheel. Steyert (1978a) has discussed designs for Carnot cycle refrigerators using $Gd_2(SO_4)_3\cdot8H_2O$ to reach temperatures of about 2 K with the upper temperature at 10 K. One of these uses time domain cycling with beryllium heat switches and the other uses a rotating wheel for space domain cycling. The wheel is made porous to allow helium gas at 10 K and liquid helium at 2 K to pass through the wheel for good heat exchange. A schematic diagram of the apparatus is shown in Fig. 11.11. In this diagram, the high field region is at the top of the wheel and the low field region is at the bottom. Calculated efficiencies are also given in Fig. 11.11 as a function of refrigeration capacity.

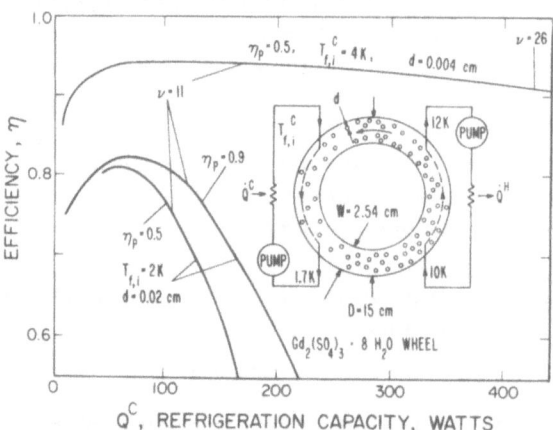

Fig. 11.11. Schematic of a $Gd_2(SO_4)_3\cdot8H_2O$ Carnot wheel refrigerator along with its estimated efficiency. Fraction of Carnot efficiency vs. refrigeration capacity is shown for various assumed values of the pump efficiency, η_p, and temperature of cold fluid at the inlet, $T_{f,i}$. Some values of rotation rate ν (in Hz) required are shown on the curves. A maximum field of 5 tesla is required for this refrigerator (after Steyert, 1978a).

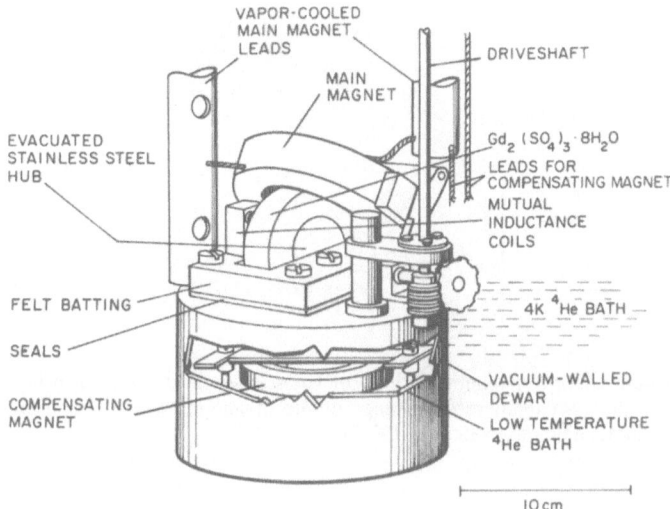

Fig. 11.12. Drawing of a $Gd_2(SO_4)_3 \cdot 8H_2O$ Carnot wheel refrigerator operating between 2 and 4 K (after Pratt *et al.*, 1977).

A prototype magnetic refrigerator using a Carnot wheel of $Gd_2(SO_4)_3 \cdot 8H_2O$ has been built and demonstrated by the Los Alamos group (Pratt *et al.*, 1977). This refrigerator, shown in Fig. 11.12, uses 4 K liquid helium as the upper bath and cools the lower superfluid bath to 2.1 K. One of the major difficulties in this refrigerator is the seal between the two helium baths. The wheel rotates at about 0.5 Hz in a field of about 3 T. Even though it is a magnetic refrigerator, the weak link in regard to reliability still is in the mechanical parts since it uses mechanical work input. The speeds are, however, quite slow. In time domain operation the work input could be in the form of electrical energy for cycling the magnetic field although large electrical losses may occur.

There are many other magnetic materials suitable for refrigeration at various temperatures. Magnetic refrigeration around room temperature utilizing gadolinium metal was proposed several years ago by Brown (1976) of the NASA Lewis Research Center.

The entropy can be changed by about $\Delta S/R = 0.3$ with a 10 T field at room temperature. We point out here that the entropies shown in Fig. 11.6 do not include the phonon entropies of these solids. The specific heat associated with this additional entropy can be large at room temperature and act as a heat load on the refrigeration cycle. Figure 11.6 shows how a field of 10 T has less and less effect on the magnetic entropy of a system as the temperature increases. This is due to the tendency of the phonons

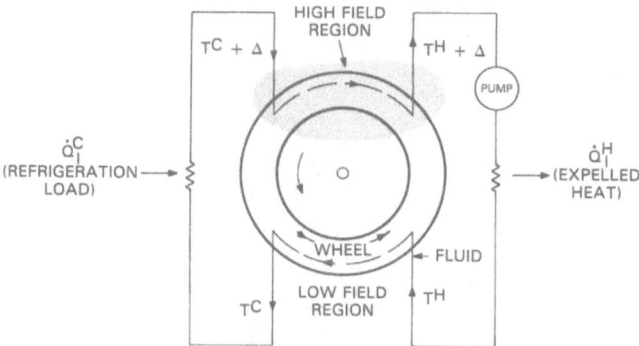

Fig. 11.13. Schematic of a magnetic Stirling- or Ericsson-cycle refrigerator using a rotating wheel. Δ is the inherent temperature change of the magnetic material upon entering and leaving the magnetic field (after Steyert, 1978b).

to disorder the magnetic spins. Carnot cycle operation with Gd would be limited to temperature differences less than 14 K. Ericsson cycle operation would follow the $B = 0$ and 10 T curves down to much lower temperatures so that larger temperature differences can be provided with that cycle or with the Stirling cycle. The high phonon specific heat places a severe requirement on the heat exchangers for such cycles. The design of a room-temperature refrigerator using a rotating wheel of Gd was described by Steyert (1978b). In this case, the cycle followed is the Ericsson cycle (constant field lines) although Steyert refers to it as the Stirling cycle. A schematic of the device is shown in Fig. 11.13. In the Carnot version in Fig. 11.12 the fluids pass through the wheel in a direction parallel to the axis of rotation. In the Ericsson version in Fig. 11.13, the fluid passes through the wheel opposite the rotation direction. A working model of the device was described by Barclay *et al.* (1979). With a 3-T magnetic field and a rotational speed of 0.1 Hz the refrigerator maintained a temperature difference of 7 K with 500 W heat input to the cold side. In this model, water was the heat exchange fluid and the 15-cm-diam wheel contains 2.3 kg of Gd.

Brown (1979) describes a laboratory model of a reciprocating magnetic refrigerator for room-temperature use which is smaller (0.9 kg Gd) than the Los Alamos rotating model but it achieves a temperature span of 80 K with 6 W heat input to the cold end. The field used is 7 T. Figure 11.14 shows the general layout for an engineering prototype of the refrigerator, which is similar to that originally proposed by Van Geuns (1966). The Gd shown in Fig. 11.14 is in the form of screens to give better heat transfer, but the experiments of Brown were performed with Gd plates. The porous

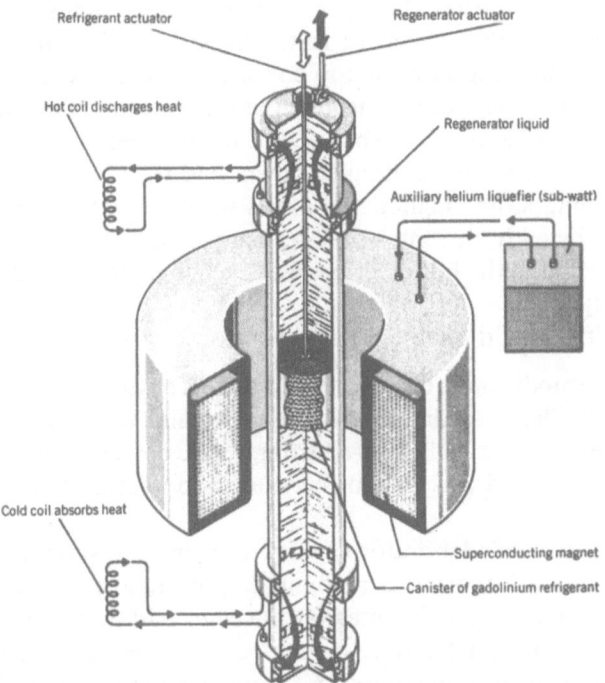

Fig. 11.14. Reciprocating magnetic refrigerator using pure gadolinium for use around room temperature. Either Stirling-cycle or Ericsson-cycle operation could be followed in this configuration (after Brown, 1979).

assembly of gadolinium plates fits inside a 1-m-long, 5-cm-diam cylinder containing the regenerator liquid (water plus ethanol) which can move up and down. The regenerator liquid flows freely through the gadolinium plates and exchanges heat with them as the cylinder moves up and down.

After a startup phase which establishes a temperature gradient in the column of regenerator fluid, the steps in the steady-state cycle are as follows:

(1) From the position shown in Fig. 11.14, the regenerator is raised while the gadolinium (Gd) assembly remains motionless in the magnet until the bottom of the regenerator reaches the Gd canister. This relative motion passes the regenerator fluid through the Gd, cooling the Gd approximately to the temperature of the bottom (cold end) of the regenerator.

(2) The regenerator continues upward, carrying the Gd out of the magnet bore. The Gd experiences a decreasing magnetic field strength, and therefore becomes cooler. In this process, the Gd absorbs heat from the cold end of the regenerator and from the external load because of the fluid circulating in the lower heat transfer loop.

(3) When the Gd and the regenerator have been completely withdrawn from the magnet, they come to rest. The Gd remains at rest while the regenerator is lowered until its top contacts the Gd assembly. The regenerator fluid thus passes down through the Gd, warming the Gd to approximately the temperature at the hot end of the regenerator.

(4) The regenerator continues down, and the Gd now moves with it. As the Gd enters the magnetic field it is warmed by the "magnetocaloric effect" (not by eddy currents), and discharges heat into the hot end of the regenerator and to the hot external heat exchanger through the upper heat transfer loop.

(5) With the Gd at rest the regenerator is raised to repeat the cycle.

The thermodynamic cycle followed by this device is the Ericsson cycle, although, like the Los Alamos group, Brown also refers to it as a Stirling-cycle device. Strictly speaking in a Stirling-cycle device the isothermal paths should be connected by constant magnetization paths instead of constant field paths.

Though Brown's laboratory refrigerator used a water-cooled magnet that consumed an enormous amount of power, any commercial device would use a superconducting magnet. It seems odd that a superconducting magnet at 4 K would be wed with a refrigerator operating near room temperature. However, in large sizes the unit could be economically feasible since the power requirement of the 4 K refrigerator would be small compared with that of the overall refrigerator. In some applications, such as air conditioning for large ships, there may already be a helium liquefier on board for a superconducting motor and generator. The principle used here for a room-temperature refrigerator could be used at any temperature with the proper choice of refrigerants and regenerator materials.

As mentioned earlier, a magnetic field will increase the entropy of an antiferromagnet until a transition to the paramagnetic phase occurs at some high field. Thus, at low fields, cooling takes place during adiabatic magnetization. The maximum entropy change will always be less than the entropy of the paramagnetic phase in zero field. Since paramagnetic systems exist which have transitions at nearly any desired temperature, it is usually best to use them just above the transition instead of an antiferromagnet below the transition.

Electrocaloric Systems

The magnetic systems just discussed have an electrical analog. Dielectric materials may have a high electric dipole entropy in the disordered, paraelectric state, but lose the entropy when the dipoles order at some low temperature to a ferro- or antiferroelectric state. An applied electric field can cause the dipoles to become ordered at some higher temperature, and so the electric field is the generalized force which can change the dipolar entropy. The entropy (Jona *et al.*, 1962; Baumgartner, 1950) of potassium dihydrogen phosphate, KH_2PO_4, in electric fields of 0 and 50 kV/cm are shown in Fig. 11.6. The order–disorder transition in zero field occurs at 122 K. The related material KH_2AsO_4 has a transition at 96 K (Blinc *et al.*, 1974), which is the lowest transition temperature of the order–disorder type of ferroelectrics known to date. These materials are analgous to the magnetic materials in the sense that they have permanent electric dipoles above and below the transition temperature.

Lower transition temperatures are possible in dilute systems such as KCl doped with OH dipoles. As shown by the entropy curves (Shepherd, 1967) for KCl:OH in Fig. 11.6, ordering can be as low as 0.1 K. The entropies shown here are on the basis of one mole of OH dipoles, but the concentration of these is only about $3 \times 10^{18} \, cm^{-3}$. Attempts to raise the concentration and provide a greater refrigeration power usually result in a clustering of the dipoles and hence a loss of their entropy. Further material studies in this area may be useful and are now going on at IBM.

Electrocaloric refrigeration has certain advantages over magnetocaloric refrigeration in the sense that high electric fields are easier to produce than high magnetic fields. High electric fields are easily produced in materials in the form of thin plates with electrodes. The electrodes can also serve as the means for transferring heat into or out of the dipolar material. Electrocaloric refrigerators could easily be miniaturized.

For a electrocaloric material the entropy change according to Eq. (11.15) can be given as

$$T \, ds = c_E \, dT + T(\partial P/\partial T)_E \, dE \qquad (11.26)$$

where s is the entropy per unit volume, c_E is the specific heat per unit volume at a constant field, and P is the polarization. The largest entropy change occurs when $(\partial P/\partial T)_E$ is largest, and that usually occurs near the transition temperature. If no remanent polarization exists in the material, the polarization can be determined from dielectric constant (ε) measurements. Thus materials with a large $\partial \varepsilon / \partial T$ would be candidates for large entropy changes. However, remanent polarizations often occur in dielectric materials due to freezing-in of impurity-vacancy dipoles (Siegwarth *et al.*, 1976) and so dc measurements of polarization are often necessary.

Even then a large $\partial P/\partial T$ at zero field does not guarantee a large total entropy change. Often $\partial P/\partial T$ decrease rapidly with increasing fields. Such behavior is found in the displacive-type materials. Unlike the order–disorder materials discussed earlier, the displacive materials have no dipole moment above the transition temperature in zero field. In these materials the transverse optic (TO) phonon mode at zero momentum (infinite wavelength) has too high an energy of excitation at high temperatures to be excited. As the temperature is lowered the excitation energy of the TO modes decreases until at the transition temperature the energy becomes zero. Because the energy, $h\omega$, is zero the frequency ω is also zero and thus the material has a permanent dipole existing throughout the sample (infinite wavelength). The dipolar entropy of the material is that of the 70 modes and can be calculated from the phonon dispersion curves. Radebaugh *et al.* (1979) did such calculations and show that the displacive materials, such as $SrTiO_3$, have rather low dipolar entropies at low temperatures. Such materials generally are not suitable for practical refrigeration at low temperatures.

The first measurements on the electrocaloric effect were performed in 1930 on Rochelle salt (Kobeko *et al.*, 1930), twelve years after the discovery of the magnetocaloric effect (Weiss *et al.*, 1918, 1926). After the ferroelectric state was better understood, several more measurements of the electrocaloric effect were made on such materials [see, for example, Thacher (1968) and references cited therein]. These measurements were all made near room temperature. Gränicher (1956) first suggested the use of $SrTiO_3$ for electrocaloric cooling at low temperatures. Experiments by Hegenbarth (1961) in $SrTiO_3$ ceramic showed a depolarization cooling of 60 mk at 17.5 K. Somewhat larger effects were seen in later measurements (Kikuchi *et al.*, 1964; Hegenbarth, 1965) on single-crystal $SrTiO_3$, but these cooling effects disappeared below about 4–5 K. Kikuchi *et al.* (1964) saw cooling effects of about 0.3 K at 12 K using fields of 7 kV/cm. Comparable effects were also seen (Radebaugh *et al.*, 1979) in $SrTiO_3$ ceramic and $KTaO_3$ single crystal for fields of about 15–20 kV/cm. Calculations showed that larger fields would not significantly increase the temperature changes.

Electrocaloric cooling at temperatures below about 1 K was first demonstrated in OH-doped KCl by Kanzig *et al.* (1964), Kuhn *et al.* (1965), and Shepherd *et al.* (1967), all working independently. Cooling to 0.36 K from a starting temperature of 1.3 K was reported by Shepherd *et al.* (1967). Cooling to as low as 0.05 K was demonstrated for CN-doped RbCl (Pohl *et al.*, 1969).

So far the only application of electrocaloric refrigeration has been the use of OH-doped KCl for thermostating crystals below 1 K while they were radiated with short light pulses (Korrovits *et al.*, 1974). Other applications would quickly open if the proper electrocaloric materials could be found. An extensive review of electrocaloric refrigeration was given by Radebaugh *et al.* (1979).

Chemical Systems

In any chemical reaction heat is either absorbed (endothermic reaction) or given off (exothermic reaction). The reactions can occur rapidly and in nonequilibrium conditions. For the reaction taking place at constant temperature and pressure the enthalpy difference gives the amount of heat transferred in the process. The enthalpy difference is simply the sum of the absolute enthalpies of the products minus the enthalpies of the reactants. The irreversible processes are, however, not very promising for an efficient refrigeration system. At equilibrium between the initial and final products the reaction occurs reversibly. The heat absorbed in reversibly converting all the initial products to the final products is simply $T\Delta S$, where ΔS is the entropy change in the reaction. When ΔS is positive the reaction absorbs heat. This entropy change can be determined for any reaction by taking the difference in the absolute entropies between the final and initial products. Absolute entropies of most elements and compounds are listed in the very comprehensive collection of data in *Tables of Selected Values of Chemical Thermodynamic Properties* (compiled by the National Bureau of Standards in conjunction with the Office of Naval Research). Unfortunately the values are generally listed only for the standard temperature of 298.16 K. For example, the entropy change ΔS^0 at 298.16 K for the reaction

$$H_2O \rightleftharpoons H_2 + \tfrac{1}{2}O_2 \qquad (11.27)$$

is calculated by

$$\Delta S^0 = S^0(H_2) + \tfrac{1}{2}S^0(O_2) - S^0(H_2O)$$

$$= 130.6 + \tfrac{1}{2} \times 205.0 - 188.7$$

$$= 44.4 \text{ J/mol K} \qquad (11.28)$$

However, in this reaction the degree of dissociation of water vapor at room temperature and atmospheric pressure is only about 10^{-27}, which means that, under these conditions, not even one molecule dissociates. The degree of dissociation is related to the equilibrium constant K. For the general reaction:

$$n_1A_1 + n_2A_2 \rightleftharpoons n_3A_3 + n_4A_4 \qquad (11.29)$$

the equilibrium constant is

$$K = \frac{x_3^{n_3}x_4^{n_4}}{x_1^{n_1}x_2^{n_2}}P^{n_3+n_4-n_1-n_2} \qquad (11.30)$$

where x_i is the fraction of the ith component at equilibrium conditions and

P is the pressure in atmospheres. The value of K at one atmosphere is related to the standard Gibbs function change ΔG^0 by

$$\Delta G^0 = -RT \ln K \qquad (11.31)$$

The tables discussed earlier for ΔS^0 also list ΔG^0. It is clear from Eq. (11.31) that for the initial and final products to be approximately equal at room temperature the term ΔG must be small. For the dissociation of water in Eq. (11.27) the term ΔG^0 is 242 K J/mol. For dissociation of nitrogen tetroxide

$$N_2O_4 \rightleftarrows 2NO_2 \qquad (11.32)$$

the term ΔG^0 is 5.39 K J/mol and so at one atmosphere and at room temperature the NO_2 concentration is about 32% at equilibrium. The entropy change ΔS^0 is 176.6 J/mol K, or $\Delta S^0/R = 21.2$, which is rather large. The transition from the ordered N_2O_4 state to the disordered NO_2 state occurs over a broad temperature interval, i.e., approximately an order of magnitude. Pressure can be used as a generalized force to vary the equilibrium concentration. The chemical reaction then adds an additional entropy change to that due to pressure changes in an ideal gas, just as a liquid–gas system adds the entropy of the phase transition. The nitrogen tetroxide system has been studied by Walker and Metwally (1977) for use in a Stirling engine. However, no chemical systems have been studied for use in refrigerators, although the magnesium chloride dihydrate system has been proposed (Gordian Associates, 1978) for use as a heat pump. An entropy decrease occurs when the system goes to a higher hydrated state in the presence of water vapor. Dehydration and entropy increase occur when the water vapor pressure is reduced.

In certain reactions an electromotive force, or potential ε, can be used to control the reaction. The electrical potential can produce a reversible chemical reaction at various temperatures. For instance, the reaction

$$Cu + ZnSO_4 \rightleftarrows Zn + CuSO_4 \qquad (11.33)$$

proceeds completely from left to right at a temperature of 273 K with a cell potential of $\varepsilon = 1.0934$ V (Zemansky, 1957). With a cell potential of 1.0481 V the reaction takes place at 373 K. In the entropy plot of Fig. 11.6 the entropy of $Cu + ZnSO_4$ is taken as zero and that of $Zn + CuSO_4$ is just the entropy increase of the reaction. A refrigeration cycle is possible by allowing the battery to discharge reversibly at room temperature and give off heat as the entropy is reduced. Heat can then be absorbed at a low temperature during the reversible charging of the battery. In some systems the charging and discharging steps are reversed. In those cases, e.g., the ordinary lead–acid battery, heat is absorbed during discharging so that it then serves as a "storage battery" for refrigeration as well as power. The first law of thermodynamics is not violated since it takes more work to charge the battery at a higher temperature. The galvanic cell has not been considered for refrigeration as far as this author knows. The rather high entropy changes possible in these systems make it appear attractive. One possible hindrance to their practicality is the Joule heat losses which accompany current flow. Another is the slowing down of chemical reactions at temperatures much below room temperature. Further studies of such systems may be useful.

Mixtures

We have already discussed the case of the He^3–He^4 solution and the dilution refrigerator. The quantum nature of that system made its behavior different from most solutions. The entropy of a solution is usually higher than that due to the sum of the components. The difference is known as the entropy of mixing, which for an ideal mixture formed at constant T and P is

$$\Delta_m S/R = -(1 - x) \ln (1 - x) - x \ln x \qquad (11.34)$$

where x is the concentration of one component. This large entropy increase is generally not available for refrigeration since the mixing is irreversible. If no work is done during the mixing, the heat absorbed or given off is just the enthalpy change. Usually there are regions in a temperature–composition diagram (phase diagram) where two phases can exist in equilibrium. Consider, for instance, a mixture of NaCl and H_2O. The phase diagram for this mixture is shown in Fig. 11.15.

The region bounded by A–E–C–A shows the region of phase separation where pure ice can be in equilibrium with a salt solution. The salt concentration in the solution is indicated by the curve A–E and the ratio of solution to ice is given by

$$n_s/n_i = w/(w + y) \qquad (11.35)$$

where w and y are the lengths shown in the phase diagram in Fig. 11.15. In the well-known ice-cream cooler, temperatures of $-21°C$ are reached when salt is added to ice at $0°C$. Since pure ice and pure salt are not in equilibrium, the ice melts and dissolves the salt to form a salt solution. The ice cannot absorb enough heat to melt immediately, and so it cools and stays on the equilibrium curve until it reaches the point where all of the salt is dissolved and forms a solution of 29% or less. If enough salt is added the cooling stops at the eutectic point E in Fig. 11.15. In a similar manner we can start with pure ice at some temperature between $0°C$ and

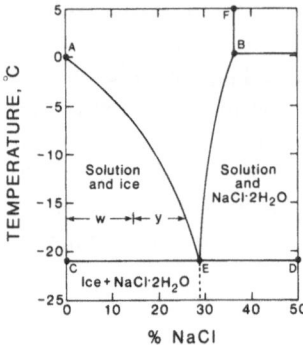

Fig. 11.15. Phase diagram for the mixture NaCl and H_2O (after Zemansky, 1957).

$-21°C$ and add salt isothermally. The ice is then converted to solution and the heat of fusion is still not related to the entropy change since the addition of pure salt, not an equilibrium phase, causes an irreversible mixing. That irreversible process would significantly decrease the efficiency of a complete refrigeration cycle. In any case we have seen an example of where the salt concentration is the generalized force controlling the phase equilibrium temperature and the system entropy. The irreversible processes are not necessarily a problem if an isolated and portable cooling system is needed for a short period of time (refrigerator "battery"), and the energy required to complete the cycle is of little concern.

In order to make a reversible cycle with this system, pure salt cannot be used to control the salt concentration since it is not in equilibrium with the system. One reversible scheme which can be used with the salt water system is like that of the dilution refrigerator. Let the ice and salt solution be contained in a low-temperature container (mixing chamber). The heavier salt solution on the bottom can pass through a heat exchanger to a still, where heat causes the water to vaporize. The water vapor is condensed and frozen. The ice then passes to the mixing chamber. In this entire process the salt remains stationary and water diffuses through the solution from the mixing chamber to the still, just as does He[3] in a dilution refrigerator. Since the process is reversible, the heat absorbed in the mixing chamber is simply $T\Delta S$, where ΔS is the entropy change of water between the solid state and the solution state. The entropy of the water in solution is known as the partial entropy. Though the system is certainly not going to make any impact on the household-refrigerator market, it does illustrate the concept of refrigeration with mixtures or solutions.

Reversible mixing of two components can be done with the use of semipermeable membranes. Such membranes allow one component to pass through it freely, but are impermeable to the other. Consider the vessel shown in Fig. 11.16. On the left side there are M_1 moles of component

Fig. 11.16. Vessel with two semipermeable membranes. Membrane G is permeable to gas A_1, but not to A_2. Membrane H is permeable to gas A_2, but not to A_1.

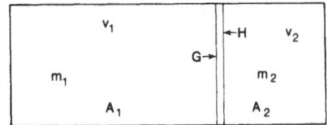

A_1, in the volume V_1. On the right side there are M_2 moles of component A_2 in the volume V_2. These components are separated by the membranes G and H. Membrane G is permeable to A_1, but impermeable to A_2. With membrane H it is the other way round. Thus A_2 passes freely through H and exerts its pressure on G, while A_1 passes freely through G and exerts its pressure on H. These pressures are known as the partial pressures of each component. Now if G moves to the left and H moves to the right in a reversible manner, i.e., doing work on the surroundings, we finally end up with a mixture of A_1 and A_2 in the entire vessel. The entropy increase is given by Eq. (11.34) where $x = m_1/(m_1 + m_2)$. Since the process was reversible, the heat absorbed is simply $Q = T\Delta S$.

We now consider a continuous refrigerator utilizing arbitrary solutions and semi-impermeable membranes as shown in Fig. 11.17. The membranes G are permeable only to A_1. At the temperature T, A_1 passes through G and is diluted by A_2 and heat $Q = T\Delta S$ is absorbed. For an ideal mixture ΔS is given by Eq. 11.34. However, in order that A_1 be in equilibrium on both sides of the membrane, the chemical potential, μ, of A_1 must be the same on both sides. For an ideal gas, this means that the partial pressures of A_1 must be equal on both sides. The total pressure on the mixture of A_1 and A_2 is then higher than that on the pure A_1 phase. This pressure difference is known as the osmotic pressure, π. The membrane G must withstand this pressure difference. The component A_1 now diffuses through the A_1–A_2 solution and leaves it through the other membrane G at the temperature T_0. The entropy reduction causes heat in the amount of $Q = T_0\Delta S$ to be given off. As for the case of the left membrane, equilibrium

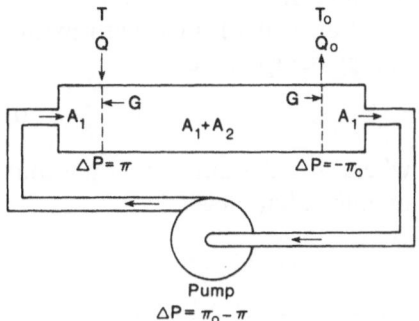

Fig. 11.17. Schematic of a solution refrigerator in which component A_1 is pumped around a cycle. Membranes G are permeable to A_1 but not to A_2. Membranes sustain an osmotic pressure π at the temperature T and π_0 at the temperature T_0.

exists when the total pressure difference across the membrane is equal to the osmotic pressure. Now for an ideal solution $\pi \propto T$. Since $T_0 > T$, π is larger at the right membrane. The total pressure on the solution must be the same at T_0 as it is at T, hence the larger π must be achieved by having the pressure on A_1 in the right side vessel less than the pressure on A_1 in the left side vessel. A pump is then necessary to complete the flow circuit for A_1. If ΔS is the same at both T and T_0, the pump work according to the first law of thermodynamics is $W = \Delta S(T_0 - T)$ and the efficiency is the Carnot value of $T/(T_0 - T)$. The maximum value of $\Delta S/R$ in Eq. (11.34) is 0.69 for $x = 0.5$. Deviations from ideality can make this value either larger or smaller. The He^4 circulating dilution refrigerator to be discussed in the next chapter uses this principle. Pure He^4 is not normally in equilibrium with a He^3–He^4 solution and so a membrane (porous plug) permeable only to the superfluid He^4 is used to provide the equilibrium.

We note here that the components A_1 and A_2 need not be molecular species. When component A_1 is electrons which are allowed to travel from one solution to another (one conductor to another), we immediately notice that the system is a thermoelectric refrigerator. The osmotic pressure in this case is simply the contact potential between the different materials and the mechanical pump acting on A_1 is replaced by a voltage source working against the thermoelectric potential of the system.

Photon Systems

All bodies emit and absorb electromagnetic radiation in the form of energy quanta of finite magnitude $h\nu$ (ν being the frequency and h Planck's constant). These quanta are called photons and have certain particle like behavior such as energy and momentum. A space filled with radiation is in many ways analogous to a space filled with gas. The only difference is that while the number of gas molecules in a closed space is independent of temperature, the number of photons in a closed space is a function of temperature.

For a space filled with monochromatic radiation of frequency ν, the energy density u_ν is

$$u_\nu = nh\nu \tag{11.36}$$

where n is the number of photons per unit volume. The pressure exerted by this radiation is

$$P_\nu = \tfrac{1}{3}u_\nu \tag{11.37}$$

Radiation from a black body consists of a wide spectrum of frequencies so that the total energy and pressure are found from a summation over all frequencies:

$$u = \sum_{\nu} u_{\nu}, \qquad p = \sum_{\nu} p_{\nu} = \tfrac{1}{3}u \qquad (11.38)$$

Heat can be transferred to the photon "gas" and work done on it by moving a piston. The radiation also possesses a certain entropy. We will not discuss the derivation here, but the energy density of photons in equilibrium with a black body is

$$u = aT^4 \qquad (11.39)$$

where $a = 7.56 \times 10^{-22}$ J/cm^3 K^4. The entropy per unit volume is given by

$$s = \tfrac{4}{3}aT^3 \qquad (11.40)$$

By way of comparison, the entropy for 1 cm^3 of He^4 gas with $P = 1$ atm and $T = 10^3$ K is 1.8×10^{-3} J/K. For photons at 10^3 K the entropy is 1.01×10^{-12} J/K for 1 cm^3. Temperatures of 10^6 K would be necessary to achieve entropy densities comparable to helium gas.

Photon radiation is produced by materials when the atoms or molecules change energy states. For an energy difference E between the states the emitted photons have the frequency $\nu = E/h$. In lasers a single energy state is used so all of the emitted photons have the same frequency. Energy flux densities of 100 W/cm^2 are not uncommon with some 104 μm wavelength CO_2 lasers and with focusing techniques the densities can be increased to as much as 10^6 W/cm^2. The energy density is found by dividing the flux density by the speed of light. Energy densities of about 3×10^{-5} J/cm^3 are then available on a continuous basis. Values up to about 10^{-2} J/cm^3 can be obtained in short pulses, which means the densities of photons are as high as 5×10^{37}/cm^3. Now for a constant-volume process $dU = T\,dS$ and so if $dU = 10^{-2}$ J/cm^3 at 10^3 K the entropy change is only 10^{-5} J/K cm^3. These low entropy values would imply that photons would be of little use for refrigeration. In addition the high-density laser beams certainly are not in equilibrium with the room-temperature surroundings and are rapidly absorbed with much heat given off. Nevertheless, a non-equilibrium system may provide useful spot cooling where overall efficiency is of little concern.

Suppose now we have a material which has an energy gap equal to the energy of the photons in the laser beam. This material absorbs a photon and an atom or molecule is shifted to the higher energy state. A photon of the same frequency and energy is then reemitted when the atom decays to the lower energy state. In this case the atom is in equilibrium with the photon beam. The use of radiation pressure for cooling was independently

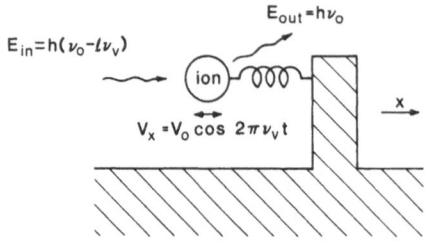

Fig. 11.18. Schematic of a photon refrigerator in which photons of energy $h(\nu_0 - l\nu_v)$ are absorbed by the harmonically bound ion but are reemitted with average energy $h\nu_0$. Energy deficit comes from the harmonic vibration which causes the ion to cool.

suggested for the case of a gas of neutral atoms (Hänsch *et al.*, 1975) and for ions bound in an electromagnetic trap (Wineland *et al.*, 1975). The method is outlined for the second case and is shown schematically in Fig. 11.18. Consider an ion with an optical transition of frequency ν_0. It is harmonically bound to the trap and constrained to move only in the x direction. Its velocity is given by $v_x = v_0 \cos 2\pi\nu_v t$, where ν_v is its vibrational frequency. This vibrational frequency causes a Doppler effect in the absorption spectrum. Thus radiation is absorbed or emitted at not only ν_0, the transition frequency, but also at $\nu_0 \pm m\nu_v$, where m is an integer. Suppose the ion is now subjected to monochromatic radiation of frequency $\nu_0 - l\nu_v$ in the x direction, where l is a specific integer. The photons are then absorbed by the ion and then reemitted at frequencies of $\nu_0 \pm m\nu_v$. The average emitted frequency is ν_0. Thus on the average the energy of photons leaving is $lh\nu_v$ higher than those entering the system. If the ions are thermally isolated from the surroundings, the net energy transfer from the ions to the photons causes the ions to cool. In experiments by Drullinger and Wineland (1979) the density of Mg II ions is only about $2 \times 10^7/\text{cm}^3$, which are achieved with pressures of about 10^{-8} Pa (10^{-10} torr). The electric repulsion of the ions limits the density. In these experiments the Mg II ion temperature has been reduced to 0.5 K using lasers of about 10 μW power. With such low densities cooling rates of about 1000 K/sec are achieved. In similar experiments on a single Ba ion, Neuhauser *et al.* (1979) have produced a temperature on the order of tens of millikelvins.

Even though the energy and entropy density of photons are very low, they are useful for cooling low-density ions or atoms. The rapid cooling occurs because of the very low specific heat of such a low-density gas. For a space of 1 cm^3 the refrigeration rate is the order of 10^{-8} W. Obviously for much higher absorber densities (possible with neutral atoms) the cooling rate is much decreased and the enhanced coupling with the room-temperature surroundings causes heat leaks which would prevent significant cooling. However, in situations where the relaxation time between an absorber and its surroundings is long, continued radiation pumping can remove a relatively large amount of energy during this time and produce significant cooling even in high-density absorbers. In Chapter 12 we discuss the use of microwave radiation to cool the nuclei in insulators.

HOW MUCH ENTROPY IS ENOUGH?

Figures 11.3–11.6 show the entropies of many systems—some commonly used for refrigeration and others never considered for refrigeration. Other workers may wish to consider other systems not shown here. A question important to both funding agencies and researchers is "how large should the entropy of some new system be before it has the potential for practicality?" There are two ways to approach this question. One way is to look at the entropy of the disordered state for existing practical systems. Generally, these entropies (S/R) lie in the range from about 10^{-1} at 1 mk to about 10 at 100 K. A straight line can be drawn between these points in the log–log plots of Figs. 11.6–11.9. This line represents the general range of S/R for present systems. Since present systems remove a large fraction of the total absolute entropy in the disordered phase, the values of $\Delta S/R$ are nearly comparable to S/R. If any new system has a molar flow rate or cycle time comparable to these systems then the controllable entropy of the discovered phase in the new system should probably not be less than $\frac{1}{10}$ that of the line discussed above. One point to consider is that for liquid or solid systems the higher density compared with gas may permit more compact refrigerators for a comparable ΔS if it is possible to transfer the same amount of heat in a more compact heat exchanger.

A second approach to determine a reasonable value of $\Delta S/R$ for a system is to consider how much entropy is required to cool itself. Such cooling is necessary with imperfect heat exchangers. For a temperature change of ΔT small compared with T, the relation

$$\frac{\Delta T}{T} = \frac{\Delta S/R}{C/R} \tag{11.41}$$

is valid, where C is the specific heat. In order to overcome inefficiencies of practical heat exchangers, $\Delta T/T$ should be at least 2×10^{-2} near room temperature. At toom temperature, C/R for most materials will be at least 3. From Eq. (11.36) an entropy of $\Delta S/R = 6 \times 10^{-2}$ would be needed to cool the sample. In order to have enough entropy left for refrigeration, a minimum value of $\Delta S/R \simeq 0.1$ seems reasonable at room temperature. At 10 K the specific heat of materials may range from $C/R \simeq 10^{-2}$ to 10^{-1}. Since heat transfer may be more of a problem at 10 K, $\Delta T/T$ would probably be at least 10^{-1}. Thus at 10 K a minimum value of $\Delta S/R \simeq 10^{-2}$ is required when a provision for some useful entropy is included. The

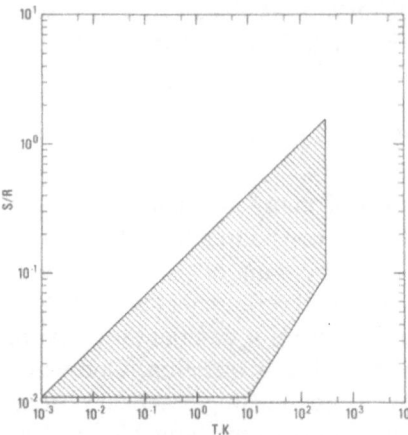

Fig. 11.19. Estimated region of reduced entropy which represents the minimum change of entropy for a system in order for that system to be a practical refrigerator.

specific heat of most active materials does not decrease much below $C/R \simeq 10^{-2}$ for temperatures down to 10^{-3} K. From this argument the minimum $\Delta S/R$ needed for a practical refrigerator is 10^{-2} from 1 mK to 10 K. From 10 K to room temperature the value should rise up to about 10^{-1}. The previous argument gave somewhat higher values. The region between these two limits is shown in Fig. 11.19. This band represents the lower limit of $\Delta S/R$ necessary for any new system to be a practical refrigerator. We emphasize that these limits are only general guidelines and will not always apply to some specific refrigeration system.

The usefulness of these entropy curves becomes clear when we consider the example of refrigeration via magnetization of superconductors. Application of a magnetic field to a superconductor will drive it into the normal state. Because the entropy of normal electrons is higher than that of the superconducting state, heat will be absorbed. However, S/R at 10 K for normal electrons is only about 10^{-3}, which is much too small for useful refrigeration.

ALTERNATE MEANS TO ELIMINATING MECHANICAL PARTS

Some of the alternate refrigeration systems discussed previously can also be adapted for use with mechanical gas–liquid systems to eliminate some of the troublesome mechanical parts. For instance the old ammonia-gas refrigerators replaced a mechanical compressor with a thermal compressor. In these refrigerators ammonia was absorbed in water at room temperature and then the solution was pumped to a high-temperature section where the ammonia gas was driven off at high pressure. The process did require a liquid pump, but these are more efficient than a gas compressor. An absorption compressor using $LaNi_5$-hydride has been described (van Mal et al., 1972) for use with a hydrogen refrigerator operating near 20 K. Other gases can be pumped and thermally compressed by using gas adsorption on zeolites. Details of such adsorption pumping are given by Hartwig (1978).

Gifford and Longsworth (1965, 1966) devised a method known as pulse tube refrigeration in which there are no moving low-temperature parts. In this method gas is pulsed in and out of a tube closed at the other end at a rate of about 30–200 pulses per minute. The gas passes through a regenerator before entering the pulse tube. Refrigeration occurs at the inlet to the tube and heat is rejected at the other end, which is closed. Two or three stages are required (Lechner et al., 1973) to reach 80 K, and it seems unlikely that a temperature as low as 10 K can be reached with this method. An advantage over the Hampson cycle is that pressures are much lower, thus a simpler compressor is required.

REFERENCES

Barclay, J. A., Steyert, W. A., and Zrudsky, D. R. (1979). *Proc. XVth Int. Congress of Refrig.*
Baumgartner, H. (1950). *Helv. Phys. Acta* **23**, 651.
Betts, D. S. (1976). *Refrigeration and Thermometry below One Kelvin.* Sussex University Press.
Blinc, R. and Zeks, B. (1974). *Soft Modes in Ferroelectrics and Antiferroelectrics, Selected Topics in Solid State Physics*, Vol. 13, North-Holland, New York.
Brown, G. V. (1976). *J. Appl. Phys.* **47**.
Brown, G. V. (1979). See *Phys. Today* **June**, 18.
Buist, R. J., Fenton, J. W., and Tuomi, D. (1971). "Low Temperature Cooler for 145 K Operation." U.S. Army Night Vision Laboratories, Fort Belvoir, Virginia, Technical Report AD888757L (Borg-Warner Thermoelectrics).
Buist, R. J. (1974). "Feasibility Study for a High Power Low Temperature Thermoelectric Cooler." U.S. Army Night Vision Laboratories, Fort Belvoir, Virginia, Technical Report ADB002643 (Borg-Warner Thermoelectrics).
Buist, R. J., Fenton, J., Lichniak, G., and Norton, P. (1976). "Low Temperature Thermoelectric Cooler for 145 K Detector Array Package." U.S. Army Night Vision Laboratories, Fort Belvoir, Virginia, Technical Report ADB008934 (Borg-Warner Thermoelectrics).
Click, P. B., Jr., and Marlow, R. (1970). "Low Temperature Thermoelectric Cooler for Operation at 145 K". U.S. Army Night Vision Laboratories, Fort Belvoir, Virginia, Technical Report AD875928 (Nuclear Systems, Inc.).
Colwell, J. H., Gill, E. K., and Morrison, J. A. (1962). *J. Chem. Phys.* **36**, 2223.
Daunt, J. G. (1970). "Preliminary Thermodynamic Data for the Inversion Curve of He^3." *Cryogenics* **10**, 473.
Drullinger, R. E., and Wineland, D. J. (1979). *Laser Spectroscopy IV.* ed. by H. Walther and K. W. Rothe, Springer-Verlag, Heidelberg, (1979), p. 66.
Dugdale, J. S., and Franck, J. P. (1964). *Phil. Tans. R. Soc.* **257**, 1.
Fisher, R. A., Hornung, E. W., Brodale, G. E., and Giauque, W. F. (1973). *J. Chem. Phys.* **58**, 5584.
Franck. A., and Clusius, K. (1937). *Z. Physik. Chem.* **B36**, 291.

Frossati, G., Godfrin, H., Hebral, B., Schumacher, G., and Thoulouze, D. (1978). In *Physics at Ultralow Temperatures*, (Physical Society of Japan, p. 205.
Giauque, W. F., and Clayton, J. O. (1933). *J. Am. Chem. Soc.* **55**, 4875.
Giauque, W. F., Fisher, R. A., Hornung, E. W., and Brodale, G. E. (1973). *J. Chem. Phys.* **58**, 262.
Gibbons, R. M., and McKinley, C. (1968). *Adv. Cryog. Eng.* **13**, 375.
Gibbons, R. M., and Nathan, D. I. (1967). AFML-TR-67-175.
Gifford, W. E., and Longsworth, R. C. (1966a). "Surface Heat Pumping". *Adv. Cryog. Eng.* **11**, 171.
Gifford, W. E., and Longsworth, R. C. (1966b). "Pulse-Tube Refrigeration Progress". *Adv. Cryog. Eng.* **10B**, 69.
Goldsmid, H. J. (1964). *Thermoelectric Refrigeration*. Plenum Press, New York.
Goodstein, D. (1979). Private Communication.
Goodwin, R. D. (1973). NBSIR 73-342.
Gordian Associated. (1978). "Heat Pump Technology." U. S. Dept. of Energy Report HCP/M2121-01.
Gränicher, H. (1956). *Helv. Phys. Acta* **29**, 210.
Hänsch, T. W., and Schawlow, A. L. (1975). *Opt Commun.* **13**, 68.
Hartwig, W. H. (1978). in *Applications of Closed-Cycle Cryocoolers to Small Superconducting Devices*, NBS Special Publication 508, p. 135.
Hegenbarth, E. (1961). *Phys. Status Solidi* **8**, 59.
Jackson, C. M., Wagner, H. J., and Wasilewski, R. J. (1972). NASA-SP5110.
Johnson, V. J. (1960). WADD Tech Tech. Report 60-56, Part I.
Jona, F. and Shirane, G. (1962). *Ferroelectric Crystals*. Macmillan, New York, p. 251,
Kanzig, W., Hart, H. R., Jr., and Roberts, S. (1964). *Phys. Rev. Lett.* **13**, 543.
Kikuchi, A., and Sawaguchi, E. (1964). *J. Phys. Soc. Jpn* **19**, 1497.
Kobeko, P., and Kurtschatov, J. (1930). *Zt. Phys.* **66**, 192.
Korrovits, V. K., Luud'ya, G. G., and Mikhkelsoo, V. T. (1974). "Thermostating Crystals at Temperatures Below 1 K by Using the Electrocaloric Effect." *Cryogenics* **14**, 44.
Kuhn, U., and Lüty, F. (1965). *Solid State Commun.* **4**, 31.
Lechner, R. A., and Ackermann, R. A. (1973). "Concentric Pulse Tube Analysis and Design." *Adv. Cryog. Eng.* **18**, 467.
London, H. (1951). Proc. of the Int. Conf. on Low Temp. Phys., Oxford, p. 157.
MacDonald, D. K. C. (1962). *Thermoelectricity: An Introduction to the Principles*. John Wiley & Sons, New York, p. 84.
McCarty, R. D. (1972). NBS Tech. Note 631.
McCarty, R. D. (1975). NASA SP-3089.
McCormick, J. E., and Brauer, J. B. (1965). *Adv. Cryog. Eng.* **10A**, 493.
Melton, R. G., Paterson, J. L., and Kaplan, S. B. *J. Low Temp. Phys.* (1980).
Nesselmann, K. (1957). *Chem. Eng. Tech.* **29**, 198.
Neuhauser, W., Hohenstatt, M., Toschek, P., and Dehmelt, H. (1979). *Laser Spectroscopy IV*. (Ed. H. Walther and K. W. Rothe), Springer-Verlag, Heidelberg, p. 73.
Pohl, R. D., Taylor, V. L., and Govban, W. M. (1969). *Phys. Rev.* **178**, 1431.
Pratt, W. P., Jr., Rosenblum, S. S., Steyert, W. A., and Barclay, J. A. (1977). "A Continuous Demagnetization Refrigerator Operating Near 2 K and a Study of Magnetic Refrigerants." *Cryogenics* **17**, 689.
Radebaugh, R. (1967). NBS Tech. Note 362.
Radebaugh, R. J. (1977). *J. Low Temp. Phys.* **27**, 91.
Radebaugh, R., Lawless, W. N., Siegwarth, J. D., and Morrow, A. J. (1979). *Cryogenics* **19**, 187.
Shepherd, I. W., and Feher, G. (1967). *J. Phys. Chem. Solids* **28**, 2027.
Siegwarth, J. D., and Morrow, A. J. (1976). *J. Appl. Phys.* **47**, 4784.
Steyert, W. A. (1978a). In *Applications of Closed-Cycle Cryocoolers to Small Superconducting Devices*, NBS Special Publication 508, p. 81; *J. Appl. Phys.* **49**, 1227.
Steyert, W. A. (1978b). *J. Appl. Phys.* **49**, 1216.
Strobridge, T. R. (1962). NBS Tech. Note 127.

Strong, C. R. (1971). *Sci. Am* **224**, 118, April.
Thacher, P. D. (1968). *J. Appl. Phys.* **39**, 1996.
Treloar, L. R. G. (1958). *The Physics of Rubber Elasticity*. Clarendon Press, Oxford.
Van Geuns, J. R., *Phillips Res. Rep. Suppl.* No. 6 (Eindhoven, Netherlands).
van Mal, H. H., and Mijnheer, A. (1972). 4th International Cryog. Eng. Conf.
Van Wylen, G. J., and Sonntag, R. E. (1965). *Fundamentals of Classical Thermodynamics*. John Wiley and Sons, New York, p. 590.
Walker, G., and Metwally, M. (1977). *Trans. A.S.M.E., J. Eng. Power* **99**, 284.
Weiss, P., and Forrer, R. (1926). *Ann. Phys. (Paris)* **5**, 153.
Weiss, P., and Piccard, A. (1918). *C. R. Acad. Sci. Paris.* **166**, 352.
Westrum, E. F., Jr. (1961). *Pure Appl. Chem.* **2**, 241.
Wilks, J. (1967). *The Properties of Liquid and Solid Helium*. Clarendon Press, Oxford.
Wineland, D. J., and Dehmelt, H. (1975). *Bull. Am. Phys. Soc.* **20**, 637.
Wineland D. J., and Drullinger, R. E. (1979). Proc. 6th Vavilov Conf. on Non-Linear Optics, Novosibirsk.
Wolfe, R., and Smith, G. E. (1962). *Appl. Phys. Lett.* **1**, 5.
Woolley, H. W., Scott, R. B., and Brickwedde, F. G. (1948). *J. Res. Nat. Bur. Stand.* **41**, 379.
Yim, W. M., Fitzke, E. V., and Rosi, F. D. (1966). *J. Mater. Sci.* **1**, 52.
Yim, W. M., and Rosi, R. D. (1972). *Solid-State Electron.* **15**, 1121.
Yim, W. M., and Amith, A. (1972). *Solid-State Electron.* **15**, 1141.
Zemansky, M. W. (1957). *Heat and Thermodynamics*. 4th ed., McGraw-Hill, New York, p. 292.

Chapter 12

Very-Low-Temperature Cooling Systems

Ray Radebaugh

INTRODUCTION

Very low temperatures can take on a variety of different meanings, depending on one's low-temperature experience. Those unfamiliar with cryogenics would usually refer to liquid nitrogen as being at very low temperatures. The adverb "very" is generally used for something out of the ordinary, and for those more familiar with cryogenics, the temperature range below 1 K might better fit this category. Such a choice merely reflects the present status of our refrigeration efforts and could therefore change with time. For this book, however, we arbitrarily define very low temperatures as those below about 1 K. Other terms, such as ultralow temperatures or millikelvin temperatures, have also been used in the past by others to describe the same region. The temperature of 1 K seems to be a good dividing line since the refrigeration techniques used below 1 K are usually quite different from those used above 1 K. The vapor pressure of He^4 becomes too small below about 1 K to be of any use in a gas refrigerator. The isotope He^3 retains a reasonable vapor pressure down to about 0.3 K, but for lower temperatures liquid–vapor equilibrium cannot be used for refrigeration. Not only are the refrigeration techniques below 1 K usually different from those above 1 K, but the temperatures are expressed in terms of millikelvin, microkelvin, or even nanokelvin.

Ray Radebaugh • Thermophysical Properties Division, National Bureau of Standards, Boulder, Colorado 80303.

Both Lounasmaa (1974) and Betts (1976) have written excellent books on the subject of very low temperatures. They cover not only refrigeration techniques but also thermometry and heat transfer. Readers interested in considerable detail about refrigeration below 1 K should refer to those books. In this chapter we do not attempt to duplicate the efforts in those books, but instead to give a brief and introductory review of the various refrigeration techniques and to show how the basic principles used are similar to those used at higher temperatures but with different refrigerants. In addition we include some of the latest developments which have occurred since the books by Lounasmaa and Betts were written. In the last few years there have been some quite interesting developments in dilution refrigeration and in nuclear demagnetization. With a dilution refrigerator continuous temperatures of 2 mK have been reached. In two-stage nuclear demagnetization cryostats, nuclear temperatures of 50 nK and electron temperatures of 50 μK have been attained.

As discussed in Chapter 11, refrigeration requires the existence of a disordered state (high entropy) at the desired temperature. Figure 12.1 shows the entropy of various systems in the range below 1 K. The ordering temperatures of these various systems indicate the lowest temperature which that system can produce. The entropies of Cu (Ehnholm *et al.*, 1979), AuIn$_2$ PrNi$_5$ (Mueller *et al.*, 1980), and solid He^3 shown in Fig. 12.1 are that of the nucleus. We note that below 1 mK (10^{-3} K) the only remaining

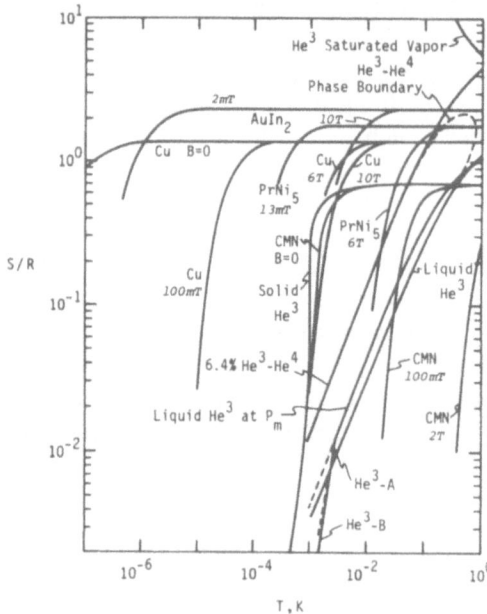

Fig. 12.1. Reduced entropy as a function of temperature for several materials used for refrigeration below 1 K.

disorder occurs in the nuclei of various materials. Even the nucleus in Cu begins to order magnetically below 10^{-6} K and so we are left with the thought that below nuclear ordering temperatures there may be nothing left with disorder to provide refrigeration. In the meantime low-temperature physicists and nuclear physicists are now making fascinating discoveries in the ordering processes taking place within the nucleus. Copper nuclei, for instance, appear to order as an antiferromagnet (Ehnholm *et al.*, 1979).

Refrigeration in the very-low-temperature range, though very common in research laboratories, has so far had few practical applications. There have been some studies recently regarding the cooling of infrared detectors to temperatures below 1 K in order to improve sensitivity for far infrared astronomy telescopes. Dilution refrigerators are now being used for cooling gravitational wave detectors (McAshan, 1976), and for cooling polarized nuclear targets but it is often difficult to say whether the use of a refrigerator is for a practical purpose or for a research purpose. Nevertheless, the number of dilution refrigerators in existence is certainly several hundred. Almost every low-temperature laboratory has at least one dilution refrigerator and some have several. The University of Leiden, site of the first dilution refrigerator, has about 15 dilution refrigerators of various types.

The organization of this chapter does not quite follow the historical sequence of events. Adiabatic demagnetization was first used in 1933 to reach temperatures considerably below 1 K. However, in this chapter we first consider the use of He^3 in various refrigeration methods (He^3 refrigerators, He^3–He^4 dilution refrigerators, and Pomeranchuk refrigerators) and then turn to the various magnetic refrigerators using different types of refrigerants (electron-spin materials, nuclear-spin materials, and hyperfine enhanced nuclear-spin materials). As discussed in Chapter 11, electrocaloric refrigeration has produced temperatures as low as 50 mK, but this type of refrigerator will not be discussed further in this chapter since there has been very little work on it at temperatures below 1 K.

He^3 REFRIGERATORS

Properties of He^3

The normal boiling point of the stable isotope He^3 is 3.19 K. Figure 12.2 compares the vapor pressures of He^3 and He^4. The smaller slope of the He^3 curve is due to the lower heat of vaporization of He^3 compared with He^4. Pumping on liquid He^3 can reduce the temperature to about

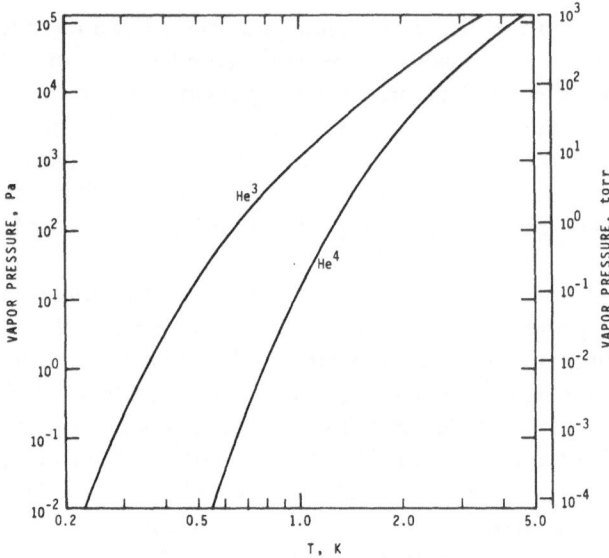

Fig. 12.2. Vapor pressure of He^3 and He^4 as a function of temperature.

0.3 K. When one liter STP of He^3 gas is condensed at 1.2 K and pumped to 0.3 K about 6% of the liquid is lost in cooling itself. The total heat of vaporization of the remaining liquid at 0.3 K is then 1.1 J. At these temperatures He^3 is still not a superfluid so there is no problem associated with superfluid film flow up the pumping tube as for the case of He^4. This means of cooling below 1 K is the simplest, but temperatures significantly below 0.3 K are not possible with it. In spite of its simplicity this technique was not used until 1954 since He^3 was not available in large enough quantities much before that time. Keller (1969) discusses in detail the history and techniques for producing He^3. Basically the process involves the two reactions

$$\text{Li}^6 + n^1 \rightarrow \text{H}^3 + \text{He}^4 \tag{12.1}$$

$$\text{H}^3 \rightarrow \text{He}^3 + \beta^- \tag{12.2}$$

The first reaction is carried out by slow neutron bombardment of Li^6. The tritium (H^3) is separated from the He^4 by selective diffusion through a heated palladium barrier. This tritium, produced in large quantities for use in thermonuclear weapons, has a half-life of 12.5 yr and decays by β^- emission to He^3. Periodically the He^3 is separated from the tritium by a diffusion process. The cost of 99.9% pure He^3 is about $150 per STP liter and this price has not changed much over the last 20 years. The relatively high cost of He^3 means that a closed system must be used. Most He^3

refrigerators, as well as dilution refrigerators and Pomeranchuk refrigerators, use only a few STP liters of He^3 and so the He^3 cost is usually negligible compared with the overall refrigerator.

Single-Cycle He^3 Refrigerators

In 1954 enough He^3 became available to study some of the thermo-dynamic properties of it. Roberts and Sydoriak (1955) then built the first He^3 cryostat in which the He^3 to be studied served as its own refrigerant. The liquid He^3 container was a 10-mm-diam sphere spun from copper sheet. This sphere was suspended in a vacuum can by a 3-mm-diam copper nickel tube 200 mm long which served as the pumping line. The top end of the copper nickel tube as well as the vacuum can were soldered to a brass flange. To maintain a reasonable pumping speed the He^3 pumping line between the flange and room temperature was a larger diameter than the line below the flange because as the He^3 gas warms the density decreases. The vacuum can was immersed in a bath of liquid He^4 pumped to a temperature of about 1.1 K. The He^3 system was then pressurized with He^3 gas to about 2000 Pa (15 torr) and the He^3 gas in contact with the 1.1 K brass flange condensed and ran down to the copper sphere. After the sphere was full the pressure on the He^3 was reduced with a vacuum pump and a minimum temperature of 0.37 K was reached. This temperature could be held until all of the liquid He^3 in the sphere evaporated. At that time the cycle could be repeated by pressurizing the system again to condense He^3.

The second He^3 refrigerator was that of Seidel and Keesom (1958) of Purdue University. This was the first He^3 refrigerator used to cool materials other than the He^3 itself. This refrigerator has been a real work-horse for Professor Keesom for the measurement of specific heats of various materials from 0.3 to 4 K. Several students have received their Ph.D. under Professor Keesom by using this refrigerator in their research. This author was one of those students, and it was with this refrigerator that he received his first experience with temperatures below 1 K. This He^3 refrigerator is still in use today after surviving two moves between different buildings. The He^3 gas used in the system was purchased while the price was still quite high and so the gas bill was $5000 for the 3 STP liters used in the system. Though the original refrigerator is still used today, it does not contain the original He^3 gas. Fortunately the price of He^3 gas had come down close to $100 per STP liter since the original purchase so the replacement charge was relatively small.

Figure 12.3 shows a diagram of the low-temperature parts of the Seidel and Keesom refrigerator. The principle of operation is nearly the same as

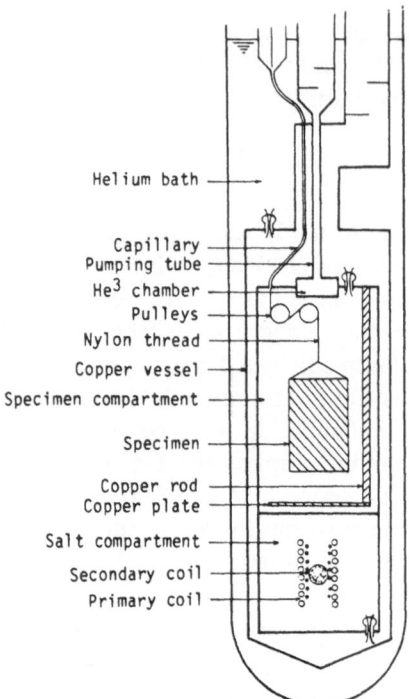

Helium bath

Capillary
Pumping tube
He³ chamber
Pulleys
Nylon thread
Copper vessel
Specimen compartment

Specimen

Copper rod
Copper plate

Salt compartment
Secondary coil
Primary coil

Fig. 12.3. Low-temperature section of the He³ refrigerator of Seidel and Keesom (1958) used for cooling specific heat samples.

that of Roberts and Sydoriak. During the condensation period liquid He³ runs down from the walls of the pumping tube into the He³ container. The refrigerator itself could reach temperatures of about 0.25 K but sample temperatures were usually limited to about 0.3 K. Temperatures are measured with a paramagnetic salt thermometer which in turn is calibrated at higher temperatures against the vapor pressure of He⁴. At 0.4 K the refrigeration capacity was about 50 μW. The 3 liters STP of He³ gas in the system (5 cm³ liquid He³ at 0.3 K) are sufficient to maintain 0.3 K for over 80 hr without warmup if so desired. Such long cycle times indicate that single-cycle He³ refrigerators are adequate for most experiments. Figure 12.4 is a diagram of the He³ pumping and storage system for the refrigerator. About 99% of the He³ gas can be recovered and stored in the two storage containers (e) by first pumping the gas into one container, then closing it off and opening the other. The 1% of the gas remaining outside the containers is only lost in the rare event that the system must be opened for repair or modifications. Temperatures are controlled by bleeding gas back through the valve (f) in Fig. 12.4 for coarse control and with a heater on the liquid He³ chamber for fine control. Since the output of the mechanical pump (d) must be sealed to keep the He³ in and air out, the entire pump is immersed in a bath of oil with the pulley shaft extending

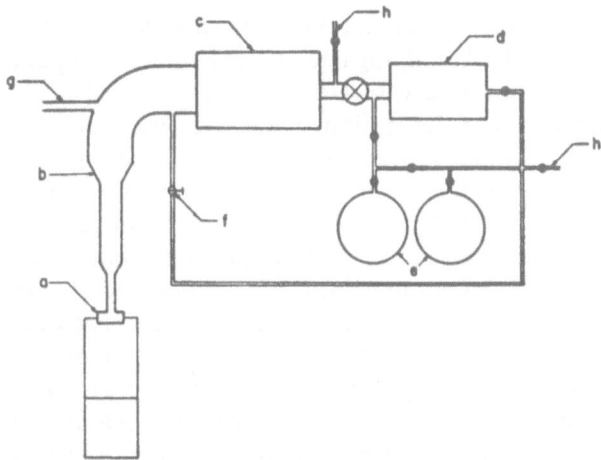

Fig. 12.4. Schematic diagram of He³ pumping and storage system for the refrigerator in Fig. 12.3. KEY: *a*, liquid He³ bath chamber; *b*, junction kept at 77 K; *c*, oil diffusion pump, *d*, rotary pump; *e*, storage containers; *f*, needle valve; *g*, line to manometers; and *h*, lines to the auxiliary vacuum system (after Seidel and Keesom, 1958).

out through a rotary seal. It is now possible to buy mechanical pumps with special oil seals on the shaft for use with He³ or other closed systems. A detailed description of He³ cryostats up to 1960 is given by Taconis (1961).

Walton (1966) shows that the ultimate temperature of a He³ refrigerator is about 0.2 K. His calculations indicate that the pumping line geometry is the limiting factor rather than the size of the pumps, although a rather large oil diffusion pump would be necessary to reach such a temperature. High pumping speeds can be achieved with the use of charcoal adsorption pumps. Several He³ refrigerators using adsorption pumping have been built and such units are also commercially available. The main advantage of the adsorption pumping is the simplicity which comes from eliminating the external pumping system. Mate *et al.* (1965) have shown that 10 g of activated charcoal is sufficient to adsorb 1.5 liters STP of He³ gas when the charcoal is held at 1.3 K. In fact the capacity of the charcoal is not significantly decreased when it is held at 4.2 K (Walton *et al.*, 1971). The He³ is desorbed when the charcoal is heated to about 40 K or above. The first adsorption pumping cryostat was that of Esel'son *et al.* (1963). In their cryostat the 30-g charcoal pump was held at 1.3 K by a pumped He⁴ bath. The pumping line to the charcoal could be closed off with a valve during the condensation of He³ into the He³ chamber. When the valve was opened pumping began and a temperature of 0.34 K could be held for about 4–5 hr. At that time the complete cryostat had to be warmed up to bring about the desorption and a repeat of the process.

In the cryostat of Mate *et al.* (1965) the charcoal cage could be moved vertically inside the He^3 pumping line by a nylon monofilament. Thus when the charcoal was saturated with He^3 it could be raised to the top of the cryostat to bring about desorption of the He^3 without warming up the whole cryostat to room temperature. The desorbed He^3 gas could then either be recovered in storage containers or recondensed in the He^3 chamber to repeat the cycle. No valve is used in the pumping line and the temperature of the He^3 chamber can be adjusted by controlling the position of the charcoal pump. Walton *et al.* (1971) eliminated all moving parts in their He^3 cryostat by using a heater wound on the charcoal container to change its temperature. The charcoal container made thermal contact with the 1.3 K helium reservoir by means of He^4 exchange gas or a copper wire. The He^4 exchange gas would be removed during the heating of the charcoal. It is now possible to buy commercial He^3 refrigerators which utilize this type of adsorption pumping.

One of the major applications for He^3 refrigerators is for cooling infrared detectors. The noise equivalent power (NEP) of semiconducting infrared bolometers varies approximately as $T^{5/2}$ (Chanin and Torre, 1976). Therefore operation at 0.33 K, rather than at 1.5 K, gives a factor of 40 improvement in performance. Generally, adsorption pumping is used for infrared applications because of the simplicity of the cryostat. Figure 12.5 shows the charcoal-pumped He^3 refrigerator of Yamamoto (1975) used

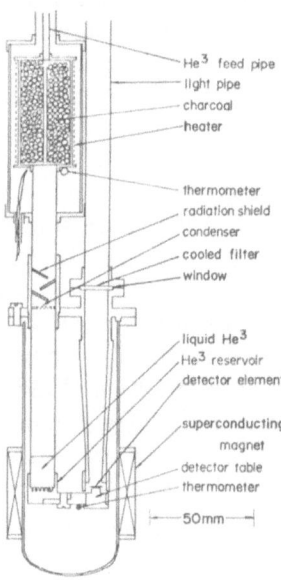

He³ feed pipe
light pipe
charcoal
heater

thermometer
radiation shield
condenser
cooled filter
window

liquid He³
He³ reservoir
detector element

superconducting
magnet
detector table
thermometer

|— 50mm —|

Fig. 12.5. Assembly of a charcoal-pumped He^3 refrigerator for cooling an infrared detector. Assembly is immersed in pumped liquid He^4 (after Yamamoto, 1975.)

for cooling an infrared detector. This refrigerator contains 9 g of activated charcoal and 1 liter STP of He^3 gas. A temperature of 0.33 K was reached with a heat load of 30 μW and could be maintained for about 7 hr. Torre and Chanin (1978) have built and tested a balloon-borne He^3 cryostat for cooling an infrared detector. The 8-kg cryostat was lifted to a height of 38 km and remained there for 3 hr. A temperature of 0.32 K was held during this time. The cryostat, with its 23 g of charcoal, could maintain such a temperature for 11 hr before running out of He^4. The He^4 bath is pumped by the low-pressure atmosphere at the high altitude reached by the balloon. Radostitz *et al.* (1978) describe an adsorption-pumped He^3 cryostat which has been used for infrared detector cooling in airplanes, in observatories, and in the laboratory.

Continuous He^3 Refrigerators

In situations where temperatures of 0.3 K are needed for long periods of time (greater than about 100 hr), the He^3 refrigerator can be made to run continuously simply by returning the high pressure He^3 to the He^3 chamber through a needle valve or a capillary. The Joule–Thomson expansion of the high-pressure He^3 (about 3000 Pa or 23 torr) from 1.2 to 0.3 K leaves a significant fraction of liquid for refrigeration purposes. Zinov'eva (1958) first described such a refrigerator. This refrigerator, shown in Fig. 12.6, was built of glass to allow the first visual observations of the phase separation in He^3–He^4 solutions (Zinov'eva and Peshkov, 1959). Temperatures of 0.35 K were reached in this cryostat but it is not clear if He^3 was being returned to the bath through the valve at that time. Reich and Garwin (1959) reached a continuous temperature of 0.5 K in a simple continuous refrigerator using only a mechanical pump. A continuous temperature of 0.26 K was reached in a He^3 refrigerator built for experiments on oriented nuclei (Ambler and Dover, 1961). This refrigerator used a commercial mercury diffusion pump with an ejector stage as the compressor. The 4.7-kPa (35-torr) output pressure of the mercury pump is high enough for condensation of He^3 at 1.2 K so that a mechanical pump is not needed. In this refrigerator the He^3 expanded into the evaporator through a porous stainless steel plug instead of a valve.

All of the He^3 refrigerators discussed so far require a pumped He^4 bath at about 1.2 K for condensation of the He^3. Daunt and Lerner (1970) describe a He^3 refrigerator which uses only 4.2 K liquid He^4. At that temperature He^3 cannot be liquefied but in their refrigerator a Joule–Thomson expansion from 4 to 1 atm produced 3.2 K liquid He^3 with a liquefaction coefficient of 59%. An oil-free diaphragm compressor was used to circulate the He^3 gas and provide continuous refrigeration at 3.2 K.

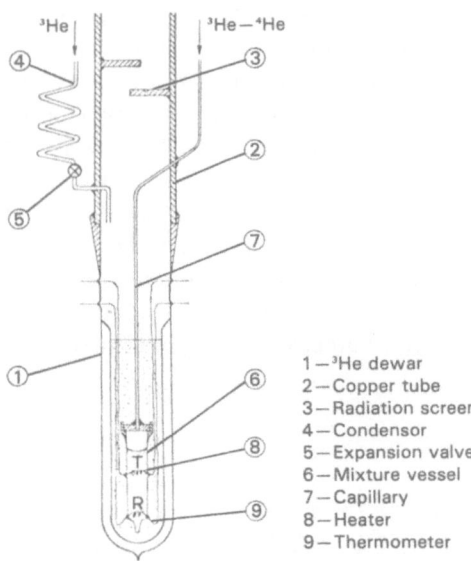

Fig. 12.6. First continuous He^3 refrigerator in which phase separation of He^3–He^4 mixtures was visually observed by the use of glass Dewars (after Zinov'eva and Peshkov, 1959).

1 — ^{3}He dewar
2 — Copper tube
3 — Radiation screen
4 — Condensor
5 — Expansion valve
6 — Mixture vessel
7 — Capillary
8 — Heater
9 — Thermometer

A heat exchanger was used to precool the gas before entering the JT valve. Temperatures down to 0.25 K could be reached and maintained for 50 hr when the return He^3 flow was stopped and the liquid He^3 reservoir was pumped on by charcoal.

Continuous temperatures below 1 K are not possible in a system with only 4.2 K liquid He^4 since all of the liquid would be vaporized during the JT expansion from 4.2 K. Thermodynamic properties of He^3, which are necessary for the design of such He^3 refrigerators, is rather scarce, although some preliminary data are given by Daunt (1970) and by Gibbons *et al.* (1967). The most recent enthalpy data are given by Kraus *et al.* (1974). Wilkes (1972) reached temperatures of 1.5 K in a system like Daunt and Lerner (1970) when a vacuum pump was placed ahead of the compressor. Another He^3 refrigerator by Daunt and Lerner (1974) replaced the He^4 bath entirely by using a commercial Gifford–McMahon refrigerator with a He^4 Joule–Thomson stage to provide a 1.2 K reservoir for liquefying He^3 at a reduced pressure. However, this refrigerator did not have a JT valve in the He^3 line so the He^3 liquid could be pumped to 0.25 K only when the return flow of He^3 was stopped. In principle the addition of a porous plug or a needle valve in the He^3 return line would permit continuous temperatures of 0.3 K or below to be maintained without the use of liquid He^4.

Except for special types of He^3 refrigerators, e.g., adsorption pumping, little work has been done in this area since the development of the dilution

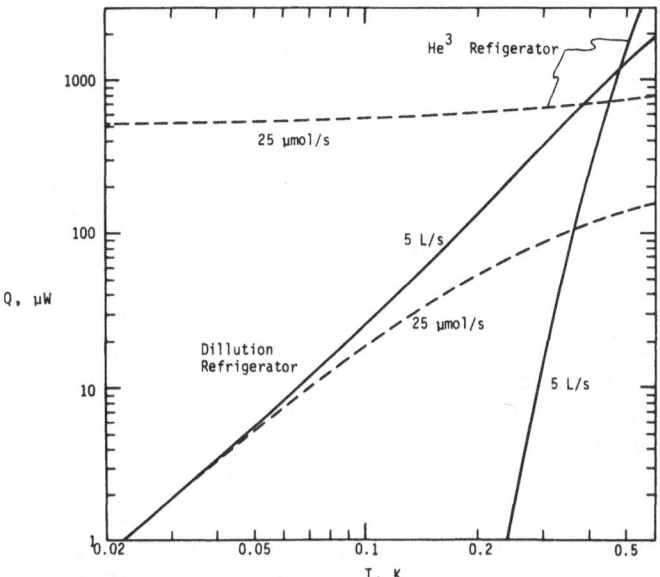

Fig. 12.7. Refrigeration powers of a He^3 refrigerator and a dilution refrigerator as a function of temperature. It is assumed that the pump used on both systems can handle 5 l/s at all relevant pressures. Also shown are the refrigeration powers for the case of constant He^3 molar flow rate of 25 μmols/s.

refrigerator in the late 1960s. The primary reason is that a dilution refrigerator can be made almost as simple as a continuous He^3 refrigerator yet be able to produce temperatures about an order of magnitude lower. Figure 12.7 compares the refrigeration rates of He^3 and He^3–He^4 dilution refrigerators when the same size pump (5 l/s) is used on both. The refrigeration rate for the He^3 refrigerator falls off rapidly below about 0.4 K because the low vapor pressure causes a low molar flow rate. According to Fig. 12.7 He^3 refrigerators have a higher refrigeration rate than that of dilution refrigerators above 0.5 K. The dashed lines in Fig. 12.7 show the refrigeration rates of the two refrigerators for a constant He^3 molar flow rate of 25 μmol/s.

He^3–He^4 DILUTION REFRIGERATORS

Properties of Liquid He^3–He^4 Mixtures

The behavior of the dilution refrigerator can only be understood after reviewing some of the properties of liquid He^3–He^4 mixtures. Below 1 K the entropy of pure liquid He^4 is extremely small and varies as T^3. This rapid ordering of He^4 is characteristic of a Bose–Einstein fluid and occurs because He^4 has zero nuclear spin. The lighter isotope He^3, on the other hand, has a nuclear spin $I = \frac{1}{2}$, so it obeys Fermi–Dirac statistics. It orders like a Fermi fluid with entropy proportional to T as shown in Fig. 12.1. The quantum-mechanical descriptions of both He^3 and He^4 are necessary at these low temperatures because the low mass of the helium atoms gives rise to rather large zero-point energy. Because the entropy of liquid He^4 below 1 K is so much smaller than that of liquid He^3, it is reasonable to expect that the He^4 would contribute a negligible amount to the entropy of a solution of He^3 and He^4. Experiments confirm such behavior. The addition of He^4 to He^3 increases the distance between He^3 atoms and weakens the already weak interaction force between them. The interaction between the He^3 and He^4 atoms causes some He^4 atoms to be dragged along with each moving He^3 atom. This dragging effect is accounted for theoretically by disregarding the presence of the He^4 atoms and substituting for each He^3 atom a He^3 quasiparticle with an effective mass m^* that is higher than the atomic mass m. The latest values of m^*/m from second sound measurements are 2.24 (van der Boog et al., 1978) and 2.28 (Brubaker et al., 1970) for zero temperature and zero concentration of He^3. An increase in the temperature or concentration causes a slight increase in m^*/m. The dilute He^3–He^4 solution is then characterized by a He^3 quasiparticle gas with weak interactions between the quasiparticles. The pressure of this gas is just the osmotic pressure of He^3 in He^4. The ideal Fermi gas model of He^3–He^4 solutions was first proposed by Landau and Pomeranchuk (1948) and the idea of weak interactions between the He^3 quasiparticles was introduced by Bardeen et al. (1967).

The entropy and other thermodynamic properties of He^3–He^4 solutions are easily calculated from the model discussed above. The specific heat and entropy are simply that of an ideal Fermi–Dirac gas with particle mass m^*. For temperatures much less than the Fermi temperature T_F the entropy of He^3 in He^4 is

$$S_3/R = 4.93(T/T_F) \qquad (12.3)$$

where

$$T_F = \frac{\hbar(3\pi^2 n)^{2/3}}{2km^*} \qquad (12.4)$$

and $\hbar$ is the reduced Planck constant, n is the density of particles, and k is Boltzmann's constant. For a He^3 concentration of $X = 6.4\%$, T_F is 0.38 K. The entropy calculated from Eq. (12.3) is on the basis of a mole of He^3. We note that the entropy per mole of He^3 increases as the concentration decreases. Radebaugh (1967), using the weakly interacting

Fermi–Dirac model discussed above and experimental data available at the time, calculated all of the thermodynamic properties of He^3–He^4 solutions for $T \leq 1.5$ K and $X \leq 0.3$. Considerably more experimental data are now available but the tabulated results of Radebaugh should still be good to about $\pm 10\%$ in most cases. Figure 12.1 shows the entropy of a 6.4% solution of He^3 in He^4 and also that of pure He^3 liquid.

Because He^3 has a higher entropy in the dilute phase than in the pure phase, an amount of heat $Q = T\Delta S$ will be absorbed for each mole of He^3 passing from the pure phase to the dilute phase in a reversible manner. The big question now remaining is how to do the dilution in a reversible manner. Pure He^3 and pure He^4 are not in equilibrium with each other and the mixing of these two components is an irreversible process which does give some cooling, but only down to about 0.2 K (Radebaugh, 1967). Fortunately nature was very cooperative in supplying a reversible mechanism for the dilution—equilibrium between a concentrated and a dilute He^3 phase as shown in Fig. 12.8. The figure shows that phase separation takes place below about 0.86 K as first seen by Walters and Fairbanks (1956). If a solution of He^3 and He^4 containing a molar fraction X of He^3 is cooled below the phase separation temperature, the solution separates into two phases. The concentrated phase (concentration X_C) is lighter, thus it floats on top of the dilute phase (concentration X_D). The amount of each phase is determined from the law of mixtures. The concentrated phase is a normal fluid whereas in the dilute phase the He^4 component is superfluid. Since the two phases are in equilibrium, He^3 can pass from the concentrated phase (low entropy) to the dilute phase (high entropy) in a reversible manner analogous to atoms passing from a liquid to a gas phase. Not only

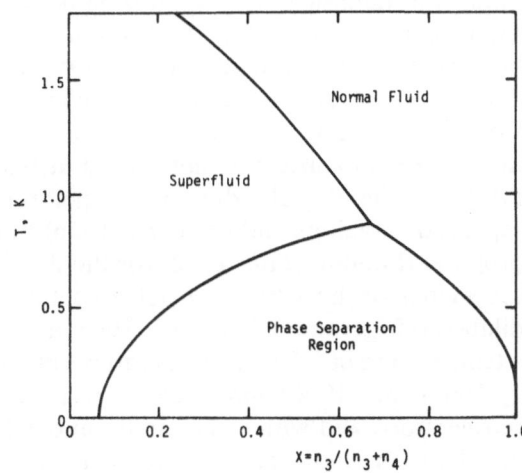

Fig. 12.8. Phase diagram of He^3–He^4 mixtures in equilibrium with their vapor. The intercept on the dilute side occurs at 6.4% He^3.

is the phase separation itself a fortunate occurrence for very-low-temperature refrigeration but so is the fact that the phase separation line on the dilute side has a finite value of X at $T = 0$ instead of the classical behavior of $X \to 0$ as $T \to 0$. In terms of refrigeration, the latter means the high entropy phase maintains a high enough particle density to permit high molar flow rates for the He^3. Calculations on a hard-sphere mixture of Bose and Fermi gases by Cohen and Van Leeuwen (1961) led to the prediction that $X_D = 0.15$ at $T = 0$. At the same time Edwards and Daunt (1961) also suggested X_D might be finite at $T = 0$ based on previous calculations of the He^3 binding energy at temperatures above 0.6 K. Ifft et al. (1967) verified experimentally that X_D is finite as $T \to 0$ and found the limiting value to be $X_D = 0.064$. Other measurements of $X_D(0)$ give values ranging from 0.064 to 0.068. The value $X_D(0) = 0.064$ is generally used since the thermodynamic properties tabulated by Radebaugh (1967) are based on that figure.

One other property of He^3–He^4 mixtures which makes dilution refrigerators possible is the behavior of the vapor–liquid equilibrium. Because He^3 has a much higher vapor pressure than He^4, the composition of the vapor in equilibrium with the liquid is close to pure He^3 for temperatures below about 1 K. Even for a liquid solution of 1% He^3 the vapor phase is 99.5% He^3 at 0.6 K but decreases to 73% at 1 K (Radebaugh, 1967). These values may need minor revisions in light of more recent data (Vvedenskii and Peshkov, 1972).

Principles of Dilution Refrigerators

After the theory (Landau and Pomeranchuk, 1949) of He^3–He^4 solutions was published in 1949, it soon became apparent to H. London that the higher entropy of He^3–He^4 solutions compared with pure He^3 could be used for refrigeration. Thus in 1951 he first proposed the He^3–He^4 dilution refrigerator in a comment to another paper at the 1st International Conference on Low-Temperature Physics (London, 1951). Phase separation in He^3–He^4 mixtures was not known at that time and so there was little interest in the idea of a dilution refrigerator. After the discovery of phase separation (Walters and Fairbank, 1956) London then published a second proposal (London et al., 1962) for the dilution refrigerator complete with schematics on how to carry out a continuous dilution of He^3. The first dilution refrigerator of Das et al. (1965) at the University of Leiden reached a temperature of 0.22 K. Temperatures of about 0.06 K were soon reached by Hall et al. (1966) and Neganov et al. (1966) in more efficient dilution refrigerators, and with some modifications Neganov et al. (1966) reached 25 mK. Many more technical refinements were contributed by Wheatley

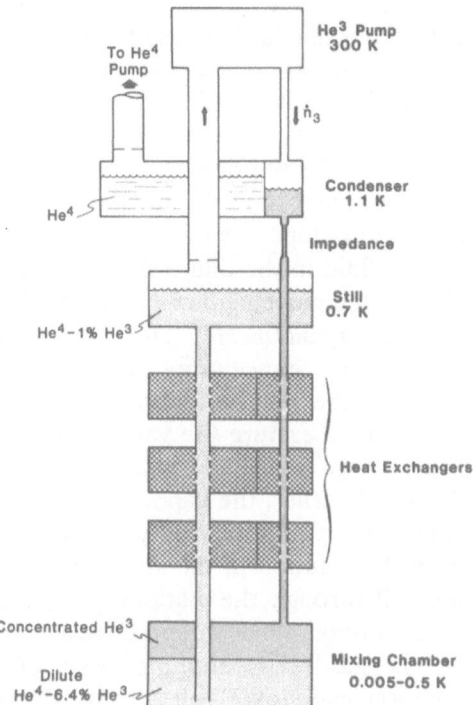

He³ Pump
300 K

To He⁴
Pump

↑   ↓ ṅ₃

Condenser
1.1 K

He⁴

Impedance

Still
0.7 K

He⁴-1% He³

Heat Exchangers

Concentrated He³

Dilute
He⁴-6.4% He³

Mixing Chamber
0.005-0.5 K

Fig. 12.9. Schematic diagram of the He³–He⁴ dilution refrigerator. He³ concentrations shown are for mixing chamber temperatures below about 40 mK.

et al. (1968, 1971), which resulted in temperatures of about 10 mK. The lowest temperature reached so far is 2.0 mK by Frossatti *et al.* (1978) at Grenoble, and commercial refrigerators are now available which reach 5 mK continuously. No other type of refrigerator has ever reached such low temperatures on a continuous basis.

The dilution refrigerator schematic, shown in Fig. 12.9, has not changed since 1966 even though many technical refinements have been made. In the description which follows it is assumed that a temperature below the phase separation line has already been reached in the mixing chamber and the proper amounts of He³ and He⁴ were initially added to the system to cause the phase boundary to be in the mixing chamber. For mixing chamber temperatures below about 0.15 K the upper concentrated phase is nearly pure He³ and the lower dilute phase is close to 6.4% He³. Because of the entropy increase, heat is absorbed when He³ passes from the concentrated phase to the dilute phase. The dilute phase has been described previously as a weakly interacting gas of He³ quasiparticles of pressure π, where π is the osmotic pressure of He³ in He⁴. The superfluid He⁴ acts only as a thermodynamic vacuum or "ether" to separate the He³ quasiparticles. The dilution process then can be described as an "evaporation" of He^3 quasiparticles from the liquid phase to the gas phase with a subsequent absorption of heat. Another analogy, that of the thermoelectric refrigerator, has already been discussed in Chapter 11.

All of the other parts in the dilution refrigerator are for the purpose of making the dilution process a continuous one. The He^3 quasiparticles will continue to "evaporate" into the dilute phase as long as these quasiparticles are pumped away from the phase boundary. The fact that the He^4 is superfluid in the dilute solution means that it does not provide viscous drag on the movement of He^3 through it. Only the container walls provide viscous drag on the He^3. Thus He^3 can be removed from the dilute solution quite some distance away from the mixing chamber. This removal of He^3 occurs in the still, held at a temperature of about 0.6–0.7 K by a heater. At this temperature the vapor pressure of the solution is high enough to permit large molar flow rates through vacuum pumps of modest sizes. As discussed earlier, the vapor composition above a dilute He^3–He^4 solution at these temperatures is very nearly pure He^3. Hence mainly He^3 is removed from the solution in the still and He^3 flows from the mixing chamber to the still through the stationary He^4. This flow of the He^3 quasiparticle "gas" through tubes will produce a pressure drop. Fortunately the pressure of this "gas," the osmotic pressure of He^3 in He^4, is rather high, 2.1 kPa (16 torr), even for $T \to 0$ and so $\Delta P/P$ can be kept reasonably small. This high osmotic pressure of He^3 is used to drive the He^3 from the low-temperature mixing chamber to the still where the temperature and He^3 vapor pressure are high enough to give high pumping rates. After being compressed by a vacuum pump at room temperature, the He^3 gas is condensed by a pumped He^4 bath at around 1.1 K or below. The pressure is reduced after passing through the capillary below the condenser. The pure He^3 entering the mixing chamber is first precooled by heat exchange with the still and then by heat exchange with the outgoing dilute He^3.

The high osmotic pressure in the dilute phase is a result of $X_D(0) = 0.064$ on the phase separation curve. The He^3 quasiparticle density is then relatively high and remains high even for $T \to 0$, unlike the particle density in He^3 vapor in equilibrium with the liquid. For temperatures below about 0.1 K the osmotic pressure of a 6.4% He^3–He^4 solution is about 2.1 kPa (16 torr) (Ebner and Edwards, 1971). Figure 12.10 shows the behavior of the osmotic pressure π as a function of temperature and composition from the calculations of Radebaugh (1967). The curves are depressed from that of an ideal Fermi–Dirac gas because of the weak interaction between the He^3 quasiparticles. The interaction chosen at the time of the calculations has turned out to be too large and so the calculated value of 1.6 kPa (12 torr) for $X = 0.064$ at $T = 0$ is smaller than the value 2.1 kPa (16 torr)

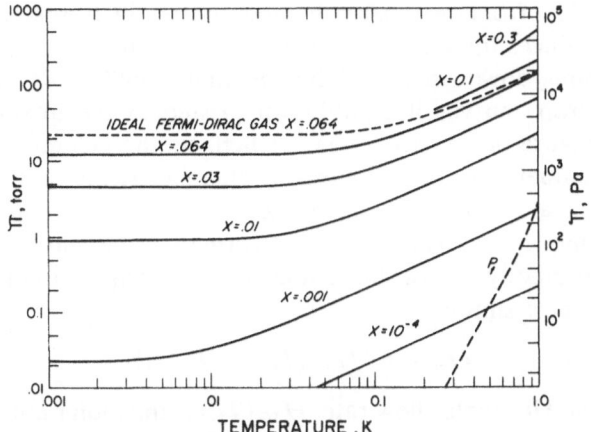

Fig. 12.10. Osmotic and fountain pressure P_f in He^3–He^4 solutions from calculations of Radebaugh (1967). Osmotic pressure may be thought of as the pressure of the He^3 quasiparticle gas in the solution.

measured later. Nevertheless, the curves in Fig. 12.10 provide us with reasonable estimates of the overall behavior of π.

For small He^3 flow rates between the mixing chamber and the still, one would expect the quasiparticle "gas" pressure to be equal at the still and mixing chamber. The exact equilibrium condition in the superfluid dilute phase is given by

$$\nabla \mu_4 = 0 \tag{12.5}$$

where μ_4 is the He^4 chemical potential. Thus μ_4 must be constant between the mixing chamber and the still. Radebaugh (1967) shows that uniformity in μ_4 is equivalent to

$$\pi + p_f = \text{const} \tag{12.6}$$

where p_f is the fountain pressure of pure He^4 as shown in Fig. 12.10. Because of the small value of p_f below 1 K it can usually be neglected in the above equation and the He^3 quasiparticle "gas" pressure is indeed constant for low flow rates. Using the condition $\pi = \text{const}$ in Fig. 12.10 shows that the equilibrium concentration in the still at 0.7 K is 0.75% He^3 for mixing chamber temperatures T_m of 20 mK or below. When the osmotic pressure curves are revised upward at low temperatures to agree with the latest experimental results the still concentration at 0.7 K becomes 1.0% He^3. The vapor pressure above the still is 11.3 Pa (0.085 torr) according to the data of Vvedenskii and Peshkov (1972) and is in reasonable agreement with the tables of Radebaugh (1967). In practice, impedance to the flow of dilute He^3 from the mixing chamber to the still causes some pressure drop which reduces the He^3 concentration in the still.

The cooling power in the mixing chamber could easily be calculated from the entropy curves if equilibrium existed in the mixing chamber. However, because of imperfect heat exchangers the incoming pure He^3 is at a higher temperature than the mixing chamber temperature T_m and so some of the ideal refrigeration power goes into cooling the incoming pure He^3 to T_m. Since no external work is performed in the dilution process the first law of thermodynamics shows that the net refrigeration power in the mixing chamber is simply

$$\dot{Q}_m = \dot{n}_3[H_{3D}(T_m) - H_3^0(T_i)] \qquad (12.7)$$

where $\dot{n}_3$ is the He^3 molar flow rate, $H_{3D}(T_m)$ is the molar enthalpy of He^3 in the dilute phase leaving the mixing chamber at a temperature T_m, and H_3^0 is the molar enthalpy of the incoming pure He^3 entering at a temperature T_i. Such enthalpy curves as calculated by Radebaugh (1967) are shown in Fig. 12.11. The H_{3D} curve is for a dilute solution along the phase separation line. As the dilute He^3 leaves the mixing chamber at constant osmotic pressure, the concentration decreases as the temperature increases and so the He^3 is no longer on the phase separation line. The specific heat of He^3 at constant osmotic pressure $C_{3\pi}$ was measured by Radebaugh and

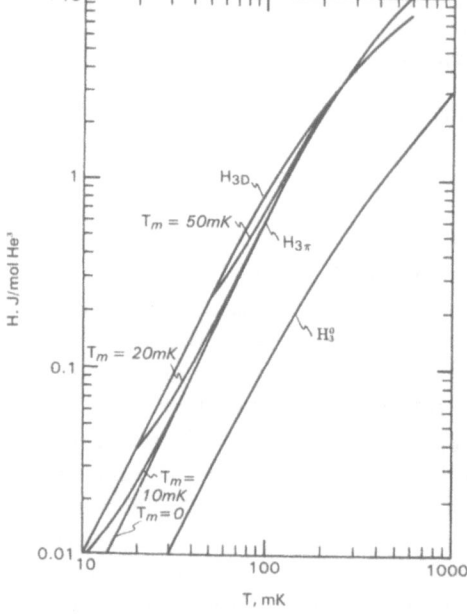

Fig. 12.11. Enthalpy as a function of temperature for He^3 in various states in a dilution refrigerator. The terms used are H$_3^0$, pure liquid He^3; H$_{3D}$, dilute He^3 along the phase separation line; H$_3\pi$-dilute He^3 at constant osmotic pressure for various mixing chamber temperatures T_m.

Siegwarth (1971) and found to be in good agreement with the calculated values (Radebaugh, 1967), especially below 0.3 K. Integration of the calculated $C_{3\pi}$ gives the $H_{3\pi}$ curves shown in Fig. 12.11. Tabulated values of $H_{3\pi}$ are given by Siegwarth and Radebaugh (1971). The $H_{3\pi}$ and H_3^0 curves are useful for heat exchanger calculations.

For temperatures below about 40 mK $H_3^0 = 12T^2$ J/mol K^2 and $H_{3D} = 94T^2$ J/mol K^2 so that Eq. (12.7) becomes

$$\dot{Q}_m = \dot{n}_3(94T_m^2 - 12T_i^2) \text{ J/mol K}^2 \tag{12.7a}$$

With perfect heat exchangers $T_i = T_m$ and the maximum refrigeration rate is then

$$\dot{Q}_m = 82\dot{n}_3 T^2 \text{ J/mol K}^2 \tag{12.8}$$

which agrees well with the value $(80 \pm 3)\dot{n}_3 T^2$ J/mol K^2 found experimentally (Gladun and Peshkov, 1972). A typical flow rate is 100 μmol/s so that at $T = 100 \, \mu$K, $\dot{Q}_m = 82 \, \mu$W. For $\dot{Q}_m = 0$ the minimum temperature occurs when

$$T_i/T_m = 2.8 \tag{12.9}$$

according to Eq. (12.7a). Exact calculations (Siegwarth and Radebaugh, 1972) show this value increases to 4.0 at 0.2 K. Equation (12.9) shows that rather large temperature differences can occur across the heat exchanger before reducing the refrigeration rate to zero. Such large values of $\Delta T/T$ relax the design requirements on heat exchangers and are not found in refrigerators operating above 1 K. The enthalpy of the dilute phase at constant osmotic pressure is given by

$$H_{3\pi} = 54T^2 \text{ J/mol K}^2 \tag{12.10}$$

for $T < 40$ mK and $T_m \ll T$. Thus for the same temperature the enthalpy in the dilute side of the heat exchanger is 4.5 times that on the concentrated side, which also greatly relaxes the design of heat exchangers. An enthalpy balance on the heat exchangers shows that for $T_m < 20$ mK and a still temperature T_S of 0.7 K the incoming pure He^3 in the heat exchanger will warm the dilute He^3 only up to a temperature of 0.18 K before entering the still. Thus a large ΔT always exists at the hot end of the heat exchanger.

In spite of the large ΔT which exists in the heat exchanger, the ultimate temperature of the dilution refrigerator, until recently, has been limited by the heat exchanger because of the very large thermal boundary resistance between helium and a solid at low temperatures. This thermal resistance, known as the Kapitza resistance, is a result of the phonon mismatch at the boundary. Most phonons then are reflected at the surface instead of being transmitted across it. Generally the Kapitza resistance is given by

$$R_K = \alpha_K/\sigma T^3 \tag{12.11}$$

where σ is the surface area of the boundary and α_K is a constant which depends on the material. Generally α_K is the order of 100 cm^2 K^4/W. Thus for temperatures around 10 mK, R_K can be extremely high. The first three dilution refrigerators discussed earlier used Cu–Ni or stainless steel tubes for the heat exchanger. Generally the dilute He^3 flowed through an annular region between tubes and the concentrated He^3 flowed inside a smaller tube or another annular region. The temperatures vary continuously along the length of the exchanger and so these are called continuous heat exchangers. Wheatley et al. (1968) soon realized that much larger surface areas were required to overcome the large Kapitza resistance and used copper powder sintered in four copper blocks as the heat exchangers to reach a temperature of 10 mK. The experimental arrangement for the parts below the 1 K pot is shown in Fig. 12.12. Each copper block had two sintered powder chambers—one for the concentrated stream and one for the dilute stream. The two liquids traveled between each block through small Cu–Ni tubes in order to thermally isolate each block. The copper powder used had particles sizes less than 44 μm and were packed to about

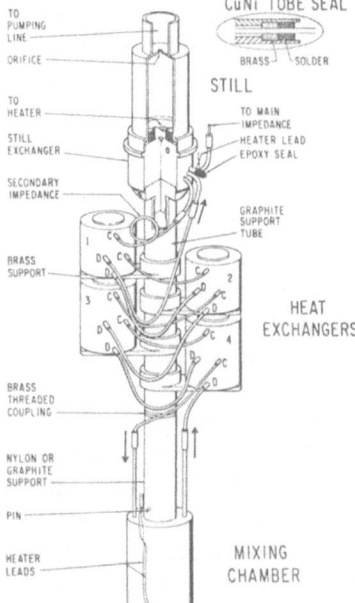

Fig. 12.12. Assembly drawing of the first dilution refrigerator with discrete heat exchangers of sintered copper powder (after Wheatley et al., 1968).

50% solid density. Generally the sintered powder chambers had length-to-diameter ratios on the order of 1. The temperature of each copper block is uniform and the liquid temperatures are changed in a stepwise fashion upon entering each block. Such exchangers are known as step or discrete heat exchangers. Discrete heat exchangers work well in a dilution refrigerator because of the large enthalpy difference between the two streams. Wheatley *et al.* (1968) and Kalvius *et al.* (1969) showed that if each block had perfect heat exchange only three are required to give $T_m = 10$ mK with a still temperature of 0.7 K, whereas four were required experimentally. Later the upper heat exchanger block was replaced with a continuous coaxial Cu–Ni tube exchanger since at the higher temperatures such large surface areas are not needed. This sintered powder heat exchanger approach pioneered by Wheatley *et al.* (1968) has been the basis for nearly all dilution refrigerators although copper foil has sometimes been used in place of the sintered powder.

Calculations to optimize the design of both the continuous and the discrete heat exchangers were made by Siegwarth and Radebaugh (1971, 1972) and by Radebaugh and Siegwarth (1971). Optimized designs would reduce the amount of He³ needed in the refrigerator and reduce the cooldown time, which often took many hours. One of the interesting results of these calculations was the detrimental effect of thermal conductance through the liquid inside a discrete heat exchanger. The thermal conductivity of both pure He³ and dilute He³–He⁴ solutions is quite high and is proportional to T^{-1} at temperatures below about 20 mK (Wheatley *et al.*, 1968). Figure 12.13 shows the calculated temperature profiles through a typical sintered copper powder heat exchanger of length 1 cm. For the case of zero liquid thermal conductivity the concentrated stream enters the exchanger at 59.9 mK and gradually cools to 27.8 mK at the other end of the exchanger. Concentrated liquid entering the mixing chamber at 27.8 mK gives $T_m = 10.0$ mK for zero external heat load. The dilute stream then enters the heat exchanger at 10.0 mK and is warmed to 25.7 mK. The temperature of the copper body is uniform throughout and is 26.5 mK. Now when the calculations are made with the proper value for the liquid thermal conductivities the temperature profiles for the two liquid streams shift dramatically. The thermal conductance prevents much of a temperature gradient in the liquid except in the small-diameter connecting tubes. The incoming concentrated stream must be at a temperature of 40.5 mK in order to give $T_m = 10.0$ mK, instead of 59.9 mK if the thermal conductivity were zero.

Obviously the thermal conductance problem could be eliminated by making the length-to-diameter ratio of the exchanger much greater than unity. Unfortunately that would make the pressure drops through the fine

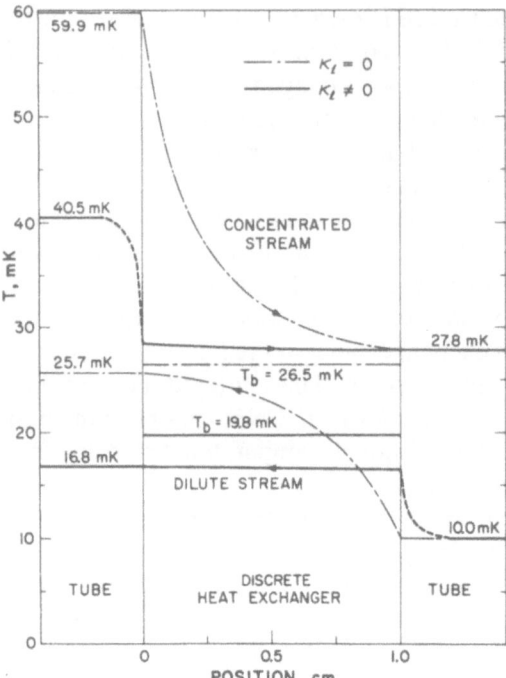

Fig. 12.13. Calculated liquid and body temperature profiles within and near a discrete heat exchanger of length 1 cm for both zero and finite liquid conductivities. For finite conductivity most of the temperature changes occur in the small tubes connected to the discrete heat exchanger (after Siegwarth and Radebaugh, 1971).

powder too large since L/D ratios of unity or even less were normally needed to make ΔP small enough. The problem could be overcome without significantly increasing ΔP by using disks of copper with a small hole in the center to partition the sintered material into several sections (Siegwarth and Radebaugh, 1971). Large temperature gradients in the liquid can then occur inside the small hole in each disk. Another approach is to use a large L/D ratio but reduce the flow impedance by placing a hole through the sintered material. Most of the liquid then flows through the hole, but Betts and Marshall (1969) showed that good radial heat transfer with the sintered material occurs because of the high thermal conductivity of the liquid. This approach was first used in a heat exchanger by Niinikoski (1971), as shown in Fig. 12.14, for a rather unique horizontal dilution refrigerator used in experiments with spin-frozen polarized proton targets. Niinikoski (1976) has also presented some recent calculations regarding the behavior of continuous heat exchangers.

The hole-through-the-sintered approach also relaxes the requirement that the particle size of the powder be large enough to allow liquid to flow through it without a large ΔP. The surface area per unit volume of a powder

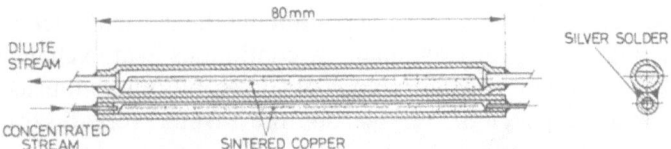

Fig. 12.14. Long and slender sintered copper heat exchanger, which reduces the liquid conductance, but yet provides a small flow impedance by the use of flow channels (after Niinikoski, 1971).

with spherical particles of diameter d is

$$\sigma/V = 6/d \tag{12.12}$$

Thus the use of smaller particles would reduce the overall Kapitza resistance in Eq. (12.11) without increasing the liquid volumes. However at 20 mK the phonon mean free path in copper is about 750 μm and the phonon wavelength is about 3 μm. Both are proportional to T^{-1}. Thus it would appear that if phonons are responsible for the heat transfer across the surface between copper and liquid helium, then particle sizes smaller than 44 μm could easily increase the constant α_K in Eq. (12.11) so that R_K would not be decreased by the increase in σ. Measurements by Radebaugh and Siegwarth (1974) of the Kapitza resistance between 2-μm-diam copper powder and dilute He^3–He^4 solutions showed that the constant α_K was about 300 cm^2 K^4/W and was the same as that for bulk copper. The results suggested that other heat transfer mechanisms were operating besides phonons.

The next step was to measure R_K for even smaller particle sizes. A sample of silver powder with a particle size of 150 nm was made available to the author on a trip to Japan in 1972. Measurements (Radebaugh *et al.*, 1974) of the Kapitza resistance on this sample gave a value for α_K of about 400 cm^2 K^4/W at 20 mK, close to the bulk copper value of 300 cm^2 K^4/W but somewhat higher than the value of 115 cm^2 K^2/W for 6-μm-diam silver powder. Thus it appears that the particle size begins to increase α_K only for sizes around 150 nm or smaller. The value of σ/V for 150-nm particles is 4×10^7 cm^{-1} and is so large that the Kapitza resistance problem is nearly eliminated. The dominant thermal resistance above about 20 mK appeared to be at the boundary between the silver powder and the solid copper cell. Another result of the measurements was that the thermal conductivity of the sintered silver powder was more than two orders of magnitude less than commercial bulk copper. At 10 mK it is also more than an order of magnitude less than the thermal conductivity of pure or dilute He^3. That meant the silver powder could be used to make a continuous heat exchanger with little axial heat conduction through the silver powder. For equal surface areas a continuous heat exchanger is always superior to a discrete heat exchanger. A continuous heat exchanger made with the 150-nm silver powder in coaxial Cu–Ni tubes was fabricated and tested by Radebaugh *et al.* (1974) but the results were disappointing. However, a discrete heat exchanger with a 0.01-cm^3 liquid volume made with 2 μm copper powder inside 2-mm-diam copper tubes was successful.

Frossati *et al.* (1978a) incorrectly identified the problem with the above silver powder and the Cu–Ni. They then solved the problem by first silver plating the Cu–Ni and then coating it with 1-μm-diam silver powder. On top of that is placed a layer of prepacked 70-nm-diam silver powder. The assembly is sintered at 200 C in a H$_2$ atmosphere for 40 min. This technique was used to construct the continuous silver powder exchanger shown pictorially in Fig. 12.15. The silver powder is bonded to both sides of a silver-plated Cu–Ni sheet which in turn is welded between formed Cu–Ni boxes. The liquids flow through the formed channels and transfer heat with the silver powder. The heat exchanger is then bent into a helix (Frossati, 1978) or semicircle (Frossati and Thoulouze, 1976) to save space. With a flow rate of 150 μmol/s through this type of heat exchanger Frossati *et al.* (1978a) were able to reach 2.0 mK continuously, the lowest temperature ever reached in a dilution refrigerator. Though this type of heat exchanger is essentially a continuous type, in practice it is usually divided into several parts so that optimum dimensions can be used for different temperature ranges and to reduce axial heat conduction even more. Frossati used sinter thicknesses on the order of 1 mm and flow channels of 12.2 mm square on the dilute side and 5.0 mm square on the concentrated side for the lowest-temperature segment. The channels decreased to 4.0 mm square and 1.6 mm square for the warmest segment. Each segment was 26 cm long and a total of four segments were used. The concentrated liquid volumes of about 16 cm^3 were used in this refrigerator and a flow rate of about

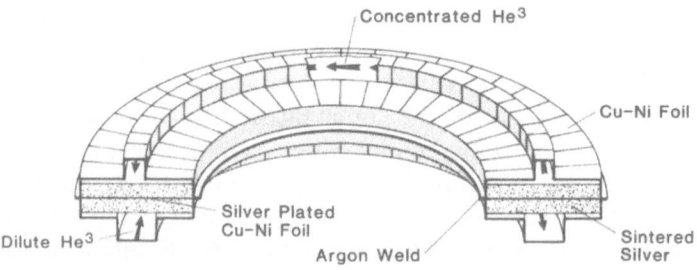

Fig. 12.15. Semicontinuous heat exchanger developed by Frossati *et al.* (1978) which uses very fine (70 nm) silver powder sintered to a silver-plated Cu–Ni foil. With several of these heat exchangers a record low continuous temperature of 2.0 mK has been reached in a dilution refrigerator.

200 μmol/s meant that 36 min were required to complete a cycle with the He^3 in the heat exchanger. The fine silver powder Frossati used was found to have a specific surface area of 1.8 m^2/g after sintering. A total surface area of about 300 m^2 was used in this refrigerator which reached 2.0 mK. These sintered silver powder heat exchangers developed by Frossati represent the present state of the art in heat exchangers for dilution refrigerators and it is this type which is used in the latest commercial refrigerators that produce temperatures of about 5 mK.

Frossati (1978) found that the ultimate temperature of 2 mK is determined mainly by viscous heating and thermal conductance in the dilute stream leaving the mixing chamber. Such behavior is in accordance with the intrinsic limit calculated by Wheatley et al. (1968), who give

$$T_m \text{ (intrinsic)} = 4(1 \text{ mm}/D)^{1/3} \text{ mK} \qquad (12.13)$$

where D is the diameter in mm of the tube for the dilute liquid leaving the mixing chamber. For $D = 10$ mm as used by Frossati, T_m (intrinsic) = 1.9 mK from Eq. (12.13). Viscous heating becomes a problem at very low temperatures because the viscosity of the dilute stream η_D varies as T^{-2}. At 2 mK, $\eta_D = 7.5 \times 10^{-3}$ Pa s (75 mpoise), which is about like that of warm olive oil, whereas at 40 mK it becomes 19×10^{-6} Pa s (190 μpoise) which is like helium gas at room temperature. Increasing the tube diameter does not help very much because the high thermal conductivity of the liquid conducts the viscous heat back to the mixing chamber.

It is possible to reach the intrinsic limit for T_m in a dilution refrigerator with less efficient heat exchangers than the silver powder type for a short period of time simply by stopping the return flow of He^3 with a valve at room temperature. That eliminates the heat load from the heat exchanger and dilution continues only until all of the He^3 in the mixing chamber is depleted. This single-cycle mode was first used by Vilches and Wheatley (1967) and is often used to find the background heat leak into the mixing chamber via Eq. (12.8) when the lower limit determined by Eq. (12.13) is not reached. Heat leaks are generally several nanowatts or more. Radebaugh (1967) analyzed the single-cycle mode and gave results useful for determining transient temperature behavior. Severijns et al. (1978) discuss the performance of an improved mixer designed specifically for single-cycle operation. They reach a temperature of 2.8 mK after starting from 32 mK.

Efficiencies of dilution refrigerators are very low but they are usually of no concern. Efficiencies are low because of the need for vacuum pumps. The minimum power required for the vacuum pumps is usually about 500 W, which does not include the pump for the 1 K He^4 bath. Such a pump could produce a refrigeration rate of about 10 μW at 0.1 K and that translates to an efficiency of 6×10^{-5} of Carnot.

Examples of Dilution Refrigerators

In this section we do not intend to describe in detail how the various refrigerators are constructed. Such details can be generally found in the book by Lounasmaa (1974) and in the original publications. Figure 12.16 shows one of the earliest really successful dilution refrigerators built by Neganov *et al.* (1966). A very large, high-speed pumping line maintained at 77 K right to the pump was used because at the time of construction they thought that the solubility X_D approached zero as $T^{2/3}$ instead of reaching the limit of 6.4% He^3. The pumping speed was 184 μmol/s and the lowest temperature reached in this refrigerator was 56 mK. The heat exchanger was made of three concentric stainless steel tubes, 22 cm long and about 20 mm in diameter. The incoming concentrated He^3 flowed in the 0.1 mm annular space between the outer two tubes and the dilute He^3 flowed through the 0.15 mm space between the two innermost tubes. After learning about the finite solubility of He^3 in He^4 at $T = 0$, Neganov *et al.* (1966) increased the surface area of the heat exchangers by a factor of 5 and reached a temperature of 25 mK with a He^3 circulation rate of 114 μmol/s. The liquid-nitrogen-cooled charcoal trap in the He^3 return

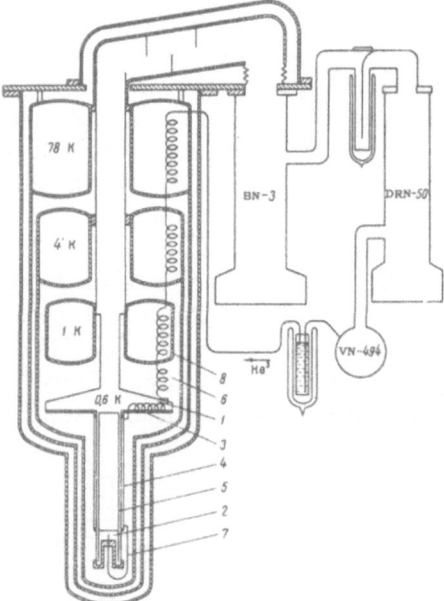

Fig. 12.16. Diagram of a pioneer dilution refrigerator in which a temperature of 25 mK was first achieved. The heat exchanger consisted of concentric stainless steel tubes (after Neganov *et al.*, 1966). KEY: 1, still, 15 cm diam.; 2, mixing chamber, 19 cm diam.; 3, still heat exchanger; 4, heat exchanger, concentrated He^3; 5, heat exchanger, dilute He^3; 6, impedance; 7, He^3 in; 8, condenser.

line is necessary to remove any gas impurities which could solidify and plug up the impedance below the 1 K pot.

Figure 12.12 shows the dilution refrigerator developed by Wheatley *et al.* (1968) that introduced the discrete heat exchange concept where sintered copper powder was used. This figure shows only the parts below the 1 K pot. Many of the refrigerators built after that time used the same concept. Figure 12.17 shows a photograph of the dilution refrigerator at the National Bureau of Standards in Boulder that was used for measurements of Kapitza resistance. This refrigerator is typical of many that were patterned after that of Wheatley *et al.* (1968), including the use of graphite rods for mechanical support because of its very low thermal conductivity at these temperatures. There are two features of this refrigerator that are unique. First the discrete heat exchangers, of which there are three, have the 44-μm copper powder partitioned by 3 copper disks with a small central hole. As discussed earlier in the section on principles, these partitions decrease the thermal conductance in the liquid and improve the efficiency of the heat exchangers. The second unique feature is the use of only the dilute stream to thermally ground various heat inputs due to thermal conduction from higher temperatures. Because of its high specific heat, the dilute stream temperature should change little with small heat inputs. The ultimate temperature depends on how cold the concentrated stream can be cooled before entering the mixing chamber, and because of the low specific heat of the concentrated stream, its temperature can be increased significantly with heat inputs. Small temperature changes in the dilute stream have a negligible effect on the heat transfer between the concentrated and dilute streams for the following reason: Because of the T^{-3} dependence in the Kapitza resistance the heat flow between the two streams is proportional to $(T_C^4 - T_D^4)$, where T_C is the concentrated stream temperature and T_D is the dilute stream temperature. As discussed earlier, an enthalpy balance on the heat exchangers shows that T_C can be 2 to 4 times T_D when the ultimate temperature is reached and so a small change in T_D has little effect on $(T_C^4 - T_D^4)$. In the refrigerator shown in Fig. 12.17 the upper thermal ground uses the dilute liquid at about 0.18 K just before it enters the still. This dilute liquid flows through a short copper tube hard soldered to a plate that intercepts heat that travels down supports from 1 K. Thus the top end of the graphite support rod is at 0.18 K instead of 0.7 K if it was attached to the still. The lower thermal ground is just above the mixing chamber and uses a small cell, filled with sintered copper powder, that is hard soldered to a plate. This thermal ground intercepts heat traveling toward the mixing chamber through the graphite. This refrigerator used a circulation rate of 20 μmol/s and could reach temperatures of 12 mK continuously. The still for this refrigerator is of the type described by

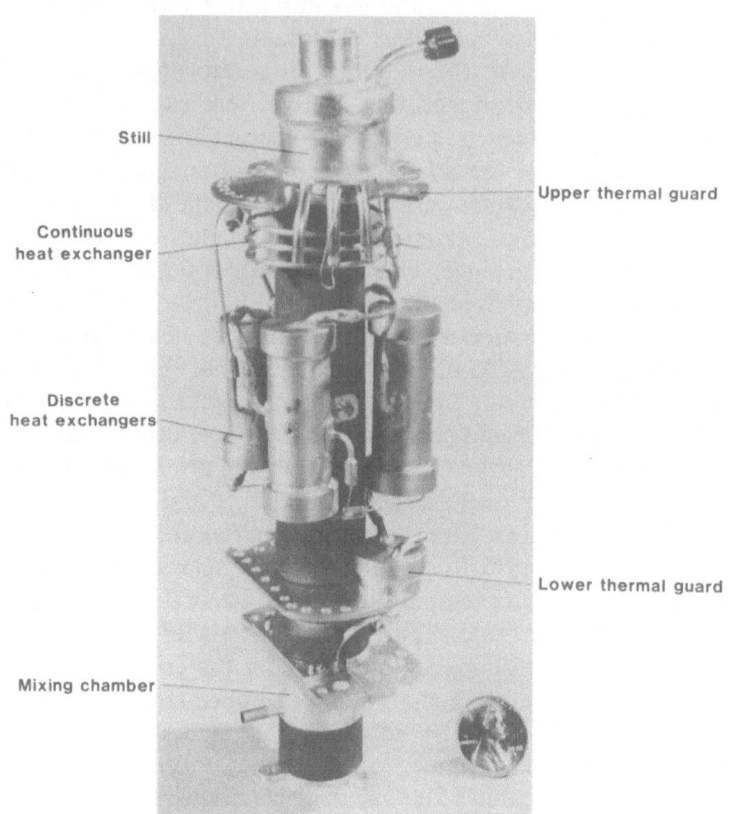

Fig. 12.17. Photograph of the dilution unit of the refrigerator used in the laboratory at the National Bureau of Standards, Boulder. A continuous heat exchanger and three discrete copper powder heat exchangers are used. Continuous temperatures of 12 mK can be maintained in this refrigerator.

Wheatley *et al.* (1971) to eliminate superfluid film flow up the pumping tube. Such film flow would increase the He^4 circulation and degrade the performance. The mixing chamber is filled with sintered copper powder to aid in the heat transfer between an externally mounted device and the liquid.

Figure 12.18a shows one of the latest commercial dilution refrigerators which employs sintered silver powder heat exchangers like that developed by Frossati. In this refrigerator five sections of the silver heat exchangers are used along with a Cu–Ni coaxial tube heat exchanger for the higher temperatures. The support rods are made of graphite-filled polyimide rods which are stronger than graphite but have almost as low a thermal conductivity (Locatelli *et al.*, 1976). Temperatures of 3.0 mK have been reached

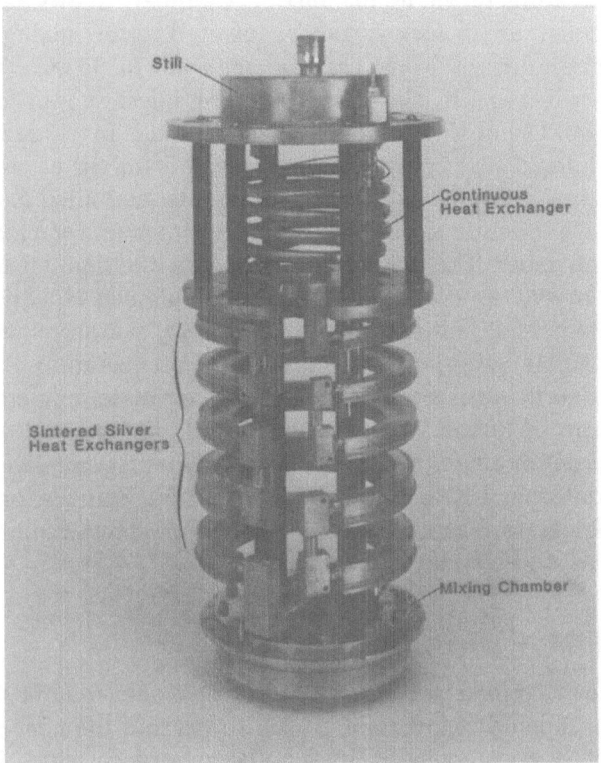

Fig. 12.18. (a) Dilution unit of a commercial dilution refrigerator that uses sintered silver heat exchangers like those in Fig. 12.15. Continuous temperatures of about 3.0 mK have been reached in such a refrigerator. (b) Entire cryostat insert for a dilution refrigerator which has a top-loading feature whereby samples can be rapidly cooled from room temperature to the mixing chamber temperature. The dilution unit is very similar to that shown in (a). [Both (a) and (b) courtesy of S.H.E. Corporation, San Diego.]

in this refrigerator. With a flow rate of 2000 μmol/s the unit can provide a refrigeration rate of at least 1.0 mW at 0.1 K. Access to the inside of the mixing chamber is achieved by the use of indium gasketed cover plates. This dilution unit sells for around $30,000 and when the necessary pumping lines and a self-regulating 1 K cold plate (DeLong *et al.*, 1971) are added to form a cryostat insert to be placed in a 4.2 K helium bath the price becomes about $50,000. A complete system, including Dewar, pumps, support structure, and temperature measurement and control instruments, sells for $100,000 to $140,000. A similar dilution unit is also available which has a top loading feature. A 1-cm-diam tube, which extends from room temperature down to the mixing chamber, allows samples to be inserted through an air lock and to be lowered to the mixing chamber. A sample temperature of 25 mK can be achieved in 80 min. This dilution unit, along with the 1 K cold plate, pumping lines, and air lock is shown in Fig. 12.18b. The mixing chamber is actually made in two separate halves, each with sintered silver powder. The top half is for the usual purpose but the bottom half is connected to the sample tube and filled partially with a liquid He^3–He^4 mixture to provide thermal contact between the sample and the mixing chamber. The feature of top loading into liquid helium was first introduced by Pavlov *et al.* (1978) and by Binnig and Hoenig (1978), and like the commercial unit shown in Fig. 12.18b, samples could be placed directly in the He^3–He^4 liquid that is used for refrigeration.

Often one is not interested in such low ultimate temperatures and if a temperature of 30 mK is satisfactory simple commercial dilution units sell for an order of magnitude less. With a 70-μmol/s circulation rate these units can cool from 1 K to below 0.1 K in 30 min. Sources for commercial dilution refrigerators as well as many other low-temperature equipment and materials are listed by Lounasmaa (1974).

Multiple Mixing Chambers

A clever technique was introduced by deWaele *et al.* (1976) whereby two or more mixing chambers are used to precool the concentrated He^3 stream directly and eliminate the Kapitza resistance which hinders heat exchanger performance. The flow pattern for a system of two mixing chambers is shown in Fig. 12.19. In this scheme the concentrated liquid is precooled in the first mixing chamber (MC-1) before going to the colder mixing chamber (MC-2). The precooling in MC-1 occurs because the heat transfer from the concentrated phase to the phase boundary occurs without

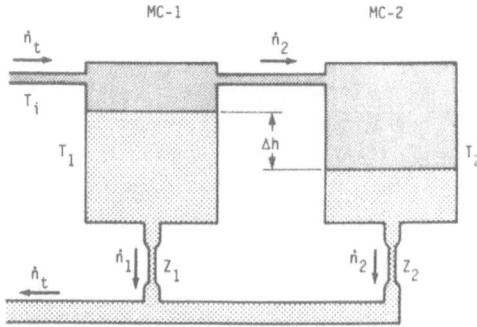

Fig. 12.19. Schematic of the arrangement used for two mixing chambers. The mixing chamber MC-1 dilutes some of the incoming He^3 and precools the remainder before it enters MC-1. The impedances Z_1 and Z_2 control the flow rates through each mixing chamber. Any number of mixing chambers can be added to this scheme to provide a lower ultimate temperature in the last mixing chamber.

the Kapitza resistance associated with a metal boundary. Dilution in MC-1 proceeds as some of the dilute He^3 is removed from it through the imped-ance Z_1. A mass balance on the He^3 shows that

$$\dot{n}_t = \dot{n}_1 + \dot{n}_2 \tag{12.14}$$

where $\dot{n}_t$, $\dot{n}_1$, and $\dot{n}_2$ are the He^3 molar flow rates as shown in Fig. 12.19. As discussed earlier an enthalpy balance on a mixing chamber leads to the condition that

$$T_2 = T_1/2.8 \tag{12.15}$$

when the heat leak to T_2 is negligible. Because $\dot{n}_2$ is less than $\dot{n}_t$, the cooling power in MC-2 is reduced from that of a single mixing chamber and the heat leak may not always be negligible. The osmotic pressure difference between MC-1 and MC-2 causes a hydrostatic head difference Δh for zero flow. For $T_1 = 18\,\text{mK}$, $\Delta h = 5\,\text{cm}$ when $T_1 = 2.8T_2$. The height difference increases with temperature and so height limitations on the mixing chamber preclude the use of multiple mixing chambers above about 20 mK. The impedance ratio Z_1/Z_2 determines the flow ratio $\dot{n}_1/\dot{n}_2$, which in turn influences the temperature reduction in MC-2. With two mixing chambers, de Waele *et al.* (1976) reached 5.5 mK in a dilution refrigerator that reached 13 mK with one mixing chamber. By adding a third mixing chamber they reached a temperature just below 4 mK. Viscous heating and thermal conduction limited the performance. Any number of mixing chambers can be added on to those shown in Fig. 12.19 and in the limit of a large number, a continuous mixing chamber evolves (de Waele *et al.*, 1978). Frossati *et al.* (1978b) describe the theoretical behavior of a double mixing chamber system and present experimental results showing a temperature reduction from 3.8 mK for a single mixing chamber to 2.8 mK for two mixing cham-bers. The experiments agree well with their calculations when all sources of heat leak are considered. The advantage of the multiple mixing chambers is that a smaller heat exchanger can be used to reach the same temperature. The disadvantage is the reduction in cooling power due to the reduction in the He^3 flow rate through the coldest mixing chamber.

He^4 Circulating Dilution Refrigerators

In order to overcome the problem of Kapitza resistance in heat exchangers, Taconis *et al.* (1971) developed a dilution refrigerator in which He^4 instead of He^3 is circulated. The very low specific heat of He^4 eliminates much of the heat exchange problem. According to the phase diagram of Fig. 12.8, pure He^4 is not in equilibrium with a He^3–He^4 solution. Therefore, the dilution process can be made reversible only if a membrane permeable just to superfluid He^4 is used in the system as discussed in Chapter 11. The semipermeable membrane in this case is a porous plug, generally made with Fe_2O_3 powder of grain size 30 nm, and known as a superleak. A schematic of the He^4 circulating dilution refrigerator is shown in Fig. 12.20. Superfluid He^4 passes through the superleak S_1 and enters the dilute He^3–He^4 solution in the mixing chamber M_1. The excess He^4 in the dilute solution causes He^3 to pass from the concentrated solution to the dilute solution in order to maintain phase equilibrium. This dilution process absorbs heat as discussed previously. The dilute solution continuously produced in this way overflows and falls, due to gravity, through the heavier concentrated solution and into the demixing chamber M_2. It is uncertain whether the dilute solution falls as droplets as shown in Fig. 12.20. Superfluid He^4 is extracted from the dilute solution in M_2 via superleak S_2 and returned to S_1. The He^3 coming into the dilute solution from the droplets then must return to the concentrated phase. The heat leak given off in this demixing process is absorbed at about 0.4 K by a He^3 refrigerator in contact with M_2. Excellent heat exchange between the ascending warm He^3 and the descending dilute He^3 occurs because of the direct contact between the two solutions.

In order to reduce thermal conduction from M_2 to M_1 through the liquid column, the length to diameter ratio of the interconnecting tube must be significantly greater than that shown in Fig. 12.20. In practice this tube is about 2.8 mm diameter and 60 cm long (Pennings et al., 1974) and coiled into a helix 15 cm high. The refrigerator does not work when the

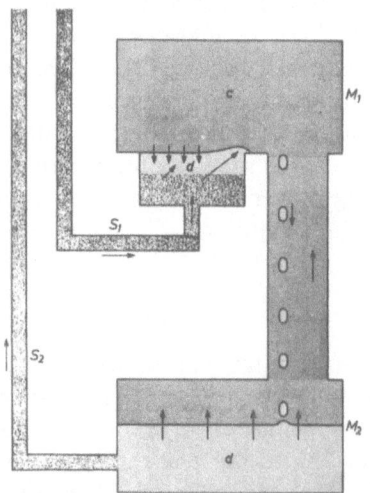

Fig. 12.20. Schematic of the He^4 circulating dilution refrigerator developed by Taconis et al. (1971) at Leiden. The He^4 is extracted from the dilute solution in the demixing chamber M_2 through the superleak S_2 and returned to the mixing chamber M_1 through superleak S_1. The heat dissipated in M_2 by the demixing process is absorbed by a He^3 refrigerator (not shown) at about 0.4 K (after Staas, 1976).

tube between M_2 and M_1 is less than about 1.5 mm diameter. Equation (12.8) for the refrigeration rate in a conventional dilution refrigerator also applies to the He^4 circulating refrigerator, except that $\dot{n}_3$ is not measured directly. In terms of $\dot{n}_3$ the He^4 flow rate is

$$\dot{n}_4 = [(1.0 - 0.064)/0.064]\dot{n}_3 = 14.6\dot{n}_3 \qquad (12.16)$$

and the maximum refrigeration rate according to Eq. (12.8) is

$$\dot{Q} = 5.6\dot{n}_4 T^2 \text{ J/mol K}^2 \qquad (12.17)$$

Thus for the same refrigeration power as a conventional dilution refrigerator, the He^4 circulating refrigerator requires much higher flow rate. However high He^4 flow rates are easily achieved because of the superfluid property of He^4. In the first experiments by Taconis et al. (1971) $\dot{n}_4$ was 400 μmol/s and the lowest temperature reached was 50 mK. After subsequent improvements the lowest temperature was reduced to 7.9 mK with a circulation rate of $\dot{n}_4 = 1200$ μmol/s (Jurriens et al., 1978). In the improved versions the He^4 circulation is provided by a He^4 fountain effect pump (thermomechanical pump) operating at about 1.3 K. The fountain pump requires only a heat input (a few mW) and has no moving parts except for the liquid. Very high flow rates are possible with the fountain pump and so these He^4 circulating dilution refrigerators are candidates for large cooling power situations. In addition, the temperature of 7.9 mK was achieved without the complex heat exchangers required in conventional dilution refrigerators.

Another advantage of the He^4 circulating dilution refrigerators is the absence of a free liquid surface. This fact means that the liquid can be put under any pressure and still maintain operation. At a pressure of about 3.4 MPa (34 atm) the He^3 begins to solidify and further cooling occurs. This cooling is known as Pomeranchuk cooling and will be discussed later. Although this Pomeranchuk cooling is not continuous, the He^4 circulating dilution refrigerator is used to precool the liquid He^3 before solidification begins.

One of the limiting factors to the ultimate temperature of the He^4 circulating refrigerator is heat conduction through the liquid from the 0.4 K demixing chamber. Obviously a reduction in the temperature of the demixing chamber would improve the performance. Thus Staas et al. (1975) and Frossati et al. (1975) used a conventional dilution refrigerator instead of a He^3 refrigerator to cool the demixing chamber. However, their ideas were more ingenious than that of attaching a conventional mixing chamber to the outside of the demixing chamber. They realized that the demixing chamber itself could be used as the mixing chamber in a conventional dilution refrigerator because both the concentrated and dilute liquids are

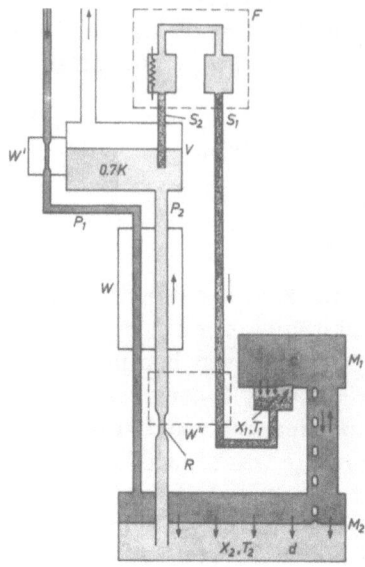

Fig. 12.21. Schematic of the double-circulating dilu-tion refrigerator in which both He^3 and He^4 are circulated. The demixing chamber M_2 also serves as the mixing chamber of a conventional He^3 circulat-ing dilution refrigerator. He^4 is extracted from the still by the fountain pump F. High circulation rates yield refrigeration powers in M_1 of 1 mW at 100 mK (after Staas, 1976).

already present. A schematic of the arrangement they used is shown in Fig. 12.21. The conventional dilution refrigerator introduces concentrated He^3 through tube P_1 into the chamber M_2 and removes dilute He^3 from M_2 through tube P_2. This dilution process in M_2 is counteracted by the demixing process that occurs because of the superfluid He^4 injection in M_1 as discussed previously. In practice (Staas *et al.*, 1975) these processes essentially cancel each other below 0.1 K so that the concentrated He^3 which enters M_2 from P_2 first detours to M_1 where it is diluted with He^4 and then returns via the droplets to M_2. The interconnecting tube between M_1 and M_2 serves as a heat exchanger without walls between the dilute and concentrated He^3. According to the acoustic mismatch model the Kapitza resistance between dilute and concentrated He^3 is about 2×10^4 times less than at a copper–helium boundary. Above 0.1 K only part of the He^3 makes the detour to M_1. The rest dilutes in M_2 and stabilizes the temperature of M_2.

The still V at 0.7 K is the source for the superfluid He^4 which is pumped into M_1 by the fountain pump F. Thus He^4 as well as dilute He^3 flows through the return line P_2. In fact the flow of He^4 exceeds the critical velocity at the constriction R and drags He^3 with it. The drag effect greatly enhances the He^3 flow rate compared with a conventional refrigerator. The first double circulation dilution refrigerator of Staas *et al.* (1975) used a He^4 flow rate of 2×10^{-2} mol/s and a He^3 flow rate of 10^{-3} mol/s. The temperature of M_1 reached about 10 mK and that of M_2 was 140 mK. The heat exchanger W was a simple coaxial tube arrangement. The maximum refrigeration rate in M_1 is given by Eq. (12.17), and with $\dot{n}_4 = 2 \times 10^{-2}$ mol/s the rate of 1 mW at 100 mK is larger than that of any existing conventional dilution refrigerator. After subsequent improvements, including the use of three discrete heat exchangers for W, this double circulating dilution refrigerator reached 4 mK in M_1 while M_2 was at 20 mK (Severijns and Staas, 1978). Further work on these double circulating dilution refrigerators will probably lead to temperatures around 2 mK before viscous heating and thermal conduction limits the lowest temperature. These refrigerators appear to be better suited for high refrigeration rates than do the conventional dilution refrigerators.

POMERANCHUK COOLING

Properties of He³ on the Melting Curve

The phase diagram of liquid–solid He³ is shown in Fig. 12.22 and the entropies of solid and liquid He³ are shown in Fig. 12.1. The entropy of the liquid at the melting pressure P_m is also indicated. These entropies are from the work of Halperin *et al.* (1978). Below 2.75 mK liquid He³ at P_m undergoes a transition to the superfluid A phase and at 2.18 mK transforms to the superfluid B phase. These ordering transitions cause the rapid decrease in liquid entropy. At zero pressure the B phase forms at 1.1 mK and no A phase exists as shown in Fig. 12.22. From a knowledge of the entropy curves for liquid and solid He³, Pomeranchuk (1950) predicted

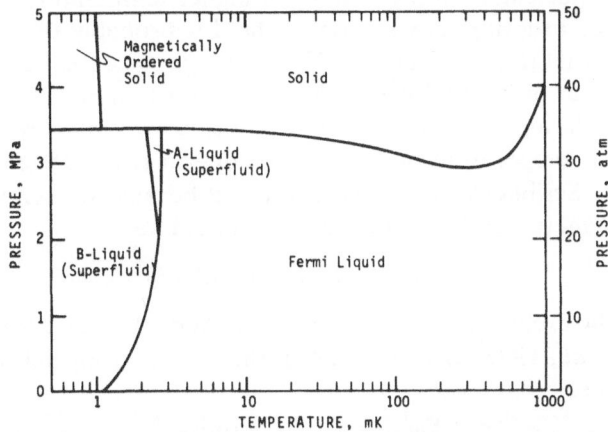

Fig. 12.22. Phase diagram of He³ showing the normal Fermi liquid, the two superfluid phases, and the ordered and disordered solid phases. Minimum in the melting curve occurs at 0.32 K.

that adiabatic solidification of He^3 would produce cooling. He also predicted the minimum in the melting curve using the Clausius–Clapeyron equation

$$\frac{dP_m}{dT} = \frac{S_l - S_S}{V_l - V_S} \qquad (12.18)$$

As shown in Fig. 12.1 the entropy of solid He^3 is greater than that of liquid He^3 below 0.32 K. This unusual set of circumstances is explained in detail by Landau and Lifshitz (1968). We make a brief explanation here. In liquid He^3 the atoms are indistinguishable from each other. Thus Fermi–Dirac statistics must be used to calculate the properties of these particles which have nuclear spin $\frac{1}{2}$. Because the atoms in the liquid are able to move about freely, their wave functions overlap considerably. The resulting interactions between atoms brings about an ordering process that lowers the entropy. In solid He^3 the atoms are distinguishable because they are frozen in a particular lattice position and do not interact with each other. Thus, classical Boltzmann statistics are used to calculate the properties. Because the atoms cannot move about there is little interaction between them and the entropy in zero field is simply that of a set of independent particles with nuclear spin $\frac{1}{2}$, or $R \ln 2$.

Pomeranchuk (1950) realized that at a sufficiently low temperature ($\sim 0.1 \mu K$) the dipole–dipole interactions would produce a nuclear ordering of the solid and a reduction in entropy. Thus Pomeranchuk thought $0.1 \mu K$ would be the lower limit for cooling by adiabatic solidification of He^3. However, it has been shown later (Nosanow and Mullin, 1965; Heterington, Mullin, and Nosanow, 1967) that because of the large zero point motion of He^3, there is a partial overlap of the wave functions that bring about nuclear ordering at a much higher temperature. The form of the exchange ordering in zero applied field is probably that of a nuclear antiferromagnet (Kummer et al., 1977) although the details of the ordering are not clear at present (Guyeer, 1978). As shown in Fig. 12.1 the nuclear magnetic ordering in solid He^3 occurs over a very narrow temperature interval at a temperature of 1.10 mK (Halperin et al., 1978). Below this temperature Scribner et al. (1969) show that because of spin waves in the antiferromagnetic solid He^3, the entropy should be

$$S_S/R = 6.87 \times 10^{-3}(kT/J)^3 \qquad (12.19)$$

where J is the exchange interaction constant. With the value $J/k = 0.70$ mK (Halperin et al., 1978; Greywall, 1976) the solid entropy below the transition becomes

$$S_S/R = 0.020T^3 \ (mK)^{-3} \qquad (12.20)$$

This temperature dependence is the same as that of the superfluid liquid entropy. Thus $S_S > S_l$ for all temperatures below 0.32 K. The lower limit for Pomeranchuk cooling is then simply the point where the solid entropy becomes very small, i.e., $T < 1.0$ mK.

In comparison with dilution refrigeration, Pomeranchuk refrigeration can offer a larger refrigeration power for the same number of moles of He[3] transferred for temperatures below about 50 mK. The entropy curves in Fig. 12.1 provide such a comparison. For *cooling* experiments along the melting line, i.e., cooling the refrigerant material, Pomeranchuk's technique is ideal. For *refrigeration* experiments, i.e., cooling other materials, problems with heat transfer through the solid formed at the warmest point offset the advantage of high cooling power. In addition, the lower limit of 1.0 mK for Pomeranchuk cooling is not much less than the 2.0 mK which can be reached with efficient dilution refrigerators. Since Pomeranchuk cooling must start at temperatures below about 100 mK, a dilution refrigerator is required anyway. Unlike the dilution refrigerator, the Pomeranchuk refrigerator cannot easily be made to operate continuously. All Pomeranchuk cooling devices to date are single-cycle devices and proceed to about 50% solid formation. Thus, it appears that applications of Pomeranchuk cooling are usually limited to studies of liquid and solid He[3] along the melting curve. Lounasmaa (1974) and Betts (1976) describe in detail various Pomeranchuk cooling devices. There have not been any significant changes or improvements in this area since the time these books were written. Consequently this section will be fairly brief.

Examples of Pomeranchuk Cooling

For many years there was little practical interest in Pomeranchuk's proposal of 1950 because of the thought that the mechanical compression of He[3] would produce too much heat. The work required to solidify He[3] is given by

$$W = P_m(V_l - V_S)n_S \qquad (12.21)$$

where V_l and V_S are the liquid and solid molar volumes along the melting line and n_S is the number of moles of solid formed. Since the heat absorbed is

$$Q = T(S_S - S_l)n_S \qquad (12.22)$$

the ratio of heat absorbed to the work input, using Eq. (12.18), is

$$\frac{Q}{W} = \frac{T(S_S - S_l)n_S}{P_m(V_l - V_S)n_S} = \frac{-T}{P_m}\frac{dP_m}{dT} \qquad (12.23)$$

This ratio ranges from a maximum of 0.077 at about 140 mK to about

0.0029 at 2.75 mK. Thus the compression of He^3 at 3 mK must be done with less than about 0.3% of the work being converted to frictional heat. With this discouraging thought, it is no wonder that 15 years elapsed before Pomeranchuk's proposal was even tried. Anufriyev (1965) was the first to try the method and succeeded in cooling the He^3 from 50 mK to about 18 mK, although the thermometers were not very accurate at temperatures below 20 mK. Adiabatic demagnetization was used to precool the He^3 since dilution refrigerators were not available then.

Compression of liquid He^3 in a cell cannot be done by pressurizing with He^3 gas at room temperature since according to Fig. 12.25 the solid first forms at a temperature of 0.32 K. That temperature would be somewhere along a capillary tube above the liquid cell, so the solid plug formed there would prevent any further pressure increase on the liquid in the cell. Anufriyev performed the compression by inserting inside the He^3 cell a second cell with flexible walls that was filled with liquid He^4. A pressure applied to the He^4 can then be transmitted to the He^3 via the flexible walls. Sydoriak *et al.* (1960) first developed this pressurization technique for He^3 studies on the melting curve above 0.3 K. However, He^4 solidifies at a minimum pressure of 2.5 MPa (25 atm), whereas He^3 requires at least 2.9 MPa (29 atm) to solidify. This difficulty is overcome by first pressurizing the He^3 cell to P_{min} with zero pressure on the He^4 cell. A subsequent pressure increase on the He^4 cell increases the He^3 pressure above P_{min} since the blocked He^3 capillary constrains the He^3 liquid in the cell.

While involved with development work on dilution refrigerators, Wheatley at La Jolla improved the Pomeranchuk cell design of Anufriyev. This improved cell, shown in Fig. 12.23a, reached temperatures somewhat below 2 mK after being precooled to 20 mK with a dilution refrigerator (Johnson and Wheatley, 1970). In this cell a flattened Cu–Ni tube with split Cu–Ni tubes soldered on each edge served as the inside flexible cell filled with He^3. The outside cell, pressurized with He^4, was mounted directly inside the mixing chamber of a dilution refrigerator. The thermal resistance at the cell walls and through the He^4 was so large that the heat leak from the mixing chamber to the Pomeranchuk cell was quite low. This same thermal resistance required 2–3 days to precool the He^3. Temperatures were measured with a powdered CMN magnetic thermometer and pressures were measured with a capacitance gauge. Figure 12.23b shows a recently developed Pomeranchuk cell of similar design to that of Fig. 12.23a in which the flexible membrane is made of plastic and a silver powder heat exchanger speeds up the precooling process. This cell was used by Chapellier *et al.* (1979) for studies of spin-polarized liquid He^3.

While work was underway at La Jolla, a similar effort was being made at Cornell. Their first Pomeranchuk device was described by Sites *et al.*

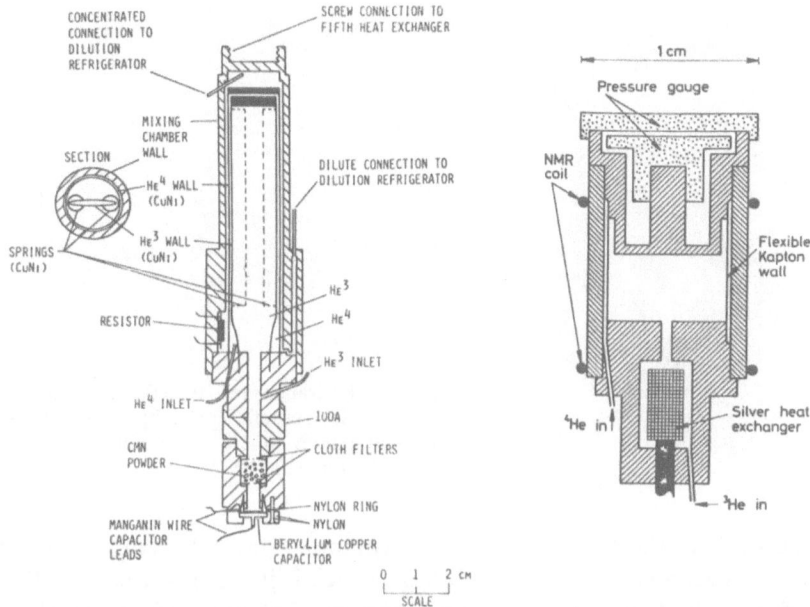

Fig. 12.23. (a) Pomeranchuk compression cell of Johnson and Wheatley (1970) which used a flattened CuNi tube for the flexible membrane to transmit pressure to He^3 inside the tube from He^4 outside the tube. (b) Miniature Pomeranchuk compression cell of Chapellier et al. (1979) which used a plastic film for the flexible wall between the He^4 and He^3 and a sintered silver heat exchanger connected to the mixing chamber of a dilution refrigerator.

(1969). Further modifications led to the apparatus described by Osheroff et al. (1972) and shown in Fig. 12.24. This arrangement employs a "hydraulic press" type of pressure amplifier which consists of two beryllium copper bellows of different diameters connected by a rigid rod. Temperatures below 2 mK were reached in this apparatus after precooling with a dilution refrigerator. Solidification rates as high as 300 μmol/s could be achieved without any noticeable irreversible heating for $T > 5$ mK. It was in this apparatus that the superfluid transitions A and B were discovered in liquid He^3 by Osheroff et al. (1972).

The Cornell group also used the same compression cell as a refrigerator to cool a liquid He^3 sample at zero pressure. The first Pomeranchuk refrigerator (Corruccini et al., 1972) used a bundle of fine copper wires to transfer heat from the liquid He^3 sample to the compression cell. This refrigerator was able to cool the He^3 sample to about 5 mK. A later version (Smith et al., 1975) used a corrugated copper foil spiral with a coating of sintered copper powder as the heat exchange surface in the compression cell. Copper wires attached to this exchanger extended through

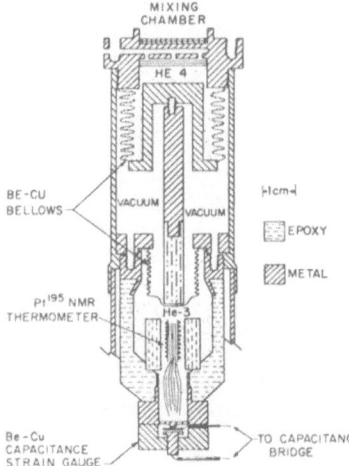

Fig. 12.24. Hydraulic press-type Pomeranchuk cell developed at Cornell University in which the superfluid phases of He^3 were first discovered (after Osheroff *et al.*, 1972).

the cell walls to the liquid He^3 sample volume. A sample temperature of 2.6 mK was achieved in this refrigerator, which is just below the superfluid transition at high pressures. It now appears that most of the work on superfluid He^3 away from the melting line must be done with magnetic refrigerators, which will be discussed in the next section.

The problem with long precooling times in Pomeranchuk cells has been overcome by using the He^4 circulating dilution refrigerator (not the double circulating dilution refrigerator) discussed earlier. Satoh *et al.* (1974) describes how the concentrated He^3 in the Leiden dilution refrigerator can be pressurized and solidified. They reach a temperature of 1.8 mK from a starting temperature of about 40 mK.

MAGNETIC REFRIGERATORS

Electron-Spin Systems

Magnetic cooling and refrigeration is the oldest technique for producing temperatures below 1 K. A fascinating account of the history of magnetic cooling is given by Hudson (1970). According to Hudson the technique was proposed independently by Giauque (1927) and by Debye (1926). Their ideas, however, were stimulated by the studies in 1918 of Weiss and Piccard on magnetocaloric heating in nickel (Weiss and Piccard, 1918) and studies in 1921 on magnetocaloric cooling in iron (Weiss, 1921). These experiments were done at room temperature and above, and the cooling produced was very small. Both Giauque and Debye realized that magnetic fields could remove large amounts of entropy in some paramagnetic materials at liquid helium temperatures but were puzzled by the source of entropy since the specific heat of materials at 1 K was very small. The concepts of entropy and the third law of thermodynamics were quite new at the time. It was Simon who pointed out to Debye that the large entropy in the paramagnetic materials in zero field must drop to zero at some low temperature and give rise to a large specific heat in that temperature range (Hudson, 1970). The temperature at which the zero-field entropy drops rapidly determines the low-temperature limit of magnetic cooling. Because of difficulties in building a helium liquefier, it was not until 1933 that Giauque and MacDougall (1933) were first able to experimentally verify the magnetic cooling concept, for which they won the Nobel Prize. They were able to cool gadolinium sulfate to 0.53 K after demagnetizing from $B = 0.8$ T at 3.5 K. Two and a half weeks later DeHass *et al.* (1933) reached 0.27 K when CeF_3 was demagnetized from 2.8 T at 1.3 K. A few months later temperatures below 0.1 K were reached in other materials and the method then became widely used for research below 1 K for the next three decades, until it was replaced in most cases by the dilution refrigerator. A lower limit of about 1 mK is possible by magnetic cooling of electron-spin systems, which, like the Pomeranchuk method, is not much less than that of a dilution refrigerator. However, magnetic cooling is quite simple and does not require a gravitational field. Thus cooling infrared detectors in space is a potential application for such refrigerators.

   Since little new work on magnetic cooling with electron-spin systems has been done in the last ten years, previous works can be consulted for more detail than is given here. The book by Hudson (1972) is the most complete treatment of this field and still up-to-date while the books by Betts (1976) and Lounasmaa (1974) are very informative. Mendoza (1961) gives some valuable experimental details and data on various paramagnetic salts. We give only a brief account here of the physics behind magnetic cooling.

   The spin associated with individual electrons in atoms couple via exchange interactions with other electron spins to form a resultant spin vector **S** whose magnitude is $\hbar[S(S + 1)]^{1/2}$, where S is the spin quantum number. Likewise the orbital motion of the electrons couple via electrostatic forces to form a resultant angular momentum vector **L** with magnitude $\hbar[L(L + 1)]^{1/2}$, where L is the quantum number. The combination of **S** and **L** gives a resultant vector **J** with magnitude $\hbar[J(J + 1)]^{1/2}$. When a magnetic field **B** is applied, **J** precesses about **B** at an angle θ and the component of **J** along **B** is given by $M\hbar$, where M is the magnetic quantum number that can take on a series of $(2J + 1)$ integrally spaced values from $-J$ to $+J$. Figure 12.25a shows a vector vodel of the relationship between M,

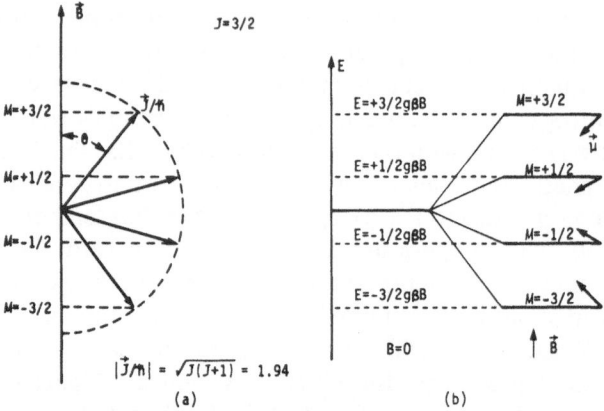

Fig. 12.25. (a) Vector model of the relationship between the magnetic field **B**, the angular momentum vector **J**, and the projection M for the case $J = 3/2$. (b) Energy level diagram of the magnetic dipole in a magnetic field.

$J/\hbar$, and **B** for $J = \frac{3}{2}$. The observable components of the angular momentum **J** are $M\hbar$. The magnetic dipole moment μ associated with **J** is in the opposite direction from **J** and given by

$$\mu = -(g\beta/\hbar)\mathbf{J} \tag{12.24}$$

where the Landé g-factor is

$$g = \frac{3}{2} + \frac{S(S+1) - L(L-1)}{2J(J+1)} \tag{12.25}$$

and the Bohr magneton is

$$\beta = e\hbar/2m_e = 9.274 \times 10^{-24}\,\text{J/T} \tag{12.26}$$

where m_e is the mass of the electron. For most paramagnetic materials $g \approx 2.0$. When a magnetic field is applied, the energy associated with each orientation of the magnetic dipole moment is

$$E = -\mu \cdot \mathbf{B} = (g\beta/\hbar)\mathbf{JB} = Mg\beta B \tag{12.27}$$

Figure 12.25b shows how the original energy level of the paramagnetic salt in $B = 0$ is split by the magnetic field into four separate levels for the case of $J = \frac{3}{2}$. This splitting is known as the Zeeman effect. For $B = 0$ and high enough temperatures each of the $(2J + 1)$ possible orientations of the dipole are equally probable and so the entropy of the paramagnetic salt is

$$S/R = \ln(2J + 1) \tag{12.28}$$

When the field is high enough such that the energy level spacing $g\beta B$ is much greater than the thermal energy kT, only the lowest energy state is occupied and μ is in the direction shown by Fig. 12.25b for that energy state. At that point the entropy has been reduced to zero. A finite value of T causes some of the higher energy levels to become populated and the distribution of spins between the various energy levels as well as the entropy are a function only of $X = g\beta B/kT$ as shown in Fig. 12.26 for three different values of J. Thus $X = \text{const}$ when $S = \text{const}$, and adiabatic demagnetization from an initial field B_i and temperature T_i to a final field B_f and temperature T_f must follow the law

$$B_f/T_f = B_i/T_i \qquad (12.29)$$

An initial condition of $B_i/T_i \approx 2T/K$ is needed to significantly reduce the entropy as is indicated in Fig. 12.26 when using the B/T axis for electron spins. A B/T axis for copper nuclear spins is also given in Fig. 12.26 and will be discussed later. From Eq. (12.29) it would appear that adiabatically reducing the applied field to zero would produce $T_f = 0$. However, due to

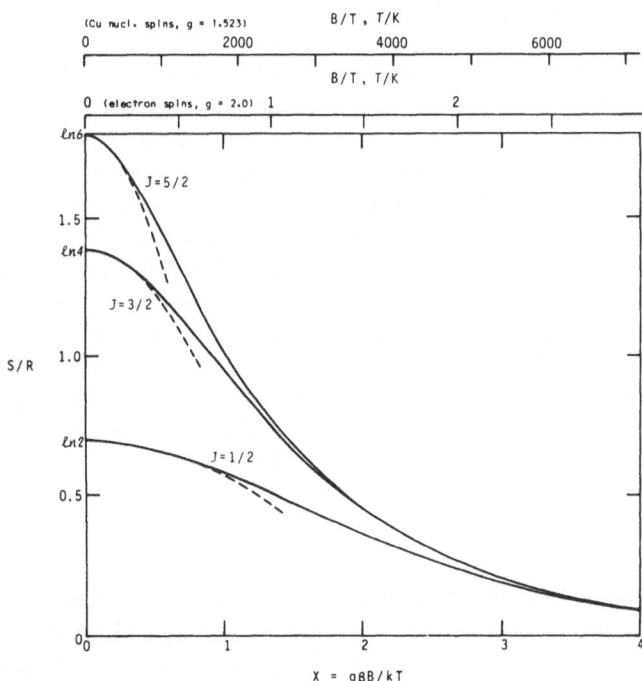

Fig. 12.26. Reduced entropy as a function of the dimensionless energy parameter $g\beta B/kT$ for different angular momentum quantum numbers J. Two top scales show the corresponding B/T values for electron spins with $g = 2.0$ and for copper nuclear spins with J substituted by I, the nuclear quantum number.

interactions between the magnetic moments there is some internal field B_{int} within the material that brings about ordering and entropy reduction at $kT \approx g\beta B_{int}$ for zero applied fields. For most commonly used paramagnetic salts B_{int} varies from 4 to 100 mT. The final temperature is determined from a modified form of Eq. (12.29) and is written as

$$T_f = (T_i/B_i)(B_f^2 + B_{int}^2)^{1/2} \qquad (12.29a)$$

Most materials have closed electron shells that give rise to $S = L = J = 0$. Such materials have no permanent magnetic moments and are diamagnetic instead of paramagnetic. These materials have no spin entropy. The main exceptions are the compounds of transition elements. In some of these materials the orbital contribution is zero ($L = 0$) so that $J = S$. Figure 12.27 shows the entropy as a function of temperature for three commonly used paramagnetic salts. These materials are cerium magnesium nitrate (CMN), chromic potassium alum (CPA), and ferric ammonium alum (FAA) and the entropies are from tabulations by Betts (1976). Except for slight upturns near 4 K, the entropy shown for these materials is due entirely to the magnetic moments. The effect of a 2-T field on the entropy of each of these materials is also shown, and is taken as that of an ideal paramagnet as shown in Fig. 12.26. These curves are easily scaled to other fields by using the fact that $S/R = $ const for $B/T = $ const, as long as $B \gg B_{int}$. For the case of CMN in a magnetic field, the entropy for powdered CMN is higher than that of a single crystal with the field parallel to the a axis because the value of g is small except in the a direction. Abel et $al.$ (1965) have calculated the entropy in a field for powdered CMN. The salt CMN has been widely used for refrigeration since it was first studied in 1953 (Daniels and Robinson, 1953) because of its low ordering temperature. Temperatures below 2 mK can be reached in CMN when it is adiabatically demagnetized from 2 T at about 1 K (see Fig. 12.27). Refrigeration of other materials by the CMN occurs as the CMN warms up at $B = 0$ and absorbs heat. Since the specific heat of CMN is

$$C = T\left(\frac{\partial S}{\partial T}\right) \qquad (12.30)$$

large amounts of heat are absorbed only where the entropy curve in Fig. 12.27 is steep, i.e., near the ordering transition. Refrigeration at a constant temperature is achieved by demagnetizing partially to reach some temperature above the ordering temperature. The material is then demagnetized to $B = 0$ at a rate such that the temperature is held constant as heat in the amount $Q = T\Delta S$ is absorbed. The volume of CMN needed to provide one mole of Ce^{+++} ions is 367 cm^3. Dilution of CMN with nonmag-

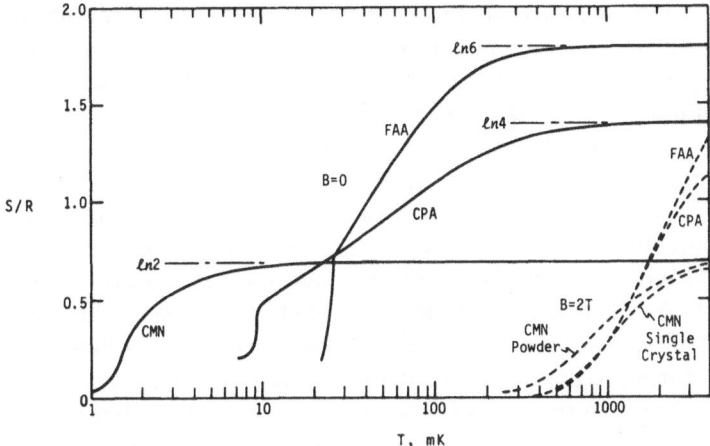

Fig. 12.27. Reduced entropy versus temperature for three electron-spin salts: CMN, cerium magnesium nitrate; CPA, chromic potassium alum; and FAA, ferric ammonium alum.

netic La reduces the ordering temperature below 1 mK but the refrigeration power per unit volume also decreases.

In cases where temperatures as low as 2 mK are not needed, other paramagnetic salts are available which provide much larger refrigeration powers. Figure 12.27 shows two common examples of materials used at higher temperatures. With CPA temperatures of about 10 mK can be reached and the $J = \frac{3}{2}$ value for CPA provides a rather large entropy of $S = R \ln 4$ at higher temperatures. In addition the molar volume of the Cr^{+++} magnetic ions in the salt is 273 cm^3/mol. If temperatures below 30 mK are not required the salt FAA provides even greater refrigeration power due to its $J = \frac{5}{2}$ value. The molar volume of the Fe^{+++} magnetic ions is 282 cm^3/mol. Figure 12.27 shows that only modest magnetic fields of 2–4 T are required to utilize all of these materials for refrigeration when starting from about 1.5 K.

As discussed in Chapter 11, refrigeration with these paramagnetic materials requires following a Carnot cycle, or some other cycle, around the entropy diagram in Fig. 12.27. In most experiments below 1 K the Carnot cycle is executed in a time domain, i.e., heat switches are used and the salt remains stationary. In most cases single-cycle operation is used in which case the material to be cooled is always in thermal contact with the salt and is cooled from about 1.5 K along with the salt. In simple cases helium exchange gas becomes the heat switch between the salt and the helium bath. The heat $Q = T_i \Delta S$ evolved from the salt during magnetization passes through the exchange gas to the helium bath. The exchange gas is pumped away to put the heat switch in the "off" condition before demagnetization is begun. Because of the difficulty in removing the exchange gas, superconducting heat switches are often used instead. At temperatures considerably below the superconducting transition temperature, the thermal conductivity in the superconducting state ($k \propto T^3$) is much less than that in the normal state ($k \propto T$) as shown in Fig. 12.28 for lead and aluminum. For lead a field of 80 mT drives it from the superconducting state to the normal state, whereas aluminum requires only 11 mT. Often the fringe field from the salt magnet is used to operate the switch. At 1 K the ratio k_n/k_s is only about 30 for lead, which is too low for an efficient heat switch between the salt and a 1 K helium bath. Thus a lower starting temperature of 0.3 K is often used with superconducting heat switches. Another possibility which has not been tried yet is the use of single-crystal beryllium as a magnetoresistive heat switch. Figure 12.28 shows the thermal conductivity of beryllium (Radebaugh, 1977) in zero field and in a 1-T transverse field as compared with lead (Childs *et al.*, 1973) and aluminum (Mueller *et al.*, 1978). The ratio $k(B = 0)/k(B)$ is about 1500 at 1 K and it remains high up to about 30 K. Note that this type of heat switch would operate in a manner opposite to a superconducting heat switch, i.e., a magnetic field is needed for the "off" state.

When CMN is to be used at about 2 mK, the $T\Delta S$ heat which can be absorbed at that temperature is quickly used up by heat leaks. Thus a

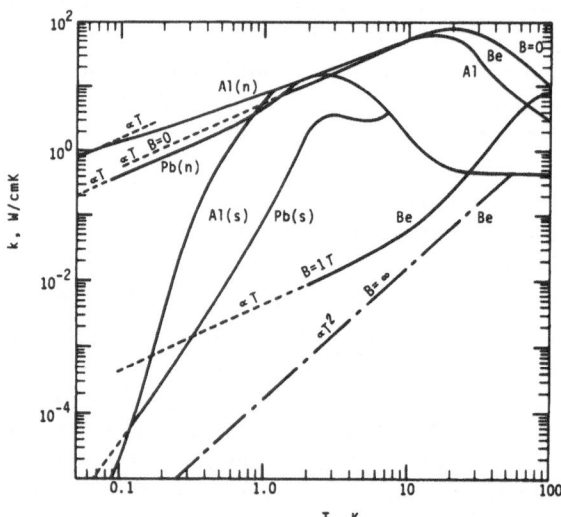

Fig. 12.28. Thermal conductivity of high-purity Al and Pb in the normal and superconducting states and of single-crystal Be in a zero field and a transverse field. Such materials are useful heat switches.

thermal guard of some other salt, e.g., CPA, must be placed between the CMN and the helium bath to absorb most of the heat leak. The use of a thermal guard does not mean the refrigerator becomes a two-stage device since the CMN is often demagnetized from the helium bath temperature along with the guard salt. A heat switch is still needed between the CMN and the guard salt. Figure 12.29 shows a schematic of an arrangement used by Wheatley (1966) where CMN and a CPA thermal guard salt are used for cooling He^3 to a few mK. In this case the upper reservoir was a 0.3 K He^3 refrigerator, which made thermal contact with the CPA through another superconducting heat switch not shown in the figure. When the very lowest temperatures were required, the CMN was demagnetized after the CPA, which makes the arrangement a two-stage refrigerator. The high surface area provided by the powdered CMN leads to excellent thermal contact with the liquid He^3. A demagnetization stage very similar to that shown in Fig. 12.29 was used in later experiments (Webb *et al.*, 1973) on superfluid He^3 away from the melting curve. At that time, however, a dilution refrigerator was used for precooling instead of a He^3 refrigerator.

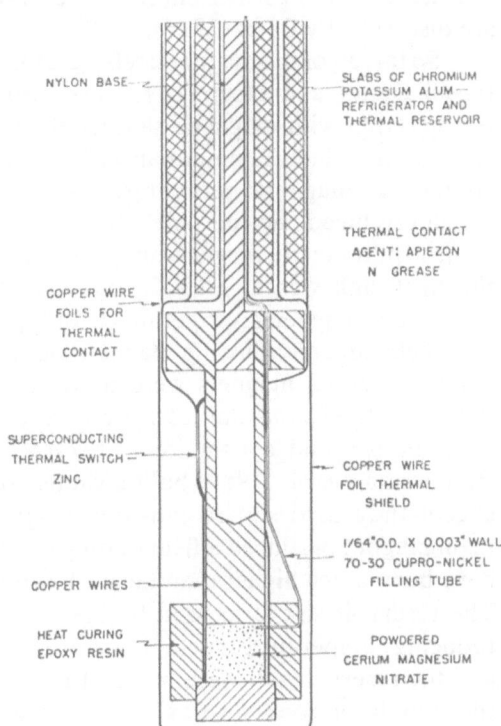

Fig. 12.29. Schematic drawing of an electron-spin magnetic refrigerator using CMN to cool He^3 to a few mK. A thermal guard is provided by slabs of CPA (after Wheatley, 1966).

Heat leaks to the low-temperature CMN refrigerant were generally on the order of a few nW.

In cases where materials other than liquid helium are to be cooled by a paramagnetic salt, heat transfer to the salt can be a problem. Thermal conductivities of these salts are low, which necessitates the use of thin sections or powder to reduce the ΔT between the inside and the surface of the salt. Sites *et al.* (1971) developed the following technique for making good thermal contact between copper wires and powdered CMN. Equal volumes of CMN powder and copper powder are mixed and packed around a bundle of copper wires. The assembly is compressed in a die using two moving punches with holes drilled in them for the copper wires to pass through. After a pressure of 275 MPa (2710 atm) is applied to the punches, the CMN–copper assembly is mechanically strong and any object to be cooled can be connected to the copper wires by soldering, welding, or other means. In later versions gold powder and gold-plated copper wires were used to eliminate corrosion of the copper by the CMN. Though the thermal contact is good, the electrical resistivity of the CMN–gold assembly is high enough to prevent excessive eddy current heating during demagnetization, at least for small CMN samples. Several other thermal contact techniques are discussed by Betts (1976).

So far, all of the magnetic refrigerators discussed have been single-cycle types. Daunt and Heer (1949) first proposed a scheme for continuous refrigeration with magnetic salts by the use of superconducting heat switches. Collins and Zimmerman (1953) built the first working model of a continuous magnetic refrigerator, but it used mechanical heat switches which produced considerable vibrational heat leak. Using a field of 0.19 T from a permanent magnet, they reached a temperature of 0.73 K while the heat sink was at 1.13 K. In 1954 Heer *et al.* (1954) built the first continuous magnetic refrigerator with superconducting heat switches. Since high-field superconducting magnets were not available at the time, water-cooled solenoid magnets were used and they were limited to a field of 0.7 T. With FAA as the refrigerant, a low temperature of 0.2 K was held continuously and a heat load of $7\,\mu W$ was absorbed at 0.26 K. Later Zimmerman *et al.* (1962) built a similar refrigerator, shown in Fig. 12.30, except they used superconducting magnets made of pure niobium. The main magnet produced a field of only 0.6 T but the overall efficiency would have been much higher than the earlier refrigerator using normal magnets. The reservoir salt, also used by Heer *et al.* (1954), is for the purpose of reducing temperature fluctuations at the cold end. Both the working salt and the reservoir salt were made of FAA and were grown in place around the bundle of copper wires. The heat switches were made of lead. This refrigerator reached a low temperature of 0.18 K and could absorb $100\,\mu W$

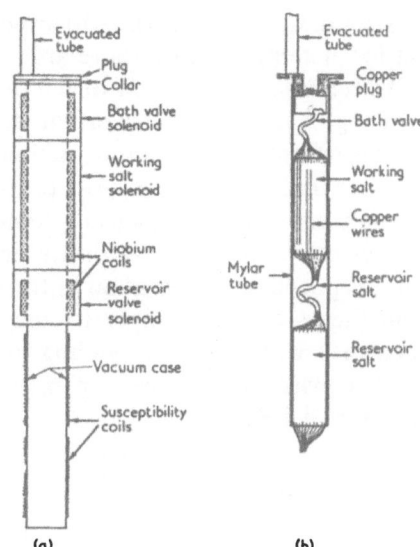

Fig. 12.30. Continuous magnetic refrigerator which uses FAA and lead heat switches, and is operated with superconducting magnets. The reservoir salt, also FAA, smooths out temperature fluctuations (after Zimmerman et al., 1962).

at 0.26 K. An electronic control system operated the salt magnet and the heat switch magnets in the appropriate sequence. A cycle time of 95 sec was used. A major source of inefficiency was in the upper lead heat switch.

A continuous magnetic refrigerator built by Rosenblum et al. (1976) with equal volumes of CMN powder and gold powder as discussed earlier maintained 10 mK while absorbing 0.1 μW of heat. In this case the upper reservoir was at about 0.1 K, which was the mixing chamber of a dilution refrigerator. Lead and tin were used for the upper and lower heat switches, respectively. The CMN operating field of 0.35 T was turned on and off with a long cycle time of about 110 min.

A quasicontinuous magnetic refrigerator designed for operation in space has maintained 0.2 K for 14 hr with a stability of ±0.5 mK (Kittel, 1980). The refrigerator is recycled to 1.3 K and back to 0.2 K in about 1 hr. A heat load of 20 μW can be absorbed at 0.2 K by the CPA refrigerant and given off to a 1.25 K helium bath. An electronic control circuit provides the proper demagnetization rate to maintain a constant temperature after a rapid demagnetization from 3 T to some finite field value. Helium exchange gas is used as the heat switch between the salt and the helium reservoir, but it is pumped away rapidly by using the salt as an adsorption pump. The ratio of heat absorbed to the theoretical capacity is reduced to about 90% because of the heat of adsorption of the exchange gas. Because the unit can operate in zero gravity, it can be used in space satellites for cooling the sensors in far infrared telescopes. Infrared telescopes cooled to 1.5 K will soon be launched by NASA as part of their Infrared

Astronomical Satellite (IRAS) program. Because of the increased sensitivity at lower temperatures, there are also plans to launch at a later date a far infrared telescope cooled to temperatures below 1 K. At the time of this writing, a magnetic refrigerator designed to operate at 0.1 K in a space environment is under construction at the NASA Goddard Space Flight Center. Dilution refrigerators probably could be made to operate in zero gravity environments by using some type of capillary confinement (Ostermeier *et al.*, 1978) for both the phase separation in the mixing chamber and vapor–liquid separation in the still. This possibility needs more study, but the associated complexity may not allow the dilution refrigerator to compete with magnetic cooling in space applications, which it has done with considerable success in most other areas of refrigeration between 2 mK and 1 K.

Nuclear-Spin Systems

One year after the successful demonstration of cooling by adiabatic demagnetization of electron-spin paramagnets, Gorter (1934) and Kurti and Simon (1935) suggested that nuclear spins could be cooled if a low enough starting temperature were used. The theory and thermodynamics are very much like that discussed in the previous section on electron-spin systems except that instead of J the appropriate quantum number I is that of the nuclear-spin system and the Bohr magneton of Eq. (12.26) is replaced by the nuclear magneton

$$\beta_n = e\hbar/2m_p = 5.051 \times 10^{-27} \, \text{J/T} \qquad (12.31)$$

where m_p is the mass of the proton. The nuclear magnetic dipole moment is

$$\boldsymbol{\mu}_n = (g\beta_n/\hbar)\mathbf{I} \qquad (12.32)$$

where the sign is opposite that for the electron moment in Eq. (12.24) and the nuclear g factor is usually about 2. The larger mass of the proton means that β_n, μ, and the energy level spacing in Eq. (12.27) are about 1836 times less than the corresponding parameter for the electron moment. Thus in order to populate mostly the lowest energy level in a magnetic field and reduce the entropy to near zero requires a much lower starting temperature than that needed for electron paramagnets. This more stringent requirement is also evident in Fig. 12.26 when β_n is substituted for β. The upper axis in Fig. 12.26 gives the B/T values for copper nuclei. For small $X = g\beta_n B/kT$ the nuclear spin entropy can be written as

$$\frac{S}{R} = \ln(2I + 1) - \frac{g^2\beta_n^2 I(I+1)}{6k^2} \frac{(B^2 + B_{\text{int}}^2)}{T^2} \qquad (12.33)$$

where the effective field is $(B^2 + B_{int}^2)^{1/2}$. The second term is often expressed in terms of the nuclear Curie constant λ, where

$$\lambda/R = g^2 \beta_n^2 I(I + 1)/3k^2 \qquad (12.34)$$

The entropy is then written as

$$\frac{S}{R} = \ln(2I + 1) - \frac{\lambda}{2R}\left(\frac{B^2 + B_{int}^2}{T^2}\right) \qquad (12.35)$$

For copper nuclei $I = \frac{3}{2}$ and $\lambda/R = 3.88 \times 10^{-7}$ K$^2/\mu^2$, which means that a B_i/T_i value of 1000 T/K would reduce the total entropy by only 14. Since $B_i = 10$ T is about the maximum field which can be reached at present, T_i must be 10 mK to give $B_i/T_i = 1000$ T/K.

On the other hand, the small value of μ_n means that the dipole–dipole interactions which bring about ordering at some low temperature are greatly reduced. This nuclear ordering temperature is generally on the order of $0.1\,\mu$K in the absence of quadrupole or exchange interactions. The value of B_{int} for copper nuclei is 0.31 mT compared with 4–100 mT for many electron paramagnets. The nuclear spin entropy of copper is shown in Fig. 12.1 for several values of applied magnetic field, including zero field. The copper nuclear molar volume is only 7.1 cm^3/mol, which means the magnetic dipoles are much more dense than those in the electron paramagnetic salts where molar volumes are on the order of 300–400 cm^3/mol.

Experimental facilities to provide the high B_i/T_i value necessary for nuclear cooling were not available at the time of the proposal in the mid-1930s. Thus about twenty years elapsed before the first experiments of Kurti et al. (1956) and Hobden and Kurti (1959) reached a nuclear temperature of about $1.2\,\mu$K in copper after demagnetizing from 3 T at 12 mK. In this experiment a bundle of copper wires (0.3 mole) was embedded at one end into a CPA precooling stage as shown in Fig. 12.31. No heat switch was used between the CPA salt and the nuclear stage. The magnetic fields for the CPA and nuclear stages were supplied by a water-cooled solenoid. The lowest temperature of $1.2\,\mu$K was that of the copper nucleus only—the copper electrons and lattice remained at the starting temperature of 12 mK. Heat leak to the nucleus from the electrons caused the nucleus to warm back up to the starting temperature in about 90 sec.

Thermal equilibrium between nuclei is established through spin–spin interactions which have a relaxation time τ_2. For copper $\tau_2 = 80\,\mu$s. Thermal equilibrium between the nuclear spins and the lattice and conduction electrons is established at the rate τ_1, the spin-lattice relaxation time. At low temperatures $\tau_2 \ll \tau_1$ and so we may speak of two distinct temperatures, the nuclear spin temperature T_n and the conduction electron temperature

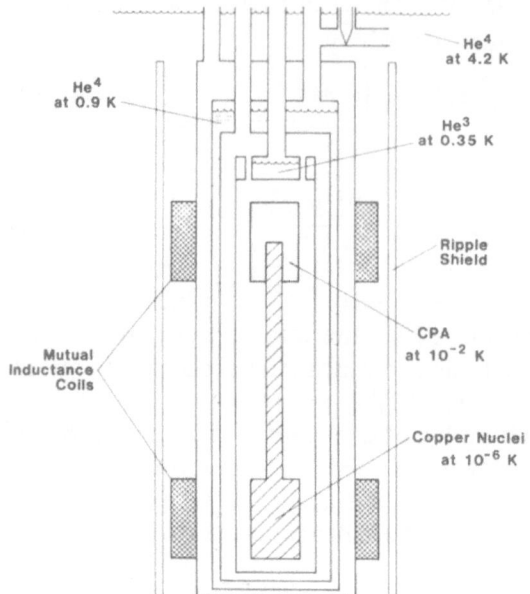

He^4 at 4.2 K

He^4 at 0.9 K

He^3 at 0.35 K

Ripple Shield

CPA at 10^{-2} K

Mutual Inductance Coils

Copper Nuclei at 10^{-6} K

Fig. 12.31. Schematic of the cryostat used by Kurti *et al.* (1956) for the first experiments on nuclear demagnetization. CPA salt precools the copper stage. CPA was also used in place of He^3 in the first experiment.

T_e. The spin-lattice relaxation time is defined by

$$\frac{d}{dt}\left(\frac{1}{T_n}\right) = -\frac{1}{\tau_1}\left(\frac{1}{T_n} - \frac{1}{T_e}\right) \qquad (12.36)$$

where τ_1 follows the Korringa (1950) relation

$$\tau_1 T_e = \kappa \qquad (12.37)$$

in which κ is the Korringa constant. For copper

$$\kappa = 1.1\left(\frac{B^2 + B_{int}^2}{B^2 + 2.7B_{int}^2}\right)sK \qquad (12.38)$$

The value of κ for copper varies from 0.4 sK at $B = 0$ up to its high field value of 1.1 sK at about B = 10 mT. Most other metals suitable for nuclear demagnetization have high field κ values that range from 1.8 sK for Al to 0.006 sK for Tl (Betts, 1976). The value of τ_1 at 10 mK is then about 100 sec or less, which permits precooling the nuclei to 10 mK by normal means. Insulators have much larger τ_1 times—the order of days or weeks at 10 mK. Thus precooling these nuclei to 10 mK is not possible by ordinary techniques. A special technique for precooling these nuclei is discussed later. For T_e held constant at T_i the warmup rate for T_n from Eqs. (12.37) becomes

$$dT_n/dt = \dot{T}_n = (T_n/\kappa)(T_i - T_n) \qquad (12.39)$$

It is obvious from Eq. (12.39) that a reduction in T_i decreases the warmup rate for T_n.

In Kurti's experiment the conduction electrons were held at 12 mK by thermal contact through the lattice to the CPA salt. Suppose now that the electrons were isolated from the precooling stage by a perfect heat switch. If the demagnetization were done slowly enough, T_e would follow T_n and the process would be reversible. The final temperature reached by T_n and T_e is then found by using entropy curves as shown in Fig. 12.1 for copper. However, the electron entropy, which has the value

$$S_e/R = 8.4 \times 10^{-5} \, (\mathrm{K^{-1}})T \qquad (12.40)$$

must be added to the nuclear entropy. For starting temperatures of about 10 mK the electron entropy is so small compared with the nuclear entropy that it has a negligible effect on the total entropy and the final temperature unless the B_i/T_i value used is so small that the nuclear entropy reduction is not much larger than the electron entropy. The demagnetization rate for reversible cooling ($T_e - T_n \approx 1 \, \mu K$) of the electrons in copper from 3 T and 12 mK is about 66 mT/s, which corresponds to a total demagnetization time of 45 sec.

For completely irreversible cooling of the electrons, i.e., the electrons start cooling from T_i after T_n has reached its lowest temperature, an energy balance is used to find the final temperature. The internal energy of the copper electrons is

$$\frac{E_e}{R} = \int (C_e/R) \, dT = 4.2 \times 10^{-5} \, (\mathrm{K^{-1}})T^2 \qquad (12.41)$$

which must be absorbed by the $T_n \Delta S$ refrigeration of the nuclear spins. For $T_n = \mathrm{const} = 1 \, \mu K$ the nuclear entropy increase in cooling the electrons from 10 mK is $\Delta S/R = E_e/RT_n = 4.2 \times 10^{-3}$, which is at least an order of magnitude smaller than the entropy change produced by typical B_i/T_i values. Thus it makes little difference whether the electrons are cooled reversibly or irreversibly for final temperatures around $1 \, \mu K$. Kurti (1967) treats in detail the reversible and irreversible cooling of electrons.

The first experiments where a heat switch was used between the precooling stage and the copper nuclear stage were performed in 1965 by March and Symko (1965) and a few months later by Goodkind's group at LaJolla (Osgood and Goodkind, 1966). These two cryostats can be called the first nuclear refrigerators since they were used to cool something other than the nuclei. In principle both refrigerators were very much the same as the first nuclear cooling apparatus of Kurti's shown in Fig. 12.31. The main difference was the use of a superconducting heat switch between the copper nuclear stage and the electron paramagnet. In the first experiment by March and Symko the electron paramagnet was CPA and lead was used for the heat switch. They demagnetized from 20 mK and 4.8 T and reached a final electron and lattice temperature of about 7 mK. The demagnetization was stopped at a finite field to keep T_n higher and provide more $T_n \Delta S$ refrigeration power. The demagnetization cryostat of Osgood and Goodkind (1966) used CMN for the precooling stage and indium for the heat switch. In their first experiments they reached an electron temperature of 5.3 mK after demagnetization from 13.2 mK and 3 T. They stopped the demagnetization at 0.75 T, which would have made $T_n = 3.3$ mK from Eq. (12.29). Later they demagnetized from 5.7 T at 13 mK and reached an electron temperature of 3 mK. In cases where more complete demagnetization was carried out, the nuclear temperature remained belw 1 mK for seven hours. The same refrigerator was used to cool liquid He^3 in contact with the 0.8 moles of copper to a temperature of 4 mK (Osgood and Goodkind, 1967). Later the He^3 and CMN stages were replaced with a dilution refrigerator and a sintered copper powder cell was attached to the nuclear stage for better thermal contact with the He^3. A liquid He^3 temperature of about 0.7 mK was achieved (Dundon et al., 1973) and solid He^3 was cooled to 0.9 mK (Dundon and Goodkind, 1974). Residual heat leaks in this apparatus were about 1 nW.

In nuclear cooling experiments where a minimum in T_n is desired the thermal resistance between the nuclear spins and the surroundings is made as large as possible so that the nuclei can remain cold for long periods of time. For nuclear refrigeration of a sample the thermal resistance between the spins and the sample is made as small as possible. Residual heat leaks are never zero, and much of this heat leak must pass through the total thermal resistance between the sample and the nuclear spins. This heat leak means that a steady state temperature difference must exist between the sample and the nuclear spins. A high thermal resistance also means a longer sample cooldown time, during which time the nucleus can warm up or increase its entropy. In cooling a sample such as liquid He^3, there are several thermal resistances in series that make up the total thermal resistance R_t between the He^3 sample and the nuclear spins: (a) the Kapitza boundary resistance R_K between the He^3 and the cell walls, (b) the electronic thermal resistance R_e between the electrons in the cell and the electrons surrounding the cooled nuclei, and (c) the thermal resistance R_{en} between the electrons and the nuclear spins.

**Table 12.1. Kapitza Resistance
between Several Metals and Pure
Liquid He^3 for $T < 10$ mK**

Metal	$a_K = R_K \sigma T$ $(\text{m}^2 \text{K}^2/\text{W})$
Copper	625
Silver	222
Gold	122
Palladium	200

The Kapitza resistance R_K is usually proportional to T^{-3}, but below about 10 mK magnetic coupling effects give rise to $R_K \propto T^{-1}$ as first seen by Avenel *et al.* (1973). For a surface area σ the Kapitza resistance can be expressed as

$$R_K = a_K/\sigma T \tag{12.42}$$

where a_K is a constant. Table 12.1 lists a_K for several metals, taken from the work of Avenel *et al.* (1973), Andres and Sprenger (1975), Edwards *et al.* (1978), and Ahonen *et al.* (1978). Silver is often used as the cell material because of its low specific heat compared with copper at these temperatures. High surface areas provided by sintered fine powder are commonly used to reduce R_K. However, there is some evidence that in very fine powders the value of a_K is increased somewhat. Harrison (1980), in an excellent review of Kapitza resistance below 100 mK, gives a table of a_K values in fine powders.

The resistance R_e is simply related to the electronic thermal conductivity k_e of the metal by the relation

$$R_e = L/k_e A \tag{12.43}$$

where L is the distance that the electrons must transfer the heat and A is the cross-sectional area. For metals below 1 K, $k_e = a_e T$, where a_e is a constant. Thus

$$R_e = L/A a_e T \tag{12.44}$$

Table 12.2 lists a_e for several metals that may be used for nuclear refrigeration. The lower limit shown is easily obtained but the high limit requires very high purity.

The resistance R_{en} is evaluated by using the definition of a thermal resistance

$$R_{en} \doteq \Delta T/\dot{Q}_n \tag{12.45}$$

Table 12.2. Properties of Metals Suitable for Nuclear Refrigeration

Metal	$a_e = k_e/T$ (W/cm K^2)	I	τ_2 (μs)	λ/R (10^{-7} K^2/T^2)	κ (s K)	κ/λ (10^5 T^2 mol/W)	V_m (cm^3/mol)	B_{int} (mT)	B_c (mT)	T_f (μK) $B_i/T_i = 10^3$ T/K
Cu	1–10	3/2	80	3.88	1.1	3.4	7.11	0.31	—	0.31
In	1–10	9/2	—	16.6	0.086	0.062	15.7	250	28	250
Al	1–20	5/2	30	8.3	1.8	2.6	9.98	—	10	10
Sn	1–10	1/2	100	0.23	0.030	1.6	16.3	—	30	30
V	0.1	7/2	—	15.2	0.79	0.63	8.34	—	141	141
Nb	0.1	9/2	—	20.8	0.19	0.11	10.9	—	206	206
Tl	1–10	1/2	30	3.5	0.006	0.021	17.2	—	18	18
Pt	1–5	1/2	1000	0.17	0.030	2.1	9.1	—	—	<1
AuIn$_2$	0.1–0.2	9/2	—	16.6	0.089	0.065	20.5	—	1.7	1.7

(Per mole In)

as long as $\Delta T/T \ll 1$. Heat flow to the nuclei $\dot{Q}_n$ is given by

$$\dot{Q}_n = nC_n\dot{T}_n \tag{12.46}$$

where n is the number of moles of nuclei, C_n is the nuclear specific heat, and $\dot{T}$ is the temperature rise rate from Eq. (12.39). Since $C_n = T\,\partial S_n/\partial T$, Eq. (12.35) gives

$$\frac{C_n}{R} = \frac{\lambda}{R}\left(\frac{B^2 + B_{\text{int}}^2}{T_n^2}\right) \tag{12.47}$$

substituting Eqs. (12.46) and (12.47) into Eq. (12.45) and using Eq. (12.39) for $\dot{T}_n$ gives

$$R_{en} = \frac{\kappa\Delta T}{nC_nT_n\Delta T} = \left(\frac{\kappa}{\lambda}\right)\Big/nT_n\left(\frac{B^2 + B_{\text{int}}^2}{T_n^2}\right) \tag{12.48}$$

For refrigeration purposes adiabatic demagnetization is carried out to a finite field, thus $(B^2 + B_{\text{int}}^2)/T_n^2 \approx (B_i/T_i)^2$ at the start of the heat absorption process. As heat is absorbed during further demagnetization at constant T_n, the term $(B^2 + B_{\text{int}}^2)/T_n$ decreases and R_{en} increases. The minimum R_{en} is

$$R_{en}(\text{min}) = (\kappa/\lambda)/nT_n(B_i/T_i)^2 \tag{12.49}$$

This resistance can be decreased by increasing B_i/T_i and using a large number of moles of a material with a minimum κ/λ. Table 12.2 lists κ/λ for several metals useful for nuclear refrigeration. For copper

$$R_{en}(\text{min}) = 3.4 \times 10^5 (\text{T}^2 \text{ mol W}^{-1})/nT_n(B_i/T_i)^2 \tag{12.49a}$$

All three resistances have a T^{-1} dependence which means that the three RT values can be added to give a R_tT value. Such a total resistance is meaningful only for $\Delta T/T \ll 1$. In practice the ΔT is often large, so the ΔT caused by each resistance must be calculated separately. To determine $T_e - T_n$ when $\Delta T/T$ is not small we substitute Eq. (12.39) into Eq. (12.46) to get

$$\dot{Q}_n = n(T_e - T_n)T_n\frac{\lambda}{\kappa}\frac{B^2 + B_{\text{int}}^2}{T_n^2} \tag{12.50}$$

Since the term $(B^2 + B_{\text{int}}^2)/T_n^2$ is a constant $= (B_i/T_i)^2$ immediately after adiabatic demagnetization, $\dot{Q}_n$ is a maximum for a given T_e when $(T_e - T_n)T_n$ is a maximum, i.e., $T_n = T_e/2$. Hence the maximum $\dot{Q}_n$ is

$$\dot{Q}_n(\text{max}) = \frac{T_e^2\lambda(B_i/T_i)^2}{4(\kappa/\lambda)} \tag{12.51}$$

This equation is used to find T_e for any given heat leak. With the maximum heat flow condition of $T_n = T_e/2$, the resistance R_{en} may be expressed in terms of T_e from Eq. (12.49) as

$$R_{en}T_e = \frac{2(\kappa/\lambda)}{n(B_i/T_i)^2} \qquad (12.52)$$

If $\Delta T/T \ll 1$ for the temperature drops across R_e and R_K, then the R_{en} in Eq. (12.52) can be added to them for a total resistance. The relationship between heat flow and resistance is simply

$$\dot{Q} = \Delta T/R \qquad (12.53)$$

When $\Delta T/T$ across R_e and/or R_K are not small, then the heat flow is determined from

$$\dot{Q} = \int_{T_L}^{T_H} R^{-1}(T)\, dT = \frac{(T_H^2 - T_L^2)}{2RT} \qquad (12.54)$$

where RT = const.

We now can calculate the temperature to which the electrons and a sample of liquid He^3 can be cooled by copper nuclei in the presence of a heat leak. Let us use a heat leak $\dot{Q} = 1$ nW and the typical conditions shown in Table 12.3. From Eq. (12.51) we calculate that $T_e = 11.7\ \mu$K for the electrons surrounding the nuclei. The electron temperature in the walls of the He^3 cell is calculated from Eq. (12.54) using $RT = 20$ K^2/W, which gives T_e(cell) $= 200\ \mu$K. The He^3 temperature from Eq. (12.54) is $T_3 = 245\ \mu$K. These numerical examples point out the importance of using materials with very high thermal conductivity. Thus the copper used in nuclear refrigeration should be purer than commercial copper wire which has $a_e = 1$ W/cm K^2. Values of at least $a_e = 5$ W/cm K^2 are feasible for small wires. Heat leaks much below 0.1 nW per mole of copper do not seem feasible at present. Such a heat leak corresponds to the sample falling at the rate of 10 nm/s with the potential energy being converted to heat instead of kinetic energy. Thus vibrations can be a serious source of heat leak. Usually nuclear refrigerators have special mountings to reduce vibrational heating.

Table 12.3. Typical Thermal Resistances between Liquid He^3 and Copper Nuclei

$n = 10$ mol, $B_i/T_i = 1000T/K$	$R_{en}T_e = 0.068$ K^2/W
$L/A = 20$ cm^{-1}, $a_e = 1$ W/cm K^2	$R_eT_e = 20$ K^2/W
$\sigma = 20$ m^2, $a_K = 200$ m^2 K^2/W	$R_KT_e = 10$ K^2/W

Because He^3 becomes a superfluid below 2.75 mK there is considerable interest in experiments on liquid He^3 below 2 mK and into the submillikelvin range. Nuclear refrigeration is better suited to this task than any other type of refrigerator, which has been the reason for much of the recent development on nuclear refrigerators. Following the pioneering work of Goodkind discussed earlier, the group at Helsinki under Lounasmaa has significantly advanced the status of nuclear refrigeration as well as nuclear cooling. The work on refrigeration of copper electrons was discussed by Gylling (1971) and by Berglund *et al.* (1972). In this work 12 moles of copper were used to cool the electrons to 0.37 mK from a starting condition of $T_i = 16$ mK and $B_i = 5$ T. With a heat leak of 12 nW temperatures below 2 mK could be maintained for 12 hr. The early Helsinki work on nuclear refrigeration of liquid He^3 was discussed by Ahonen *et al.* (1976). This group achieved a He^3 temperature of 0.7 mK using 22 moles of copper wires as the refrigerant with initial conditions of $T_i = 17$ mK and $B_i = 7.5$ T. Sintered copper powder was used in the He^3 cell. The total heat leak was 5 nW, of which only 0.5 nW was to the He^3. The sum of electron and Kapitza resistances $R_e T + R_K T$ was 160 K^2/W.

In order to reduce the heat leak to the nuclear refrigerator, the Helsinki group built a "double-bundle" nuclear refrigerator in which a second nuclear stage surrounds the first and acts as a thermal guard to intercept heat leaks as was commonly done in electron-spin refrigeration systems. This work is described by Veuro (1978) and Ahonen *et al.* (1978). A general view of this refrigerator is shown in Fig. 12.32a and details of the "double bundle" are shown in Fig. 12.32b. A tin heat switch separates both bundles from the mixing chamber and a lead heat switch separates the inner-bundle refrigerant (10 moles) from the outer-bundle thermal guard (15 moles). Both bundles are demagnetized simultaneously and the lead heat switch is operated by the main magnet. Heat leak to the inner bundle was reduced to 1.2 nW, of which about 20% entered the He^3 sample. The He^3 cell was made of silver and contained 13 m^2 of sintered silver powder. The high-purity copper wires used for the bundle had $a_e \approx 2$ W/cm K^2. With a value of $(R_e + R_K)T = 90$ K^2/W they were able to cool the He^3 sample to 0.38 mK from starting conditions of $T_i = 18$ mK and $B_i = 5.3$ T (rms applied field). The temperature stayed below 1 mK for 50 hours.

Lower He^3 temperatures were achieved at Orsay (Varoquaux, 1978) using a single 30 mole nuclear stage of very high-purity copper wires ($a_e = 7$ W/cm K^2). With $(R_K + R_e)T = 50$ K^2/W they reached a He^3 temperature of 0.35 mK in the presence of a 3-nW heat leak (1 nW to the He^3). The lowest He^3 temperature reached to date (0.21 mK) was performed at the Bell Laboratories by Osheroff and Sprenger (1980). Their refrigerator, shown in Fig. 12.33, used 15 moles of copper and starting conditions of $T_i = 17\,\text{mK}$ and $B_i = 6.4\,\text{T}$ rms. The success of their refrigerator is due to the low total heat leak of 1.5 nW to the refrigerator and the low thermal resistance of $(R_K + R_e)T = 20\,\text{K}^2/\text{W}$. The low R_e is obtained by using many pure silver ($\rho = 2 \times 10^{-9}\,\Omega\,\text{cm}$ at 4.2 K) strips between the copper nuclear bundle (11) and the sample chamber (6). The entire assembly is made demountable by using squeeze-type heat plugs to connect the copper bundle, silver strips, and sample chamber together. These heat plugs are a modified version of that described by Boughton *et al.* (1967) where the thermal contraction of a nylon compression collar

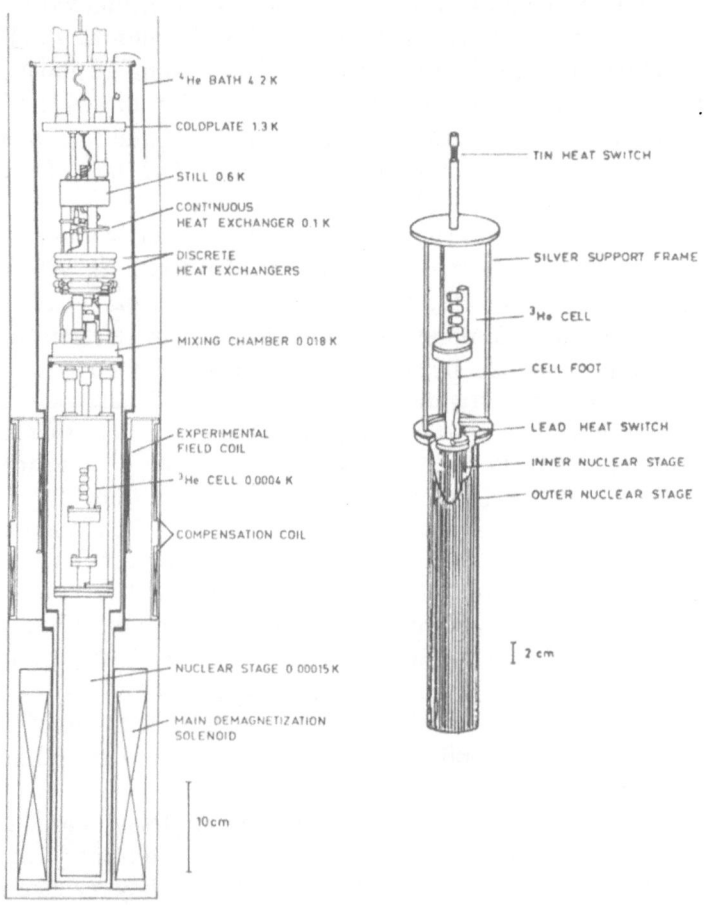

Fig. 12.32. (a) General view of the nuclear refrigerator at Helsinki (Veuro, 1978) used to cool He^3 to 0.38 mK. (b) Detailed view of the nuclear stages, known as a double bundle. The two stages are demagnetized simultaneously, but the outer stage acts as a thermal guard.

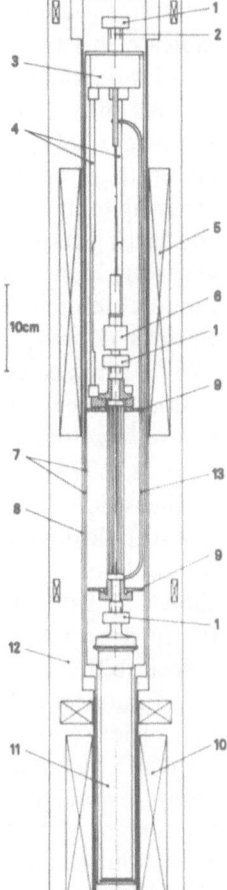

Fig. 12.33. Schematic of the nuclear refrigerator built by Osheroff and Sprenger (1980) of Bell Laboratories to cool He^3 to a record low temperature of 0.21 mK. 1, Compression collar for a heat plug; 2, coldfingers from 10 m^2 heat exchanger in mixing chamber; 3, mixing chamber with annular hole down central axis; 4, phenol fiber thermal isolation rods; 5, NMR magnet; 6, one of many interchangeable sample chambers; 7, 1 K and mixing chamber thermal shields; 8, exchange gas can; 9, graphite positioning flange; 10, demagnetization magnet; 11, copper nuclear bundle; 12, He^4 bath; 13, 3-mm-diam silver rod for thermal contact to heat switch.

squeezes a split outer member against the inner member. The outer and inner member are generally gold plated to enhance the heat transfer. The modified version used by Osheroff and Sprenger uses the low thermal contraction of tungsten to provide a force from within the inner member. Tungsten segments held between the outer member and a BeCu collar also apply force from outside. These all-metal heat plugs have electrical resistances of $(1-2) \times 10^{-8} \, \Omega$ at 4 K, which is an order of magnitude lower than previous designs. With this refrigerator Osheroff and Sprenger have been able to cool pure He^3 at saturated vapor pressure to 0.21 mK and a 5%

He^3–He^4 solution to 0.59 mK. Superfluidity of the He^3 in a He^3–He^4 solution should occur at a low enough temperature and some calculations indicate this transition may occur as high as about 10^{-4} K. Thus there is now interest in cooling He^3–He^4 solutions as well as pure He^3.

As shown in Table 12.2 there are several metals which may be used as nuclear refrigerators. The refrigeration powers of several of these are larger than copper but they have certain disadvantages such as higher ordering temperatures and manufacturing problems. Those that are super-conducting require a final field above the critical field B_c since τ_1 becomes very long in the superconducting state. The last column in Table 12.2 shows the final temperature reached after initial conditions of $B_i/T_i = 10^3$ T/K. Besides copper, only indium has been studied experimentally (Symko, 1969; Hunik *et al.* 1978a, b). Because of a nuclear quadrupole interaction, indium nuclei order at about 0.5 mK. However, the refrigeration capacity per unit volume above this temperature is about twice that of copper. Andres (1978) has suggested the use of the compound $AuIn_2$ for nuclear refrigeration. It has the cubic CaF_2 structure, which means that the nuclear quadrupole effects seen in pure In are absent in this compound. The volume per mole of In nuclei is increased only somewhat over that of pure In as shown in Table 12.2. This material thus appears to be a good candidate for nuclear refrigeration but so far it has been used only as a nuclear susceptibility thermometer (Andres and Bucher, 1972).

Equation (12.39) shows that in order to reduce the warmup time of the nuclei in nuclear *cooling* experiments, the electron temperature must be made as small as possible. Also if one is interested in nuclear tem-peratures much below 1μK, the initial entropy reduction must be a large fraction of $R \ln 4$ (see Fig. 12.1). To accomplish such a large entropy reduction would require B_i/T_i values of about 5000 T/K. For $B_i = 7$ T it would be necessary to have $T_i = 1.4$ mK. Only nuclear refrigeration can be used to provide such low starting temperatures since all other refrigerators have too small refrigeration powers at 1 mK. This two-stage nuclear approach was recently attempted by Lounasmaa's group at Helsinki to achieve a record low nuclear temperature of 50 nK (Ehnholm *et al.*, 1979). Their two-stage nuclear demagnetization cryostat is shown in Fig. 12.34. Details of this cryostat are given by Ehnholm *et al.* (1980). A brief review of this work as well as refrigeration below 1 K is given by Lounasmaa (1979). Both nuclear stages are precooled with a powerful dilution refrigerator (He^3 flow rate of 0.4 mmol/s) to about 10 mK. A tin heat switch separates the first nuclear stage and the mixing chamber. Both nuclear stages are made of a bundle of insulated high-purity copper wires. The first stage contains 10 moles of copper and the second stage contains 0.031 moles (2 g) of copper. No heat switch is used between the two stages

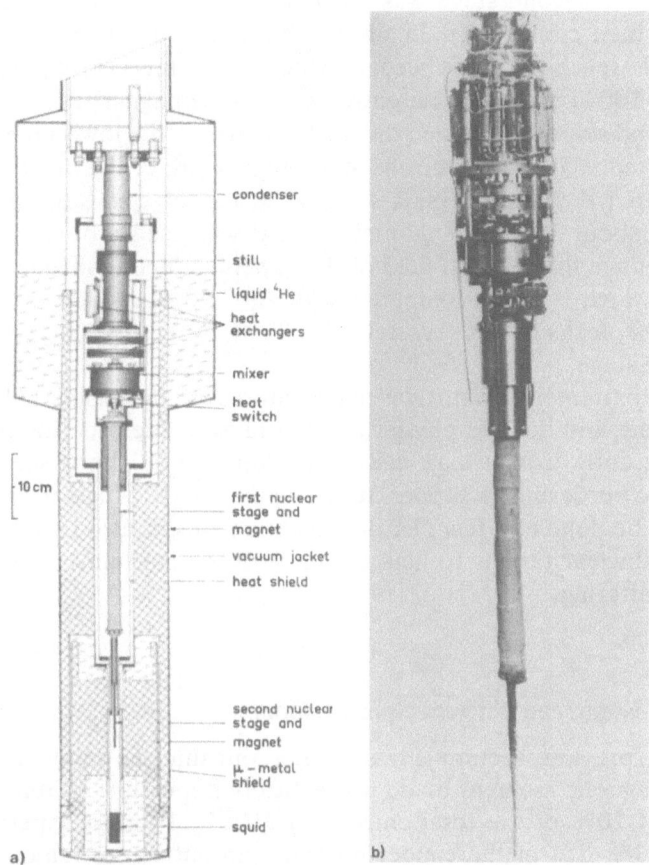

Fig. 12.34. (a) Schematic of the two-stage copper nuclear demagnetization cryostat at Helsinki (Enholm *et al.*, 1980) used to cool copper nuclei to a record low temperature of 50 nK. (b) Photograph of same apparatus.

so that a heat leak to the electrons in the lower stage can be transferred to the first stage.

The experimental procedures used to cool the nuclei are as follows. With a field of 8 T on the first stage and the heat switch turned on the two stages were cooled to 10 mK by the dilution refrigerator. With a refrigeration rate of about 0.6 μW at 10 mK it takes the dilution refrigerator about 6 hr to remove the 10 mJ heat of magnetization $T \Delta S$. After the heat switch was turned off the first stage was demagnetized to 0.1 T at the

same time the second stage was magnetized to 7 T. The second nuclear
stage was then cooled to 0.25 mK by the first stage. The value $B_i/T_i =$
28,000 T/K reached in this second stage is then high enough to remove
practically 100% of the nuclear entropy. Rapid demagnetization ($\sim$10 min)
of the second stage then cooled the nuclei to 50 nK, the lowest temperature
ever reported. With the electrons at a temperature of 0.2 mK the nuclear
warmup rate from Eq. (12.39) is 0.025 nK/s and the spin-lattice relaxation
time τ_1 is about 30 min. A μ-metal shield was used around the second
stage to reduce the residual field of the superconducting magnet to below
0.02 mT, which is small compared with $B_{int} = 0.31$ mT for copper. The
residual heat leaks in this cryostat were 3 nW to the first stage and 0.3 to
the second stage.

These extremely low nuclear temperatures are of considerable interest
to solid state and nuclear physicists for studies of the ordering process in
nuclei. Recently Lounasmaa and co-workers (1983) have seen antiferromagnetic bordering of copper nuclei below 63 nK. Nuclear orientation,
which can be done at a few mK in magnetic fields, is also used to provide
polarized nuclear targets to help understand the mechanisms involved in
nuclear scattering.

Hyperfine Enhanced Nuclear-Spin Systems

In the previous section it was pointed out that the small value of the
nuclear magnetic moment made it practically impossible to remove more
than about 10% of the total entropy at 10 mK. In most experiments to
date only 1%–2% of the copper nuclear-spin entropy is removed with a
dilution refrigerator precooling stage. Because so little entropy is removed
before demagnetization, the total refrigeration capacity of copper is limited.
With 1 mole of copper a heat leak of 1 nW can use up the available
refrigeration at 10 μK in about 30 min. Much higher entropy changes
would be possible if the nuclei could be subjected to fields in excess of
10 T. Such large entropy changes would require long precool times at
10 mK with a dilution refrigerator since the refrigeration capacity at such
a temperature is quite low. Thus higher precooling temperatures would be
desirable. As indicated by Eq. (12.35) the entropy reduction is proportional
to B^2. In atoms with unfilled electron shells the electronic magnetic
moments can produce a very large magnetic field at the site of the nucleus.
This hyperfine field, often the order of several hundred teslas, is high
enough to remove most of the nuclear entropy at 100 mK or higher.
However, in most of these materials the hyperfine field is permanent and
cannot be removed or controlled by some external means.

It was Al'tshuler (1966) who first recognized that nuclear cooling should be possible in materials where the hyperfine field is induced by an applied field. In such materials, known as Van Vleck paramagnets, the electronic magnetic moment is proportional to the applied field, which leads to a hyperfine field that is proportional to the applied field, i.e.,

$$B_{hf} = KB \qquad (12.55)$$

where the constant K is known as the Knight shift. The total field then becomes

$$B_t = B_{hf} + B = (1 + K)B \qquad (12.56)$$

and the term $(1 + K)$ is called the hyperfine enhancement (hfe) factor, which can be the order of 10 to 100 in useful materials. All of the equations presented in the previous section on nuclear-spin systems can be used for the Van Vleck paramagnetic materials by substituting $(1 + K)B$ for B. Another approach to the explanation of these materials is that the apparent magnetic moment of the nucleus is one to two orders of magnitude larger than the nuclear magneton β_n but one to two orders of magnitude less than the Bohr magneton β. Dipole–dipole interactions between these magnetic dipoles would then cause spontaneous nuclear ordering at about 10–$100 \, \mu K$. When exchange interactions are present, the ordering occurs around 1 mK.

Van Vleck paramagnetism occurs only in materials that contain ions with an even number of electrons. The paramagnetic salts discussed previously contain ions with an odd number of electrons. Other considerations discussed by Andres and Bucher (1968, 1971, 1972), limit the number of useful materials to the intermetallic compounds of praseodymium and thulium. In the case of metals the spin-lattice relaxation time is short and the thermal conductivity is good. Table 12.4 lists the useful properties of some of the materials studied or considered to date. Most of the data are based on many years of work by Andres (1978) and co-workers (Bucher et al., 1971) at Bell Laboratories.

The values for T_{min} represent the lowest temperatures reached with that material to date (the value for $AuIn_2$ is estimated) and T_c is the ordering temperature. The materials Cu, $AuIn_2$, and CMN are shown only for comparison.

The first experiments on hyperfine enhanced nuclear cooling were completed in 1967 and 1968 by Andres and Bucher (1968). The first material tested was TmSb, but only heating occurred when it was demagnetized from 40 mK and 1.6 T. It was concluded that irreversible heating canceled the nuclear cooling effect. A slight cooling effect was first seen in PrSb, but the first successful experiment was done on PrBi, where a final

Table 12.4. Properties of Van Vleck Paramagnets Used for Hyperfine Enhanced Nuclear Refrigeration

Material	$1+K$	I	λ/R $(10^{-7}\,K^2/T^2)$	V_m (cm^3/mol)	S/V_m $(J/cm^3\,K)$	$\Delta S/V_m$ $(J/cm^3\,K)$ $(B_i/T_i = 300\,T/K)$	$\Delta S/S$ %	T_{min} (mK)	T_c (mK)
TmSb	86	1/2	0.0976	33.9	0.170	0.161	95	—	—
TmBi	79	1/2	0.0976	35.7	0.161	0.149	93	10	—
TmCd	157	1/2	0.0976	—	—	—	99.9	—	—
PrSb	10.8	5/2	11.4	38.8	0.384	0.287	75	—	—
PrBi	12	5/2	11.4	40.6	0.367	0.290	79	10	10
PrMg	7.1	5/2	11.4	35.3	0.422	0.238	56	—	—
PrTl$_3$	20	5/2	11.4	64.4	0.231	0.217	94	1.5	1
PrIn$_3$	5.7	5/2	11.4	61.4	0.243	0.112	46	1.7	—
PrPt$_5$	23.5	5/2	11.4	65.5	0.227	0.219	96	3	—
PrNi$_5$	12.2	5/2	11.4	51.0	0.292	0.247	85	3.19	0.4
PrCu$_6$	12.5	5/2	11.4	62.8	0.237	0.207	87	2.2	2
PrBe$_{13}$	8.7	5/2	11.4	83.9	0.178	0.117	66	0.85	—
Cu	1.0	3/2	3.88	7.1	1.62	0.0204	1.3	0.00005	0.0001
AuIn$_2$	1.0	9/2	16.6	20.5	0.934	0.0303	3.2	(0.0017)	—,
CMN	1.0	1/2	3.81×10^6	367	0.0157	0.0157	100	0.6	1.6

temperature of 10 mK was reached with starting conditions of 40 mK and 2.0 T. After considerable persistence and patience in developing and testing new materials, Andres has advanced the state of hyperfine enhanced nuclear refrigeration to the point where lattice temperatures below 1 mK can now be achieved. Figure 12.35a shows the demagnetization apparatus Andres and Darack (1974) used to cool $PrTl_3$ to a temperature of 1.6 mK from starting conditions of 25 mK and 2.0 T. The lattice temperature is measured with the $AuIn_2$ nuclear susceptibility thermometer located some distance from the $PrTl_3$ sample. Figure 12.35b shows a new insert to the refrigerator in Fig. 12.35a in which $PrNi_5$ rods are used to cool a sample of liquid He^3 to a temperature of 1.0 mK (Andres and Darack, 1977). The $PrNi_5$ reached a temperature of 0.8 mK from initial conditions of 17 mK and 2.3 T. Further work on testing and improving materials for hyperfine enhanced nuclear refrigeration is now continuing at several laboratories. The lowest temperature reached so far in these materials is 0.19 mK in $PrNi_5$ by Mueller *et.al.* (1980) from starting conditions of 10 mK and 6 T. With those conditions almost 100% of the $R \ln 6$ entropy is removed by the field. Because

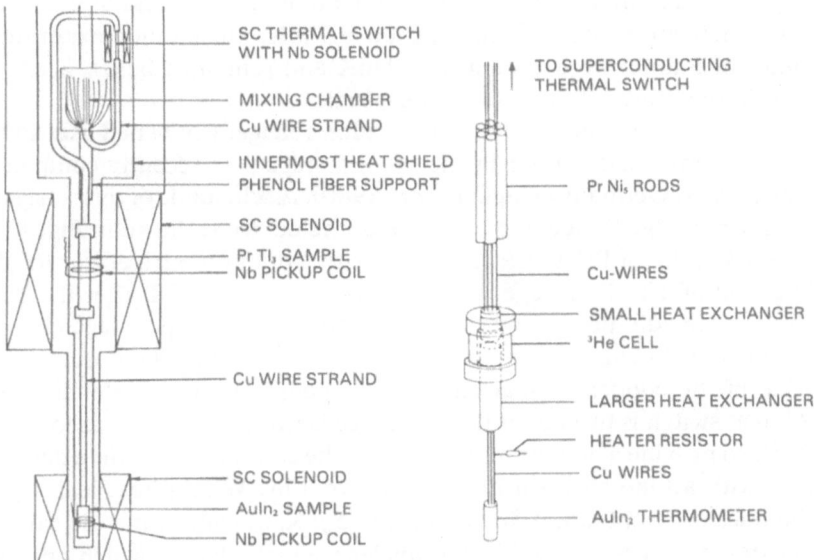

Fig. 12.35. (a) Hyperfine enhanced nuclear refrigerator of Andres and Darack (1974) at Bell Laboratories which uses the Van Vleck paramagnet $PrTl_3$. The $AuIn_2$ nuclear susceptibility thermometer recorded a low temperature of 1.6 mK. (b) New hyperfine enhanced nuclear insert of $PrNi_5$ used by Andres and Darack (1977) in the refrigerator of (a) to cool He^3 to 1.0 mK.

of the large cooling power per unit volume of these Van Vleck paramagnets, they are now being used as the first stage of a two-stage nuclear refrigerator to cool copper electrons below 50 μK. These combined systems are discussed in the next section. To illustrate just how large the refrigeration power of these materials are, we consider the case of 100 cm^3 PrNi$_5$ which has had 85% of its available entropy removed at 20 mK by a 6-T field. This material can then absorb a 1-nW heat leak at 1 mK for nearly 8 months!

Combined Systems

The precooling of nuclear refrigerants to temperatures of about 1–5 mK appears to be best done by using a hyperfine enhanced nuclear stage which in turn is precooled by a dilution refrigerator. This temperature is then low enough to remove a significant fraction of the copper nuclear entropy. According to Table 12.4 all of the copper nuclear entropy could be removed with a volume of PrNi$_5$ that was 6.6 times that of the copper stage. A copper precooling stage would require a volume 79 times the second copper stage.

Hunik *et al.* (1978a, b) first used an enhanced nuclear system of PrCu$_6$ to precool a 2 mole indium nuclear stage to 8 mK in a field of 6.3 T. After demagnetization the indium reached 0.7 mK. In another experiment 8 moles of indium were cooled to 0.9 mK and remained below 2 mK for 3 days in the presence of a 16-nW heat leak.

A two-stage nuclear demagnetization refrigerator which uses PrNi$_5$ in the first stage and copper in the second stage was recently completed in Jülich, West Germany (Mueller *et al.*, 1980). A schematic of this refrigerator is shown in Fig. 12.36a and a picture of the device is shown in Fig. 12.36b. The 4.3 moles of PrNi$_5$ are precooled with a dilution refrigerator to 25 mK in a field of 6 T. This stage is then demagnetized to 0.2 T after the Al heat switch to the dilution refrigerator is turned off. The copper stage (10 moles) is cooled in a field of 8 T by the PrNi$_5$ stage to 5.5 mK. At these conditions 23% of the copper nuclear entropy has been removed. After the second Al heat switch is turned off the first stage is further demagnetized to below 2 mK to provide a better thermal guard. The copper stage is then demagnetized with a time constant of the order of 1 hr. With a heat leak of 1 nW the nuclear temperature finally reaches 5 μK, at which time the calculated electron temperature within the nuclear stage is 9 μK, which appears to be the lowest electron temperature reached to date. Thermometers mounted in the experimental space are in electronic thermal contact with the nuclear stage through a resistance $R_e T = 0.93$ K^2/W. The calculated electron temperature in the Pt NMR thermometer is then 44 μK, which agrees well with the measured value of 48 μK. Electron temperatures below

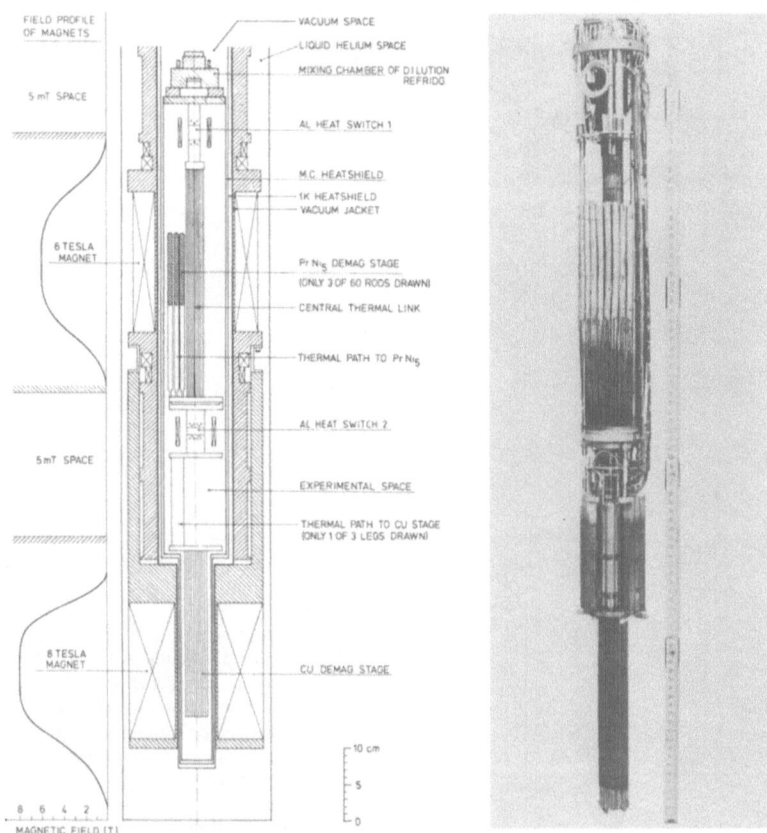

Fig. 12.36. (a) Schematic of the two-stage nuclear refrigerator of Mueller *et al.* (1980) at Jülich which uses a PrNi₅ hyperfine enhanced stage to precool a 640-g copper nuclear stage. Electron temperatures below 50 μK were measured in the experimental space. (b) Photograph of same refrigerator.

60 μK can be maintained for a few days because of the slow, 2-μK/day, rise of the nuclear temperature.

Figure 12.37 shows another two-stage nuclear demagnetization refrigerator which used $PrCu_{7.2}$ for the first stage. The nonstoichiometric material $PrCu_{7.2}$ in less brittle than $PrCu_6$. This refrigerator built by Ono *et al.* (1980) of the University of Tokyo uses only 13 g of $PrCu_{7.2}$ (0.022 moles) and 1.2 g of copper (0.019 moles). The $PrCu_{7.2}$ stage is precooled by a dilution refrigerator to 15 mK in a field of 5.5 T through a lead heat switch. Demagnetization of this stage precooled the copper stage to about 3 mK in a field of 5.5 T. Thus the copper entropy was reduced by more

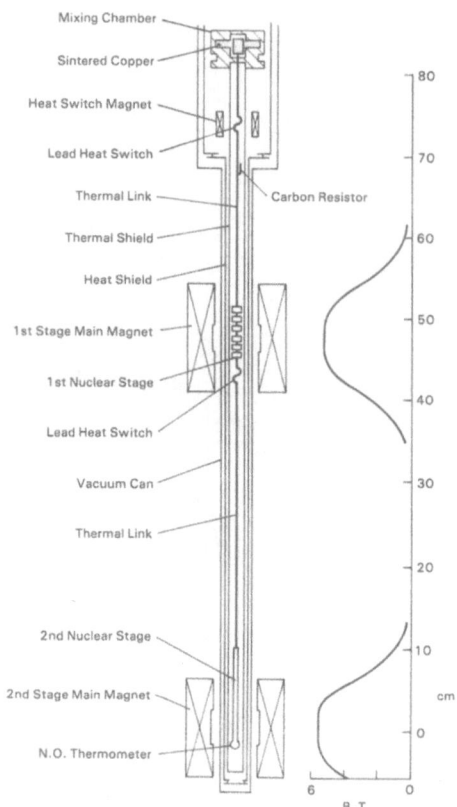

Fig. 12.37. Schematic of the two-stage nuclear refrigerator of Ono *et al.* (1980) at Tokyo which uses a PrCu$_{7.2}$ hyperfine enhanced stage to precool a 1.2-g copper nuclear stage. Electron temperatures below $50\,\mu$K were achieved in the nuclear orientation (N.O.) thermometer.

than 30%. Demagnetization of the copper stage to 30 mT produces a calculated nuclear temperature of $22\,\mu$K. The electron temperature was estimated to be $(37 \pm 14)\,\mu$K. The thermometer used was an AlMn$_{54}$ nuclear orientation thermometer (N.O. thermometer) located within the main magnetic field. It is pointed out by Ono *et al.* (1980) that this thermometer is actually sensing the temperature of the Al nuclear spins, which are demagnetized along with the copper sample. The measured temperatures are then always less than the electron temperature in the copper. The value $T_e = 37\,\mu$K is based on some calculations involving the heat leak and the $22\,\mu$K measured by the thermometer. A temperature rise rate as low as $0.3\,\mu$K was observed in some cases and heat leaks to the copper nuclear stage were as low as 0.03 nW, which corresponds to 1.6 nW/mol.

The results obtained in these nuclear refrigerators at Jülich and Tokyo indicate that electron temperatures of the order of $10\,\mu$K can be reached with electrons in close proximity to the nuclei. Such low electron temperatures are of interest in studies of superconductivity. So far all the existing superconductors have the simplest type of ordering—ordinary S-wave pairing. The much more sophisticated and interesting P-wave pairing has never been observed, but it may occur in some metals below 1 mK.

Dynamic Nuclear Polarization

The spin-lattice relaxation time τ_1 in insulators is very long—the order of days or weeks at 10 mK. Such long relaxation times mean that these materials in bulk form cannot be used for refrigeration purposes. However, Gates and Potter (1975) have shown that in fine powders τ_1 is shorter and have proposed the use of such powders for nuclear refrigeration of He^3. For studies of nuclear ordering in insulators the long τ_1 can be a great advantage once the nuclei are polarized at about 10 mK. A subsequent demagnetization then cools the nuclei to the μK temperature range where they begin to order. The long τ_1 gives experimenters a long time to study the ordered nuclei before they warm up significantly. The problem with conventional demagnetization experiments, sometimes called the brute force technique, is that the long τ_1 at 10 mK prevents the magnetized nuclei from being cooled to this temperature in a reasonable time.

An elegant technique, originally called dynamic nuclear polarization by the solid effect (Abragam and Proctor, 1958), but now better known as just dynamic nuclear polarization (Abragam and Goldman, 1978a) exists whereby the nuclei in insulators can be substantially polarized in a magnetic field at temperatures as high as 0.5 K. Paramagnetic electronic impurities are introduced into the sample at concentrations of the order of 10^{-4} and because of the short electron spin lattice relaxation time they have no difficulty in reaching the temperature of the lattice (and of the refrigerator), which may be about 0.3 K and obtained with a He^3 refrigerator. When a dc field B_i is applied to the sample the electronic magnetic moments are polarized to nearly 100% whereas the polarization of the nuclear magnetic moments is practically zero. According to Eq. (12.27) the energy levels of the electronic and nuclear magnetic moments split as $E_e = Mg\beta B_i$ and $E_n = Mg\beta_n B_i$. Figure 12.38 shows a diagram of the energy levels as a function of the field B_i for the case of $M = \pm\frac{1}{2}$ in both the electron system since $\beta \gg \beta_n$. Nearly all of the electronic moments are in the low-energy state ($S_e = 0$), whereas both levels of the nuclear moments are equally populated ($S_n = R \ln 2$). In order for a nuclear moment to change from the higher to the lower energy state the nuclear Zeeman energy $g\beta_n B_i = \hbar\omega_n$ must be absorbed by some other system. The lattice does not absorb this

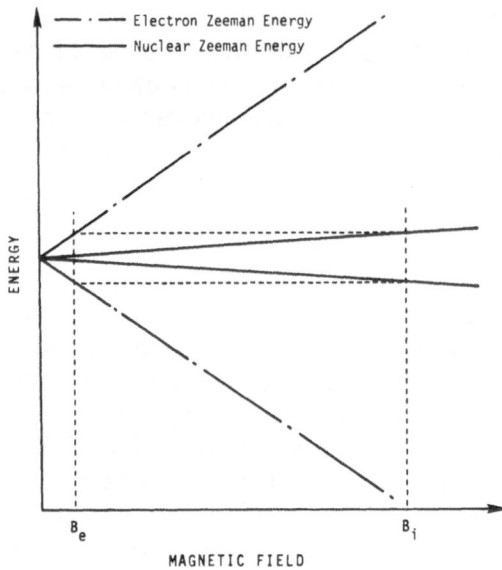

Fig. 12.38. Diagram which shows the electron and nuclear Zeeman energies as a function of magnetic field. With an applied field of B_i the electron and nuclear levels strongly interact if the effective field on the electrons is reduced to B_e.

energy easily because of the long τ_1. In addition the electron spins will not absorb this energy since they require a much greater energy $g\beta B_i = \hbar\omega_e$ to change states. The electron spins would absorb the nuclear Zeeman energy if the effective field B_e on just the electron spins were reduced to the point where their level splitting was equal to the splitting of the nuclear spins in the field B_i as shown in Fig. 12.38. The field reduction on the electron spins is accomplished by applying to the sample a small magnetic field of amplitude B_0 rotating in a plane perpendicular to B_i with an angular frequency ω_0 which is the same order of magnitude as ω_e. In a frame of reference rotating at ω_0, the Larmor frequency of the electron spins is reduced to $\omega_e - \omega_0$ and the spins behave as if they were seeing a small dc field only: a dc field parallel to B_i but of amplitude $\hbar(\omega_e - \omega_0)/g\beta$ and a dc field of amplitude B_0 perpendicular to B_i. This technique is called adiabatic demagnetization in the rotating frame (ADRF). For B_0 small enough the effective field is parallel to B_i, and when $\omega_0 = \omega_e - \omega_n$, the effective field B_e seen by the electron spins is such that their level splitting shown in Fig. 12.38 equals the level splitting of the nuclear spins. The energy released by a nuclear spin going from the high-energy state (spin opposite to the field direction) to a low-energy state (spin parallel to the field direction) can then be absorbed by the electron spin system and transferred to the lattice. The simultaneous change of spin direction in the electron and nuclear moments is known as a flip–flop transition.

The explanation given above for dynamic nuclear polarization can also be explained in a manner similar to the photon refrigerator discussed in

Chapter 11. First we consider the upper half of Fig. 12.39. The initial condition at (a) shows an electron magnetic moment and several nuclear magnetic moments in the field B. The sample is then irradiated with microwaves (~ 100 GHz). The energy of the incoming photons is $\hbar(\omega_e - \omega_n)$. As shown in position (b) each photon causes an electron moment to flip from the parallel to the antiparallel direction and a nuclear moment to flop from the antiparallel to the parallel direction. The electron moment then quickly comes back to equilibrium with the lattice in (c) by emitting a photon of energy $\hbar\omega_e$. The energy difference $\hbar\omega_n$ between the outgoing photons and the incoming photons comes from the nuclear spins, which results in a reduction of their entropy and a polarization in the field direction (positive polarization). For $I = \frac{1}{2}$ the equilibrium nuclear polarization is given by

$$P_n = \tanh{(g\beta_n B/2kT)} \tag{12.57}$$

This polarization is related to the fraction of spins n_1 in the lower energy level and the fraction of spins n_2 in the higher energy level by

$$P_n = n_1 - n_2 \tag{12.58}$$

The fractions n_1 and n_2 are

$$n_1 = e^x/(e^x + e^{-x}) \tag{12.59}$$

$$n_2 = e^{-x}/(e^x + e^{-x}) \tag{12.60}$$

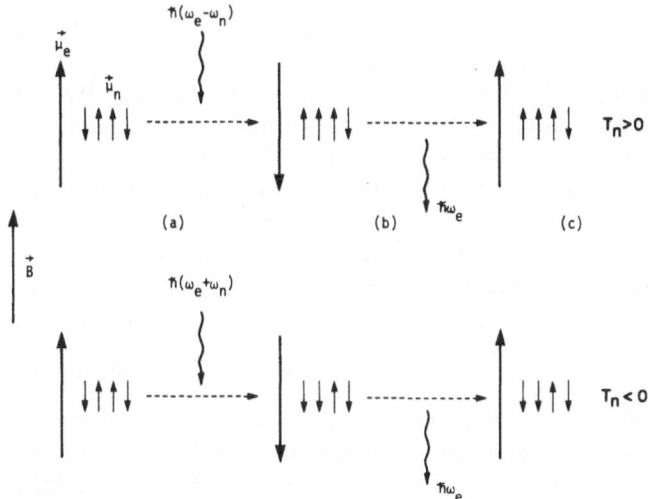

Fig. 12.39. Sequence of events, (a), (b), and (c) in which positive or negative nuclear polarizations and temperatures are produced by microwave irradiation. Process is known as dynamic nuclear polarization.

where $x = g\beta_n B/kT$. From Eq. (12.57) the degree of polarization yields some corresponding nuclear spin temperature. For instance, for the case of protons where $I = \frac{1}{2}$ and $g = 2$, a polarization of 50% in a field of 2.5 T yields a spin temperature of 5 mK, which is sufficiently low for a starting temperature for adiabatic demagnetization.

A conventional demagnetization of the nuclei can be carried out to give final nuclear temperatures on the order of 1 μK. However, in low fields the energy difference between the electron spin system and the nuclear spin system is small, which leads to a short nuclear spin relaxation time. Thus demagnetization is best accomplished in the rotating frame. Thus a rotating magnetic field is applied to the sample again, but this time rotating at the nuclear Larmor frequency ω_n, which is of the order of 100 MHz. The effective final applied field in this demagnetization is the magnitude of the rotating field, which may be the order of or smaller than the internal field B_{int} of the nucleus and yield final temperatures of about 1 μK or lower.

An interesting situation occurs if in the dynamic nuclear polarization (DNP) of the sample, a microwave frequency of $\omega_e + \omega_n$ is used instead of $\omega_e - \omega_n$, as shown in the lower half of Fig. 12.39. The radiation is then absorbed when both the nuclear and electronic spins are flipped from a direction parallel to the field to a direction opposite to the field, i.e., from the low- to high-energy state. This is known as a flip–flop transition. Continued radiation will cause the fraction of spins n_2 in the higher energy state to be greater than the fraction of spins n_1 in the low-energy state. According to Eq. (12.58) the polarization is negative and by Eq. (12.57) the absolute temperature must be negative also. When such a system comes to equilibrium with the surroundings it passes through the point where $n_1 = n_2$ or $P_n = 0$, which means that $T = \infty$. Thus negative absolute temperatures are "hotter" than infinite temperatures, and the third law of thermodynamics is not violated since the system never passes through $T = 0$ K. Infinite and negative temperatures are only possible in systems with a finite number of energy levels, such as electron or nuclear magnetic moments in a magnetic field. The input of a finite amount of energy can then bring the system to infinite temperatures and beyond to negative temperatures. For $I = \frac{1}{2}$ the entropy is related to the polarization by

$$S/R = \ln 2 - \tfrac{1}{2}[(1 + P_n) \ln (1 + P_n) + (1 - P_n) \ln (1 - P_n)] \quad (12.61)$$

This equation shows that the entropy does not depend on the sign of the polarization. A change in polarization from 0 to -1 then reduces the entropy from $R \ln 2$ to 0. Since T is negative, the term $T\Delta S$ is consistent with the fact that energy is added to the system. Just as in the case of positive polarization, a polarization of -50% in a field of 2.5 T yields a spin temperature of -5 mK and demagnetization from there gives a final nuclear temperature of about $-1 \mu K$. Experiments have shown that the nuclear ordering around $\pm 1 \mu K$ is different at negative temperatures than it is at the same positive temperature.

Negative polarization and temperatures can also be achieved without the microwave pumping by starting with a metal sample that has been polarized in a magnetic field at about 10 mK. A sudden rotation of the field or sample by 180° would give a negative polarization and temperature, but it would rapidly decay at the rate τ_1 back to a positive polarization and temperature.

Most of the theoretical studies and all of the experimental studies on nuclear magnetic order using dynamic nuclear polarization followed by demagnetization in the rotating frame have been performed at the Laboratory of Nuclear Magnetism at Saclay. Recent reviews of their work are given by Abragam et al. (1978a, b) and Goldman (1977). The first observation of nuclear magnetic ordering was seen by this group in 1969 (Chapellier et al., 1969; 1970). The nuclear spin system studied was that of F^{19} nuclei in Ca_2F. Polarizations of about $\pm 60\%-70\%$ were produced at 0.7 K in a field of 2.7 T after 3 hr of irradiation with microwaves with $B_0 = 5 \mu T$. After adiabatic demagnetization in the rotating frame spin temperatures of about $\pm 1 \mu K$ or less were reached. For positive spin temperatures the nuclei remained in the paramagnetic state up to the highest polarization studied. For negative spin temperatures the F^{19} nuclei underwent a transition to the antiferromagnetic state above polarizations of $|P_n| = 29\%$ (Chapellier, 1971). More recently the same group has studied LiH (Roinel et al., 1978). With starting conditions of 50 mK and 6.5 T they achieved Li^7 polarizations of 80% and H^1 polarizations of 95% after 3 days of microwave irradiation. After demagnetization in the rotating frame to $T_n = \pm 1 \mu K$ nuclear antiferromagnetism was observed for both $T_n > 0$ and $T_n < 0$, but the antiferromagnetic structure was different for the two cases. These experiments on nuclear magnetic ordering in CaF_2 and LiH are a striking example of the fact that negative absolute temperatures are real and material can behave differently at negative temperatures compared with positive temperatures.

REFERENCES

Abel, W. R., Anderson, A. C., Black, W. C., and Wheatley, J. C. (1965). *Physics* **1**, 337.

Abragam, A., Bouffard, V., Goldman, M., and Roinel, Y. (1978b). *Proc. LT15, J. Phys. (Paris) Colloq. C6, Suppl. No. 8* **39**, 1436.

Abragam, A., and Goldman, M. (1978a). *Rep. Prog. Phys.* **41**, 395.

Abragam, A., and Proctor, W. G. (1958). *C.R. Acad. Sci. Paris* **246**, 2253.

Ahonen, A. I., Berglund, P. M., Haikala, M. J., Krusius, M., Lounasmaa, O. V., and Paalanen, M. A. (1976). *Cryogenics* **16**, 521.

Ahonen, A. I., Gully, W. J., Lounasmaa, O. V., and Veuro, M. C. (1978). *Proc. LT15, J. Phys. (Paris) Colloq. C6, Suppl. No. 8* **39**, 1153.

Ahonen, A. I., Lounasmaa, O. V., and Veuro, M. C. (1978). *Proc. LT15, J. Phys. (Paris) Colloq. C6, Suppl. No. 8,* **39**, 265.

Al'tshuler, S. A. (1966). *JETP Lett.* **3**, 112.

Ambler, E., and Dover, R. B. (1961). *Rev. Sci. Instrum.* **32**, 737.

Anderson, A. C. (1970). *Rev. Sci. Instrum.* **41**, 1446.

Andres, K. (1978). *Cryogenics* **18**, 473.

Andres, K., and Bucher, E. (1968). *Phys. Rev. Lett.* **21**, 1221.

Andres, K., and Bucher, E. (1971). *J. Appl. Phys.* **42**, 1522.

Andres, K., and Bucher, E. (1972). *J. Low Temp. Phys.* **9**, 267.

Andres, K., and Darack, S. (1974). *Phys. Rev. B* **10**, 1967.

Andres, K., and Darack, S. (1977). *Physica* **86–88** B + C, 1071.

Andres, K., and Sprenger, W. O. (1975). *Proc. 14th Int. Conf. on Low Temp. Phys.,* North-Holland Publishing Co., Amsterdam. Vol. 1, p. 123.

Anufriyev, Yu. D. (1965). *Zh. Eksp. Teor. Fiz. Pis'ma* Red. (Engl. transl.) *JETP Lett.* **1**, 155.

Avenel, O., Berglund, M. P., Gylling, R. G., Phillips, N. E., Vetleseter, A., and Vuorio, M. (1973). *Phys. Rev. Lett.* **31**, 76.

Bardeen, J., Baym, G., and Pines, D. (1967). *Phys. Rev.* **156**, 207.

Berglund, P. M., Ehnhom, G. J., Gylling, R. G., Lounasmaa, O. V., and Sovik, R. P. (1972). *Cryogenics* **12**, 297.

Betts, D. S. (1976). *Refrigeration and Thermometry Below One Kelvin.* Sussex University Press.

Betts, D. S., and Marshall, R. (1969). *Cryogenics* **9**, 460.

Binnig, G., and Hoenig, H. E. (1978). *Proc. LT15, J. Phys. (Paris) Colloq. C6, Supp. No. 8* **39**, 1148.

Boughton, R. I., Brubaker, N. R., and Sarwinski, R. J. (1967). *Rev. Sci. Instrum.* **38**, 1177.

Brubaker, N. R., Edwards, D. O., Sarwinski, R. E., Seligman, P., and Sherlock, R. A. (1970). *J. Low Temp. Phys.* **3**, 619; *Phys. Rev. Lett.* **25**, 715.

Bucher, E., Andres, K., Maita, J. P., Cooper, A. S., and Longinotti, L. D. (1971). *J. Phys. (Paris) Colloq. C-1* **32**, 114.

Chanin, G., and Torre, J. P. (1976). *6th Int. Cryog. Eng. Conf. IPC* Science and Technology Press, Guildford, England, p. 96.

Chapellier, M. (1971). *Proc. 12th Int. Conf. on Low Temp. Phys.,* Academic Press of Japan, p. 637.

Chapellier, M., Frossati, G., and Rasmussen, F. B. (1979). *Phys. Rev. Lett.* **42**, 904.

Chapellier, M., Goldman, M., Vu Hoang Chau, and Abragam, A. (1969). *C.R. Acad. Sci. Paris* **268**, 1530.

Chapellier, M., Goldman, M., Vu Hoang Chau, and Abragam, A. (1970). *J. Appl. Phys.* **41**, 849.

Childs, G. E., Ericks, L. J., and Powell, R. L. (1973). *Thermal Conductivity of Solids at Room Temperature and Below.* National Bureau of Standards Monograph 131.

Cohen, E. G. D., and Van Leeuwen, J. M. J. (1961). *Physica* **27**, 1157. See also Cohen, E. G. D., and Van Leeuwen, J. M. J. (1960). *Physica* **26**, 1171; and Van Leeuwen, J. M. J., and Cohen, E. G. D. (1967). *Phys. Lett.* **26A**, 89.

Collins, S. C., and Zimmerman, F. J. (1953). *Phys. Rev.* **90**, 991.

Corruccini, L. R., Osheroff, D. D., Lee, D. M., and Richardson, R. C. (1972). *J. Low Temp. Phys.* **8**, 229.

Daniels, J. M., and Robinson, F. N. (1953). *Phil. Mag.* **44**, 630.

Das, P., de Bruyn Ouboter, R., and Taconis, K. W. (1965). *Proc. 9th Int. Conf. on Low Temp. Phys.,* Plenum Press, New York, p. 1253.

Daunt, J. G. (1970). *Cryogenics* **10**, 473.

Daunt, J. G., and Heer, C. V. (1949). *Phys. Rev.* **76**, 985.

Daunt, J. G., and Lerner, E. (1970). *Cryogenics* **10**, 476.

Daunt, J. G., and Lerner, E. (1974). *Proc. 5th Int. Cryog. Eng. Conf. IPC* Science and Technology Press, Guildford, p. 238.

Debye, P. (1926). *Ann. Phys. (Leipzig)* **81**, 1154.

de Haas, W. J., Wiersma, E. C., and Kramers, H. A. (1933). *Nature* **131**, 719; *Physica* **1**, 1 (1934).

De Long, L. E., Symko, O. G., and Wheatley, J. C. (1971). *Rev. Sci. Instrum.* **42**, 147.

de Waele, A. J. A. M., Coops, G. M., and Gijsman, H. M. (1978). *Proc. LT15, J. Phys. (Paris) Colloq.* C6, *Suppl. No.* 8 **39**, 1150.

de Waele, A. Th. A. M., Reekers, A. B., and Gijsman, H. M. (1976). *Physica* **81B**, 323; *Proc. 6th Int. Cryog. Eng. Conf.*, IPC Science and Technology Press, Guildford, p. 112.

Dundon, J. M., and Goodkind, J. M. (1974). *Phys. Rev. Lett.* **32**, 1343.

Dundon, J. M., Stolfa, D. L., and Goodkind, J. M. (1973). *Phys. Rev. Lett.* **30**, 843.

Ebner, C., and Edwards, E. O. (1971). *Phys. Rep.* **2C**, 77.

Edwards, D. O., and Daunt, J. G. (1961). *Phys. Rev.* **124**, 640.

Edwards, D. O., Feder, J. D., Gully, W. J., Ihas, G. G., Landau, J., and Muething, K. A. (1978). *Proc. LT15, J. Phys. (Paris) Colloq.* C6, *Suppl. No.* 8 **39**, p. 260.

Ehnholm, G. J., Ekstrom, J. P. Jacquinot, J. F., Loponen, M. T., Lounasmaa, O. V., and Soini, J. K. (1979). *Phys. Rev. Lett.* **42**, 1702.

Ehnholm, G. J., Ekstrom, J. P., Jacquinot, J. F., Loponen, M. T., Lounasmaa, O. V., and Soini, J. K. (1980). *J. Low Temp. Phys.*, to be published.

Esel'son, B. N., Lazarev, B. G., and Shvets, A. D. (1963). *Cryogenics* **3**, 203.

Frossati, G. (1978). *Proc. LT15, J. Phys. (Paris) Colloq.* C6, *Suppl. No.* 8 **39**, 1578.

Frossati, G., Godfrin, H., Hebral, B., Schumacher, G., and Thoulouze, D. (1978a). *Physics at Ultralow Temperatures*. Physical Society of Japan, p. 205.

Frossati, G., Hebral, B., Schumacher, G., and Thoulouze, D. (1978b). *Cryogenics* **18**, 277.

Frossati, G., Schumacher, G., and Thoulouze, D. (1975). *Proc. 14th Int. Conf. on Low Temp. Phys.* North-Holland Publishing Co., Amsterdam, Vol. 4, p. 13.

Frossati, G., and Thoulouze, D. (1976). *Proc. 6th Int. Cryog. Eng. Conf.*, IPC Science and Technology Press, Guildford, p. 116.

Gates, J. V., and Potter, W. H. (1975). *Proc. 14th Int. Conf. on Low Temp. Phys.*, North-Holland Publishing Co., Amsterdam, Vol. 4, p. 5.

Giauque, W. F. (1927). *J. Am. Chem. Soc.* **49**, 1870.

Giauque, W. F., and MacDougall, D. P. (1933). *Phys. Rev.* **43**, 768.

Gibbons, R. M., and McKinley, C. (1967). *Adv. Cryog. Eng.* **13**, 375; Gibbons, R. M., and Nathan, D. I. (1967). Technical Report AFML-TR-67-175, Wright–Patterson Air Force Base.

Gladun, A., and Peshkov, V. P. (1972). *Zh. Eksp. Teor. Fiz.* **62**, 1853 [Engl. Transl. (1972) *Sov. Phys. JETP* **35**, 965].

Goldman, M. (1977). *Phys. Rep.* **32C**, 1.

Gorter, C. J. (1934). In a remark following a paper by P. Debye, *Phys. Z.* **35**, 923.

Greywall, D. S. (1976). *Phys. Rev. Lett.* **37**, 105.

Guyeer, R. A. (1978). *J. Low Temp. Phys.* **30**, 1.

Gylling, R. G. (1971). *Acta Polytechnica Scandinavica* No. Ph81.

Hall, H. E., Ford, P. J., and Thompson, K. (1966). *Cryogenics* **6**, 80.

Halperin, W. P., Rasmussen, F. B., Archie, C. N., and Richardson, R. C. (1978). *J. Low Temp. Phys.* **31**, 617.

Harrison, J. P. (1980). *J. Low Temp. Phys.* **37**, 467.

Heer, C. V., Barnes, C. B., and Daunt, J. B. (1954). *Rev. Sci. Instrum.* **25**, 1088.

Heterington, J. H., Mullin, W. J., and Nosanow, L. H. (1967). *Phys. Rev.* **154**, 175.

Hobden, M. V., and Kurti, N. (1959). *Phil. Mag.* **4**, 1092.

Hudson, R. P. (1970). *Proc. 1970 Ultralow Temperature Symposium*, NRL Report 7133, p. 3.

Hudson, R. P. (1972). *Principles and Application of Magnetic Cooling*. North-Holland Publishing Co., Amsterdam.

Huiskamp, W. J., and Lounasmaa, O. V. (1973). *Rep. Prog. Phys.* **36**, 423.

Hunik, R., Bongers, E., Konter, J. A., and Huiskamp, W. J. (1978b). *Proc. LT15, J. Phys. (Paris) Colloq.* C6, *Suppl. No.* 8 **39**, 1155.

Hunik, R., Konter, J. A., and Huiskamp, W. J. (1978a). *Proc. ULT Hakone Symposium, Physics at Ultralow Temperatures*. Physical Society of Japan, p. 287.

Ifft, E. M., Edwards, D. O., Sarwinski, R. E., and Skertic, M. M. (1967). *Phys. Rev. Lett.* **19**, 831; Edwards, D. O., Ifft, E. M., and Sarwinski, R. E. (1969). *Phys. Rev.* **177**, 380.

Johnson, R. T., and Wheatley, J. C. (1970). *J. Low Temp. Phys.* **2**, 423.

Jurriens, R. G., Pennings, N. H., Satoh, T., Taconis, K. W., and de Bruyn Ouboter, R. (1978). in *Physics at Ultralow Temperatures*, Physical Society of Japan, p. 226.

Kalvius, G. M., Katila, J. E., and Lounasmaa, O. V. (1969). *Mössbauer Effect Methodology* (I. Gruverman, ed.) Plenum Press, New York, Vol. 5, p. 231.

Keller, W. E. (1969). *Helium-3 and Helium-4*. Plenum Press, New York.

Kerley, N. W. (1978). in *Advances in Refrigeration at the Lowest Temperatures*. International Institute of Refrigeration Comm. Al-2, p. 159.

Kittel, P. (1980). *Cryogenics*, to be published.

Korringa, J. (1950). *Physica* **16**, 601.

Kraus, J., Uhlig, E., and Wiedemann, W. (1974). *Cryogenics* **14**, 29.

Kummer, R. B., Mueller, R. M., and Adams, E. D. (1977). *J. Low Temp. Phys.* **27**, 319.

Kurti, N. (1967). *Contemp. Phys.* **8**, 21.

Kutri, N., Robinson, F. N. H., Simon, F. E., and Spohr, D. A. (1956). *Nature* **178**, 450.

Kurti, N., and Simon, F. E. (1935). *Proc. Roy. Soc.* **A149**, 152.

Landau, L. D., and Lifshitz, E. M. (1968). *Statistical Physics*, Pergamon Press, Oxford.

Landau, L. D., and Pomeranchuk, I. (1948). *Dokl. Akad. Nauk. SSSR* **59**, 669; Pomeranchuk, I. (1949). *Zh. Eksperim. Teor. Fiz.* **19**, 42.

Locatelli, M., Arnaud, D., and Routin, M. (1976). *Cryogenics* **16**, 374.

London, H. (1951). *Proc. Int. Conf. Low Temp. Phys.*, Oxford, Clarendon Lab., p. 157.

London, H., Clarke, G. R., and Mendoza, E. (1962). *Phys. Rev.* **128**, 1992.

Lounasmaa, O. V. (1974). *Experimental Principles and Methods Below 1K*. Academic Press, London.

Lounasmaa, O. V. (1979). *Physics Today* **32**(12), 32.

Lounasmaa, O. V. (1983). Private Communication.

March, R. H., and Symko, O. G. (1965). *Proc. Grenoble Conf.*, International Inst. of Refrigeration, Annexe 1965, 2, p. 27.

Mate, C. F., Harris-Lowe, R., Davis, W. L., and Daunt, J. G. (1965). *Rev. Sci. Instrum.* **36**, 369.

McAshan, M. S. (1976). *6th Int. Cryog. Eng. Conf.* IPC Science and Technology Press, Guildford, England, p. 3.

Mendoza, E. (1961). *Experimental Cryophysics*. (ed. F. E. Hoare, L. C. Jackson, and N. Kurtii), Chap. 8.

Mueller, R. M., Buchal, Chr., Folle, H. R., Kubota, M., and Pobell, F. (1980). *Cryogenics* **20**, 395.

Neganov, B., Borisov, N., and Libury, M. (1966). *Zh. Eksp. Teor. Fiz.* **50**, 1445 [Engl. transl. *Sov. Phys. JETP* **23**, 959 (1966)].

Niinikoski, T. O. (1971). *Nucl. Instrum. Methods* **97**, 95.

Niinikoski, T. O. (1976). *6th Int. Cryog. Eng. Conf.*, IPC Science and Technology Press, Guildford, p. 102.

Nosanow, L. H., and Mullin, W. J. (1965). *Phys. Rev. Lett.* **14**, 133.

Ono, K., Kobayasi, S., Shinohara, M., Asahi, K., Ishimoto, H., Nishida, N., Imaizumi, M., Nakaizumi, A., Ray, J., Iseki, Y., Takayanagi, S., Tenuri, K., and Sugawars, T. (1980). *J. Low Temp. Phys.* **38**, 737.

Osgood, E. B., and Goodkind, J. M. (1966). *Cryogenics* **6**, 54.

Osgood, E. B., and Goodkind, J. M. (1967). *Phys. Rev. Lett.* **18**, 894.

Osheroff, D. D., Richardson, R. C., and Lee, D. M. (1972). *Phys. Rev. Lett.* **28**, 885.

Osheroff, D. D., and Sprenger, W. O. (1980). Private Communication from D. D. Osheroff.

Ostermeier, R. M., Nolt, I. G., and Radostitz, J. V. (1978). *Cryogenics* **18**, 83.

Pavlov, V. N., Neganov, B. S., Konicek, J., and Ota, J. (1978). *Cryogenics* **18**, 115.

Pennings, N. H., Taconis, K. W., and de Bruyn Ouboter, L. (1974). *Cryogenics* **14**, 53.

Pomeranchuk, I. (1950). *Zh. Eksp. Teor. Fiz.* **20**, 919.

Radebaugh, R. (1967). NBS Tech. Note 362.

Radebaugh, R. (1977). *J. Low Temp. Phys.* **27**, 91.

Radebaugh, R., and Siegwarth, J. D. (1971). *Cryogenics* **11**, 368.

Radebaugh, R., and Siegwarth, J. D. (1971). *Proc. 12th Int. Conf. on Low Temp. Phys.*, Academic Press of Japan, p. 163.

Radebaugh, R., and Siegwarth, J. D. (1974). *Proc. 13th Int. Conf. on Low Temp. Phys.*, Plenum Press, New York, Vol. 1, p. 401.

Radostitz, J. V., Nolt, I. G., Kittel, P., and Donnelly, R. J. (1978). *Rev. Sci. Instrum.* **49**, 86.

Reich, H. A., and Garwin, R. L. (1959). *Rev. Sci. Instrum.* **30**, 7.

Roberts, T. R., and Sydoriak, S. G. (1955). *Phys. Rev.* **98**, 1672.

Roinel, Y., Bouffard, V., Bacchella, G. L., Pinot, M., Meriel, P., Roubeau, P., Avenel, O., Goldman, M., and Abragam, A. (1978). *Phys. Rev. Lett.* **41**, 1572.

Rosenblum, R. S., Sheinberg, H., and Steyert, W. A. (1976). *Cryogenics* **16**, 245.

Satoh, T., Jurriens, R. G., Taconis, K. W., and de Bruyn Ouboter, R. (1974). *Physica* **77**, 523.

Scribner, R. A., Panczyk, M. F., and Adams, E. D. (1969). *J. Low Temp. Phys.* **1**, 313.

Seidel, G., and Keesom, P. H. (1958). *Rev. Sci. Instrum.* **29**, 606.

Severiyns, A. P., and Staas, A. P. (1978). in *Advances in Refrigeration at the Lowest Temperatures*, International Institute of Refrigeration, Comm. Al-2, p. 175.

Severijns, A. P., Staas, F. A., and Cense, W. A. (1978). *Cryogenics* **18**, 87.

Siegwarth, J. D., and Radebaugh, R. (1971). *Rev. Sci. Instrum.* **42**, 1111.

Siegwarth, J. D., and Radebaugh, R. (1972). *Rev. Sci. Instrum.* **43**, 197.

Sites, J. R., Osheroff, D. D., Richardson, R. C., and Lee, D. M. (1969). *Phys. Rev. Lett.* **23**, 836.

Sites, J. R., Smith, H. A., and Steyert, W. A. (1971). *J. Low Temp. Phys.* **4**, 605.

Smith, E. N., Bozler, H. M., Truscott, W. S., Richardson, R. C., and Lee, D. M. (1975). *Proc. 14th Int. Conf. Low Temp. Phys.* North-Holland Publishing Co., Amsterdam, Vol. 4, p. 9.

Staas, F. A. (1976). *Philips Tech. Rev.* **36**, 104.

Staas, F. A., Severijns, A. P., and Van Der Waerden, H. C. M. (1975). *Phys. Letters* **53A**, 327.

Sydoriak, S. G., Mills, R. L., and Grilly, E. R. (1960). *Phys. Rev. Lett.* **4**, 495; Mills, R. L., Grilly, E. R., and Sydoriak, S. G. (1961). *Ann. Phys. (N.Y.)* **12**, 41.

Symko, O. G. (1969). *J. Low Temp. Phys.* **1**, 451.

Taconis, K. W. (1961). *Progress in Low-Temperature Physics*, Vol. III (ed. C. J. Gorter), North-Holland Publishing Co., Amsterdam, Chapter 5.

Taconis, K. W., Pennings, N. H., Das, P., and de Bruyn Ouboter, R. (1971). *Physica* **56**, 168.

Torre, J. P., and Chanin, G. (1978). *Adv. Cryog. Eng.* **23**, 640.

van der Boog, A. G. M., Husson, L. P. J., and Kramers, H. C. (1978). *Physics Lett.* **66A**, 305.

Varoquaux, E. (1978). *Proc. LT15, J. Phys. (Paris) Colloq. C6, Suppl. No. 8*, **39**, 1605.

Veuro, M. C. (1978). *Acta Polytechnica Scandinavica* No. Ph. 122.

Vilches, O. E., and Wheatley, J. C. (1967). *Phys. Lett.* **24A**, 440.

Vvedenskii, V. L., and Peshkov, V. P. (1972). *Zh. Eksp. Teor. Fiz.* **63**, 1363 [Engl. Transl. *Sov. Phys. JETP* **36**, 721 (1973)].

Walters, G. K., and Fairbank, W. M. (1956). *Phys. Rev.* **103**, 262.

Walton, D. (1966). *Rev. Sci. Instrum.* **37**, 734 (1966).

Walton, D., Timsuk, J., and Sievers, A. J. (1971). *Rev. Sci. Instrum.* **42**, 1265.

Webb, R. A., Greytak, T. J., Johnson, R. T., and Wheatley, J. C. (1973). *Phys. Rev. Lett.* **30**, 210.

Weiss, P. (1921). *J. Phys. (Paris)* Ser. 6, No. 2, 161.

Weiss, P., and Piccard, A. (1918). *C.R. Acad. Sci. (Paris)* **166**, 352.

Wheatley, J. C. (1966). in *Proceedings of the Sussex University Symposium on Quantum Fluids* (ed. D. T. Brewer), North-Holland, Amsterdam, p. 183.

Wheatley, J. A., Rapp, R. E., and Johnson, R. J. (1971). *J. Low Temp. Phys.* **4**, 1.

Wheatley, J. A., Viches, O. E., and Abel, W. R. (1968). *Physics* **4**, 1.

Wilkes, W. R. (1972). *Cryogenics* **12**, 180.

Yamamoto, J. (1975). *Japan. J. Appl. Phys.* **14**, 1807.

Zinov'eva, K. N. (1958). *Zh. Eksp. Teor. Fiz.* **34**, 609 [Engl. Transl. *Sov. Phys. JETP* **7**, 421].

Zinov'eva, K. N., and Peshkov, N. P. (1959). *Zh. Eksp. Teor. Fiz.* **37**, 33 [Engl. Transl. *Sov. Phys. JETP* **10**, 22].

Zimmerman, J. E., McNutt, J. D., and Bohmn, H. V. (1962). *Cryogenics* **2**, 153.

Chapter 13

Cryogenic Engineering and Cryocooler Development in the USSR

Evgeny Ivanovich Mikulin

INTRODUCTION

Cryogenics began in 1877 when L. Kalete observed the transition of oxygen from the gaseous to the liquid state. Small systems for liquefaction of air were developed by Linde and Hampson in 1895, working independently, but the broad development of cryocoolers came much later and especially after World War II.

Significant developments in cryogenic engineering have also occurred in the USSR. Many Russian scientists have contributed to cryocooler development, but undoubtedly the main contribution has been by Academician P. Kapitza. The achievements of Kapitza in the field of low temperatures are well known to scientists around the world. His interests are very broad, ranging from fundamental investigations of the properties of liquid helium to designing new cryogenic plants, expansion engines, and apparatus. In the field of low-temperature physics his classical work is well known on the fluidic effects of liquid helium including the discovery of superfluidity, thermal "Kapitza resistance," and other phenomena in helium II. In the field of cryogenic engineering Kapitza (1935) designed the first helium liquefier with an expansion engine (Claude-type liquefier) instead of the alternative Joule–Thomson systems which needed liquid hydrogen.

Evgeny Ivanovich Mikulin • Baumann Institute, Moscow Technical High School, Moscow, USSR.

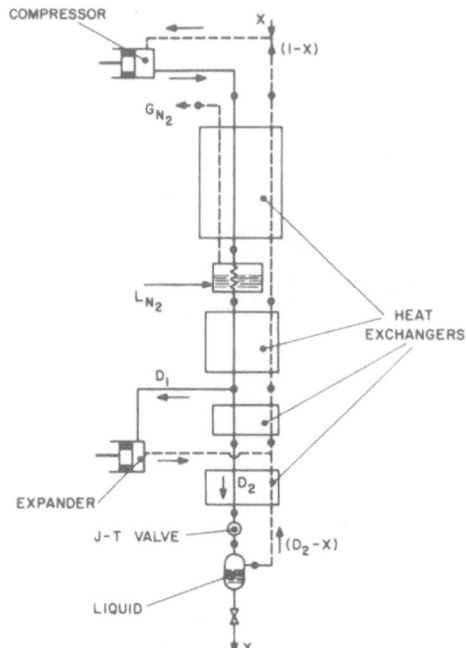

Fig. 13.1. Flow diagram for Kapitza helium liquefier.

The general ideas of Kapitza were later developed by Collins and others in designing modern liquid helium facilities. Again, Kapitza (1939) designed a new type radial-inflow expansion turbine having an isentropic efficiency over 80%. Such expansion turbines are in regular use today for modern cryogenic systems. In 1937 Kapitza suggested the low-pressure cycle with turboexpander for air-liquefaction plants. Today all large air-separation systems employ the Kapitza low-pressure cycle.

The flow diagram of Kapitza's helium liquefier is shown in Fig. 13.1. This cycle employs liquid nitrogen for precooling, with one expansion engine and a Joule–Thomson expansion valve. A schematic cross section of Kapitza's liquefier has been given earlier, Chapter 7. The flow diagram of Kapitza's cycle for liquefaction of air is shown in Fig. 13.2.

Kapitza is presently head of the Institute of Physical Problems, Academy of USSR. Numerous advances in cryophysics have been accomplished in this institute under his guidance. It is first of all necessary to mention the theoretical work of L. Landau on the properties of liquid helium II. Ginzburg, Landau, Abrikosov, and Gorkov developed the phenomenological theory of superconductivity (the GLAG theory). Khalatnikov has given a theoretical explanation of the "Kapitza resistance," while Peshkov investigated the properties of helium-3. A new method of cooling

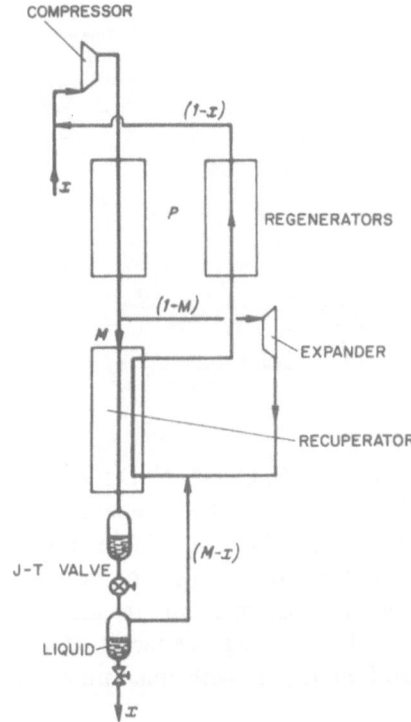

Fig. 13.2. Flow diagram for Kapitza low-pressure cycle with turboexpander for liquefaction of air.

below 5 mK, based on the adiabatic compression of solid helium-3, was developed by Pomeranchuk in 1950, and Anufriev reduced the process to practice. Pomeranchuk cooling is shown on the temperature–entropy diagram, Fig. 13.3. Adiabatic compression of helium-3 from liquid to solid state (process $A–B$) decreases the temperature from T_H to T_k.

In the Kharkov Physico-Technical Institute, Esselson, Lasarev, and others (1973) have investigated properties of liquid helium and the properties of solutions of helium-3 and helium-4. One of the first effective working dilution refrigerators employing helium-3/helium-4 solution was designed by Neganov, Borisov, and Liburg (1966). They achieved a temperature as low as 25 mK. Later Neganov improved the cooling system and achieved a minimum temperature of 5.5 mK. A schematic view of this refrigerator is shown in Fig. 13.4. Further details of other Russian work in cryophysics are given by Arkharov et al. (1978).

With regard to air separation plants, cryocoolers, and other cryogenic equipment, continuous progress has been made since the 1930s. The first air separation plants producing oxygen were built in 1932–1933 with capacities in the range $100/250 \, \text{m}^3/\text{hr}$. These plants were designed by

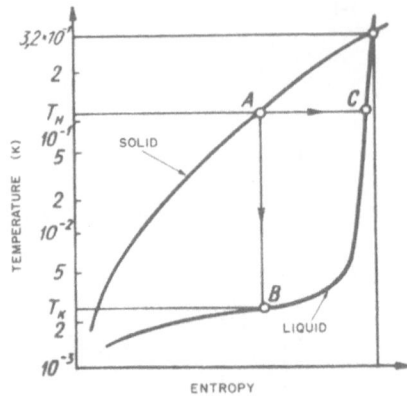

Fig. 13.3. Process of cooling by adiabatic compression of helium-3 (Pomeranchuk process).

Professors S. Gersh, N. Dollezal, and S. Semichatov. Today there are many different types of air separation plants operating in the USSR. Some of these plants have very large capacities and provide a wide range of products including oxygen (gas or liquid), pure nitrogen liquid and gas (perhaps compressed), and the rare gases.

The average capacity of these large air separation plants is about 30,000 m³/hr, with maximum capacity ranging up to 70,000 m³/hr. The

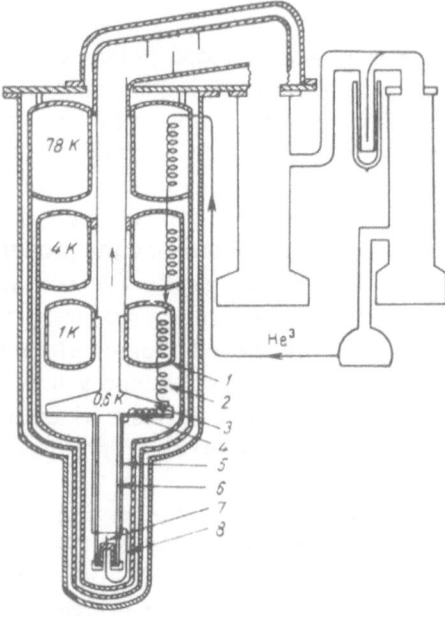

Fig. 13.4. Neganov–Borisov–Liburg dilution refrigerator. KEY: 1, condenser; 2, coil; 3, evaporator; 4,5,6, heat exchangers; 7, dilution chamber; 8, inlet He³ in dilution chamber.

operating principle of the large plants is based on the low-pressure Kapitza cycle. A recent trade catalog (Anon., 1978) listed 13 types of large air separation plants available for export. The cryogenic industry in Russia also produces different types of hydrogen and helium liquefiers, cooling systems (refrigerators), the cryogenic pressure chambers, and other cryogenic equipment (Anon., 1975) (Belyacov *et al.*, 1977).

These plants are principally different versions of the low-pressure Kapitza cycle for air separation and for helium liquefiers or refrigerators. Among these low-temperature cycles it is necessary to mention the multistage process of cooling based on the mixtures of several gases as a working fluid. This process, shown in Fig. 13.5, is basically a development to improve the efficiency of the classical Linde–Hampson process, including, of course, a Joule–Thomson expansion. Klimenko proposed this process in 1952, first for liquefying natural gas, and later for cryogenic refrigerators. The principal advantage of the cycle is that mixtures of gases can be handled using only one compressor in a cascade process. Another advantage lies in the increase in the Joule–Thomson effect for mixtures compared with a single component working fluid. Temperature differences in the heat exchangers are much less than with the usual Hampson process. As a result, the reversibility of the throttling process increases; the efficiency is more than twice the Hampson process using only nitrogen as the working fluid. The mixed-flow multistage process is becoming increasingly popular in many countries for liquefying natural gas and for cryogenic minicoolers (Gresin *et al.*, 1977).

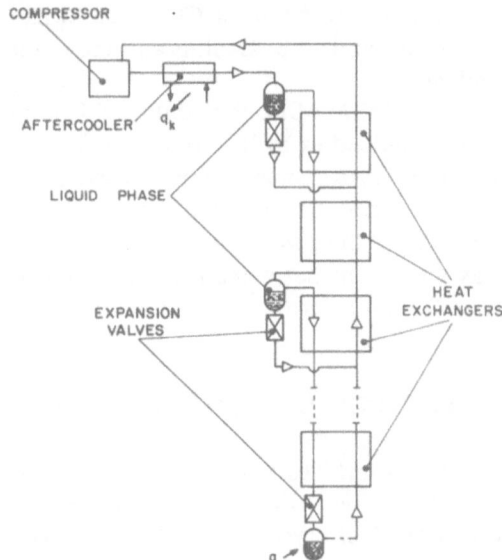

Fig. 13.5. Flow diagram of the Klimenko process for cooling mixtures of gases.

The first Russian version of the Brayton cycle cryocoolers was developed by Dubinski and Martynovski (Socolov *et al.*, 1968). In this machine the inlet air enters the regenerator at atmospheric pressure, is cooled by counterflow heat exchangers, and then enters the turboexpander, where it expands to 0.5 atmospheres to produce refrigeration at the minimum cycle temperature. On leaving the turboexpander the gas flows back through the counterflow heat exchanger and regenerator to cool the incoming inlet flow stream. It is finally pumped from the system by an axial vacuum pump. The cycle does not require a compressor and has demonstrated quite high efficiencies at temperature levels in the range 120–150 K.

About the same time, a cryogenic system for low-temperature separation of hydrogen producing deuterium was developed under the direction of Malkov (1961).

CRYOGENIC RESEARCH CENTERS IN THE USSR

In the USSR there are several large cryogenic centers, mainly at Technological Institutes. Leading the field is the Baumann Institute of the Moscow High Technological School (MHTS), where S. Gersh began his teaching of cryogenics in 1933. Today at MHTS experimental studies of cryogenic problems are carried out on a broad range and are regularly reported (MHTS 1967–1976).

Teaching of students and research in cryogenics is carried out also in the Moscow Institute of Chemical Machine Building, the Moscow Power Institute, and the Leningrad and Odessa Institutes of Refrigeration Industry.

VNIIKIMASH, the center of Russian cryogenic research and design, was organized in 1946. Today the leading center in the field of cryogenics is NPO Cryogenmash. Most Russian research and cryogenic engineering work is concentrated in Cryogenmash. The Cryogenmash regularly publish accounts of their work, i.e., Belyakov (1971). Additional work on cryogenics has also been performed at the Kharkov Physico-Technical Institute of Low Temperatures.

PRINCIPAL CRYOGENIC PUBLICATIONS IN THE USSR

Many articles and books on cryogenics have been published in the USSR. In this review we give first a description of the general books on the subject of cryocoolers. Books concerned with particular problems of cryogenics are described later.

The earliest Russian book on cryogenic fundamentals was written by Professor S. Gersh under the title of *Glubokoe ochlazdenie* (*Deep Cooling*). This was published in 1937 and reprinted several times (Gersh, 1957; 1960). The book covered all international experience in low-temperature technology and treated many cryogenic problems such as the properties of gases, processes of cooling, separation of gases, description of the heat exchangers, and expansion engines. Many original methods of analysis developed by Gersh were included.

A popular and useful publication has been the *Spravotchnik po phizico-technicheskim osnovam cryogenici* (*Handbook on Physico-Technical Fundamentals of Cryogenics*) (Malkov, 1973). This handbook was first published in 1947 and has been reprinted twice. It includes a very wide range of information on different aspects of cryogenic technology. The handbook includes about 1200 references and covers such topics as thermodynamic fundamentals of cryogenic cooling cycles, physical properties at low temperatures, heat transfer and insulation, separation of gases, purification of gases, transport and storage of cryogenic liquids, low temperature thermometry, cryogenic engines, apparatus, and other equipment.

The text *Technica Nizkich Temperatur* (*Cryogenic Engineering*), edited by E. Mikulin, I. Marfenina, and A. Arkharov was published in 1964 and subsequently revised in 1975 (Mikulin *et al.*, 1975). It represents the most complete and up-to-date publication on the subject. The book includes chapters on the history of cryogenics, theory of low-temperature processes, separation of gases, low-temperature heat exchangers, expansion engines, and turboexpanders and contains many examples to illustrate some of the procedures for analyzing cryogenic cycles, expansion engines, heat exchangers. The book was translated into English by NBS in 1968 (Arkharov *et al.*, 1968).

It is also necessary to mention a two-volume book, *Razdelenie vosducha metodom glubokogo ochlazdenia* (*Separation of Air by Low-Temperature Process*) (Epifanova, 1973). This book was first published in 1964 and again in 1973. It is the most complete publication on the subject in the world literature. The book includes details and descriptive procedures for the computation of air separation processes including three-component systems, i.e., oxygen–nitrogen–argon, and methods for the design of air separation plants.

An album of drawings entitled *Ustanovki, mashiny i apparaty cryogennoi techniki* (*Cryogenic Plants, Machines, and Apparatus*) was published under the editorship of Ousikin (1975). An earlier edition of the album was translated into English and published by Pergamon Press in 1967.

The principal textbook on cryogenics for students is *Teoria i raschet criogennich system* (*Theory and Computation of Cryogenic Systems*)

(Arkharov *et al.*, 1978). The textbook includes fundamental problems of cryogenic systems, such as different processes of cooling, analysis of cryogenic cycles, properties of fluids and solids at low temperatures. This text also describes basic principles involved with the separation of gases, absorption processes, and heat transfer at low temperatures. Other general texts of a similar nature have also been published (Socolov *et al.*, 1968; Fastovsky *et al.*, 1974; and E. Mikulin, 1969).

Problems of minicoolers have been described in another text, *Mikrocriogennia technica* (*Microcryogenic Engineering*) (Gresin *et al.*, 1977). Minicoolers of the Linde–Hampson type have been discussed extensively in a volume by Suslov *et al.* (1978).

In the USSR there is no special journal on cryogenics. Many articles concerned with cryogenics and cryogenic engineering are published in various technical journals, principally *Chemichescoe i Neftianoe Mashino-stroenie* (*Chemical and Oil Industry*). Some cryogenic centers regularly publish transactions of their work (Belyakov *et al.*, 1977; Belushcin *et al.*, 1972; Butkevitch *et al.*, 1968; and Kalitin, 1977).

Next we shall consider some specific types of problems in the cryogenic field. We will limit our survey to contemporary problems researched over the past decade, including the properties of substances at low temperatures, the analysis, computing, and optimization of cryogenic systems, heat transfer problems at low temperatures, problems of designing expansion engines, turboexpanders, and Stirling engines. Finally there are brief descriptions of typical modern Russian cryogenic systems.

PROPERTIES OF SUBSTANCES AT LOW TEMPERATURES

Information covering the properties of fluids and solids at low temperatures is required for computing and designing cryogenic systems. Information is needed about thermodynamic and transport properties and also the mechanical, electrical, magnetic, and other properties at low temperatures for gases, liquids, and structural materials.

There is considerable interest in the properties of materials among scientists in the USSR. Of the many different publications on this subject the most substantial are the works of Vasserman and Rabinovich *et al.* (1966, 1968). They have published two books on the properties of air and components of air—nitrogen, oxygen, argon. In the first book *Teplophizicheskie svoistva zidkogo vosducha i ego componentov* (*Thermophysical Properties of Air and Components of Air*), they describe a new equation of state, suggested by Kazavchinsky, now widely used in the USSR. This book also contains detailed data on the thermodynamic and transport properties of these fluids in the gaseous state up to a pressure of

1000 bar and from saturation temperature to 1300 K. The latter text includes complete information on the properties of air and its components in the liquid state up to a pressure of 500 bar. More recently, Rabinovich and Vasserman (1976) have published *Teplophizicheskie svoistva neona, argona, kriptona i xenona* (*Thermophysical Properties of Neon, Argon, Krypton, and Xenon*). This book summarizes experimental data and presents equations of state for these inert substances in gaseous, liquid, and solid state covering a temperature range from 0 to 1300 K and pressures up to 1000 bar. These authors also suggest temperature–pressure equations for different equilibrium states: solid–liquid, liquid–vapor, solid–vapor. This book includes tables of thermodynamic and transport properties, 27 tables in all. The properties of inert gases are considered in great detail and the book contains considerable new information including experimental data of authors previously unpublished.

Helium has been studied by many Russian scientists, first by Kapitza and his co-workers and after that in cryogenic centers of Kharkov and Tbilisy. Zelmanov developed the first reliable temperature–entropy chart for helium. A more recent publication of the book by Tzederberg *et al.*, *Thermodynamical and Thermophysical Properties of Helium*, contains additional information on properties of helium.

Properties of the two isotopes of helium—helium-4 and helium-3—and their mixtures are thoroughly discussed by Esselson *et al.* (1976) in the book *Solutions of Quantum Liquids*. This book contains extensive international data on the experimental and theoretical research concerning solutions of these unique quantum liquids. There are detailed descriptions of equilibrium states of He^3–He^4 mixtures, phase transitions in this system, thermodynamical properties of solutions, and transport properties. The book includes reference data including 600 bibliographic entries.

There are also many publications on the mechanical properties of structural materials—metals, alloys, plastics at low temperatures. Much useful design information is included in a handbook prepared by P. Kosehlev and S. Belyaev (1967), *Prochnost i plasyichnost constructzionnich materialov pri nizkich temperaturach* (*Mechanical Properties of Structural Materials at Low Temperatures*). This handbook contains information concerning such properties as yield strength, tensile strength, hardness, impact resistance, brittle behavior, and plastic resistance. The handbook provides data for different kinds of steels and alloys of aluminum, titanium, nickel, copper, magnesium. Properties of the pure metals are also included. The handbook summarizes experimental data of many Russian and foreign scientists in the temperature region 20–300 K.

The most complete bibliography on the properties of substances at low temperatures was presented by Malkov (1973).

COMPUTING AND ANALYZING CRYOGENIC
PROCESSES AND CYCLES

In the USSR great attention has been devoted to thermodynamic analysis, optimization, and the evaluation of cryogenic cycles and processes. For optimal designs of cryogenic systems it is necessary to define many parameters, such as temperatures, pressures in different parts of the system, parameters of heat exchangers, efficiency of expansion engines, and so on. All these parameters have to be carefully matched to achieve maximum efficiency of the cryogenic system. Modern plants consume a great deal of energy and thus efficiency is a critical factor. There are many different approaches to the problem of optimal design of cryogenic systems. Optimization is based on the solution of many complex problems including computation of the cycle with maximum thermal efficiency and designing all the apparatus and machines to be as effective as possible.

Methods for analyzing multistaged cryogenic liquefaction and refrigeration cycles have been presented by Arkharov *et al.* (1978) and by Mikulin *et al.* (1975). Analysis begins by computing the number of cooling stages. As a rule, the initial model of multistage cycle systems is based on the ideal Claude cycle with several expansion engines—the cycle shown in Fig. 13.6. Results of an analysis of this cycle used for liquefaction of a monatomic gas are shown in Fig. 13.7. The work required for liquefaction of one mole of gas P_Σ depends on temperature of condensation T_x and on the number of stages of the expansion engine, n. When temperature T_x decreases, the required work P_Σ sharply increases and therefore we need more stages of

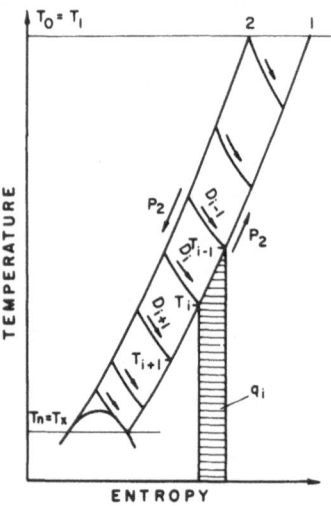

Fig. 13.6. *T–S* chart for multiple expansion cycle.

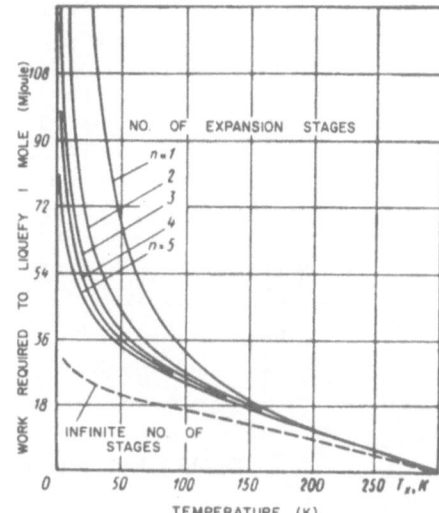

Fig. 13.7. Work required to liquefy 1 mol of gas in multistage cycle.

cooling to keep P_Σ at a low level. The dotted line in Fig. 13.7 represents the work of liquefaction of the ideal cycle.

For this method of analysis, based on known temperatures after each stage of cooling T_i from ambient T_0, Kapitza suggested (1959) that

$$T_i = \left[T_0^{n-1} T_x^i \frac{(\alpha_1, \alpha_2, \ldots, \alpha_n)^i}{(\alpha_1, \alpha_2, \ldots, \alpha_i)^m} \right]^{1/n} \tag{13.1}$$

where the parameter α_1 reflects the process of expansion of gas in the expansion engine for each stage with number i and efficiency η_i for n stages is

$$\alpha_i = \frac{\eta_i \left[1 - \left(\frac{1}{\pi} \right) \left(\frac{K-1}{K} \right) \right]}{1 - \eta_i \left[1 - \left(\frac{1}{\pi} \right) \left(\frac{K-1}{K} \right) \right]} \tag{13.2}$$

where K is a constant related to the heat transfer. All streams of gas in the multistage cycle can be determined by making a heat balance around the separate stages of cooling.

The optimal pressure ratio $P_2/P_1 = \pi$ generally corresponds to the condition of maximum thermal efficiency

$$\eta_t = \frac{l_\omega}{l} = f(\pi) \tag{13.3}$$

where l is the required work and l_ω is the work required in the ideal cycle.

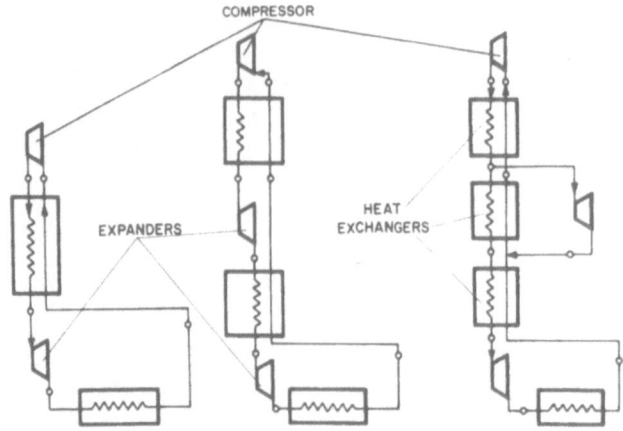

Fig. 13.8. Different versions of Brayton-type cycles.

Detailed analyses of Brayton-type cycles were developed by Marfenina (see Mikulin, 1975). Figure 13.8 shows some cycles of this type. Analysis and optimization of these cycles takes into account such factors as temperature levels T_0, T_i, T_2, pressure ratio $\pi = P_2/P_1$, efficiency of compressor η_K and expanders η_i, heat leak from surroundings q, temperature difference at warm end of each heat exchanger $\alpha_i = \Delta T_i/T_i$ relative drop of pressure for both streams of gas, the mass flow of gas, and also the full characteristics of each heat exchanger (heat transfer coefficient, type of surface, and so on). All of these variables depend upon each other and all affect the optimum efficiency.

The next step of analysis consists of establishing a system of initial equations and developing correlations between all parameters and the maximum thermal efficiency η_t. Some of the results of such analyses are shown in Figs. 13.9 and 13.10. Figure 13.9 shows how the thermal efficiency η_t and the optimal pressure ratio π depend on the temperature of cooling T_α and efficiency of expander η_i. Figure 13.10 shows how π depends on relative temperature difference at the warm ends of the heat exchanger. These results make it possible to select the best correlations among different parameters of cycles. More advanced analyses such as presented by Arkharov *et al.* (1975) take into account the change in efficiency of the compressor η_K and the expanders η_i as a function of the mass flow of gas. The efficiencies η_K and η_i generally decrease as the mass flow decreases resulting in changes of the optimum h_t and π.

Because of the need to account for many variables, the problem of optimization of cryogenic cycles is complicated. The above method of cycle analysis is the basis for contemporary design of cryogenic systems. Complete

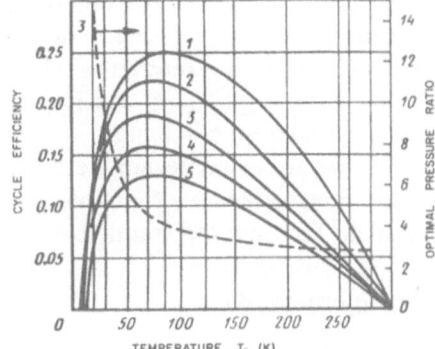

Fig. 13.9. Efficiency of Brayton cycle with one expander η_t and optimal pressure ratio π opt (broken line) as a function of temperature of cooling T_x for different conditions.

analyses have to include a mathematical model of the cryogenic system for all processes, apparatus, and machines with procedures for computing the system of equations. The analysis must include not only thermodynamic, but also economical correlations for optimized cryogenic system design.

In addition to the method of analysis and optimization described above a great deal of attention is also paid to thermodynamic analysis of cryogenic processes based on the second law of thermodynamics. This approach is

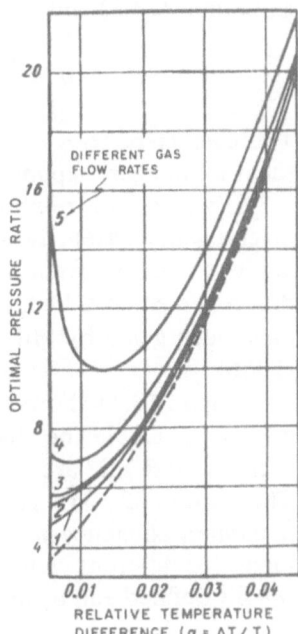

Fig. 13.10. Optimal pressure ratio π for Brayton cycle with one expander as a function of relative temperature difference $\alpha = \Delta T/T$. Temperature of cooling $T_x = 50$ K; 1–5 represent different flow rates of gas (flow increasing from 1 to 5).

most effective for comparing the real and the ideal thermodynamical process. The goals of this analysis are (a) to determine the degree of reversibility for different parts (elements) of the cryogenic system; and (b) to delineate steps which can improve the reversibility of these elements and increase their efficiency.

There are two different methods of thermodynamic analysis: one based on the thermodynamic function "exergy" (available energy) and the other based on the thermodynamic parameter "entropy."

Entropy Method

The basic equation for this method was developed by Clausius. It was first applied by Keesom to cryogenic systems. The basic equation is

$$l = l_{u\partial} + T_0 \sum_\Delta S_i \qquad (13.4)$$

and is the relationship between the work of an ideal reversible cycle $l_{u\partial}$ and the real cycle l. The losses of energy because of nonreversibility of the real processes may be represented by

$$T_0 \sum \Delta S_i = \alpha_i$$

Here $\sum \Delta S_i$ is the total change of entropy of all elements of the system. Equations (13.3) and (13.4) give us

$$\eta_t = \frac{l - T_0 \sum \Delta S_i}{l} = 1 - \frac{\sum \alpha_i}{l} = 1 - \sum \Omega_i \qquad (13.5)$$

The coefficient $\Omega_i = \alpha_i/l$ is the efficiency for each element of the system (heat exchanger, expansion engine, throttling valve, and so on). Equation (13.5) provides the relation between the efficiency of the individual elements of cryogenic system Ω_i and the efficiency of system as a whole, η_t. The analysis consists essentially of calculating the change of entropy for the different elements. Detailed descriptions of this method for cooling systems have been given by Mikulin (1969), Mikulin et al. (1975), and Gochstein (1966).

Figure 13.11 shows some results of an entropy analysis for the Brayton cycle (Fig. 13.8). The change in efficiency of the separate elements of this cycle Ω_i and $\sum \Omega_i$ are given as a function of the cooling temperature T_x. The losses $\sum \Omega_i$ are summarized at $T_x = 80\,\text{K}$ and correspond to the maximum efficiency $\eta_t = 1 - \sum \Omega_i$. Figure 13.11 shows the distribution of losses for different elements. This allows the ineffective components to be readily defined and by improving these to effectively improve the efficiency of the whole system.

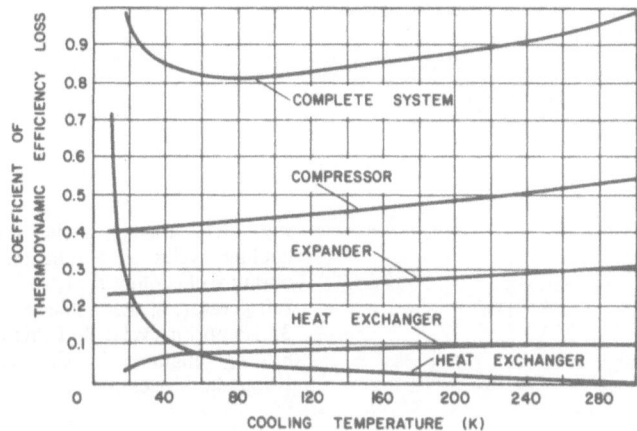

Fig. 13.11. Thermodynamic efficiency loss for the elements of a Brayton cycle with one expander as a function temperature of cooling T_x. Pressure ratio $\pi = 8$.

Exergy Method

This method is based on the calculation of the exergy flow. Exergy is the available energy (useful energy) which can be transferred from one form to another—for example from heat to electric energy. There are losses of exergy because of the nonreversibility of real processes.

The basic equation for an exergy analysis is

$$\sum E_{in} = \sum E_{out} + \sum D \tag{13.6}$$

Where E_{in} is the inlet exergy, E_{out} is the outlet exergy, and D are the losses of exergy. Exergy efficiency can be expressed as

$$\eta_e = \frac{\sum E_{out}}{\sum E_{in}} \tag{13.7}$$

η_e can be calculated for the separate elements of the system and also for the cooling system as a whole.

The change of exergy can be calculated with the help of exergy-enthalpy diagrams $(E-H)$ (Brodiansky, 1973). The change of exergy during a given process 1–2 can also be found from the expression

$$\Delta E = \int_{1}^{2} \left(\frac{T - T_0}{T} \right) dq \tag{13.8}$$

This is the change in exergy with heat flow. Alternatively

$$\Delta E = (h_1 - h_2) - T_0(S_1 - S_2) \tag{13.9}$$

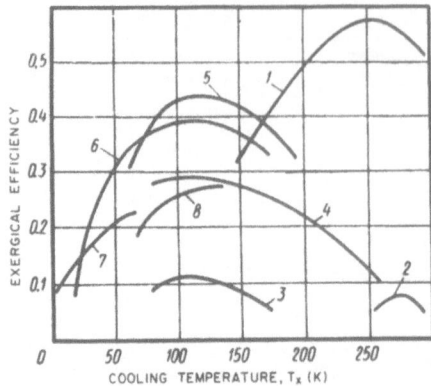

Fig. 13.12. Exergy efficiency η_l for various cooling cycles as a function of the temperature of cooling T_x: 1, Ammonia or Freon refrigerator; 2, thermoelectric cooling cycle; 3, Hampson cycle; 4, Brayton cycle; 5, 6, Stirling engines; 7, Claude process; 8, Klimenko process.

This is the change in exergy with mass flow. A detailed description of this method of analysis for cooling systems was given by Brodiansky (1973). Figure 13.12 shows the exergy efficiency of some cooling cycles as a function of cooling of the temperature T_x (Gresin et al., 1977).

EXPANDERS AND STIRLING ENGINES

Reciprocating expansion engines and turboexpanders are widely used in Claude cycle cooling systems because they provide the most efficient process of cooling. Stirling engines are also effective coolers and are in general use in the USSR. Here we shall describe some research work on engines and several types of expanders and Stirling engines used in the USSR.

Expansion Engines

Modern expansion engines have changed considerably from the first engines designed by Claude at the beginning of the century. Contemporary reciprocating piston and cylinder expansion engines are similar in principle to the Claude engine but have a relatively high efficiency, high speed, and advanced systems of lubrication and valve action. Many different types of expansion engines exist. The most complete book on this subject was published by Arkharov (1974), with the title *Criogennie porshnevie detanderi* (*Cryogenic Expansion Engines*). The book covers all problems concerning expansion engines and includes thermodynamic analysis of the engines, computation of losses, internal heat transfer between gas and cylinder walls, and many other problems. The book includes a detailed description of different types of engines, and details problems of valve action and sealing

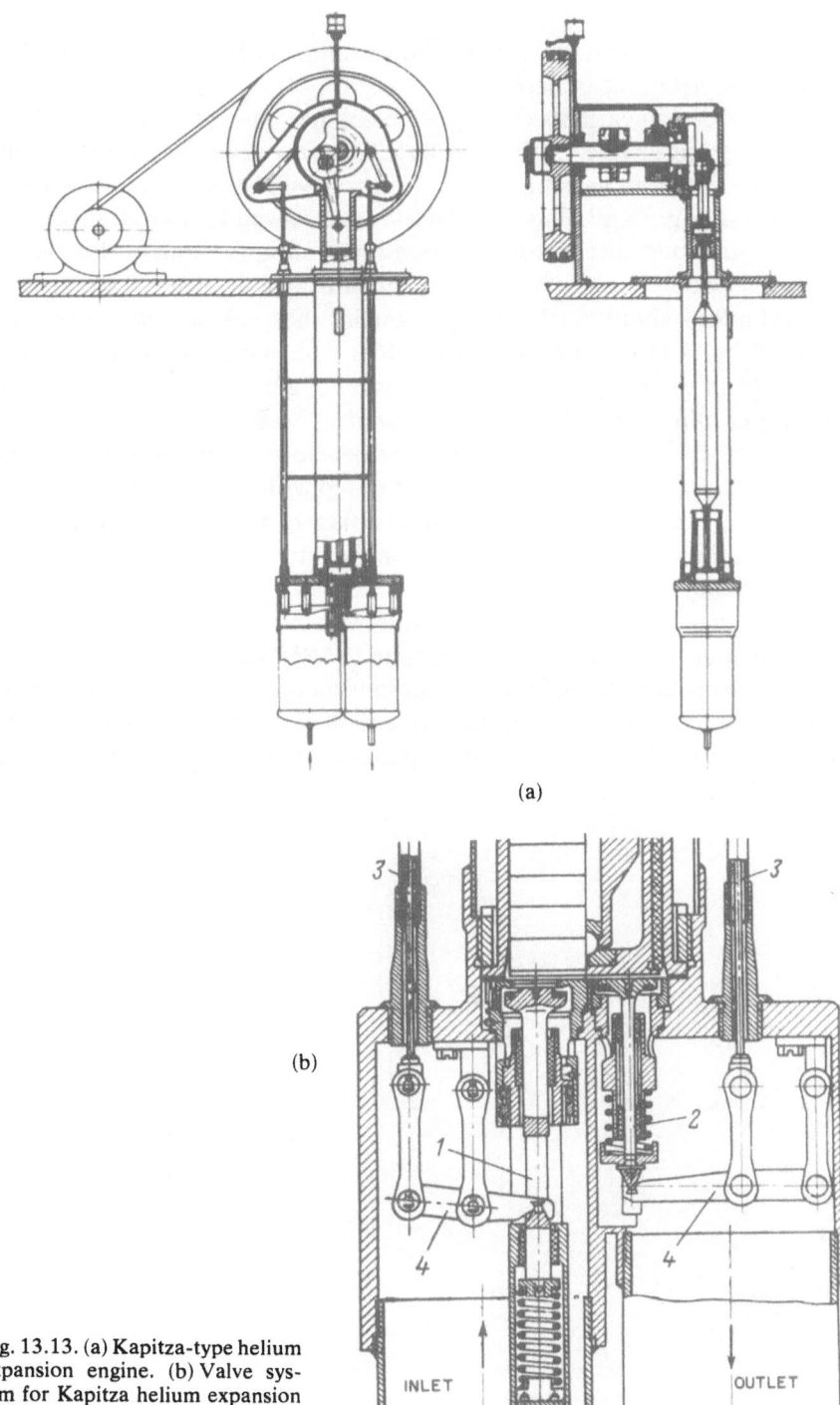

(a)

(b)

Fig. 13.13. (a) Kapitza-type helium expansion engine. (b) Valve system for Kapitza helium expansion engine.

INLET

OUTLET

of the piston and so on. It contains many examples of computing procedures and a large list of references.

Many design variants of expansion engines have been developed. The first expansion engine for helium liquefiers was developed by P. Kapitza in 1934. A modern variation of the Kapitza-type helium expansion engine is shown in Fig. 13.13. The piston rod is a thin-wall tube from nickel steel and the surface of the piston is covered by Teflon. Sealing of the piston is accomplished by the use of a small radial distance between the piston and the cylinder—about 0.01 mm. The engine has self-sealing valves, also covered by Teflon. Motion of the valves is provided by long, thin rods, driven from the crankshaft. The expansion engine runs at 300/400 revolutions per minute, and has an efficiency of 75%–80%.

Butkevich and Dobrov (1968) have developed expansion engines with the pistons sealed with leather rings saturated with paraffin. Steel expansion rings press the leather rings onto the surface of the cylinder. The heat of friction is absorbed by liquid nitrogen on the internal surface of the cylinder. The efficiency of engine is about 85%.

The first uniflow expansion engine with no outlet valve but a piston controlled port was developed by Gridin in 1958 (Strachovich, 1966). This system was sufficiently effective to permit increase in the engine speed to 1500 revolutions per minute. A cross section of the uniflow engine is shown in Fig. 13.14. The outlet flow of expanded gas takes place through ports

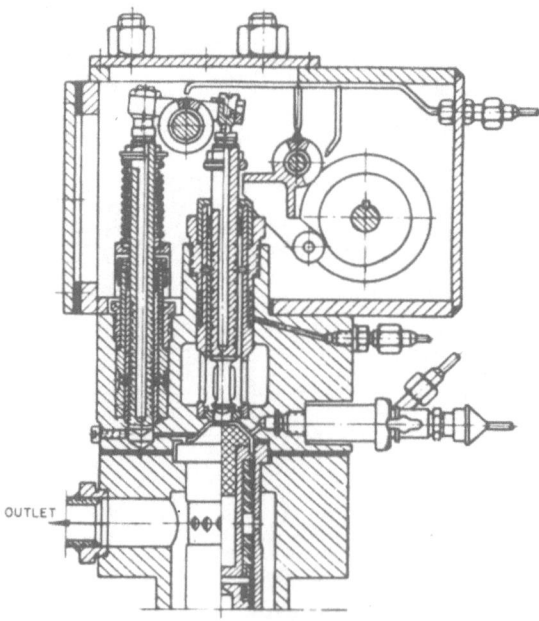

OUTLET

Fig. 13.14. Expansion engine.

in the wall of the cylinder (instead of through an exhaust valve), opened by descent of the piston.

Figure 13.15 shows an expansion engine with internally operated valves (Belushcin *et al.*, 1972). This engine was designed for the expansion of hydrogen. Both inlet and outlet valves are operated by special pins mounted on the piston and cylinder. The piston is sealed by leather rings installed in the warm zone of the cylinder. The expansion engine runs at 400 rpm and has an efficiency of 80% with partial liquefaction of gas during expansion.

A radically new system is presented in Fig. 13.16. It is an expansion engine with an electromagnetic inlet valve (Arkharov, 1974). The ball valve is opened and closed solely by electromagnetic means with no complicated valve gear. The engine is of the uniflow type with ported exhaust.

For regions of moderate temperature, the rotor-type expansion engines developed by A. Suslov and Y. Frolow are satisfactory. A cross section of

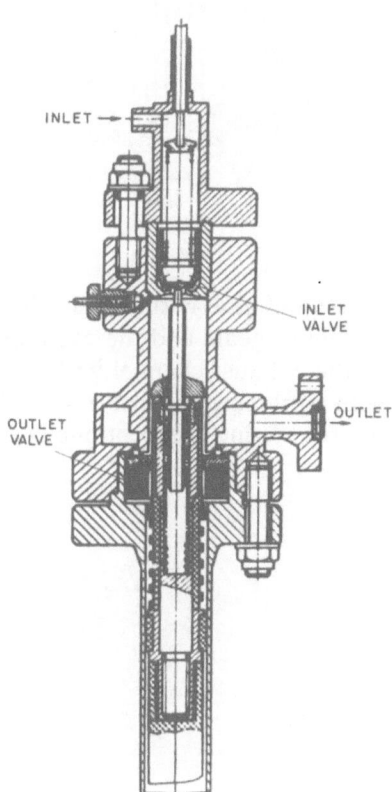

Fig. 13.15. Piston–cylinder assembly of the expansion engine with internal self-acting system of valves.

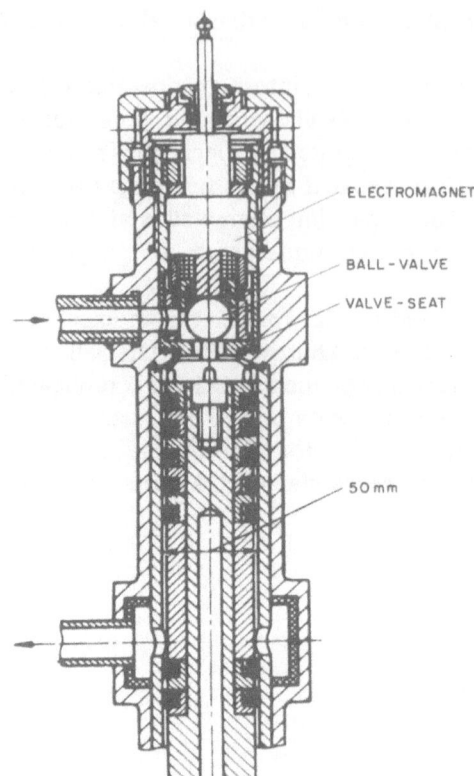

ELECTROMAGNET

BALL - VALVE

VALVE - SEAT

50 mm

Fig. 13.16. Expansion engine with electromagnetic inlet valve.

such an engine is shown in Fig. 13.17. Two rotors are synchronized by a system of gears and the expanded gas (shaded area) moves from the inlet to the exhaust channel without valves. The engine operates at 4000 rpm with a pressure ratio of $\pi = 5$. The efficiency is not high, only about 60%, but the engine is exceptionally simple and reliable.

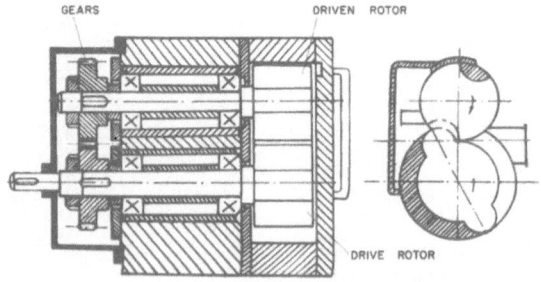

GEARS   DRIVEN ROTOR

DRIVE ROTOR

Fig. 13.17. Schematic diagram of rotary expansion engine.

Turboexpanders.

Expansion turbines are becoming more widely used in applications previously the domain of piston and cylinder expansion engines, i.e., high-pressure turboexpanders with inlet pressures of 10–20 MPa for helium liquefiers and refrigerators. Kapitza designed the first effective turboexpander for air separation plants in 1939. Research in this field continues today in the USSR. The most complete book on the subject was published by V. Epifanova (1974) with the title *Nizkotemperaturnie radialnie turbodetandery* (*Low Temperature Radial Turboexpanders*). It is a classic and is probably the most complete work on turboexpanders. The book describes basic principles for designing expansion turbines and includes complete thermodynamic and gas dynamic analyses of these engines. All of the large air-separation plants in the USSR are provided with turboexpanders, having efficiencies of 85%–90%. The turbines have regulated flow nozzles to maintain high efficiency when flow is changed. There are several types of turboexpander for inlet pressures as high as 20 MPa. These expanders can replace the huge piston-type expansion engines with an order of magnitude saving in weight.

Figure 13.18 shows a turboexpander for expansion of air from 4 MPa (inlet temperature 170 K) to 0.6 MPa (exhaust temperature 110 K). This is a single-stage inward-flow turbine with semishrouded mixed flow wheel. Between the cold and warm parts of the turbine there is an insulating gasket. The energy produced by the turbine is absorbed in the oil-filled hydraulic brake.

High-pressure turboexpanders have more or less the same cross section. One large helium plant generating 250 W of cooling at 4.5 K employs

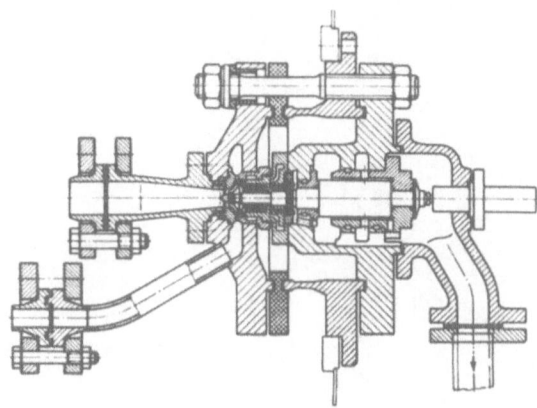

Fig. 13.18. Intermediate pressure turboexpander.

two turboexpanders for expansion of helium from 1.5 MPa to 0.15 MPa at inlet temperature levels of 70 K and 15 K (inlet pressure 0.8 MPa). These are radial-inflow turbines with gas bearings working at 180,000 rpm. Several other types of turboexpanders exist. Some have a shaft diameter as small as 10 mm and speeds up to 400,000 rpm.

Stirling Engines

Stirling engines operating as cooling systems are popular because of their high efficiency and other advantages—small weight, reliability, and so on. Much research has been done on these engines in the Russian cryogenic centers and many articles have been published on this subject. The book *Nizcotemperaturnie gasovine mashini* (*Low-Temperature Gas Machines*) (A. Arkharov, 1969) describes the various types of reciprocating regenerative cryogenic cooling engines and, especially, the Stirling engines. The book outlines a theoretical description of the process for machines with a closed regenerative thermodynamic cycle.

The proceedings of MHTS (1967–1976) contain very complete experimental and theoretical studies on Stirling machines. These articles discuss such problems as computation of losses and their influence on the efficiency of the machines, the real process in the internal spaces of the Stirling engine, the function of the regenerator, and system optimization. A recently published book by A. Gresin and V. Zinoviev (1977), *Mikrocriogennia*

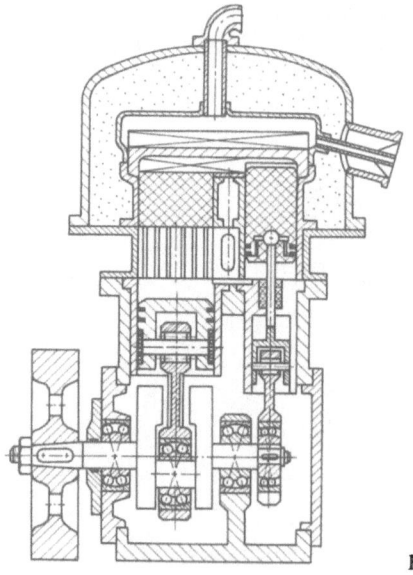

Fig. 13.19. Stirling engine.

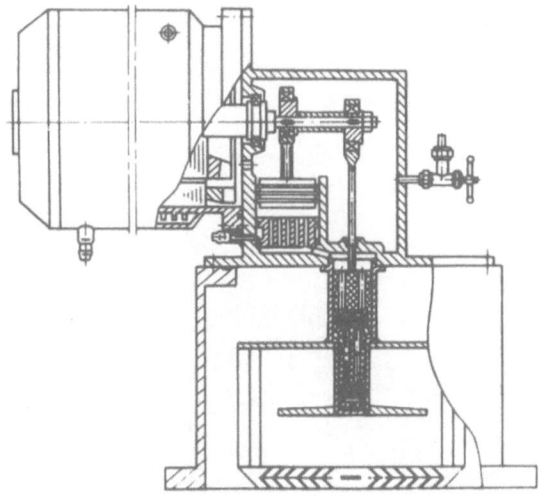

Fig. 13.20. Two-stage Stirling engine for cryopumping vacuum applications.

technica (*Microcryogenic Engineering*), outlines a general approach to optimization of Stirling engines, based on nonlinear programming.

The industry produces several types of Stirling engines having refrigerating capacities of 700 and 1000 W. These machines are based on the classical Philips design. Another variant with piston and displacer in separate cylinders is employed for the engine shown in Figure 13.19. Figure 13.20 shows a two-stage engine with separate cylinders intended for cooling at two temperature levels, 100 K (10 wt) and 25 K (4 wt). This unit is used for cryopumping. The regenerators of the engine are located in the displacer (Minaichev, 1976). Drawings of two small Stirling minicoolers are shown in Fig. 13.21, a two-stage engine and a single-stage engine reproduced from Gresin *et al.* (1977). The heat generated in the compression space is absorbed by the surrounding air. Both machines have regenerators placed inside the displacers and have unlubricated piston and displacer rings.

HEAT TRANSFER PROCESSES AND HEAT EXCHANGERS

In addition to the cooling processes of throttling and adiabatic expansion, the process of heat transfer is very important in cryogenic systems. Heat exchangers are perhaps the most widely used devices in cryogenics. The heat transfer process is important because of the significant change in the properties of materials and working fluids at low temperatures,

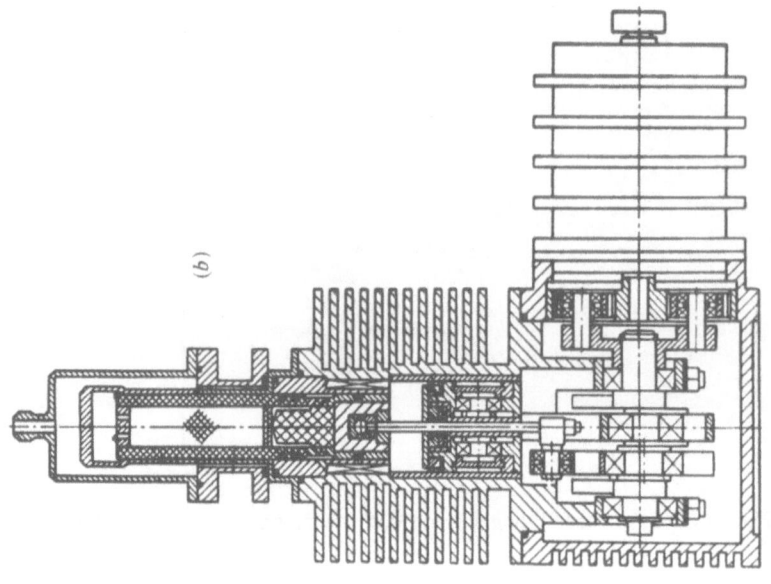

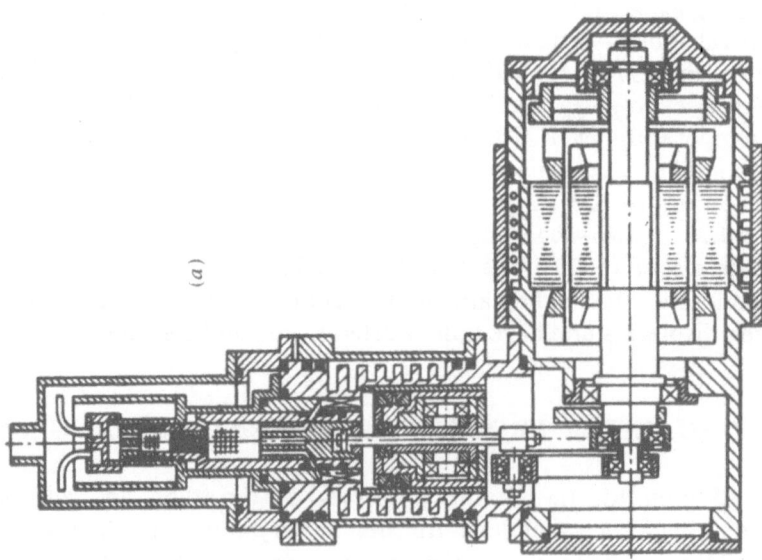

Fig. 13.21. (a) Single stage and (b) two-stage Stirling minicooler.

particularly when a fluid changes phase during cooling or heating. This section will mention some of the research and design work concerning heat transfer problems in cryogenic cooling systems in progress in the USSR.

Boiling of Cryogenic Liquids

Boiling heat transfer has been studied by many scientists and many experimental data exist. Unfortunately, the data do not correlate well. The results of experimental work depend very much on conditions of the experiments. The most comprehensive publication on the subject in the Russian literature is the book by Grigoriev *et al.* (1977), *Kipenie criogennich zidkostei (Boiling of Cryogenic Liquids)*. In this text the generalized data of many Russian and foreign researchers in this field are presented for the three regions of boiling: nucleate, film, and transient. The authors introduce and develop a new point of view on the problem, emphasizing the very strong dependence of the physical properties of the heating surface on the process of heat transfer. Grigoriev does not accept the contention that there is no difference between the boiling of cryogenic liquids and conventional liquids. The influence of the heating surface properties is especially significant for cryogenic liquids because they have a very small angle of wetting and because the physical properties of surface change very sharply in dependence of temperature in the region of low temperatures.

The experimental data, reproduced from Grigoriev *et al.* (1977), shown in Fig. 13.22 tend to support this point of view. It is clear from the figure that the intensity of heat flux for a copper surface is 10 times greater than that for a nickel surface. It is thought the heat flux for different surfaces depends upon a combination of the thermal conductivity, heat capacity, and density, $\lambda C_p \rho$. The boiling heat transfer coefficient, h, for nitrogen and helium appears to be proportional to $\lambda C_p \rho$ such that

$$h \sim [(\lambda C_p \rho)^{1/2}]^{0.5} \tag{13.10}$$

These experiments were carried out for the pool boiling at the end of a rod, having a diameter 8–10 mm, at atmospheric pressure and with a surface roughness about 0.01–0.05 mm. Most other authorities agree that boiling is a surface phenomenon but few include the properties of surface in the correlation. Grigoriev introduced the theoretical distribution of this phenomenon for calculating the radius of bubbles in connection with surface properties.

Other Russian authors who have extensively studied boiling of cryogenic liquids include Kirichenco *et al.* (1970) and Vishnev (1973).

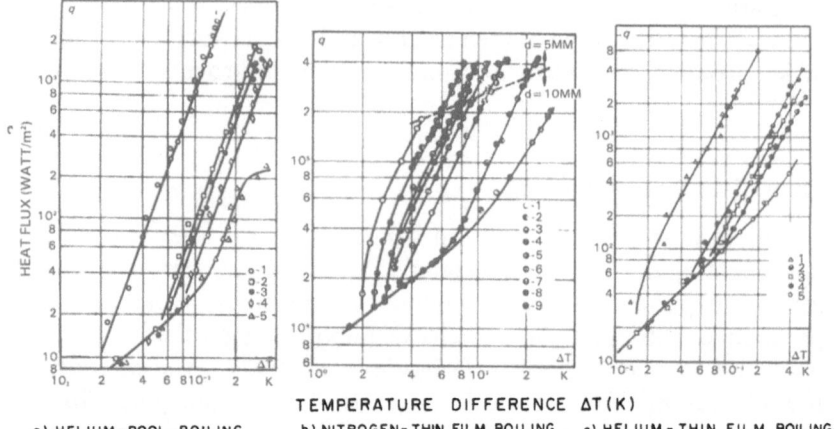

o) HELIUM-POOL BOILING   b) NITROGEN-THIN FILM BOILING   c) HELIUM-THIN FILM BOILING

Fig. 13.22. Heat flux boiling cryogenic liquids as a function of the temperature difference ΔT for nucleate boiling on the surfaces of different materials. (a) Pool boiling of helium. 1, copper; 2, bronze; 3, nickel; 4, brass; 5, stainless steel. (b) Boiling of nitrogen in thin films. 1, silver; 2, copper; 3, brass; 4, wood alloy; 5, nickel; 6, ferrum; 7, bronze; 8, stainless steel; 9, Teflon. (c) Boiling of helium in thin films. 1, copper; 2, bronze; 3, nickel; 4, brass; 5, stainless steel.

Convective Heat Transfer

Processes of convective heat transfer are commonly found in cryogenic cooling systems. For many purposes, especially for cooling superconducting systems, it is preferable to use cryogenic fluids at supercritical states rather than fluids experiencing a phase change. Study of convective heat transfer at supercritical states is, therefore, an important problem of cryogenic heat transfer.

Maliznev *et al.* (1972) presented experimental data on the heat transfer of helium in the supercritical state. They found there are two different kinds of heat transfer processes at this state: "normal" and "inhibited." The inhibited process of convective heat transfer near the critical state occurs because of the substantial difference in the properties of the fluid, especially heat capacity, in the cross section of the flow. When the fluid is cooling a duct with warm walls the density and heat capacity of the fluid near the wall of the channel is less than at the center of the channel so the heat transfer is reduced. "Inhibited" heat transfer usually occurs when the temperature T_m corresponding to peak heat capacity is between the wall temperature T_w and the temperature in the center of the flow T_b, i.e.,

$$T_b < T_m < T_w$$

Maliznev *et al.* (1972) gave correlations for both kinds of supercritical helium heat transfer inside a tube: For "normal" heat transfer his correlation is

$$\mathrm{Nu}_m = 0.037\mathrm{Re}^{0.8}\mathrm{Pr}\left(\frac{T_B}{T_m}\right)^{-0.6}\left(\frac{P}{P_c}\right)^{-0.2} \tag{13.11}$$

where P_c is the critical pressure and Nu, Re, and Pr are the Nusselt, Reynolds, and Prandtl numbers, respectively. For "inhibited" heat transfer this is

$$\mathrm{Nu} = \mathrm{Nu}_m K^{-1.4} \tag{13.12}$$

where

$$K = 687\left(\frac{q}{\rho wh}\right)\left(\frac{P}{P_c}\right)^{0.3}\left(\frac{T_B}{T_m}\right)^{0.9} \tag{13.13}$$

A transition from "normal" to "inhibited" heat transfer occurs when $K > 1$.

Heat Exchangers and Regenerators

Much work has been done designing and researching cryogenic heat exchangers. Compact regenerators are important devices used in Stirling engines and other types of cryogenic cooling engines. The packing utilized for Stirling engine regenerators is often constructed of thin wires or screens with a wire diameter of 0.03–0.04 mm. The heat transfer coefficient for such surfaces has been studied by many investigators but the data do not correlate well. Mikulin *et al.* (1972) found that the packing of regenerators works as a stabilizer of turbulent oscillations. As a result the heat transfer coefficient for long packed beds was smaller than for short beds. Figure 13.23 shows the results of this experimental work. When the number of screens exceeded 200, no further change in heat transfer coefficient was observed. The correlation for heat transfer in screen regenerators when the number is greater than 200 is

$$\mathrm{Nu} = 0.05\mathrm{Re}^{0.85} \tag{13.14}$$

Pron'ko *et al.* (1976) used adsorbents for packing Stirling regenerators operating at temperatures less than 20 K. These materials adsorb gas (helium) and therefore the heat capacity of such packing remains high at low temperatures, whereas the heat capacity of metals tends to zero. It has been found that silica gel coated with silver is the best material for such regenerator packing.

Classical forms of the Hampson coil heat exchanger have been improved by new design features. One very effective measure is the heat exchanger where the tubes are finned by wire coils. It is important to note

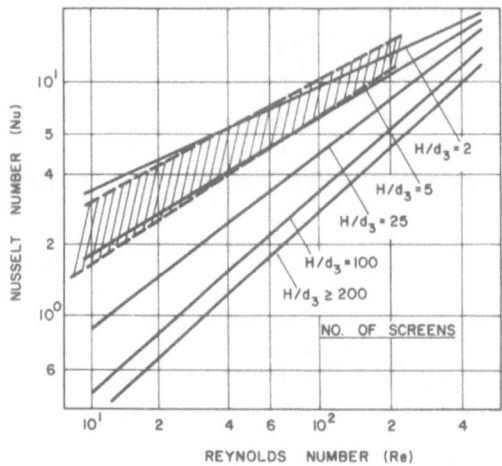

Fig. 13.23. Nusselt number as a function of Reynolds number for Stirling engine screen regenerator packing. H/d_3 is the number of screens in the packing.

that the coil does not need to be soldered to the surface of the tubes (Ousikin, 1972). This heat exchanger is shown in Fig. 13.24. The wire coil serves two purposes: first, it guarantees uniform spacing between the tube coils, thereby distributing the flow equally across the flow section; secondly, the wire acts as a fin and increases the external surface of the tubes. Typical dimensions are tube diameter 4 mm, wire diameter, 0.8 mm. The heat transfer coefficient is generally about the same as for Collins-type heat exchangers. Modern Russian cryogenic cooling systems are usually provided with such types of heat exchanger.

Cryogenic Insulation

There are many heat transfer problems related to insulation with systems for storing and transporting liquefied gases. In this field the most comprehensive, fundamental book has been presented by Kaganer (1972), *Teplovaia isoliatia v technike nizkich temperatur* (*Heat Insulation in Cryogenic Engineering*). In this book Kaganer gives analytical methods for many problems of cryogenic insulation, including contact heat transfer, radiation heat transfer, residual gas heat transfer, and the efficiency of shields. This reference includes tables of the properties of different insulating materials: mechanical, physical, humidity, structure, and other properties. There are detailed analyses of evaluated porous insulation and multilayer superinsulation. The book has been translated into English.

Kaganer has also published many articles concerning problems of storing liquefied gases, including the influence of the level of the liquid on

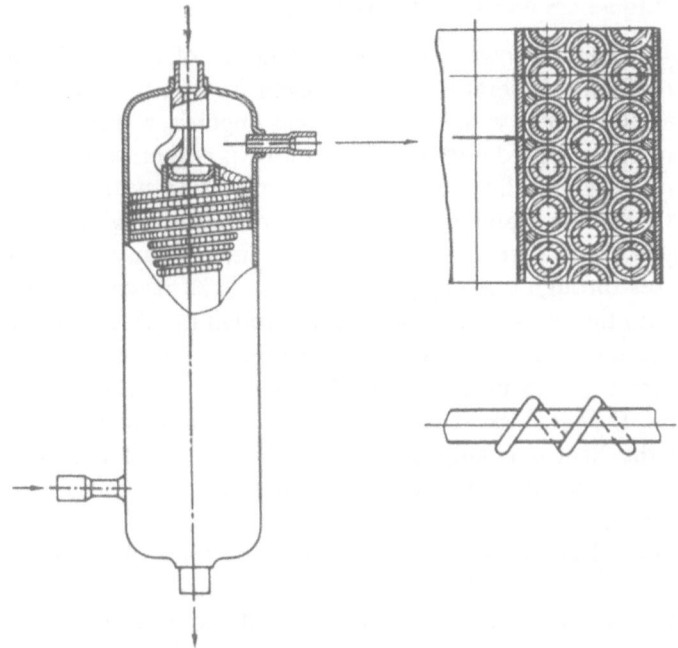

Fig. 13.24. Coiled heat exchanger with tubes finned by wire coils.

losses in cryogenic Dewars, thermal resistance of multilayer support system, and others (see Belyakov *et al.*, 1971). Gorbachev (see Arkharov *et al.*, 1975) has given a detailed analysis of the well-known problem of complicated heat transfer in vapor-cooled vent tubes. He has also analyzed the efficiency of vapor-cooled shields in cryogenic insulation.

There are many other research studies concerning insulation and handling of liquefied gases. These investigations have resulted in design improvements to many types of cryogenic equipment including stationary and transport Dewar vessels, cryostats, transfer lines, and cryogenic pressure vessels.

CRYOGENIC COOLING SYSTEMS AND SOME TYPES OF CRYOGENIC EQUIPMENT

In this section of the review, we include a description of several types of commercial cryogenic systems and equipment available in the USSR.

Helium Liquefiers and Cooling Systems

Figure 13.25 shows a schematic diagram of a small helium cooling system (helium refrigerator) having a capacity of 1.5 W at 4.2 K. This is a Claude-cycle system with two expansion engines, a throttling valve, and no precooling of liquid nitrogen. The helium is compressed by a reciprocating compressor and then passed to the purification system before entering the cold box. The mainstream of helium passes through counterflow heat exchangers and is divided into several streams. Twenty-five percent of the flow passes through one expansion engine, 40% passes through the other engine, and the remaining 35% passes through the throttling valve, where it is expanded to 0.7 MPa. After passage through further heat exchangers, the helium expands in a second JT valve and the vapor–liquid mixture enters the Dewar flask. The low-pressure gas returns to the compressor through the heat exchangers. The initial pressure of the helium is 2.0–2.5 MPa. The plant is simple and reliable, and the compressors are not oil-lubricated.

Figure 13.26 shows the flow diagram of the well-known and widely used general-purpose helium plant having a refrigerating capacity of 250 W at 4.5 K. This can be increased to 400 W if required. It can operate as a liquefier producing about 100 liters per hour of liquid helium or alternatively as a cold-gas refrigerating system. The operation of the plant is based on the Collins-type Claude cycle with precooling by liquid nitrogen. It has two turboexpansion engines and one stage of throttling. The helium is compressed to 1.6 MPa in the compressor at the rate of 1200 m^3/hr. Purification of the mainstream from oil contamination is accomplished by

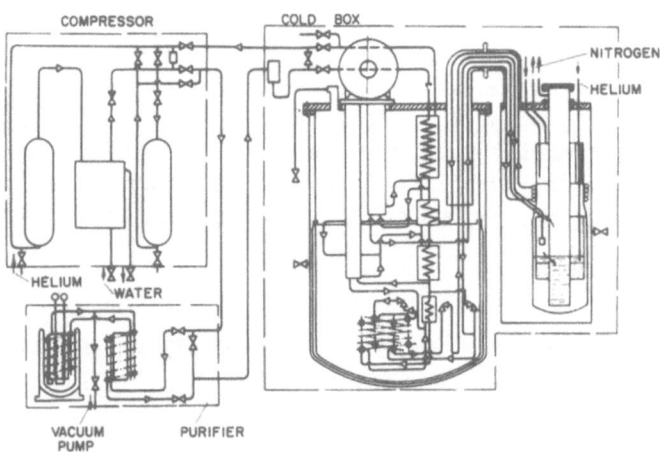

Fig. 13.25. Schematic diagram of small helium cooling system of 1.5 W capacity at 4.2 K.

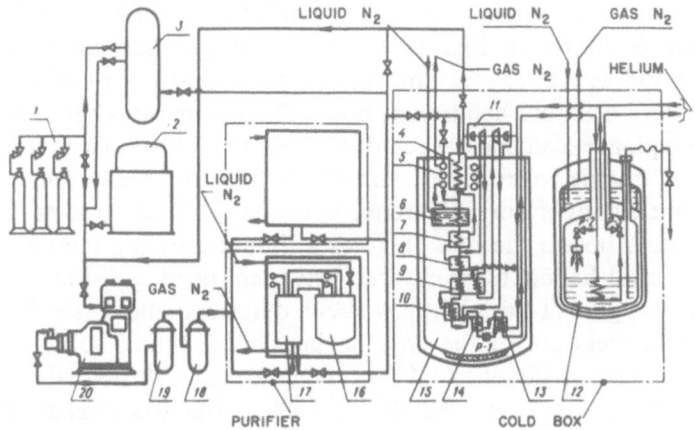

Fig. 13.26. Schematic diagram of the general-purpose plant having a capacity of 250 W at 4.2 K. 2, helium gas receivers; 4, 5, heat exchangers; 6, liquid nitrogen bath; 7–10, heat exchangers; 11, turboexpanders; 12, Dewar; 13, 14, heat exchangers; 15, cold box; 16, 17, purification system; 18, 19, oil filters; 20, compressor.

filters and purification from other impurities in the adsorption block at liquid-nitrogen temperatures. The helium then enters the cold box and passes through a series of recuperative heat exchangers. Liquid nitrogen is used only when the plant works as a liquefier. After the upper stage heat exchanger, part of the helium stream (74%) passes to the first turboexpander and expands to 0.8 MPa and cools to a temperature of 38 K. The same stream then passes through the intermediate heat exchanger to a second turboexpander where it expands to near atmospheric pressure, cooling to 10 K. This stream then returns back through the heat exchangers. The remaining 26% of the mainstream at 1.6 MPa passes through a further recuperative heat exchanger and expands through the throttling valve, to a pressure of 0.66 MPa. A second stage of expansion takes place in the Dewar container. The flow then passes across the heat load at a pressure somewhat above atmospheric and returns to the compressor inlet through the heat exchangers.

The turboexpanders of this system have gas bearings. The heat exchangers are of the coil type with tubes finned by the wire as discussed above. The plant is equipped with vacuum multilayer superinsulation. Consumption of power is 200 kW. The plant as a liquefier requires 120 liters per hour of liquid nitrogen. General-purpose plants such as this one are increasingly popular since they can be used as liquefiers or refrigerators.

Another general-purpose plant based on the cycle introduced by Kapitza in 1934 (see Arkharov, 1975) with liquid-nitrogen precooling, has a single expansion engine and one throttling stage. The flow diagram of this cycle is shown in Fig. 13.1. This plant has the same rate of gas consumption as the 250-W unit but produces 500 W of cooling at 4.3 K or 150 liters of liquid helium per hour. These superior characteristics are the consequence of an increased initial pressure (2.4 MPa), a high-efficiency expansion engine (over 80%), and highly effective heat exchangers with only a small temperature difference between the fluid streams. The plant has two cold boxes; the first operates at liquid-nitrogen temperature, the other at liquid-helium temperatures. The reciprocating piston and cylinder expansion engine of this plant have been described above (see Fig. 13.14). Heat exchangers are of the type having coiled, finned tubes. When the plant operates as a liquefier, 64% of the flow passes through the expansion engine and the rate of liquefaction is 8.6% of the flow stream. The liquid nitrogen in the precooling bath is maintained at low temperatures at a pressure of 0.02 MPa under vacuum.

Cryogenic Vessels and Associated Apparatus

Storage and transportation of liquefied gases at cryogenic temperatures pose a number of engineering problems including the provision of highly efficient insulation to ensure a small loss by evaporation. There are many different types of cryogenic containers, from small Dewar vessels having the capacity of a few liters of liquid to very large stationary vessels with capacities of hundreds of cubic meters. Figure 13.27 shows a stationary vertical storage vessel for liquefied gases available commercially (Anon, 1979). The vessel is of double-wall metal construction with the inner vessel fabricated of stainless steel and the external vessel of carbon steel. The space between the vessels is filled with highly efficient multilayer superinsulation. The overall capacity of the vessel is 63 m^3. The internal pressure is 0.4 MPa. The losses per day amount to 0.25% for oxygen and 1.2% for hydrogen. Similar vessels are available in capacities up to 225 m^3.

A transport vessel for liquid oxygen or nitrogen is shown in Fig. 13.28. The low weight and high strength make the vessel suitable for transporation by aircraft. The volume of vessel is 1155 liters, the internal pressure is 0.25 MPa, and the empty weight is 1160 kg. The insulation of this vessel is evacuated powder, the rate of evaporation is 1.3% per day. The vessel has a system for increasing the pressure when evacuating liquid from it. Other vessels of this type are available to a capacity up to 7000 liters.

Figure 13.29 shows a typical Dewar vessel for the storage of liquid helium (Anon, 1975). The insulation is vacuum, multilayer superinsulation with two copper shields cooled by vent vapor. Capacity of the Dewar is 520 liters. The rate of evaporation loss is 1% per day. The weight of the

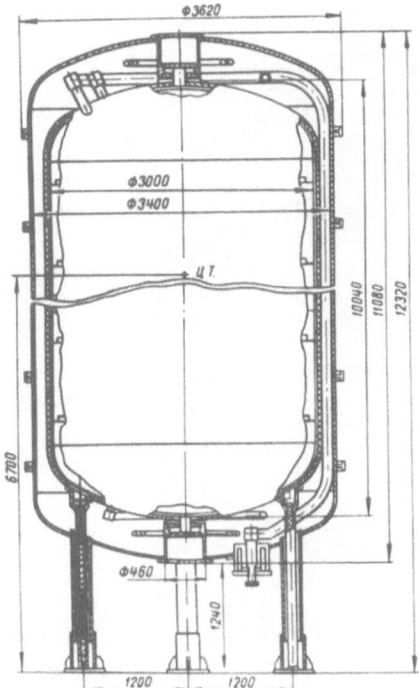

Fig. 13.27. Storage vessel for liquefied gases.

Fig. 13.28. External view of the transport vessel for liquid oxygen or nitrogen.

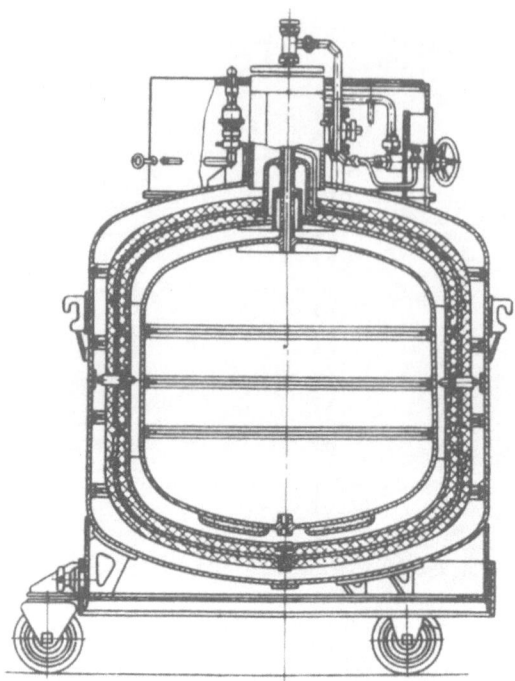

Fig. 13.29. Dewar storage vessel for liquid helium.

Dewar filled with liquid is 500 kg. It does not require liquid nitrogen. The inner vessel is constructed of stainless steel and the internal pressure is limited to 0.17 MPa.

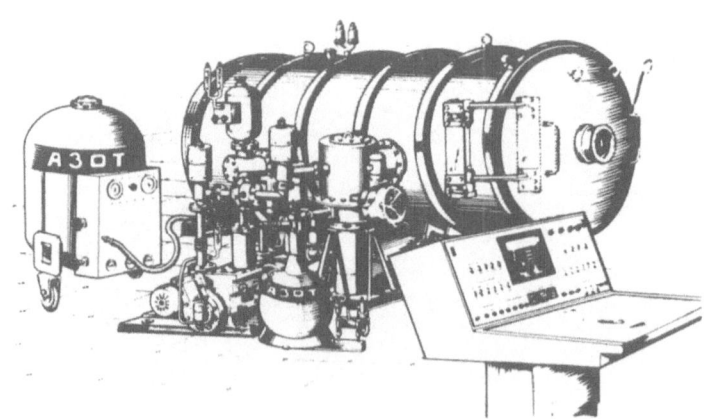

Fig. 13.30. Cryogenic pressure chamber.

Experimental work at low temperature and high vacuum conditions may be carried out with the cryogenic pressure chamber shown in Fig. 13.30 (see Anon, 1979). The principal units are the cryogenic pressure chamber, the liquid-nitrogen supply system, and the vacuum system. The pressure chamber has a working volume of $1 \, m^3$. Liquid-nitrogen-cooled shields are available to cool the internal space and also serve as cryogenic vacuum pumps to reduce the pressure inside the chamber to 10^{-6} mm Hg. The liquid-nitrogen supply system is based on the cryogenic container shown in Fig. 13.28. The system consumes 4.8 kg of liquid nitrogen per hour and requires a power output of 8 kW.

REFERENCES

Anon. (1971). "*Technica nizkich temperatur*, Proceedings of Conference, Leningrad.

Anon. (1975). *Cryogennoe oborudovanie*, Catalog p-1 TZINTICHIMNEFTEMASH, M., p. 55.

MHTS (1967–1976). "*Glubokyi cholod i condizionirovanie*", Proceedings of MHTS No. 124, 1967, p. 370; No. 132, 1969, p. 368; No. 138, 1972, p. 270; No. 149, 1974, p. 192; No. 193, 1974, p. 242; No. 239, 1976, p. 150.

Anon. (1978). *Criogennoe i kislorodnoe oborudovanie*, Catalog. Techmash-export 130224, USSR. Moscow, p. 212.

Arkharov, A., *et al.* (Nov. 1968). *Cryogenic Engineering* (Two Vols.), Foreign Technology Division, US Department of Commerce, NBS AD 685-948.

Arkharov, A. M. (1969). *Nizcotemperaturnie gasovine mashini*, M. Mashinostroenie, p. 222.

Arkharov, A. M. (ed.) (1974). *Criogennie porshnevie detanderi*, A. M. Mashinostroenie, p. 216.

Arkharov, A., Belyakov, B., Malkov, M., and Vneshtorgizdat, M. (1975). *Voprocy sovremennoi criogenici.*

Arkharov, A. M., Marfenina, J. V., and Mikulin, E. I. (1979). *Teoria i raschet criogennich system*, M. Mashinostroenie, p. 415.

Belushkin, V. A. *et al.* (1971) "Ozizitel vodaroda c detanderom na vsem potoke szatogo gasa." *Technica nizkich temperatur*. Proceedings of conference, Leningrad, p. 49–56.

Belushcin, V. A., and Gotvanskyi, N. F. (1972) "Novyi porshnevoi detander s vnutrenim privodom klapanov dlia vodorodnogo ozizitelia." *Zh. Khim. Neftianoe Mashinostroenie*, No. 1, pp. 36–39.

Belyakov, V. (ed.) (1971) "Apparaty i mashiny kislorodnich i criogennich ustanovok," *Proceedings of VNIICRYOGENMASH* No. 13, M. Mashinostroenie.

Belyakov, V. P., and Shein, G. F. (1977). "Razvitie cryogennogo mashinostroenia v USSR.," *Zh. Khim. Neftianoe Mashinostroenie*," No. 10, pp. 36–39.

Brodiansky, V. M. (1973). *Exergetichesky metod termodinamicheskogo analiza*, M. Energia, p. 296.

Butkevich, I. K., and Dobrov, V. M. (1968). "Helievyi porshnevoi detander s manzetnim uplotneniem porshnia," *Zh. Khim. Neftianoe Mashinostroenie*," No. 8, pp. 4–5.

Epifanova, V., and Akselrod, L. (1973). *Razdelenie vosducha metodom glubokogo ochlazdenia*, Mashinostroenie M., Vol. I, p. 472; Vol. II, p. 568.

Epifanova, V. I. (1974). *Nizkotemperaturnie radialnie turbodetandery*, M. Mashinostroenie, p. 445.

Esselson, B. N., *et al.* (1973). *Rastvory kwantovich zidcostei*, M. Nauka publishers, p. 421.

Fastovsky, V. G., Petrovsky, Ju. V., and Rovinsky, A. E. (1974). *Criogennaia Tehnica*, M. Energia, p. 496.

Gersh, S. J. (1957). *Glubokoe ochlazdenie*, Vol. 1, Gosenergoizdat, p. 392.
Gersh, S. J. (1960). *Glubokoe ochlazdenie*, V. II, Gosenergoizdat, p. 495.
Gochstein, D. P. (1966). *Sovremenie metody termodinamichescogo analiza energeticheskich ustanovok*, M. Energia, p. 365.
Gresin, A. K., and Zinoviev, V. C. (1977). *Mikrocriogennia technica*, M. Mashinostroenie, p. 226.
Grigoriev, B. A., Pavlov, Y. M., and Ametistov, E. V. (1977). *Kipenie criogenich zidkostei*, M. Energia, p. 288.
Kaganer, M. G. (1972). *Teplovaia isoliatia v technike nizkich temperatur.*
Kalitin, P. P., Pron'kov, V. G., and Davidenkov, I. A. (1977). "Osnovnie napravlenia razvitia criogennogo helievigo oborudovania," *Zh. Khim. Neftianoe Mashinostroenie, No.* 7, *pp.* 5–8.
Kapitza, P. H. (1935). "Adiabaticheskyi metod ozizenia helia," *Usp. Fiz. Nauk* **16**, (2).
Kapitza, P. H. (1939). "Turbodetander dlia poluchenia nizkich temperatur i ego primenenie dlia ozozenia vosducha," *Zh. Tekh. Fiz.* **IX**, (22).
Kapitza, P. H. (1959). "Raschet helievogo ozizitelnogo tzikla s cascadnim vklycheniem detandera," *Zh. Tekh. Fiz.* **XXIX**, (4), 427–432.
Kirichenco, V., and Dolgoi, M. (1970). "Issledovanie kipenia v ploskich naklonnich conteinerach modeliruyshich stable gravitationnie polia", *Zh. Teplophizica Visokich Temperatur* **1**, (1), 130–155.
Koshelev, P. F., and Belyaev, S. B. (1967). *Prochnost i plasyichnost constructzionnich materialov pri nizkich temperaturach*, *M*. Mashinostroenie, p. 315.
Maliznev, G., and Pron'o, V. G. (1972). *Osobennosti teploobmena pri turbulentnom techenyi helia*, v. sverchkritichescom sostoianyi., *Inzh.-Fiz. Zh.* **XXII**, (5).
Malkov, M. P., et al. (1961). *Videlenie deiteria is vodaroda metodom glubokogo ochlazdenia*, Cosatomizdat, M., p. 150.
Malkov, M. P. (1973). "*Spravotchnik po phizico-technicheskim osnovam cryogenici*," M. Energia, p. 390.
Mikulin, E. E. (1969). *Criogennaia technica*, M. Mashinostroenie, p. 270.
Mikulin, E. I., and Shevich, V. A. (1972). "Experimentalnoe issledovanie teploobmena v setchatich matrizach," *Inzh.-Fiz. Zh.* **XXII**, (6).
Mikulin, E. I., Marfenina, J. V., and Arkharov, A. M. (1975). *Technica nizkich temperatur*, Second edition, Energia, p. 392.
Minaichev, V. E. (1976). *Vacuumnie cryonasosi*, M. Energia, p. 150.
Neganov, N., Borisov, N., and Liburg, M. (1966). *Zh. Eksp. Teor. Fiz.* **50**, 1045.
Ousikin, I. P. (ed.) (1972). *Ustanovki, mashiny i apparaty cryogennoi techniki*, Album of drawings p. I, M. Pishevaia promishlennost, p. 172.
Pron'ko et al. (1976). "Some problems of using adsorbents as matrix material for low-temperature regenerators of cryogenic refrigerators," *Proc. Sixth International Cryogenic Engineering Conference*, Grenoble, France, IPC Press, Guildford, U.K.
Rabinovich, V. A., and Vasserman, A. A., et al. (1976). *Teplophizicheskie svoistva neona, argona, kriptona i xenona*, M. Standards p. 636.
Socolov, E., and Brodiansky, V. (1968). *Energeticheskie osnovy transformatzyi tepla i processov ochlazdenia*, M. Energia, p. 336.
Strachovich, K. M. (ed.) (1966). *Rasshiritelnie mashini*, Mashinostroenie, p. 336.
Suslov, A. D. (1969). "Metody issledovania i rascheta mashin so vstroennymy teploobmennimi apparatani," Proceedings of MHTS No. 132, M., p. 80–86.
Suslov, A. D., Gorshcov, A. M., and Maslakov, V. B. (1978). *Drosselnie microochladitely*, Mashinostroenie, M., p. 140.
Vasserman, A. A., Kazavchinskyi, J. Z., and Rabinovich, V. D. (1966). *Teplophizicheskie svoistva zidkogo vosducha i ego componentov*," M. Hauka, p. 376.
Vasserman, A. A., and Rabinovich, V. D. (1968). *Teplophizicheskie svoistva zidkogo vosducha i ego componentov*, M. Standards, p. 240.
Vishnev, I. P. (1973). "Vlianie orientatzyi poverchnosti nagreva v gravitatzionnom pole na crisis puzirkovogo kipenia zidkostey," *Inzh.-Fiz. Zh.* **24**(1), 59–64.

Chapter 14

Cryogenic Engineering and Cryocooler Development in Japan

Yoshihiro Ishizaki

HELIUM LIQUIFIERS AND REFRIGERATORS

The first helium liquefier in Japan was an ADL Collins liquefier installed at Tohoku University in 1952. The first successful helium liquefier built in Japan was a cascade-type machine which produced liquid helium in 1954. The liquefier was constructed at the University of Tokyo under the supervision of K. Oshima with the support of the Japan Oxygen Co. This was the start of a period during which significant advances were made in the field of solid state physics and its applications. In connection with the rapidly growing activities, many helium liquefiers were imported not only for basic research at universities but also in industry. During this period about fifty liquefiers of the Collins type were built at several companies in Japan.

In 1974, there were ninety-nine liquefiers operating in Japan; of these forty-eight units were made in Japan and fifty-one were imported units (Oshima *et al.*, 1974). The units installed since then have been mostly imported machines because of their low cost, high performance, and low maintenance requirements. In 1977, a 750-l/h hydrogen liquefier was imported from France in connection with the development of H_2/O_2 rocket engines.

Although several score helium liquefiers and cryogenic refrigerators have been in operation for several years in Japan, they have been used mainly for research in laboratories. Problems related to the reliability of

Yoshihiro Ishizaki • Yamanouchi 253–5, 247 Kamakura, Japan.

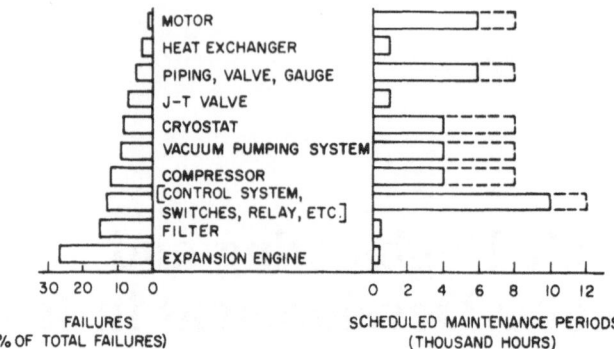

Fig. 14.1. Failure record and scheduled maintenance periods for cryogenic refrigerator components.

refrigeration systems have now arisen with the advent of large development projects on MHD power generation, superconducting magnetic levitation of trains, superconducting power cables, superconducting generators, and, more recently, developments in application to high-energy physics and large-scale superconducting magnets for the experimental fusion reactor.

The first investigation of reliability in this country was conducted by K. Kasamatsu and his collaborators (1971). Figure 14.1 shows the relation between elements, maintenance period, and percentage of failure in reciprocating-expander cryogenic refrigerators and helium liquefiers. It can be seen that the failure rate of the reciprocating expanders was the highest and, conversely, the maintenance period was the shortest. Therefore, with the object of developing reciprocating expanders with improved reliability, they point out the importance of decreasing the number of components and the simplification of the control mechanism.

A similar investigation was conducted by a committee of the Cryogenic Association of Japan chaired by K. Yasukochi (1978). According to this investigation reviewing a total of a hundred and eleven refrigerators and liquefiers below 20 K, 492 malfunctions were reported as shown in Table

Table 14.1. Refrigerator Class, Number Installed, and Failures Frequency

Class	Refrigerator	Number installed	Failures reported	Failures per unit installed
I	Medium size with turboexpander	10	53	5.3
II	Medium size with reciproexpander	58	356	6.1
III	Small size (4.2 K)	7	10	1.4
IV	Small size (20 K)	28	39	1.4
V	Hydrogen liquefier	8	34	4.3
Total		111	492	4.4

**Table 14.2. Location and Frequency of Failure for Five Different Classes of Machine
(See Table 14.1)**

Location of failure	Class of refrigerator					
	I	II	III	IV	V	Total
Small refrigerator	0	0	0	19	2	21
Reciprocating compressor	9	70	0	7	7	93
Screw compressor	9	0	0	5	0	7
Other compressors	0	0	0	0	1	1
Reciproexpansion engine	0	116	0	1	2	119
Turbo expander	13	0	0	0	1	14
JT Valve	2	9	1	0	1	13
Roots blower	0	1	0	0	0	1
Cryogenic ejector	0	0	0	0	0	0
Heat exchanger	1	29	2	0	0	32
Regenerator	0	0	0	2	0	2
Transfer tube	3	13	2	1	3	22
Dewar vessel	3	7	0	0	1	11
Recovery compressor	3	28	0	0	0	31
Oil separator	0	2	0	1	0	3
Oil absorber	0	2	0	0	0	2
Purifier	2	11	1	0	1	15
Inner absorber	2	4	0	0	0	6
Gas holder	1	18	0	0	0	19
Balloon	1	2	1	0	0	4
Vacuum insulation	3	5	0	0	1	9
Vacuum line	3	6	0	0	2	11
Piping	0	14	2	3	1	20
Valve	4	11	1	0	1	17
Cooling tower	0	4	0	0	0	4
Control system	0	4	0	0	10	14
Drier	1	0	0	0	0	1
Total failures	53	356	10	39	34	492

14.1. Table 14.2 shows the location of the malfunction. It can be seen that the number for reciprocating expanders is the largest (119 cases) followed by the reciprocating compressor.

Table 14.3 shows the laboratories where research related to the refrigeration and liquefaction systems are presently being conducted in Japan.

COMPONENT DEVELOPMENT

Dry Helium Compressor

The compressor is a key component affecting both the efficiency and reliability of a refrigeration system. Figure 14.2 shows an experimental

Table 14.3. Research Groups in Japan for Cryogenic Refrigeration and Liquefaction Systems

Institute	Research being conducted	Purpose	Leader
Aisin Seiki Co.	R. L. system, components	Maglev	T. Tani
Electrotechnical Laboratory	R. L. system, distribution, heat transfer	Fusion, MHD, power cable	Y. Akiyama
Hitachi Ltd	R. L. system, components	Maglev, fusion	S. Shimamoto
Japan Atomic Energy	R. L. system, cryopump	Fusion	H. Nakajima
Japanese National Railways	R. L. system	Maglev	I. Yamashita
Mechanical Engineering Inst.	Cryoengine	Power generation	O. Ogino
Mitsubishi Electric Co.	R. L. system	Maglev, synchronous condenser	
National Laboratory for High Energy Physics	Conceptional design	Bending magnet	M. Kobayashi and H. Hirabayashi
Nihon University	Heat transfer, instability	Fusion	Y. Matsubara
Sumitomo Heavy Industries	R. L. system, components	General	K. Kikuchi
Tohoku University	Mini R. L. system	SQUID	T. Ohtsuka
Toshiba Corp.	R. L. system, components	Maglev	S. Nakaya
University of Tokyo Mechanical Eng.	Distribution, dynamics	Fusion	T. Saito
Nuclear Eng.	System design, heat transfer	Fusion	M. Akiyama
Nuclear Eng.	System design, components	General	Y. Ishizaki

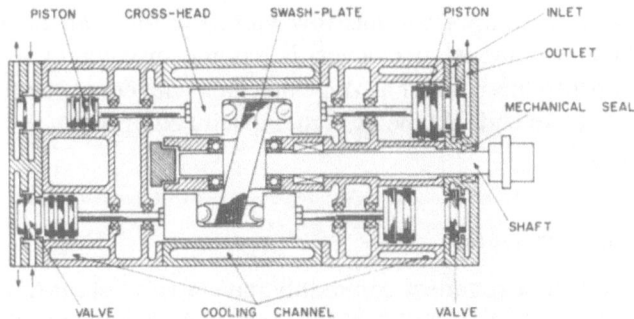

Fig. 14.2. Cross section of three-stage dry-lubricated helium compressor with swash-plate drive.

three-stage dry helium compressor constructed at the University of Tokyo (Ishizaki, 1976). A swash-plate piston driving mechanism located at the center of the cylinder block drives four pistons, two on each side of the swash plate and moving counterwise to each other. The first compression stage consists of four 105-mm-diam pistons. The second and third stages consist of two pistons with diameters of 86 and 56 mm, respectively. Figure 14.3 shows the experimental performance of the prototype. The isothermal compression efficiency is relatively good for such a low-capacity machine. The compressor is quiet with the noise level below 78 dB at 1 m distance at 1000 rpm.

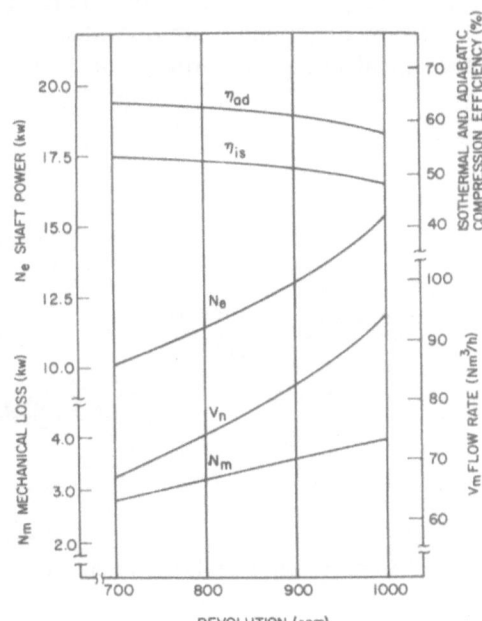

Fig. 14.3. Performance characteristics of three-stage dry-lubricated helium compressor with swash-plate drive. (After Ishizaki, 1976.)

This type of compressor has low vibration and can be made both light-weight and compact. However, it has the disadvantage of a larger number of parts than a crankshaft-type machine with consequent difficulties in maintenance. The ideal type of high-capacity compressor is probably of the rotary type.

Reciprocating Expanders

Reciprocating expanders commonly have a valve stem to operate the inlet and outlet valves. Expanders having a high expansion efficiency can be constructed with the cam driving the valve stems and designed so that the opening and closing of the valve is optimum. The disadvantage of this is that the expander tends to become complicated and somewhat heavy because of the number of components required in the crank case. Heat leaks through the valve stem and sheath are also significant. To obviate these problems, a double-acting-type expander with no valve stem as shown in Fig. 14.4 was constructed in 1972. In this expander, temperatures of 10 K are produced in one expansion space, 70 K in the other. Temperatures in the buffer and the middle region of the piston are about 50–60 K due to helium gas leakage from top to the bottom.

Figure 14.5 shows one of the expanders with self-acting valves con-structed by the JNR (Japanese National Railways) in 1977. It is presently being used as the on-board helium refrigerator (30 W at 4.4 K) for the Maglev (magnetic levitation) test vehicle (Sekiguchi *et al.*, 1978). The valves in this type of expander are operated by a combination of helium pressure, mechanical springs and motion of the piston.

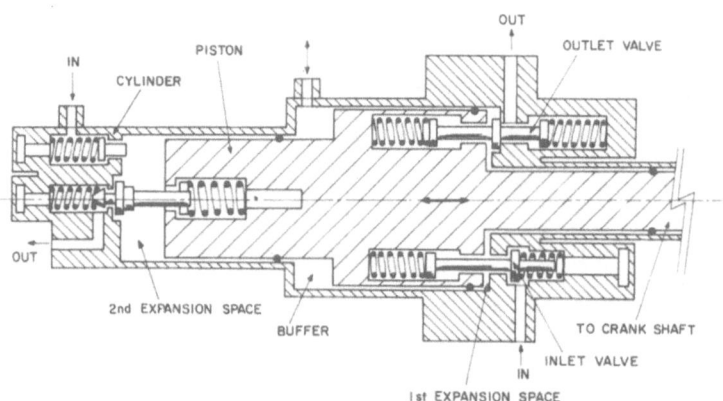

Fig. 14.4. Cross section of the double-acting expansion engine with self-acting inlet and outlet valves.

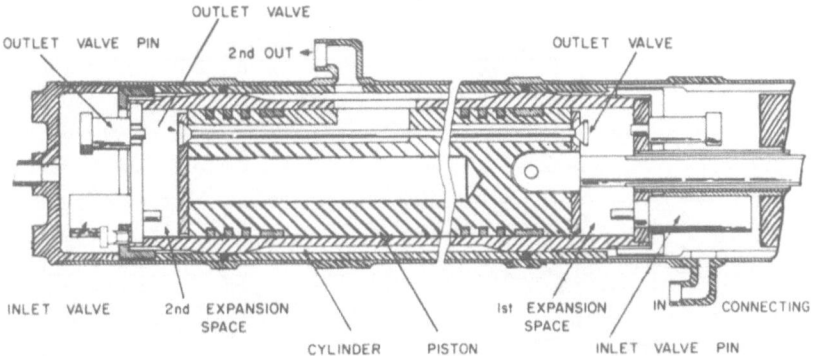

Fig. 14.5. Cross section of the double-acting engine expander with self-acting valves (after Sekiguchi, 1978).

Further developments of this type of expander are foreseen for different temperature regimes and expansion pressure ratios. The number of design parameters involved is relatively large. Further, simplifications are necessary to reduce the number of components involved. This type of machine offers the possibility of drastic simplification of the crank case mechanism and should lead to lighter, compact expanders of high performance.

Final Liquefaction Process and Efficiency

In general, three methods are used for liquefaction after high-pressure helium gas has been cooled below its inversion temperature: (a) isenthalpic expansion by a JT valve, (b) adiabatic expansion by an expansion engine from the supercritical region, (c) simultaneous adiabatic expansion and compression by an ejector. These various methods are illustrated in Fig. 14.6. Of the three methods, (b) has the highest liquefaction rate. However,

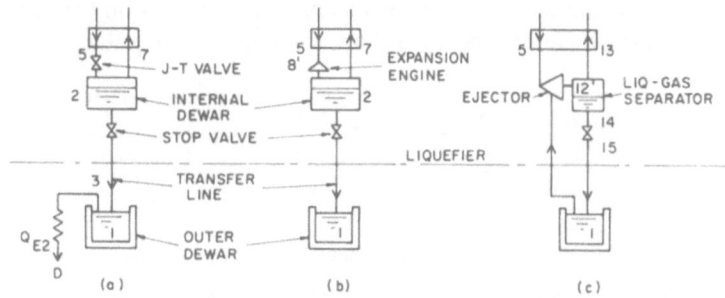

Fig. 14.6. Three methods for helium liquefaction.

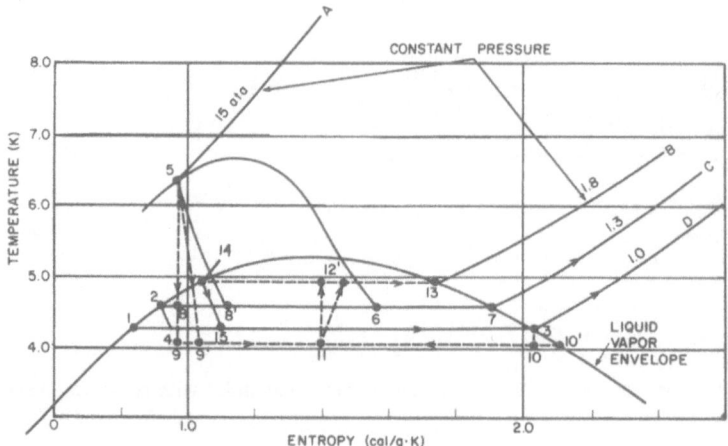

Fig. 14.7. *T–S* diagram for three helium liquefaction methods by expansion.

(c) is more favorable when considering the efficiency and reliability of the refrigerator or liquefier as a whole. Figure 14.7 illustrates the temperature–entropy diagram for the three methods. Starting from the same pressure and temperature before expansion (point 5, Fig. 14.7), the sequence of flow diagram for the three methods when operating as a liquefier is shown in Fig. 14.8.

A comparison of the liquefaction rate at 1 atm, 4.2 K, and the input power per unit liquefaction will be made for the three methods starting from point 5 at 15 atm and 6.5 K. The pressure after expansion in the expansion engine or the JT valve, is assumed to be 1.3 atm so as to allow for the pressure loss in the low-pressure side of the heat exchanger. The suction pressure of the compressor is assumed to be 1 atm, the pressure at the ejector (point 13) 1.8 atm, and the suction pressure of the ejector 0.7 atm.

In method (a), helium gas at 15 atm (point 5) reaches a gas–liquid mixture state at 1.3 atm (point 6) via isenthalpic expansion through the JT valve. The gas phase enters the JT heat exchanger at point 7 and passes

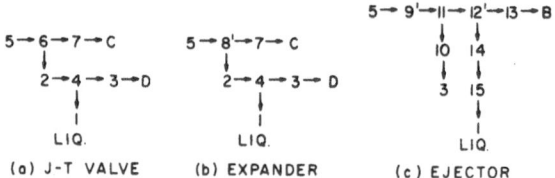

Fig. 14.8. State-flow diagrams for the three methods of helium liquefaction illustrated in Fig. 14.7.

to the suction side of the compressor at 1 atm. The liquid phase at state 2 passes, via a stop valve, to an external storage Dewar at state 1. There is a flash loss caused by the pressure difference of 0.3 atm in transferring to the Dewar, so that a portion of the liquid evaporates to state 3. The liquefaction rate for this process may be obtained from the enthalpy values at the various states as follows:

$$\varepsilon_{\mathrm{JT}} = \left(\frac{H_7 - H_5}{H_7 - H_2}\right) \cdot \left(\frac{H_3 - H_4}{H_3 - H_1}\right) \tag{14.1}$$

In method (b), the gas expands adiabatically from point 5 in an expansion engine to 1.3 atm (point 8) producing a gas–liquid mixture. Due to inefficiences in the expansion process, the state actually reached is point 8′. From there on, the fluids follow routes similar to method (a). The gas flows from 7 through the JT heat exchanger and the liquid flows from 2 to 4 to 1 with some loss at state 3. The liquefaction rate is as follows:

$$\varepsilon_{\exp} = \left(\frac{H_7 - H_{8'}}{H_7 - H_2}\right) \cdot \left(\frac{H_3 - H_4}{H_3 - H_1}\right) \tag{14.2}$$

In method (c), the gas at state (5) expands to a gas–liquid mixture 9′. The portion of gas not liquefied is at state 10′ and is mixed with some of the liquid to mixture state 11. The mixture is then adiabatically compressed to 12′ (4.95 K, 1.8 atm). Here gas returns to the compressor from state 13 via the heat exchanger B. Due to pressure losses of about 0.1 atm, the pressure of the gas will be reduced to 1.7 atm in passage through the exchanger. The liquid component of the mixture after compression to 12′ is at state 14. This expands to state 15 in a mixture of liquid at state (1) and gas at state (3). In calculating the liquefaction rate, the flow rate of gas circulating from 5 to 9′ without being liquefied was added to the flow rate between 11 to 12′. The liquefaction rate is then

$$\varepsilon_{Ej} = \frac{H_{13} - H_5}{H_{13} - H_1} \tag{14.3}$$

Both calculation and experiments show that the $\varepsilon_{\exp} > \varepsilon_{\mathrm{JT}} < \varepsilon_{Ej}$ without taking into account heat leak and pressure loss. One must, however, take into account the input power when evaluating the efficiencies as a liquefier. The work input by the compressor W_{ac} is

$$W_{ac} = \frac{mRT \ln (P_2/P_1)}{\eta_{\mathrm{is}}} \tag{14.4}$$

For methods (a) and (b), the suction pressure of the compressor is the same, $P_1 = 1$ atm. Therefore, for $P_2 = 15$ atm, $\ln (P_2/P_1) = 2.71$. For

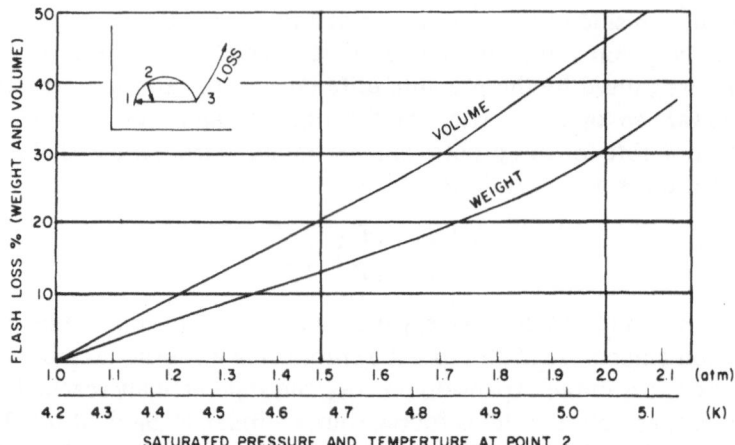

Fig. 14.9. Flash loss of liquid helium (after McCarty, 1972).

method (c), however, $P_1 = 1.7$ atm and, hence, $\ln (P_2/P_1) = 2.18$. Further-more, as the compression ratio is reduced from 15 to 8.8, the isothermal compression efficiency of the compressor η_{is} is increased. Therefore, the input power required for the ejector method is reduced by about 30% when compared to the JT method. The relation between the input powers for liquefaction of a unit amount of liquid is thus $W_{ej} < W_{exp} < W_{JT}$.

Three idealized methods for the final expansion process for helium liquefaction have been described. In the design of liquefiers using these methods, it is important to reduce the pressure loss in the piping between the bath and compressor inlet to a minimum so as to reduce the flash loss when transferring the liquid helium. Figure 14.9 shows the relation between the saturation pressure of the liquid-helium temperature, and the flash loss.

INSTABILITY IN FORCED COOLING SYSTEMS

Problems related to instability phenomena of circulating supercritical helium have attracted attention, particularly in connection with the forced cooling of large scale superconducting magnets and power transmission cable, etc.

J. W. Dean *et al.* (1978) reported density wave type oscillations in experiments related to cooling of superconducting cables. According to their report, density wave oscillations occurred in a copper tube with an inner diameter of 4.8 mm and a length of 500 m when the inlet temperature was about 10 K and the flow rate less than 3 g/s. A gradual rise in the temperature of the test section was observed due to this oscillation. They concluded that when the required cooling temperature was appreciably higher than liquid-helium temperature, as in the case of cooling Nb₃Sn, it was absolutely necessary to control the temperature at the inlet of the test section to avoid this phenomenon.

Tamada and Tomiyama (1978) reported that the generation of density wave oscillations induces temperature oscillations of higher frequency. Their analysis, based on the model cooling system in Fig. 14.10, was as follows: If, for some reason, the gas temperature at point a in Fig. 14.10 increases by ΔT_a, the evaporation rate of liquid helium in the heat exchanger will increase. This increase in evaporation rate $\Delta \dot{m}$ will, in turn, change the temperature at point a which causes a further variation of $\Delta \dot{m}$ in such a manner that temperature oscillation sets in. Now if the transfer function of the heat exchanger expressing the effect of the temperature rise ΔT_a on the change in evaporation $\Delta \dot{m}$ is denoted by $G_{Tm}(s)$, and that of the effect of $\Delta \dot{m}$ on the temperature change at point a by $G_{mT}(s)$, then the temperature loop on the low-temperature side may be expressed as follows:

$$G_{Tm}(s)G_{mT}(s) = 1 \qquad (14.5)$$

Here (s) is a Laplace operator. $G_{Tm}(s)$ and $G_{mT}(s)$ may be calculated by assuming the thermal efficiency of the liquid-helium heat exchanger to be

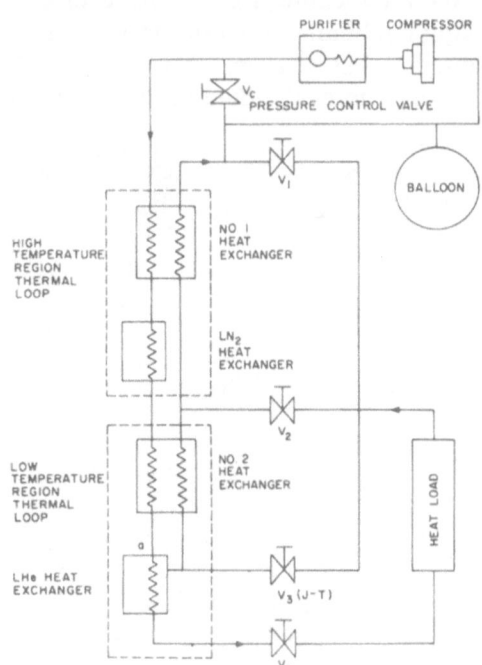

Fig. 14.10. Line diagram of refrigeration system for analysis of density wave oscillation (after Tamada et al., 1978).

large. Equation (14.5) may then be expressed as follows:

$$s^3 + a_2 s^2 + a_1 s + a_0 = 0 \tag{14.6}$$

Expressions for the coefficients a_0, a_1, and a_2 are given in Appendix 14.1. From inspection of Eq. (14.6), it can be seen that an oscillatory solution exists when the circulation rate exceeds a critical rate $\dot{m}$, i.e., $\dot{m} > \dot{m}^*$. The expression for $\dot{m}$ is given also in the Appendix 14.1. Tamada *et al.* concluded that oscillations are more easily generated in large-scale cooling systems, and, further, that the temperature difference between the hot and cold ends of the second heat exchanger should be made as small as possible so as to stabilize the system. In order to check their analysis, they conducted experiments on a test tube with an inner diameter of 5 mm and a length of 100 m. Figure 14.11 is an example of the flow rate oscillation observed at an operating pressure of 7 kg/cm^2. Figure 14.12 shows the calculated relation between the oscillation period and mass flow together with the observed data.

Matsubara *et al.* (1979) have reported studies made on the forced cooling of superconducting magnet systems, such as the medical pion beam transfer system. Using copper test tubes 10 and 30 m long with an inner diameter of 3 mm, they have observed various types of oscillations with flow rates below 1 g/s and the tube inlet temperature held at 4.3 K by liquid helium. No density wave oscillation was observed for a 10-m-long test tube. Density wave oscillations with periods about equal to the residence time of the fluid in the test loop were observed but the oscillation

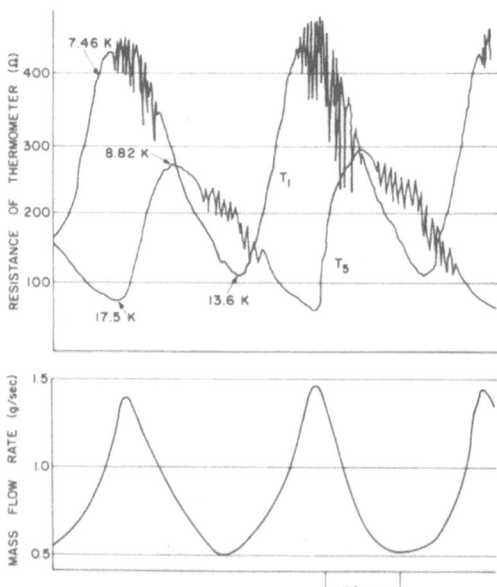

Fig. 14.11. An example of mass flow oscillation observed at an operating pressure of 7 kg/cm^2 (after Tamada *et al.*, 1978).

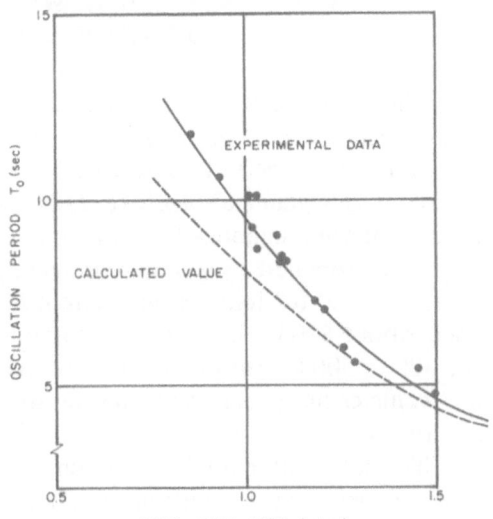

Fig. 14.12. Relation between oscillation period and mass flow (after Tamada *et al.*, 1978).

could not be sustained and damped out in approximately 15 cycles. No temperature instabilities were encountered. However, pressure oscillations were observed with frequencies of approximately 2 and 0.3 Hz for the 10- and 30-m loops, respectively. These pressure oscillations were believed to be resonant Helmholtz mode oscillations in the test channels. They were found to agree well with calculations based on single mass-spring model. It was further shown that this oscillation can be suppressed effectively by local heating of the tube wall. Figure 14.13 shows the effect of local heating.

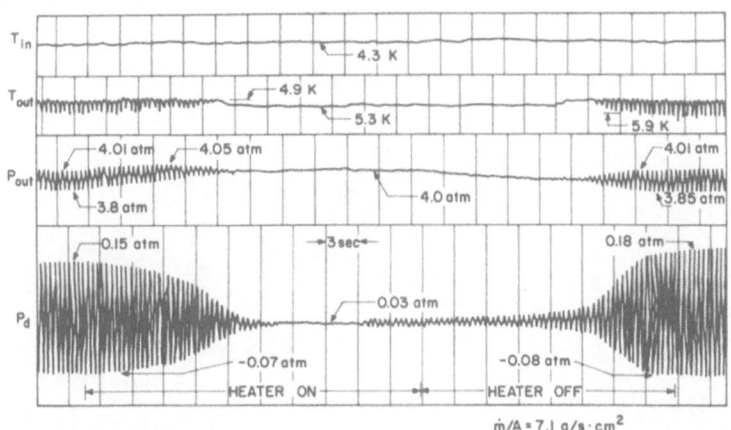

Fig. 14.13. Effect of local heating on the temperature and pressure oscillations in low-temperature tube flow (after Matsubara *et al.*, 1979).

SUPERCONDUCTING MAGNETIC LEVITATION
OF TRAINS (MAGLEV)

The Japanese National Railways (JNR) is now operating Superexpress trains (Shinkansen) between Tokyo and Hakata, Kyushu (distance: 1177 km) at a speed of 210 km/hr. Lines are now being extended north of Tokyo to Nigata and Morioka. JNR is planning to operate, in the future, trains that are magnetically levitated and guided by superconducting magnets and propelled by a superconducting linear synchronous motor. The route selected for initial development is between Tokyo and Osaka (distance: about 500 km), where the traffic is heaviest. Train speeds will be in the 300 to 500 km/hr range. Development of magnetically levitated (Maglev) vehicles has progressed considerably in the past ten years (Ohtsuka *et al.*, 1979).

The test center for Maglev vehicle experiments is located near Miyazaki, Kyushu. Test speeds up to 458 km/hr have been achieved on a test track 7 km long. Further development to speeds up to 500 km/hr are planned in the near future. Table 14.4 lists the test vehicle specifications and test center facilities (Kyotani, 1979). The vehicle is levitated approximately 20 cm and has no mechanical contact with the guideway, so that vibration and noise levels are low and maintenance of the guideway is minimal.

On-board refrigeration to liquid-helium temperatures is necessary to cool and operate the superconducting magnets. Figure 14.14 is a schematic drawing on the on-board refrigeration system consisting of cryostats, refrigerators, liquid-helium tank, etc. Several hundred of these systems will be necessary as Maglev trains are put into service.

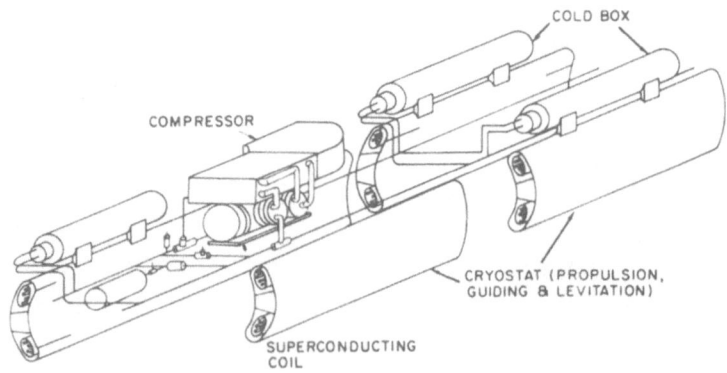

Fig. 14.14. Schematic of on-board refrigeration system for Maglev train.

Table 14.4. Miyazaki Test Facility for Maglev Trains

Vehicle

Dimensions, length × width × height	13.5 × 3.7 × 2.6 m
Weight	10 tons
Speed	500 km/hr
Levitation height	250 mm
Supporting	Opposed-coil type
Guiding	Null-flux type
	(used in combination LSM)
Propulsion	LSM
Brake	Regenerative brake
	Rheostatic brake
	Mechanical brake

Superconducting magnets

Number of cryostats	4
Cryostat dimensions, length × width × height	4.15 × 0.7 × 1.1 m, L-type
Number of coils	4 in one cryostat
Coil pitch:	
Supporting	2.1 m
Guiding	2.1 m
Coil dimensions:	
Supporting, length × width	1.65 × 0.3 m
Guiding, length × width	1.65 × 0.5 m
Ampere turns:	
Supporting	250 kAT
Guiding	450 kAT
Persistent switch	Thermal type

Test track

Total track length	7 km
Track coil pitch:	
Supporting	0.7 mm
Guiding	1.4 m
Track coil dimensions:	
Supporting, length × width	0.45 × 0.33 m
Guiding, length × width	1.1 × 0.7 m

Power distribution

Frequency changer:	
Motor	60 Hz, 10 MW
Generator	120 Hz, 25 MVA
Cryoconverter:	
Input	120 Hz
Output	0–33.1 Hz
Cryogenic equipment	Helium liquefier

In the vehicle (ML-500) now being tested, 8 superconducting magnets are enclosed in 4 sealed cryostats. The heat leak of each cryostat is a few watts at 4.5 K, and several score watts at 77 K. On-board refrigeration systems of three refrigeration cycles, Claude, Gifford–McMahon, and

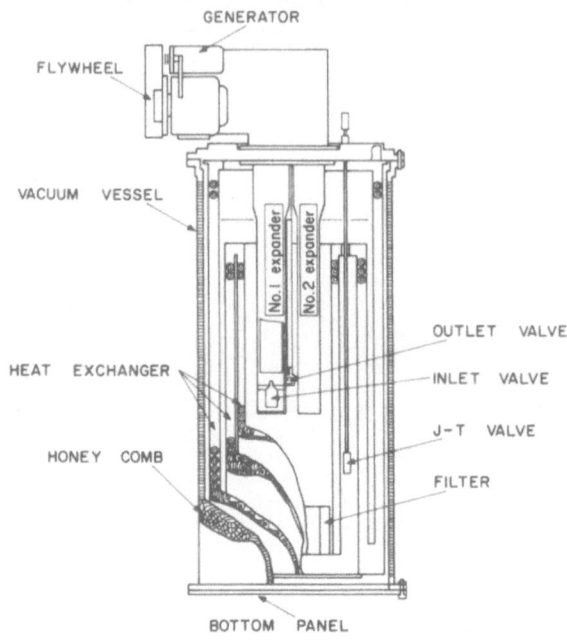

Fig. 14.15. Cold box for the Claude cycle on-board-refrigerator for the Maglev train.

Stirling cycles, are under development. Competitive evaluation of weight, compactness, COP, reliability, exchangeability, ease of maintenance, to match the requirement of a mass transportation system, is in progress.

Figure 14.15 shows the Claude cycle on-board refrigeration system now being evaluated. The expansion engines and heat exchangers are sealed inside the cryostat. The refrigeration power is about 30 W at 4.5 K and the weight, not including the compressor, is 163 kg.

PULSED REFRIGERATION SYSTEM

In superconducting energy storage systems for fusion, accelerator, laser, maglev, etc., the superconducting magnet is used in pulsed operation where the stored energy is discharged in time intervals ranging from a few microseconds to a few tens of seconds (Ishizaki, 1977). This results in unavoidable losses to the cryogenic environment because of eddy current and hysteresis losses generated in the magnet during field change. These losses in turn develop pulsating variation in the heat load, as shown in Fig. 14.16. Difficulties could be encountered in attempting to operate the

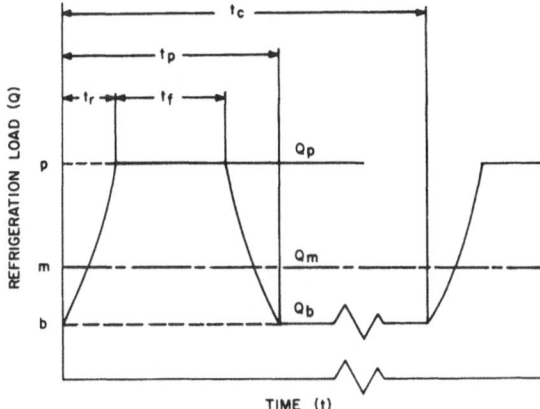

Fig. 14.16. Relation between refrigeration load and time in pulsed superconductors magnet operation (after Ishizaki, 1977). t_c, cycle time; t_f, flat-top time; t_p, pulse time; t_r, pulse rise time; Q_b, base load; Q_p, peak load.

refrigerator at high efficiency if measures are not incorporated to respond to this pulsating heat load.

The heat load developed when the magnet is pulsed depends on the ramp rate, maximum field, magnet configuration, and the characteristics of the superconducting wire used. Eddy current and hysteresis loss can, in principle, be calculated. However, to ascertain what magnitude of heat load was actually generated, an experiment was conducted using a magnet with the specifications listed in Table 14.5. Figure 14.17 shows the apparatus used. The carbon resistor was maintained at about 10 K during steady-state conditions by passing a constant current through it. This was used to sense the change in the flow rate of evaporating gas at 4.2–5 K during pulsed operation. An electric heater was placed as a monitor near the magnet. The magnet was encased in a plastic container so as to eliminate a contribution to losses arising from eddy currents induced in the metal Dewar wall.

Table 14.5. Specifications of Experimental Magnet for Pulsed Refrigeration Study

Coils		Wires	
Size:		Material:	Nb–Ti
Inner diameter	56 mm	Outer diameter:	76 mm
Outer diameter	100 mm	Diameter of filaments:	30 μm
Length	110 mm	Number of filaments:	132
Central field:	5.5 T at 154 A	Cu/Nb–Ti ratio:	2.6
Inductance:	47 H	Critical current:	256 A at 5 T
Number of turns:	3824		
Wire length:	950 m		

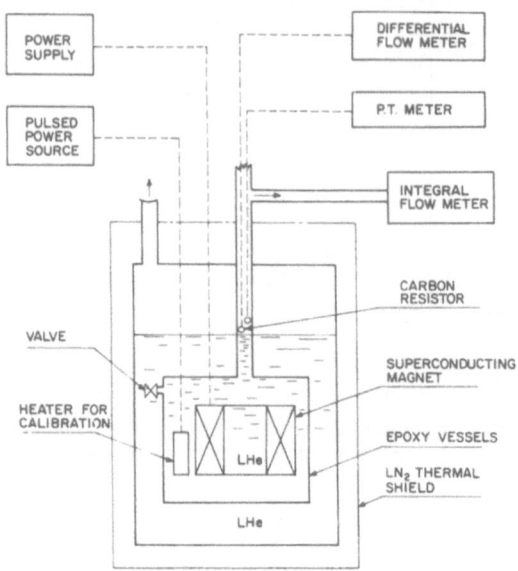

Fig. 14.17. Experimental apparatus for superconducting energy storage studies (after Ishizaki, 1977).

The heater was then pulsed with known rates of heat generation and served to calibrate the carbon resistor sensor. An example of the heat loss variation observed is shown in Fig. 14.18. It can be seen that the evaporation rate of helium increased abruptly with change in the coil current. The recovery to steady-state conditions takes place more slowly with a characteristic time of about 30 sec. If $\eta_R = 0.9$ (which can be easily achieved), this

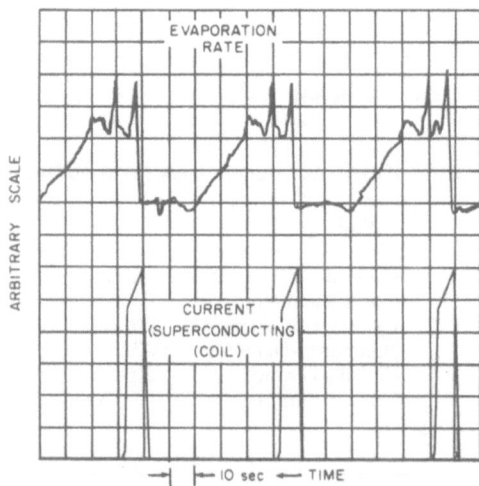

Fig. 14.18. An example of the heat loss variation observed in superconducting energy storage studies (after Ishizaki, 1977).

temperature will be $T_0 \approx 30$ K. This design not only allows leveling off the pulsating gas flow but also increases the overall efficiency of the refrigeration system by partially recovering the intermittently released cold.

Conceptual Design

The effect of the pulsating heat load on the refrigeration system will depend primarily on the ratio of the peak heat load Q_p to the standby base load Q_b, and ratio of the cycle time τ_c to the pulse time τ_p (see Fig. 14.16). A refrigeration system matched to the peak load Q_p will provide excessive cooling power when τ_c/τ_p becomes much larger than unity as in fusion systems where $\tau_c/\tau_p = 10^2$–10^3. In such a case, methods for load leveling must be devised so as not to impair the operation of the refrigeration system. With this problem in mind, a refrigeration system has been designed to accomplish the necessary heat load leveling by effectively utilizing the intermittently released cold from the helium bath due to the pulsating heat load (Ohtsuka *et al.*, 1979).

The system is shown schematically in Fig. 14.19. During a heat load pulse the excessive evaporating gas developed in the liquid-helium Dewar (4.2 K, 1 atm) passes through valve I to a cold buffer tank which is held

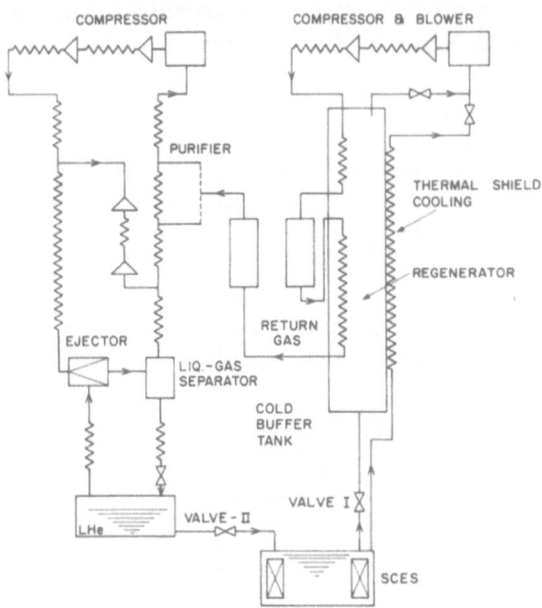

Fig. 14.19. Schematic flow diagram of the pulsed refrigeration system.

at about 6 K. The cold gas then flows through a regenerator. In this process, the gas absorbs heat from the regenerator and warms up to room temperature. This warm gas is pressurized by a blower and flows back through heat exchanger tubes embedded in the regenerator. The gas is thus recooled and enters the low-pressure side of the liqufier at temperatures depending on the thermal efficiency η_R of the regenerator.

An estimate has been made of the increase in efficiency expected for cooling a magnet for fusion use with 3 GJ stored energy and a cycle time of 360 s or 10 pulses per hour. It was assumed that the loss per cycle was 0.01% of the stored energy of 300 kJ/pulse. The required capacity for the liquefier was 1150 l/hr. Assuming a 14% Carnot efficiency for a liquefier of this size, the required input power would be about 1500 kW. The power requirement for the improved system is shown in Fig. 14.20 as a function of the thermal efficiency of the regenerator. When $\eta_R = 0.9$, the power required will be about 1000 kW, a 33% reduction in the system input power.

In recent years, superfluid helium at 1.8 K, about 1 atm or more has been favored for cooling large superconducting energy storage coils because of the enhanced heat transfer characteristics. The large heat capacity of superfluid helium near the lambda point (2.17 K) also serves as a heat reservoir for load leveling. Based on our previous experience with a 1.8 K refrigerator, we are presently considering the design of an efficient cycle for producing such environments as described by Bon Mardion (1976) for use with large-scale superconducting coils.

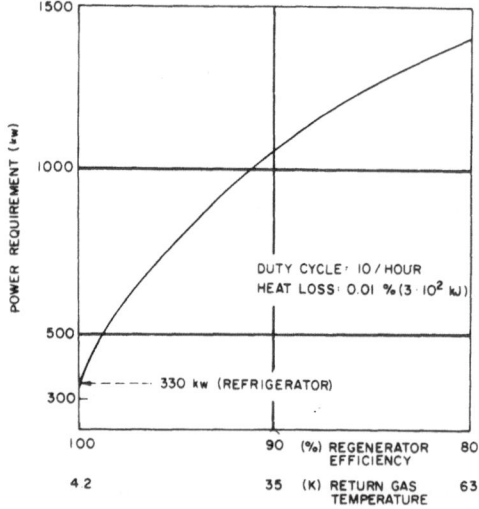

Fig. 14.20. Power requirements of the pulsed refrigeration system as a function of regeneration efficiency.

APPENDIX 14.1

$$a_2 = \frac{Cp\dot{m}^2}{\lambda^* \rho_{he} V_{he}} \ln \frac{\phi_{he}}{1 - \phi_{he}} + \beta l\left(\frac{1}{Cp_1 V\rho_1} + \frac{1}{Cp_2 V_2 \rho_2}\right)$$

$$a_1 = \frac{Cp_2 \beta l \dot{m}^2}{\lambda^* \rho_{he} V_{he}} \ln \frac{\phi_{he}}{1 - \phi_{he}} \left(\frac{1}{Cp_1 V_1 \rho_1} + \frac{1}{Cp_2 V_2 \rho_2}\right)$$

$$a_0 = \frac{Cp_2 \beta l^2 G}{Cp_1 V_1 \rho_1 V_2 \rho_2} \frac{\dot{m}^2}{\lambda^* \rho_{he} V_{he}} \ln \frac{\phi_{he}}{1 - \phi_{he}}$$

$$\dot{m} = \frac{1}{\frac{1}{2}\left(\frac{1}{\rho_{he} V_{he}} \ln \frac{\phi_{he}}{1 - \phi_{he}}\right)^2} \left[\frac{\beta l}{4}\left(\frac{1}{Cp_1 V_1 \rho_1} + \frac{1}{Cp_2 V_2 \rho_2}\right)\right.$$

$$\left. \times \frac{1}{\rho_{he} V_{he}} \ln \frac{\phi_{he}}{1 - \phi_{he}} - \frac{Cp\beta l}{\lambda^* Cp_1 V_1 \rho_1 V_2 \rho_2}(T_H - T_L)\right]$$

Here Cp is the specific heat at constant pressure; V is the inner volume of No. 2 heat exchanger; ρ is the density; β is the heat transfer coefficient; l is the No. 2 heat exchanger length; G is the temperature gradient of the low-pressure side at steady state $[= (T_H - T_L)/l]$; T is the hot end temperature of the low-pressure side; T_L is the cold end temperature of the low-pressure side; $\dot{m}$ is the flow rate; ϕ is the enthalpy efficiency; and λ is the latent heat. The subscripts are 1, high-pressure side of No. 2 heat exchanger; 2, low-pressure side of No. 2 heat exchanger; and he, liquid-helium heat exchanger.

REFERENCES

Bon Mardion, G., *et al.* (1976). "Superfluid Helium Bath for Superconducting Magnets." *Proc. 6th Int. Cryog. Eng. Conf., Grenoble*, pp. 159–162, IPC Science & Technology Press, London.

Dean, J. W., and Stewart, W. (1978). "Performance of a Liquid Helium Refrigerator Operated above the Critical Temperature." *7th Int. Cryog. Eng. Conf.*, London, pp. 629–634, IPC Science and Technology Press, London.

Ishizaki, Y. (1974). "Refrigeration," *Bussei* (in Japanese). Vol. 15, No. 12, pp. 683–97, Maki Shoten, Tokyo.

Ishizaki, Y. *et al.* (1974). "Sealed Cryostat System for Magnetically Levitated Vehicles." *5th Int. Cryog. Eng. Conf.*, Kyoto, pp. 102–105, IPC Science & Technology Press, London.

Ishizaki, Y. *et al.* (1976). "Dry Helium Compressor for Refrigeration Systems." *Adv. Cryog. Eng.* **21** 219–223.

Ishizaki, Y. (1977). "Pulsed Refrigeration System for the Superconducting Energy Storage." 1st Superconductive Energy Storage Mtg. (Nat. Lab. for High Energy Physics), Tsukuba, Proc., pp. 75–80.

Kasamatsu, K. *et al.* (1971). "An Experiment and Consideration on the Reliability of the Cryogenic Refrigerator," *Cryogenic Engineering* (in Japanese) 6(5), 182–189.

Kyotani, Y. (1979). "Maglev JNR." Int. Symp. on Traffic and Transp. Tech., Hamburg, Proc. (to be published).

Matsubara, Y. *et al.* (1979). "Experimental Study of Flow Instabilities in Forced Helium Cooling Channels." Cryogenic Engineering Conference.

McCarty, R. D. (1972). "Thermophysical Properties of Helium-4 from 2 to 1500 K with Pressures to 1000 Atmospheres," NBS Tech. Note No. 631 (US Government Printing Office).

Ohtsuka, T., and Ishizaki, Y. (1979). "Refrigeration System for the Superconducting Energy Storage." 1st Int. Symp. on Superconductive Storage, Kobe, Proc. (to be published).

Ohtsuka, T., and Kyotani, Y. (1979). "Superconducting Maglev Tests." 2nd Joint Intermag. MMM-Conf., New York, Proc. (to be published).

Oshima, K., and Ishizaki, Y. (1974). Refrigeration Technology in Japan." *Proc. 5th Int. Cryog. Eng. Conf.*, Kyoto, Japan, pp. 357–360, IPC Science and Technology Press, London.

Sekiguchi, H. *et al.* (1978). Helium Refrigerator On-board." 21st Semi-Annual Mtg. of the Cryog. Assoc. of Japan, Proc., p. 63.

Tamada, N., and Tomiyama, S. (1978). "Temperature Oscillation Caused by Coupling of Heat Exchangers," *Cryogenic Engineering* (in Japanese) 13(4), 198–204.

Yasukochi, K. *et al.* (1978). "Reliability Research of the Low-Temperature Refrigeration System," *Cryogenic Engineering* (in Japanese) 13(5), 257–271.

Bibliography

GENERAL READING

Abadzic, E. E., and Scholz, H. W. (1973). "Coiled Tube Heat Exchangers." *Adv. Cryog. Eng.* **18**, 42–51.

Abe, H. (1974). "ESR Measurements at Temperatures around 0.1K." *Jap. J. Appl. Phys.* **13**(7), 1145–1150.

Aberle, J. L., and Westbrook, A. J. (1963). "Liquid Helium and Nitrogen Supply Systems for Space Simulators." *Adv. Cryog. Eng.* **8**, 190–198.

Abraham, B. M., and Falco, C. M. (1976). "Demountable (4) He for Dilution Refrigerators," *Rev. Sci. Instrum.* **47**, 253–254.

Abraham, B. M., Ketterson, J. B., Roach, P. R., and Pfeiffer, E. R. (1974). "Demagnetization Experiments on Some Promising New Compounds for Very-Low-Temperature Refrigeration." *J. Low Temp. Phys.* **14**(3–4), 387–396.

Ackermann, R. A. (1971). "Veuilleumier Cycle Refrigeration Analysis." *Prog. Refrig. Sci. Technol.* (Proc. XIII Int. Cong. of Refrig., Washington) AVI Publishing Co., Inc., Westport, Connecticut, pp. 55–64.

Ackermann, R. A., and Gifford, W. E. (1971). "A Heat Balance Analysis of a Gifford–McMahon Cryorefrigerator." *Adv. Cryog. Eng.* **16**, 221–229.

Adamovitch, N. I., Zuev, V. I., and Pavlovskaya, T. F. (1970). "A Cryostat for a Neutron Scintillation Spectrometer." *Cryogenics* **10**, 440.

Afgan, N., and Schlünder, E. U. (1974). *Heat Exchangers, Design and Theory Sourcebook.* Scripta Book Co./McGraw-Hill Book Co., New York.

Agarwal, K. L., and Betterton, J. O. (1974). "On Low-Temperature Indium Seals." *Cryogenics* **14**, 520.

Agsten, R. (1974). "Optimum Cooling of Cryoelectrical Connections." *Luft Kaeltetech.* **10**(6), 294–299.

Ahonen, A. I., Berglund, P. M., Haikala. M. T., Krusius, M., Lounasmaa, O. V. and Paalanen, M. A. (1976). "Nuclear Refrigeration of Liquid He^3." *Cryogenics* **16**, 521.

Akiyama, Y., Togo, S., and Ishii, H. (1971). "Helium Turbo-Expander with an Alternator." *Prog. Refrig. Sci. Technol.* (Proc. XIII Int. Cong. of Ref., Vol. 1, Washington). AVI Publishing Co. Inc., Westport, Connecticut, pp. 159–164.

Almond, D. P. and Lea, M. J. (1974). "Co-Axial Transmission Line for Use in a Dilution Refrigerator," *Cryogenics* **14**, 226.

Alvesalo, T. A., Anufriyev, Yu. D., Lund, P. B., *et al.* (1974). "A Cryogenic System for Studying the Properties of He(3) between 1 and 15 MK." *Cryogenics* **14**(7), 384–390.

Amstutz, L. I., and Eaton, R. III, (1973). "Comparative Analysis of Methods for Cooling High Current Leads." Proc. Closed-Cycle Cryogenic Cooler Conf. USAF Academy, Colorado, October, 12 pp. (AFFDL-TR-73-149, Vol. 1, AD No. 918234).

Anashkin, O. P., Danilov, I. B., and Krivenko, V. G. (1966). "Helium Dewar Vessel without Nitrogen Cooling." *Cryogenics* **6**, 106.

Anashkin, O. P., Keilin, V. E., and Patrikeev, V. M. (1976). "Compact High-Efficiency Perforated-Plate Heat Exchangers." *Cryogenics* **16**, 437.

Angrist, S. W. (1968). "Perpetual Motion Machines." *Sci. Am.* **218**, 114–122.

Anon. (1965). "Neue Tieftemperatur Kälte-Satze im Philips-Programm für den Bereich von −200 bis −261 C. (New Refrigerating Units at Low Temperature for the Range −200 to −261 Degrees C)." *Kältetechnik* **18**(3), 132.

Anon. (1967). "New Plant Aids Decentralization of Helium Conversion." *J. Refrig. (London)* **10**(4), 82–83.

Anon. (1969). "Technical Product Analysis. Compressors, Pumps and Expanders. Aerospace Cutback—Blessing in Disguise." *Cryogenic Technol.* **5**(4), 172–176.

Anon. (1973). "Specialist Heat Transfer Service Aids Heat Exchanger Designers." *Cryogenics* **13**, 637.

Anon. (1974). "Meeting on Technology Arising from High-Energy Physics." European Organization for Nuclear Research, Geneva, Report No. CERN 74-9, Vol. 2, (June) pp. 153–202.

Anon. (1974). "Miniturbine for Very Low Temperatures." *Rev. Polytech.* No. 1322, (February) 131.

Anzin, V. B., Kosichkin, Y. V., Kotlov, Y. N., and Nadezhdinsky (1976). "Portable Optical Helium Cryostat with Cold Windows for IR Investigations." *Cryogenics* **16**, 375.

Appleton, A. D. (1969). "Motors, Generators and Flux Pumps." *Cryogenics* **9**, 147.

Arkharov, A. M. (1969). *Low Temperature Gas Machines (Cryogenerators)*. Mashinostroenie, Moscow, USSR, 224 pp.

Arkharov, A. M., Bondarenko, L. S., and Kuznetsov, B. G. (1970). "Analysis of Grouping Plans for Cryogenerators, Operating without Consumption of Mechanical Work." *Deep Cold and Conditioning* (Ed. G. I. Voronin), Trudy MVTU No. 138, Moscow, USSR, pp. 12–20.

Arkharov, A. M., and Desyatov, A. T. (1975). "Improvement of Microcryogenic Throttling Systems." *Chem. Pet. Eng. (USSR)* **11**(7–8), 752–758.

Arkharov, A. M., Gridin, V. B., and Mirkin, A. Z. (1971). "Designing and Investigation of the Free Piston Expander." *Prog. Refrig. Sci. Technol.* (Proc. XIII Int. Cong. of Refrig., Washington) AVI Publishing Co., Inc., Westport, Connecticut, pp. 143–151.

Arkharov, A. M., and Kuznetsov, B. G. (1972). "Cybernetic Modeling of the Working Process in Regenerative Heat Exchangers of Stirling–Philips-Type Gas Machines." *Deep Cold and Conditioning* (Ed. G. I. Boronin), Trudy MVTU No. 149, Moscow, USSR, pp. 65–72.

Arregger, J. E., and Haselden, G. G. (1967). "Cold-Compression Cascades." *Prog. Refrig. Sci. Technol.* (Proc. XII Int. Cong. of Refrig., Madrid. Vol. 1. Paper 1.41), pp. 217–222.

Ashworth, T., Bunting, J. G., and Smith, S. J. (1968). "Cryogenic Lead Seals." *Cryogenics* **8**, 167.

Aslenian, J., and Weil, L. (1960). "The Cooling of Specimens for Calorimetry at Very Low Temperatures." Cryogenics **1**, 117.

Astrov, D. N., and Belyanski, L. B. (1967). "A High-Vacuum Seal for Low Temperatures." *Cryogenics* **7**, 111.

Augustynowicz, S. (1971). "Cryogenic Refrigerators." *Chlodnictwo* **6**(1), 2–7.

Avenel, O., Der Nigohossian, G., and Roubeau, P. (1976). "A Liquid Helium Saver." *Int. Cryog Eng. Conf., Proc. 6th*, Grenoble, France. IPC Sci. and Technol. Press, Guildford, England, pp. 163–165.

Avenel, O., and Roubeau, P. (1976). "Experimental Set-up for Ultralow Temperatures." *Bull. Inf. Sci. Technol.*, Commis. Energ. At. (Fr.) No. 214, (May) 59–63.

Babcock, G. H. (1885). "Substitutes for Steam." *ASME Trans.* **7**, 680–741.

Babiichuk, V. P., Golub, A. A., Esel'son, B. N., and Serbin, I. A. (1975). "Continuous Adsorption Refrigerator for Producing Temperatures Below 1 K." *Cryogenics* **15**, 254–256.

Baehr, H. D. (1963). "On the Thermodynamics of the Cold-Air Cycle with Throttling." *Prog. Refrig. Sci. Technol.* (Proc. XI Int. Cong. of Ref., Munich. Vol. 1. Paper No. 11-11, pp. 319–328), Pergamon Press, Oxford.

Bagatskii, M. I., Manzhelii, V. G., Minchina, I. Ya., and Popov, V. A. (1976). "Investigation of the Physical Mechanism of Obtaining Temperatures below 2 K by Desorption of Helium." *Sov. J. Low Temp. Phys.* **2**(4), 212–214.

Bahnke, G. D., and Howard, C. P. (1964). "The Effect of Longitudinal Heat Conduction on Periodic-Flow Heat Exchanger Performance." *J. Eng. Power* **A86**, 105–120.

Balas, C. Jr. (1973). "An Acoustically Quiet, Low Power, Minimum Vibration, Stirling Cycle Refrigerator." Proc. Closed Cycle Cryogenic Cooler Conf. USAF Academy, Colorado, October, 12 (AFFDL-TR-73-149, Vol. 1, AD No. 918234).

Balas, C., Leffel, C. S., and Wingate, C. A. (1977). "The Stirling Cycle Cooler," *Adv. Cryog. Eng.* **23**, 411–419.

Balas, C., Jr., and Wingate, C. A., Jr. (1975). "An Efficient, Long-Life Cryogenic Cooling System for Spacecraft Applications." Int. Astronautical Cong., 26th, Lisbon, Portugal, Paper No. 75-170, (September), 9 pp.

Baldus, W. (1963). "Liquefication of Hydrogen and Helium for Nuclear Applications." *Prog. Refrig. Sci. Technol.* (Proc. XI Int. Cong. of Ref. Munich, Vol. 1, Paper No. 1, pp. 115–119), Pergamon Press, Oxford.

Baldus, W., Kneuer, R., and Stephen, A. (1975). "On-Board Cryogenic System for Magnetic Levitation of Trains: Cryogenic System of EET," St. Asztalos.

Baldus, W., and Sellmaier, A. (1968). "A Continuous Helium II Refrigerator." *Adv. Cryog. Eng.* **13**, 434–440.

Bannister, J. D. (1968). "Heat Exchanger for Cryogenic Service." *Appl. Cryog. Technol.* **1**, 81–96.

Baranov, G. M. (1959). "The Main Trend in the Designing of Large Gaseous Oxygen Plants." *Prog. Refrig. Sci. Technol.* (Proc. X, Int. Cong. in Ref., Copenhagen., Vol. 1. Paper 1-a-7. pp. 19–23), Pergamon Press, Oxford.

Barron, R. (1966). *Cryogenic Systems.* McGraw-Hill Book Co., Toronto.

Barthelemy, R. R., and Oberly, C. E. (1966). "Optimum Magnets for MHD Generators." Int. Symp. on Generation, Salzburg, Austria (July). Report. No. AFAPL-CONF-67-2, CFSTI AD-643 832. WPAFB, Dayton, Ohio.

Batrakov, B. P., and Dravchenko, V. A. (1976). "A Neon Refrigerator for Cooling Magnetic Systems." IEEE, New York, pp. 722–723.

Batrakov, B. P., and Kravchenko, V. A. (1977). "Neon Liquefier." *Chem. Pet. Eng.* (*USSR*) **12**(5), 474–475.

Baumgartner, J. P., and Wapato, P. G. (1977). "Reliability and Repair Policy Assessment for Long-Duration Operation of Helium Refrigeration Systems." *Adv. Cryog. Eng.* **23**, 397–410.

Beale, W. (1969). "Free-Piston Stirling Engines—Some Model Tests and Simulations." S.A.E. Paper 690230, S.A.E. Auto Eng. Congress, Detroit, Michigan.

Beasley, S., Ruhemann, M., and Seddon, W. L. (1959). "A Single Column Plant with Expansion Turbine for Producing Pure Nitrogen under Pressure." *Prog. Refrig. Sci. Technol.* (Proc. X Int. Cong. in Ref., Copenhagen, Vol. 1. Paper No. 1-a-1, pp. 29–33), Pergamon Press, Oxford.

Becker, H., Doll, R., and Eder, F. X. (1968). "Measurement of the Efficiency of the Valveless Doll–Eder Expansion Engine." *Proc. Second Int. Cryog. Eng. Conf.* Iliffe Scii. and Tech. Pub., Guildford, England, pp. 9–11.

Bejan, A. (1975). "Discrete Cooling of Low Heat Leak Supports to 4.2 K." *Cryogenics* **15**, 290–292.

Bejan, A. (1976). "Refrigeration for Rotating Superconducting Windings of Large AC Electric Machines." *Cryogenics* **16**, 153–159.

Bejan, A. (1977). "Refrigerator–Recirculator Systems for Large Forced-Cooled Superconducting Magnets." *Cryogenics* **17**, 97–105.

Beliakov, V. P., Epifanova, V. I., Baikov, V. S., Rosenoer, T. M., and Usanov, V. V. (1971). "Helium Microejectors." *Prog. Refrig. Sci. Technol.* (Proc. XIII Int. Cong. of Refr. Vol. 1., Washington). AVI Publishing Co. Inc., Westport, Connecticut, pp. 65–69.

Baliakov, V. P., Ermakov, A. P., and Semiletov, V. L. (1974). "Multiconnected Systems for Automatic Stabilization of Temperature in Regenerators of Air-Separating Units." *Chem. Pet. Eng.* (*USSR*) **10**(3–4), 420–422.

Beliakov, V. P., Narinsky, G. B., Orlina, I. A., and Pronko, V. G. (1971). "Determination of Optimum Parameters of the Helium Cooling Plant at Temperature Levels Ranging From 12 to 30 K with Regard to Heat Exchange Surface." *Prog. Refrig. Sci. Technol.* (Proc. XIII Int. Cong. of Ref. Vol. 1, Washington). AVI Publishing Co., Inc., Westport, Connecticut, pp. 91–94.

Beliakov, V. P., Pronko, V. G., Yelukhin, N. K., Zhuravleva, I. N., and Krasnikova, O. K. (1971). "Compact Heat Exchangers for Helium Refrigeration Systems." *Prog. Refrig. Sci. Technol.* (Proc. XIII Int. Cong. of Ref. Vol. 1, Washington). AVI Publishing Co. Inc., Westport, Connecticut, pp. 101–105.

Belonogov, A. V., Butkevich, I. K., and Povarov, Yu. I. (1972). "Experimental Investigation of a Piston Helium Expander." *Chem. Petrol. Eng.* (*USSR*) **8**(7–8), 62–23 [Transl. of *Khin. Neft. Mashinostr.* No. 7, 14–16 (1972)].

Bennings, M. A., Kunkle, E. B., and Singleton, A. H. (1969). "Development of a Practical Thermodynamic Cycle for a Space-Borne Hydrogen Reliquefier." *Adv. Cryog. Eng.* **14**, 378–386.

Berchowitz, D. M., Rallis, C. J., and Urieli, I. (1977). "A New Mathematical Model for Stirling Cycle Machine." Proc. 12th I.E.C.E.C., Washington, D.C., August 28–September 2, 1522–1527.

Berglund, P. M., Ehnholm, G. J., Gylling, R. G., Lounasmaa, O. V., and Sovik, R. P. (1970). "A Powerful Dilution Refrigerator." Annexe 2, *Cryophysics and Cryoengineering.* IIR. Paris. pp. 49–55.

Berry, R. L. (1973). "Modular Cryogenic Refrigerators." Proc. Closed Cycle Cryogenic Cooler Conf. USAF Academy, Colorado, October, 18 pp. (AFFDL-TR-73-149, Vol. 1, AD No. 918234).

Bertinat, M., Brewer, D. F., and Butterworth, G. J. (1973). "Observations on the Use of Sintered Copper Heat Exchangers in Experiments with Superfluid Helium." *Cryogenics* **13**, 48.

Betts, D. S. (1974). "Pomeranchuk Cooling by Adiabatic Solidification of Helium-3." *Contemp. Phys.* **15**(3), 227–247.

Betts, D. S. (1974). "Refrigeration and Thermometry below One Kelvin." Crane, Russak and Co., Inc., New York, 293 pp.; and *Cryogenics* **17**, 583.

Betts, D. S., and Marshall, R. (1969). "Design and Construction of Sintered Copper for Use in Dilution Refrigerator Heat Exchangers." *Cryogenics* **9** 460.

Bevan, T. (1946). *The Theory of Machines.* Longmans Green and Co. Ltd., London.

Birmingham, B. W., Goodman, B. B., Hartwig, W. H., Kamper, R. A., and Kunzler, J. E. (1968). "A Report on the 1967 Applied Superconductivity Conference." *Cryogenics* **8**, 176–179. *CFSTI PB*-180 899. National Bureau of Standards, Boulder, Colorado.

Birmingham, B. W., and Smith, C. N. (1976). "A Survey of Large Scale Applications of Superconductivity in the U.S.," *Cryogenics* **16**, 59.

Black, W. C., Kirschkoff, E. C., Mota, A. C., and Wheatley, J. A. (1969). "He^4 Film Flow Suppressing Evaporator for a Dilution Refrigerator." *Rev Sci. Instrum.* **40**, 846–848.

Blackford, J. E. (1972). "The Use of a Computer for the Determination of Gas State in the Design of Low-Temperature Machines." *Cryogenics* **12**, 170.

Blaisse, B. S. (1970). "A Proposal to Decrease the Heat Resistance of Heat Exchangers in He^3–He^4 Dilution Refrigerators." Annexe 2, *Cryophysics and Cryoengineering.* IIR. Paris, pp. 79–82.

Blinkov, Ye. L. (1977). "An Analysis of the Operating Conditions of Superconducting Power Transmission Lines." Proc. Applied Superconductivity, Conf., *IEEE Trans. Mag.* **MAG-13**(1), 144–148.

Block, R. F. (1973). "The Design, Development and Test of a Heat Pipe Interface for the Hot Cylinder of a VM Refrigerator." Proc. Closed Cycle Cryogenic Cooler Conf. USAF Academy, Colorado, October, 18 pp. (AFFDL-TR-73-149, Vol. 1, AD No. 918234).

Boddeker, K. W., and MacWood, G. E. (1962). "Unconventional Refrigeration Techniques." Report. No. RADC-TDR-62-341. Ohio State Univ., Research Foundation, Columbus, Ohio.

Bodio, E. (1975). "Refrigerating Micro-Units for Nitrogen and Argon Liquefaction." *Chlodnictwo* **10**(7), 18–21.

Bogner, G. (1976). "Research at Siemens Laboratories on Large Scale Applications of Superconductivity." *Cryogenics* **16**, 259.

Boiko, A. A., and Shmyt'ko, I. M. (1975). "Cryostat for X-Ray Reflection Photography with a Widely Divergent Beam." *Cryogenics* **15**, 35.

Boom, R. W., Haimson, B. C., McIntosh, G. E., *et al.* (1975). "Superconductive Energy Storage for Large Systems." Proc. Applied Superconductivity Conf., *IEEE Trans. Mag.* **MAG-11**(2), 475–481.

Borovik, E. S., Batrakov, B. P., and Kobzev, P. M. (1965). "Helium Liquefier with Liquid Flow Heat Exchangers." *Cryogenics* **5**, 338.

Borovik, E. S., Mikhailov, I. F., and Kosik, N. A. (1964). "Hydrogen Liquefiers with Efficient Heat Exchangers." *Cryogenics* **4**, 358.

Brandt, N. B., Svistova, E. A., and Semenov, M. V. (1971). "A Miniature Apparatus for Obtaining Intermediate Temperatures." *Cryogenics* **11**, 59.

Bratsberg, H. G., Soevik, R. P. (1976). "He^3/He^4 Dilution Methods," *Fra Fep. Verden* **38**(1), 9–10, 17–20.

Brechna, H. (1974). "Superconducting Magnet Systems." *Cryogenics* **14**, 357.

Breckenridge, R. W., Jr. (1969). "A 3.6°K Reciprocating Refrigerator." *Adv. Cryog. Eng.* **14**, 387–393.

Breckenridge, R. W., Jr., (1973). "Summary of Rotary-Reciprocating Refrigerator Technology." Proc. Closed-Cycle Cryogenic Cooler Conf. USAF Academy, Colorado, October 20 pp. (AFFDL-TR-73-149, Vol. 1, AD No. 918234).

Breckenridge, R. W., Jr. (1975). "Cryogenic Coolers for IR Systems." *Opt. Eng.* **14**(1), 57–62.

Bretherton, A., Granville, W. H., and Harness, J. B. (1971). "Performance of Regenerators at Low Temperatures." *Adv. Cryog. Eng.* **16**, 333–341.

Brewer, D. F., Edwards, D. O., Howe, D. R., and Whall, T. E. (1966). "Simple Helium-3 Cryostat and the Specific Needs of Nylon and an Epoxy Resin." *Cryogenics* **6**, 49.

Brezzi, F. (1976). "Technique and Problems of Hydrogen Liquefaction." *Quad. Ing. Chim. Ital.* **12**(1), 26.

Brickwedde, F. G. (1958). "The Thermodynamic Theory of a Liquid Nitrogen Generator Using a Norelco Refrigerator–Liquefier." *Adv. Cryog. Eng.* **3**, 101–105.

Brickwedde, F. G. (1959). "Fifty Years of Progress in Low Temperature Research and Development Since Onnes' First Liquefaction of Helium." *Prog. Refrig. Sci. Technol.* (Proc. X Int. Cong. In Ref., Copenhagen, Vol. 1, Paper No. 8. pp. 3–19), Pergamon Press, Oxford.

Brodyansky, V. M. (1963). "Thermodynamic Analysis of Gas Liquefaction Processes." *Inzh.-Fix. Zh. Akad. Wauk. Belorus. SSR* **6**(7), 36–42.

Brodyansky, V. M. (1968). "Calculation and Optimization of Expander Cryogenic Cycles on Electronic Digital Computers." *Proc. First Int. Cryog. Eng. Conf.*, Heywood-Temple Ind. Pub., London, Eng., pp. 198–201.

Brodyansky, V. M., and Grachev, A. B. (1966). "Thermodynamic Analysis of Real Gas Type Refrigerating Machines with Displacer." *Izv. Vyssh. Ucheb. Zaved. Energ.*, No. 2, 57–63.

Brodyansky, V. M., Gresin, A. K., Gromov, V. M., Yagodin, V. M., Nicolsky, V. A., and Alpheev, V. N. (1971). "The Use of Mixtures as the Working Gas in Throttle Joule–Thomson Cryogen Refrigerators." *Prog. Refrig. Sci. Technol.* (Proc. XIII Int. Cong. of Refrigeration. Vol. 1. Washington). AVI Publishing Co., Inc. Westport, Connecticut, pp. 43–45.

Brown, G. V. (1971). "The Practical Use of Magnetic Cooling," *Prog. Refrig. Sci. Technol.* (Proc. XIII Int. Cong. of Refrig. Vol. 1., Washington). AVI Publishing Co., Inc. Westport, Connecticut, pp. 587–592.

Browning, C. W., Potter, V. L., Miller, W. S., and Gasser, M. G., (1973). "Developments Toward Achievement of Long-Life Cryogenic Vuilleumier Refrigeration Systems." Proc. Closed-Cycle Cryogenic Cooler Conf. USAF Academy, Colorado, October, 20 pp. (AFFDL-TR-73-149, Vol. 1, AD No. 918234).

Buchal, C., Mueller, R. M., Oversluizen, T., and Pobell, F. (1978). "Some Remarks on a Nuclear Demagnetization Apparatus to be Built in Jülich." Int. Symp., Hakone, Japan, 284–286.

Buchter, H. H. (1979). *Industrial Sealing Technology.* Wiley-Interscience, New York.

Budnevich, S. S., and Gurin S. P. (1975). "Thermodynamic Analysis of Helium Ejector–Throttle Refrigeration Cycles. *Izv. Vyssh. Uchebn. Zaved, Energ.* **18**(2), 127–131.

Budnevich, S. S., Kondryakov, I. K., Akulov, L. A., and Golovko, G. A. (1963). *Prog. Refrig. Sci. Technol.* (Proc. XI Int. Cong. of Refrig., Munich, Vol. 1. Paper 1-12, pp. 137–140), Pergamon Press, Oxford.

Budnevich, S. S., Novotelnov, V. N., and Davydov, I. A. (1970). "Analysis of Energetic and Weight Indices of Cryogenic Systems." *Azv. Vyssh. Ucheb. Zaved., Mashinostr.* No. 6, 111–114. (Transl. by John Crerar National Translations Center, Chicago, Illinois).

Bukshpan, I., Cederbaum, M., and Eckstein, Y. (1974). "Pomeranchuk Cooling to Millikelvin Temperature without Forming Bulk Solid (3)He." *Proc. European Phys. Soc. Topical Conf.*, Haifa, Israel; John Wiley and Sons, New York, pp. 141–144.

Buller, J. S. (1971). "A Miniature Self-Regulating Rapid-Cooling Joule–Thomson Cryostat." *Adv. Cryog. Eng.* **16** 205–213.

Buschow, K. H. J., Olijhoek, J. F., and Miedema, A. R. (1975). "Extremely Large Heat Capacities between 4 and 10 K." *Cryogenics* **15**, 261.

Buttefield, A. W. (1971). "Design of a Simple Liquid Nitrogen Cooled Photomultiplier Housing." *Phys. E: Sci. Instrum.* **5**, 518–519 (NTIS AD-754 137. Defence Stds. Labs., Marebyrnong, Australia).

Byrns, R., and Green, M. A. (1975). "The Escar Helium Refrigeration System. *IEEE Trans. Nucl. Sci.* **NS-22**(3), 1168–1171.

Cairelli, J. E., and Thieme, L. G. (1977). "Initial Test Results with a Single Cylinder Rhombic Drive Stirling Engine." Proc. ERDA, Highway Veh. Syst. Cont. Coord. Mtg., Dearborn, Michigan, October 4–6.

Campbell, D. N. (1967). "A Miniature Closed Circuit 28 K Claude Cycle Refrigerator." *Prog. Refrig. Sci. Technol.* (Proc. XII Int. Cong. of Refrig., Madrid, Vol. 1., Paper 1.22) pp. 79–83.

Campbell, D. N. (1967). "Miniature Closed Circuit Single State Claude Cycle Refrigerator for 28 K." *Proc. First Int. Cryog. Eng. Conf.*, Heywood Temple Ind. Pubs., London, pp. 197–198.

Carbonnell, E. (1974). "Large Scale Refrigerators with Turbines." Proc. *Fifth Int. Cryog. Eng. Conf.*, Kyoto, pp. 353–356, IPC Sci. and Technol., Guildford, U.K.

Carbonnell, E., Chovet, P., Johannes, C., Marinet, D., and Solente, P. (1972). "Refrigerator without Moving Parts at Low Temperature Able to Cooldown to 90–100 K." Proc. 4th Int. Cryo. Eng. Conf., Eindhoven, pp. 68–70.

Carbonell, E., and Solente, P. (1974). "Refrigeration and Cryoelectricity." *Entropie* **56**, 40–46.

Carpetis, C. (1971). "Circulation of Cold Gas in a Closed Cycle and Its Use for Experiments in the Region of 15 to 80 K." *Deut. Luft. Baumfahrt Mitt.*, No. 9, 47.

Carter, C. N. (1973). "Superconductive DC Transmission Lines: Design Study and Cost Estimates." *Cryogenics* **13**, 207.

Charkey, E. S. (1972). *Electro Mechanical System Components.* Wiley-Interscience, New York.

Chellis, F. F. (1973). "The Future of the Stirling Cycle Refrigerator in Airborne I-R Applications." Proc. Closed-Cycle Cryogenic Cooler Conf. USAF Academy, Colorado, October, 8 pp. (AFFDL-TR-73-149, Vol. 1, AD No. 918234).

Chellis, F. (1977). "Design Compromises in the Selection of Closed Cycle Cryocoolers." Proc. Appl. of Closed-Cycle Cryocoolers to Small Superconducting Devices, pp. 109–122, Boulder, Colorado, October (NBS Spec. Publn. 508).

Chellis, F. F., and Hogan, W. H. (1964). "A Liquid-Nitrogen-Operated Refrigerator for Temperatures below 77°K. "*Adv. Cryog. Eng.* **9**, 545–551.

Chellis, F. F., Hosmer, T. P., and Keller, E. (1971). "Closed-Cycle Refrigeration for an Airborne Illuminator." *Adv. Cryog. Eng.* **16**, 214–220.

Chernetskii, V. C., and Epifanova, V. I. (1968). "Idealized Cycle of a Gas-Expansion Machine with a Built-In Regenerator." *Vses. Nauchn. Issled. Inst. Kriogennogo Machin.*, *Trudy*, No. 12, 172–189.

Chernysheva, E. A., Tumanov, A. I., and Platonova, S. N. (1975). "Calculation of the Regenerators of the Second Stage of Gas-Coolers, Using Nomograms." *Chem. Pet. Eng.* (*USSR*) **11**(8), 715–717 (Transl. of *Khim. Neft. Mashinostr.* No. 8, 19–20).

Chovet, P., and Solente, P. (1970). "200 Litres per Week Helium Liquefier." *Proc. Third Int. Cryog. Eng. Conf.*, Illife Sci. and Tech. Pubs., Guildford, England., 251–254.

Claeson, T., and Wingbro, T. (1974). "Unusual Features of an He³ Refrigerator. *Cryogenics* **14**, 468–469.

Clark, A. F., and Kropschot, R. H. (1970). "Low Temperature Specific Heat and Thermal Expansion of Alloys." Annexe 2, *Cryophysics and Cryoengineering.* IIR. Paris, pp. 249–254.

Clarke, A. F. (1968). "Low-Temperature Expansion of Some Metallic Alloys." *Cryogenics* **8**, 282.

Clarke, M. E., and Brittaine, L. J. (1973). "Developments in Helium Gas-Bearing Expansion Turbines." *Prog. Refrig. Sci. Technol.* (Proc. XIII Int Cong. of Refrig. Vol. 1, Washington.) AVI Publishing Co. Inc., Westport, Connecticut, pp. 153–158.

Clarke, M. E., and Gardner, J. B. (1968). "New Developments in Expansion Machinery for Low-Temperature Refrigerators." *Proc. First Int. Cryog. Eng. Conf.*, Heywood-Temple Ind. Pubs., London, pp. 273–274.

Clarke, W. D. III. (1973). "Reliability and Maintainability of USAF Cryogenic Coolers." Proc. Closed-Cycle Cryogenic Cooler Conf. USAF Academy, Colorado, October, 10 pp. (AFFDL-Tr-73-149, Vol. 1, AD No. 918234).

Class, C. R., Spero, R. P., and McIntosh, G. E. (1960). "Efficient Utilization of Ortho-Para Catalyst." *Adv. Cryog. Eng.* **3**, 64–72.

Claude, G. (1913). "*Liquid Air, Oxygen and Nitrogen.* P. Blakiston Son and Co., Philadelphia, Pennsylvania.

Claudet, G., Dif, P., and Lacaze, A. (1976). "Analysis of Some Processes Derived from the Claude Refrigeration Cycle." *Int. Cryog. Eng. Conf., Proc. 6th*, Grenoble, France (May). IPT Sci. and Tech. Press, Guildford, England, pp. 82–85.

Claudet, G., and Verdier, J. (1972). "Simplified Cryogenic Reciprocating Expansion Engine." *Proc. Fourth Cryog. Eng. Conf.*, IPC Sci. and Tech. Press, Guildford, England, p. 50–52.

Cleaver, A. V. (1974). "Cryogenic Fluids in the Aerospace Industry." Cryotech 73-British Cryogenics Council, Proc. Conf., Brighton, England (November). IPC Sci. and Tech. Press, Guildford, England, pp. 107–111.

Cobourne, M. H., and Williams, W. T. (1975). "A Cryostat and Discharge Tube for Investigating Electron Emission from Superconducting Surfaces in Ultra-High Vacuum." *Cryogenics* **15**, 37.

Cohen, E. G. D. (1977). "Toward Absolute Zero." *Am. Sci.* **65**(6), 752–758.

Colangelo, J. W., Fitzpatrick, E. E., Rea, S. N., and Smith, J. L., Jr. (1968). "An Analysis of the Performance of the Pulse Tube Refrigerator." *Adv. in Cryog. Eng.* **13**, 494–504.

Collins, S. C. (1947). "A Helium Cryostat." *Rev. Sci. Instrum.* **18**(3).

Collins, S. C. (1952). "Helium Liquefiers." *Science* **116**, 289.

Collins, S. C. (1956). "Helium Liquefiers and Carriers." *Handbuch der Physik*, Vol. XIV, pp. 113–135, Springer-Verlag, Berlin.

Collins, S. C. (1958). *Expansion Engines for Gas Liquefaction.* Oxford University Press.

Collins, S. C. (1958). "Hydrogen-Helium Liquefier." *Adv. Cryog. Eng.* **2**, 8–11.

Collins, S. C. (1964). "Practical Means for Maintaining Very Low Temperatures." *Metals Eng. Quart.* **4**(3), 157, 157–167.

Collins, S. C. (1966). "Helium Refrigerator and Liquefier." *Adv. Cryog. Eng.* **11**, 11–15.

Collins, S. C. (1968). "Early History and Development of Cryogenics." *Cryog. Technol.* **4**(4), 144–145.

Collins, S. C. (1968). "Liquefaction Techniques." Monograph 111, Chap. 3, National Bureau of Standards, Boulder, Colorado, pp. 55–76.

Collins, S. C. (1969). "Early History and Development of Cryogenics." *Proc. Seminar Cryog. Technol.* (Ed. H. Weinstock), Boston Technical Publishers, Boston.

Collins, S. C. (1969). "Refrigeration at Temperatures Below the Boiling Point of Helium." Proc. 1968 Summer Study on Superconducting Devices and Accelerations, Part 1. Brookhaven National Lab., Upton, New York, pp. 59–66.

Collins, S. C., and Cannaday, R. L. (1958). *Expansion Machines for Low-Temperature Processes.* Oxford University Press, Oxford, England.

Collins, S. C., and Doherty, P. R. (1971). "Freeze-Out Type Helium Make-up Gas Purifier." *Prog. Refrig. Sci. and Technol.* (Proc. XIII, Int. Cong. of Ref. Vol. 1. Washington). AVI Publishing Co., Inc. Westport, Connecticut, pp. 95–100.

Collins, S. C., and Hughes, R. W. (1961). "New Refrigeration Cycle for the Production of Liquid Nitrogen." *Cryogenics* 2, 43.

Collins, S. C., Robinson, G. Y., and Selldorff, J. T. (1960). "Design and Performance Data on a Laboratory Size Liquid Nitrogen Plant." *Adv. Cryog. Eng.* 3, 106–113.

Collins, S. C., and Streeter, M. H. (1968). "Refrigerator for 1.8° K." *Proc. First Int. Cryog. Eng. Conf.*, Heywood-Temple Ind. Pubs., London, pp. 215–217.

Collins, S. C., Stuart, R. W., and Streeter, M. H. (1967). "Closed-Cycle Refrigeration at 1.85°K. *Rev. Sci. Instrum.* 38, 1654–1657.

Colyer, D. B. (1967). "Long-Life Miniature Refrigeration Systems." *Prog. Refrig. Sci. and Technol.* (Proc. XII Int. Cong. of Ref., Madrid, Vol. 1, Paper No. 1.31, pp. 85–95).

Colyer, D. B., and Bjerklie, J. W. (1962). "Cryocycle Runs and Cools Space Equipment." *J. Soc. Automot. Eng.* 70(9), 90–95.

Colyer, D. B., Hurley, J. D., and Oney, W. R. (1973). "Long-Life Turbo Refrigerators for Unattended Space Applications." Proc. Closed-Cycle Cryogenic Cooler Conf. USAF Academy, Colorado, October, 48 pp. (AFFDL-TR-73-149, Vol. 1, AD No. 918234).

Conte, R. R. (1969). "Irradiations at Low Temperatures. Part 1—Device Designs." Report No. CEA-R-3910, CFSTI N70-13879. Commissariat à l'Energie Atomique, Fontenay, Aux-Roses, France.

Cooke-Yarborough, E. H., and Yeats, F. W. (1975). "Efficient Thermomechanical Generation of Electricity from the Heat of Radioisotopes." Proc. 10th I.E.C.E.C., Paper No. 759150, pp. 1003–1011, Newark, New Jersey.

Coppage, J. (1952). "Heat-Transfer and Flow-Friction Characteristics of Porous Media." Ph.D. Thesis, Stanford University, Stanford, California.

Coppage, J. E., and London, A. L. (1953). "The Periodic-Flow Regenerator—A Summary of Design Theory." *Trans. Am. Soc. Mech. Eng.* 75, 779–787.

Coppage, J. E., and London, A. L. (1956). "Heat-Transfer and Flow Friction Characteristics of Porous Media." *Chem. Eng. Prog.* 52(2), 56–57.

Cosier, J., and Croft, A. J. (1970). "A New Form of Finned-Tube Heat Exchangers" *Cryogenics* 10, 239.

Cowans, K. W. (1974). "A Countercurrent Heat Exchanger that Compensates Automatically for Maldistribution of Flow in Parallel Channels." *Adv. Cryog. Eng.* 19, 437–444.

Cowans, K. W., and Walsh, P. J. (1965). "Continuous Cryogenic Refrigeration for Three to Five Micron Infrared Systems." *Adv. Cryog. Eng.* 10, 468–476.

Crabtree, L. F. (1973). "Engineering—Art and Science." *Aerospace* 3(7), 22–25.

Crawford, A. H. (1970). "Specifications of Cryogenic Refrigerators." *Cryogenics* 10, 28–37.

Crawford, A. H. (1971). "Cryogenic Cooler." *Kaelte* 24(12), 525–527.

Creswick, F. A. (1957). "A Digital Computer Solution of the Equation for Transient Heating of a Porous Solid, Including the Effects of Longitudinal Conduction." *Ind. Math,* 8, 61–68.

Croft, A. J. (1964). "The New Hydrogen Liquefier at the Clarendon Laboratory." Cryogenics 4, 143.

Croft, A. J. (1971). "Liquefiers and Refrigerators." *Advanced Cryogenics* (Ed. C. A. Bailey) Plenum Press, New York, pp. 183–224.

Croft, A. J., and Robertson, C. W. (1969). "A 50-W Hydrogen Refrigerator." *Cryogenics* 9, 365.

Croft, A. J., and Tebby, P. B. (1970). "The Design of Finned-Tube Cryogenic Heat Exchangers." *Cryogenics* 10, 236.

Croft, A. J., and Thomas, J. O. (1969). "Small-Scale Liquid-Helium Cryostat." *Cryogenics*, **9**, 57.

Crouch, J. N. (1965). "Cryogenic Cooling for Infrared." *Electro-Technol.* (**May**), 96–106.

Crouthamel, M. S., and Shelpuk, B. (1973). "A Combustion Heated, Thermally Activated, Vuilleumier Refrigerator." *Adv. Cryog. Eng.* **18**, 339–351.

Cryogenic Technology Incorporated—Anon. (1978). "Standard Helium Liquefier Model No. 1410." Tech. Mem. 91477, C.T.I., Waltham, Massachusetts.

Currie, R. B. (1967). "A Joule–Thomson Laboratory Expander." *Adv. Cryog. Eng.* **12**, 557–563.

Daly, E. F., and Dean, T. J. (1970). "A New Heat Pump Cycle and Its Application to a 3.5 K Refrigerator." *Cryogenics* **10**(2), 123–135.

Danby, G. T., and Powell, J. R. (1971). "The Central Role of Cryogenics in Magnetically Levitated High Speed Trains." *Prog. Refrig. Sci. Technol* (Proc. XIII, Int. Cong. of Refrig. Vol. 1., Washington). AVI Publishing Co. Inc., Westport, Connecticut, pp. 605–609.

Daney, D. E., McConnell, P. M., and Strobridge, T. R. (1972). "Low-Temperature Nitrogen Ejector Performance." (Final Rept.), *Cryog. Eng. Conf.*, Boulder, Colorado (August) Paper L-4, V18, 476–485. NTIS COM-73-50712/1. National Bureau of Standards, Washington, D.C.

Daniels, A., and DuPre, F. K. (1971). "Miniature Refrigerators for Electronic Devices." *Philips Tech. Rev.* **32**(2), 49–56.

Daniels, A., and DuPre, F. K. (1971). "Triple Expansion Stirling Cycle Refrigerator. *Adv. Cryog. Eng.* **16**, 178–184.

Daniels, A., and DuPre, F. K. (1973). "Miniature Refrigerators for Electronic Devices." *Cryogenics*, **13**, 134–140.

Danilov, I. B. (1964). "Textolite Piston for a Helium Expansion Engine." *Cryogenics* **4**(2) 93–94.

Danilov, I. B. (1975). "Combined Piston for a Helium Gas-Expansion Machine." *Chem Pet. Eng. (USSR)*. **11**, (9–10), 806–807. (Transl. of *Khim. Neft. Mashinostr.* No. 9, 16–17).

Danilov, I. B., and Andrianov, V. P. (1963). "An Electrical Indicator for Adjusting and Checking the Working of a Piston Expansion Engine." *Cryogenics* **3**(1), 31–32.

Danilov, I. B., and Kovatchev, V. T. (1972). "Two-Stage Expansion Engine with Differential Piston." *Proc. Fourth Int. Cryog. Eng. Conf.* IPC Sci. and Tech. Press, Guildford, England 87–89.

Darinskaya, E. V., and Roshanskii, V. N. (1970). "Device for Deep Refrigeration of Objects in an Electron Microscope." *Cryogenics* **10**, 505.

Daunt, J. G. (1956). "The Production of Low Temperatures down to Hydrogen Temperatures." *Handbuch der Physik* Vol. IV, Springer-Verlag, Berlin.

Daunt, J. G. (1967). "Desorption Cooling in Miniaturized Systems at Temperatures below 20 K." *Prog. in Refrig. Sci. and Technol.* (Proc. XII Int. Cong. of Ref., Madrid, Vol. 1, Paper No. 1.02), pp. 97–102.

Daunt, J. G. (1970). "Preliminary Thermodynamic Data for the Inversion Curve of He^3 " *Cryogenics* **10**, 473.

Daunt, J. G. (1972). "Present and Future Cryogenic Refrigerators for the Temperature Range 1–20 K." *Appl. Cryog. Technol.* **4**, 365–371.

Daunt, J. G., and Lerner, E. (1970). "A Closed-Cycle Joule–Thomson Liquefier and Cryostat for He^3." *Cryogenics* **10**, 475.

Daunt, J. G., and Lerner, E. (1970). "Joule–Thomson Liquefaction of He^3." Annexe 2, *Cryophysics and Cryoengineering* IIR., Paris, pp. 83–87.

Daus, W., and Ewald, R. (1975). "A Refrigeration Plant with 300 W Capacity at 1.8 K." *Cryogenics* **15**, 591.

Davies, M. (1949). *The Physical Principles of Gas Liquefaction and Low-Temperature Rectification.* Longmans Green and Co. Ltd., London.

Davies, S. J., and Singham, J. R. (1951). "Experiments on a Small Thermal Regenerator." *General Discussion on Heat Transfer.* I. Mech. E., London, 434–435.

Davydov, A. B., Kobulashvili, A. Sh., and Antipenkov, B. A. (1975). "High-Speed Turboexpanders with Gas Bearings for Cryogenic Helium Plants." *Chem. Pet. Eng. (USSR)* **11**(9–10), 819–822. (Transl. of *Khim. Neft. Mashinostr.* No. 9, 23–25).

Dean, J. W. (1974). "The Thermal Efficiency of a Pumped Supercritical Helium Refrigeration System Operating below 6 K." *Cryogenics*, **14**(6), 307–312.

Dean, J. W. (1976). "Calculating the Recovery Time of a Supercritical Helium Refrigerator after a Thermal Upset." *Proc. 6th Int. Cryog. Eng. Conf.*, Grenoble, France, IPC Sci. and Technol. Press, Guildford, England, pp. 80–81.

Dean, T. J., and Daly, E. F. (1968). "A Closed-Cycle 3.5° K Refrigerator." *Proc. Second. Int. Cryog. Eng. Conf.* Illife Sci. and Technol. Pub., Guildford, England, pp. 16–18.

De Brey, H., Rinia, H., and van Weenan, R. L. (1968). "Fundamentals for the Development of the Philips Air-Engine." *Philips Tech. Rev.* **9**, 97–104.

de Jange, A. K. (1979). A Small Free-Piston Stirling Refrigerator." Proc. 14th I.E.C.E.C., Boston, Massachusetts, Paper No. 799245, pp. 1136–1141.

Delage, C., Coron, N., and Grenier, P. (1976). "Versatile High-Performance Helium Cryostat with Vapour-Cooled Shields." *Cryogenics* **16**, 483.

Deland, R. E., Haroldsen, O. O., Helley, L. R., and Porter, R. N. (1970). "Task 6 Storable Propellant Module Environmental Control Technology." Report No. NASA-CR-111082. NASA.

Deom, W., and Schreurs, D. (1975). "Problems of the Optimization of Refrigerating and Cryogenic Cycles." *Mem. Ing. Univ. Catholic (Louvain) Rev. M* **2**(2), 212.

Depierre, Y., Verdier, J., Artiguebieille, P., Parine, D., and Solente, P. (1970). "Helium Refrigerator Producing 130 Watts at 4.5 K Connected with an Electron Irradiation Dewar." Annexe 2, *Cryophysics and Cryoengineering*. I.I.R. Paris. pp. 223–230.

De Waele, J. (1963). "Reciprocating and Turbo-Expanders for Low Temperature Refrigeration." *Prog. Refrig. Sci. Technol.* (Proc. XI Int. Cong. of Refrig., Munich, Vol. 1. Paper No. 111-11., pp. 569–579)., Pergamon Press, Oxford.

De Waele, J. (1963). "Reciprocating and Turbo-Expanders for Low Temperature Refrigeration." Proc. 11th Intern. Congr. of Refrig., Munich, I.I.R. Paper No. 11.

De Waele, J. (1964). "Les Moto-Detendeurs et Les Turbo-Detendeurs au Service de L'Industrie Frigorifique a Tres Basse Temperature. Journees des Basses et Tres Basses Temperatures." ("Reciprocating Expanders and Turbo-Expanders in the Refrigeration Industry at Very Low Temperature.") Meeting on Low and Very Low Temperatures. Paris, June.

De Waele, A. Th. A. M., Reekers, A. B., and Gijsman, H. M. (1976). "A (3)He Circulating Dilution Refrigerator with Two Mixing Chambers." *Physica (Utrecht) B + C* **81**(2), 323–326.

De Waele, A. Th. A. M., Reekers, A. B., and Gijsman, H. M. (1976). "A (3)He Circulating Dilution Refrigerator with More Than One Mixing Chamber." *Int. Cryog. Eng. Conf.*, Proc. 6th, Grenoble, France. IPC Sci. and Technol. Press, Guildford, England, pp. 112–115.

De Waele, A. Th. A. M., Reekers, A. B., and Gijsman, H. M. (1977). "A Dilution Refrigerator without Pumped He[4] Bath." *Cryogenics* **17**, 175.

Dobrov, V. M., and Orlov, A. V. (1974). "The Valveless Piston Expansion Engine with Six-Phase Gas Distribution Cycle." *Proc. Fifth Cryog. Eng. Conf. Kyoto*, IPC Sci. and Technol. Press Ltd., Guildford, England.

Doll, R., and Eder, F. X. (1964). "Gas-Lubricated Low-Temperature Piston Expansion Engine without Control Valves." *Adv. Cryog. Eng.* **9**, 561–564.

Draine, B. T., and Sievers, A. J. (1974). "A Far-Infrared Bolometer at 0.10 K." *Int. Conf. On Submillimeter Waves and Their Applications*, Georgia Inst. of Technol., Atlanta (June). Inst. of Elec. and Electronic Engs. Inc., New York, pp. 84–85.

Dros, A. A. (1965). "Industrial Gas Refrigerating Machine with Hydraulic Piston Drive." *Philips Tech. Rev.* **26**(11–12), 353–367.

Dros, A. A. (1965). "Large-Capacity Industrial Stirling Machine." *Adv. Cryog. Eng.* **10**, 7–12.

Dros, A. A., and Roosendaal, K. (1959). "Gas Circulatory System for the Transport of Cold from the Philips Gas Refrigeration Machine to a Cold Box." *Prog. Refrig. Sci. Technol.* (Proc. 10 Int. Cong. of Ref.) Copenhagen, Vol. 1, Paper 6B40, pp. 595–600. Pergamon Press, Oxford.

Duminil, M. (1969). "Refrigeration Cycles for the Production of Low Temperatures." *Rev. Gen. Froid.* **60**(6), 819–837.

Dupart, J. M., Bodin, C., and Pech, T. (1975). "Reliable Point Contact Preparation and Adjustment for SQUIDs" *Cryogenics* **15**, 227.

DuPlessis, P. de V. (1977). "Heat Transport in a Cryo-Tip Refrigerator via a Metal-Filled Epoxy." *Cryogenics* **17**, 707.

Dupre, F. K., and Daniels, A. (1973). "Gap-Regeneration Method for Stirling and Similar Cycles." *Prog. Refrig. Sci. Technol.* (Proc. XIII Int. Cong. of Refrig. Washington) AVI Publishing Co., Inc., Westport, Connecticut, pp. 137–141.

Eatwell, A. J., and Clarke, M. E. (1976). "A 4.4 K Refrigeration System for an Industrial Magnetic Separator." *Proc. 6th Int. Cryog. Eng. Conf.*, Grenoble, France. IPC Sci. and Technol. Press, Guildford, England., pp. 62–65.

Eatwell, A. J., Clarke, M. E., and Gardner, J. B. (1974). "U.K. Developments in High-Reliability Helium Refrigerators for Cryoelectric Applications." Amer. Soc. of Mech. Engs. Winter Ann. Mtg., New York, Paper No. 74-WA/PI0-12.

Edel'man, V. S. (1972). "A Dilution Refrigerator with Condensation Pump." *Cryogenics* **12**, 385.

Edelsack, E. A., Kropschot, R. H., Olien, N. A., and Olsen, J. L. (1973). "A Directory of European Low Temperature Research." *Cryogenics*, **13**, 132.

Eder, F. X. (1958). "Indicator for Helium Expansion Engines." *Bull. IIR Annexe* 1958-1, pp. 33–37.

Eder, F. X. (1968). "Methoden zur Erzeugung Tiefster Temperaturen." ("Methods of Producing Very Low Temperatures"). *Electrotech. Z.* **A89**(13), 304–307.

Edwards, D. O., Feder, J. D., and Gully, W. J. (1977). "A Nuclear Demagnetization Cryostat for Cooling He(3)." Proc. Int. Symp., Hakone, Japan (September). Phys. Soc. of Japan, Tokyo (1978), pp. 280–283.

Ehnholm, G. J., and Gylling, R. G. (1971). "A Dilution Refrigerator with Large Cooling Power." *Cryogenics* **11**, 39.

Ehnholm, G. J., Katila, T. E., Lounasmaa, O. V., and Reivari, P. A. (1968). "A He^3/He^4 Dilution Refrigerator for Mössbauer Experiments at Very Low Temperatures." *Cryogenics* **8**, 136.

Elfving, T. M. (1963). "The Practical Use of Thermoelectric Refrigeration." *Prog. Refrig. Sci. Technol.* (Proc. XI Int. Cong. of Ref., Munich. Vol. 1. Paper No. 7., pp. 23–38), Pergamon Press, Oxford.

Endig, M., and Lange, F. (1968). "Metal Cryostats for Superconducting Solenoids." *Cryogenics* **8**, 215.

Epifanova, V. I. (1959). "Low-Temperature Expansion Turbines." *Prog. Refrig. Sci. Technol.* (Proc. X Int. Cong. in Ref., Copenhagen. Vol. 1. Paper No. 1-a-3, pp. 24–29), Pergamon Press, Oxford.

Epifanova, V. I. (1965). "Analyzing the Gas Regenerative Refrigerating Cycles." *Tr. Vses. Nauchn. Isslfd. Inst. Koslorodn. Mashinostr.* No. 9, 3–35.

Erb, J., Heinz, W., Hofmann, A., Koefler, H. J., and Komarek, P. (1975). "Comparison of Advanced High-Power Underground Cable Designs." Inst. für Experimentelle Kernphysik, NTIS KFK-2207, 171 pp.

Ermakov, A. P., and Semiletov, V. L. (1974). "New Device for Measuring and Monitoring Temperature in Regenerators of Air Separation Plants." *Chem. Pet. Eng. (USSR)* **10**(1–2), 174–175 [Transl. of *Khim. Neft. Mashinostr.* No. 2 (Feb) 38–39].

Esel'son, B. N., Lazarev, B. G., and Shvets, A. D. (1962). "Production of Temperatures below 1 K by Pumping Liquefied Helium Vapour Using Adsorption Pumps." *Cryogenics* **2**, 279.

Esel'son, B. N., Shvets, A. D., and Bereznyak, N. G. (1962). "An Apparatus for Obtaining Temperatures down to 0.3 K by Using Helium-3." *Cryogenics* **2**, 361.

Fagaly, R. L., and Boh, R. G. (1976). "Simple Method for Determining Molar Flow Rates in Dilution Refrigerators." *Rev. Sci. Instrum.* **47**, 1307–1308.

Fern, A. G., and Nau, B. S. (1976). *Seals.* Engineering Design Guides, No. 15, Oxford University Press, Oxford, England.

Finegold, J. G., and Sterrett, R. H. (1978). "Stirling Engine Regenerators—Literature Review." Report No. 5030-230, Jet Propulsion Laboratory, California Inst. of Tech., Pasadena, California.

Finegold, J. G., and Vanderbrug, T. G. (1977). "Stirling Engines for Undersea Vehicles." Final Report No. 5030-63, Jet Propulsion Laboratory, Pasadena, California.

Finkelstein, T. (1953). "Self-Acting Cooling Cycles." D.I.C. Thesis, Imperial College, London.

Finkelstein, T. (1959). "Air Engines." *Engineer,* **207**, 492-497, 522-527, 568-571, 720-723.

Finkelstein, T. (1959). "Development and Testing of a Stirling Cycle Machine with Characteristics Suitable for Domestic Refrigeration." English Electric Report W/M/3A.

Finkelstein, T. (1960). "Generalized Thermodynamic Analysis of Stirling Engines." S.A.E. Paper No. 118B, January.

Finkelstein, T. (1970a). "Thermocompressors, Vuilleumier and Solvay Machines." Proc. 5th I.E.C.E.C., Las Vegas, Nevada.

Finkelstein, T. (1975). "Computer Analysis of Stirling Engines." *Adv. Cryog. Eng.* **20**, 269-282.

Finkelstein, T. (1978). "Modeling Design and Optimization of Stirling Engines." Proc. 13th I.E.C.E.C., San Diego, California.

Finkelstein, T. (1978). "Pressure Compounding of Stirling Engines." Proc. 13th I.E.C.E.C., San Diego, California.

Fleming, R. B. (1962). "An Application of Thermal Regenerators to the Production of Very Low Temperatures." Sc.D. thesis, MIT.

Fleming, R. B. (1967). "The Effect of Flow Distribution in Parallel Channels of Counterflow Heat Exchangers." *Adv. Cryog. Eng.* **12**, 352-362.

Fleming, R. B. (1969). "A Compact Perforated Plate Heat Exchanger." *Adv. Cryog. Eng.* **14**, 197-204.

Fleming, R. B. (1973). "Turbo Refrigerators for Superconductive Mobile Power Equipment." Proc. Closed Cycle Cryogenic Cooler Conf. USAF Academy, Colorado, October, 6 pp. (AFFDL-TR073-149, Vol. 1, AD No. 918234).

Fleming, R. B., Wheeler, D. B., and Kerr, D. L. (1973). "Evaluation of Long-Life Turbo Refrigerators Applied to Superconductive Ship Propulsion Systems." Proc. Closed-Cycle Cryogenic Cooler Conf. USAF Academy, Colorado, October, 12 pp. (AFFDL-TR-73-149, Vol. 1, AD No. 918234).

Flood, D. J. (1974). "Suitability of the Rare-Earth Compounds Dy(2)Ti(2) D(7) and Gd(3)Ak(5)O(12) for a Low-Temperature (4-20 K) Magnetic Refrigeration Cycle. *Proc. 19th Annual Conf.,* Boston, Massachusetts (November). 1973. American Institute of Physics, New York, Conf. Proc. No. 18, PT. 2, 1345-1348.

Forsyth, E. G. (1977). "Cryogenic Engineering for the Brookhaven Power Transmission Project." *Cryogenics* **17**, 3.

Forsyth, E. G., and Gibbs, R. J. (1977). "The Brookhaven Superconducting Cable Test Facility." *Applied Superconductivity,* Proc. Conf., *IEEE Trans. Mag.* **MAG-13**(1), 172-176.

Foster, W. G., and Murray, D. O. (1973). "Development Program for a Liquid Methane Heat Pipe." *Adv. Cryog Eng.* **18**, 96-102.

Fredkov, A. B. (1962). "Helium and Hydrogen Cryostats without Additional Liquid Nitrogen Cooling." *Cryogenics* **2**, 177.

Fredkov, A. B., and Troitskii, V. F. (1965). "Liquefier with Two-Stage Conversion to Obtain 98 per cent Parahydrogen." *Cryogenics* **5**, 136.

Frossati, G., Godfrin, H., and Hebral, B. (1977). "Conventional Cycle Dilution Refrigeration down to 2.0 MK." *Physics at Ultralow Temps.,* Proc. Int. Symp., Hakone, Japan. Phys. Soc. of Japan, Tokyo (1978), pp. 205-225.

Frossati, G., and Thoulouze, D. (1976). "A Dilution Refrigerator with Sintered Silver Powder Heat Exchangers for Use in the Millikelvin Range." *Int. Cryog. Eng. Conf.,* Proc. 6th, Grenoble, France. IPC Sci. and Technol. Press, Guildford, England, pp. 116-118.

Fujii, G., and Nagano, H. (1971). "High-Pressure Piston–Cylinder Apparatus at Low Temperature." *Cryogenics* **11**, 142.

Fujii, Y., Kodama, T., Shigi, T., and Okuda, T. (1970). "An He³ Refrigerator for a Long-Time Operation." Annexe 2, *Cryophysics and Cryoengineering*. IIR. Paris. pp. 89–93.

Furihara, A., Kikuchi, K., and Nagano, H. (1967). "Closed-Cycle Helium Refrigerator for Laser Applications." *Proc. First Int. Cryog. Eng. Conf.*, Heywood-Temple Ind. Pubs., London, England, pp. 213–214.

Furnas, C. (1932). "Heat Transfer from a Gas Stream to a Bed of Broken Solids." *Bull. U.S. Bur. Mines*, No. 361.

Fushimi, K. (1970). "Applications of Superconducting Magnets in Japan." *Cryogenics*, **10**, 116.

Gardner, J. B., and Smith, K. C. (1960). "Power Consumption and Thermodynamic Reversibility in Low-Temperature Refrigeration and Separation Processes. *Adv. Cryog. Eng.* **3**, 32–46.

Gasser, M. G., Yoshikawa, D. K., and Browning, C. W. (1972). An Approach to Long-Life VM Cryogenic Refrigerators for Space Applications." *Appl. Cryog. Technol.* **4**, 416–434.

Gaule, G. K., Breslin, J. T., Ross, R. L., and Logan, R. S. (1964). "Superconductors in Advanced Electronics." Army Science Conf., United States Military Academy, West point, N. Y., Vol. I, AD-611 432. Army Elect. Labs., Fort Monmouth, New Jersey.

Gauthier, M., and Varoquaux (1973). "Flow Impedance Construction for Liquid-Helium Continuous Refrigerators." *Cryogenics* **13**, 272.

Gavrish, I. G., Permyakov, V. V., and Bogomolov, A. G. (1975). "Cooling of Superconducting Magnetic Systems." Akad. Nauk UKR. SSSR. Lenin Physico-Technical Inst., Kharkov, Rept. No. KHFTI 75-33. COPR. AT. NAUK. TEKH., Ser. Fundam. PRIKL. Sverkhprovodimost Vol. 1, No. 3, 34–35.

Gedeon, D. (1978). "Optimization of Stirling Cycle Machines. Proc. 13th I.E.C.E.C., San Diego, California.

Geist, J. M., and Lashmet, P. K. (1960). "Miniature Joule–Thomson Refrigeration Systems." *Adv. Cryog. Eng.* **5**, 324–331.

Geist, J. M., and Lashmet, P. K. (1961). "Compact Joule–Thomson Refrigeration Systems 15–60 K." *Adv. Cryog. Eng.* **6**, 73–81.

Gerhold, J. (1976). "Optimization of Refrigerator–Cooled Current Leads for Superconducting Devices." *Cryogenics* **16**(7), 401–408.

Gessner, R. L., and Colyer, D. B. (1968). "Miniature Claude and Reverse Brayton Cycle Turbomachinery Refrigerators." *Adv. Cryog. Eng.* **13**, 474–484.

Giarratano, P. J. (1973). "Supercritical Helium Heat Transfer." (Final Rept.) Pub. in Proceedings of CRYO-72 Conf., Chicago, Illinois, NTIS COM-74-50419/2. National Bureau of Standards, Washington, D.C.

Giarratano, P. J., and Jones, M. C. (1975). "Deterioration of Heat Transfer to Supercritical Helium at 2.5 Atmospheres." (Final Rept.) NTIS COM-75-50309/4ST. National Bureau of Standards, Washington, D.C.

Gifford, W. E. (1956). "Low-Temperature Heat Exchanger Usage." *Adv. Cryog. Eng.* **2**, 276–281.

Gifford, W. E. (1957). "Low-Temperature Heat Exchange," *Chem. Eng. Prog.* **53**(6), 278–281.

Gifford, W. E. (1960). "Closed Cycle Helium Refrigeration." Proc. of Superconductive Techniques for Computing Systems, Washington, D.C.

Gifford, W. E. (1960). "Fundamentals of Hydrogen Liquefaction." *Adv. Cryog. Eng.* **2**, 1–7.

Gifford, W. E. (1961). "Novel Refrigeration Cycles and Devices." *Prog. Cryog.* **3**, 49–74.

Gifford, W. E. (1963). "Pulse Tube Refrigerator." Technical Report No. 2, Syracuse Univ., Research Institute, New York.

Gifford, W. E. (1964). "The Thermal Check Valve—A Cryogenic Tool." *Adv. Cryog. Eng.* **9**, 551–555.

Gifford, W. E. (1966). "The Gifford–McMahon Cycle." *Adv. Cryog. Eng.* **11**, 152–159.

Gifford, W. E. (1966). "Study of Novel Refrigeration Methods." Report No. NASA-CF-80513. Syracuse Univ. Research Institute., New York.

Gifford, W. E. (1968). "Refrigeration at Very Low Temperatures." *Proc. First Int. Cryog.*

Eng. Conf., Tokyo and Kyoto, Japan. Heywood Temple Ind. Pubs., London, pp. 221–224.

Gifford, W. E. (1970). "Refrigeration to Below 20 K." *Cryogenics* **10**, No. 1, 23–27.

Gifford, W. E. (1973). "A New Performance Parameter for Heat Exchangers in Transient Operation." *Prog. Refrig. Sci. Technol.* (Proc. of XIII Int. Cong. of Ref., Vol. 1., Washington, D.C.), AVI Publishing Co. Inc., Westport, Connecticut, pp. 223–233.

Gifford, W. E. (1975). "Miniature Cryogenic Refrigeration." Proc. XIV Int. Cong. of Refrig., Moscow, USSR, September.

Gifford, W. E., and Acharya, A. (1968). "Optimization of a Cryogenic Refrigerator Heat Exchanger." *Adv. Cryog. Eng.* **13**, 599–606.

Gifford, W. E., and Acharya, A. (1970). "Low-Temperature Regenerator Test Apparatus." *Adv. Cryog. Eng.* **15**, 436–442.

Gifford, W. E., Kadaikkal, N., and Acharya, A. (1970). "Simon Helium Liquefaction Method Using a Refrigerator and Thermal Valve." *Adv. Cryog. Eng.* **15**, 422–427.

Gifford, W. E., and Ackermann, R. A. (1969). "Small Cryogenic Regenerator Performance." Trans. of ASME, Winter Annual Mtg., New York, December.

Gifford, W. E., and Ackermann, R. A. (1971). "A Heat Balance Analysis of a Gifford–McMahon Cryorefrigerator." *Adv. Cryog. Eng.* **16**, 221–229.

Gifford, W. E., Ackermann, R. A., and Acharya, A. (1969). "Compact Cryogenic Thermal Regenerator Performance." *Adv. Cryog. Eng.* **14**, 353–360.

Gifford, W. E., and Deabler, H. E. (1959). "A New Refrigeration Process and Its Application to Cooling of Infrared Detectors." Proc. Third Nat. Mtg. of IRIS, Pasadena, California.

Gifford, W. E., and Fuller, P. D. (1961). "Liquid Hydrogen from Refinery Waste Gases." *Adv. Cryog. Eng.* **6**, 177–184.

Gifford, W. E., and Hoffman, T. E. (1961). "A New Refrigeration System for 4.2 Degrees K." *Adv. Cryog. Eng.* **6**, 82–94.

Gifford, W. E., and Kyanka, G. H. (1967). "Reversible Pulse-Tube Refrigeration. *Adv. Cryog. Eng.* **12**, 619–630.

Gifford, W. E., and Longsworth, R. C. (1963). "Pulse-Tube Refrigeration." Paper No. 63-WA-290. ASME Winter Ann. Meeting, Philadelphia, 8pp.

Gifford, W. E., and Longsworth, R. C. (1964). "Pulse-Tube Refrigeration." *Eng. Dig.* (*London*) **25**(9), 109–112.

Gifford, W. E., and Longsworth, R. C. (1964). "Pulse-Tube Refrigeration." *J. Eng. Ind.* **B66**(3), 264–268.

Gifford, W. E., and Longsworth, R. C. (1964). "Pulse-Tube Refrigeration." *Trans. ASME, Ser. B*, August.

Gifford, W. E., and Longsworth, R. C. (1965). "Pulse-Tube Refrigeration Progress." *Adv. Cryog. Eng.* **10**, 69–79.

Gifford, W. E., and Longsworth, R. C. (1966). "Surface Heat Pumping." *Adv. Cryog. Eng.* **11**, 171–179.

Gifford, W. E., and McMahon, H. O. (1959). "A Low-Temperature Heat Pump." *Prog. Refrig. Sci. Technol.* (Proc. 10th Int. Cong. in Refrig., Copenhagen Vol. 1, Paper 1-e-6, pp. 100–104), Pergamon Press, Oxford.

Gifford, W. E., and McMahon, H. O. (1959). "A New Low-Temperature Gas Expansion Cycle," *Adv. Cryog. Eng.* **5**, 354–372.

Gifford, W. E., and McMahon, H. O. (1959). "A New Refrigeration Process." *Prog. Refrig. Sci. Technol.* (Proc. X Int. Cong. of Refrig., Copenhagen, Vol. 1. Paper 1-e-7, pp. 105–109). Pergamon Press, Oxford.

Gifford, W. E., and McMahon, H. O. (1960). "Closed-Cycle Helium Refrigeration." *Solid-State Electron.* **1**(4), 273–278.

Gifford, W. E., and McMahon, H. O. (1960). "A New Low-Temperature Gas Expansion Cycle." Parts I and II. *Adv. Cryog. Eng.* **5**, 368–372.

Gifford, W. E., and Withjack, E. M. (1969). "Free-Displacer Refrigeration." *Adv. Cryog. Eng.* **14**, 361–369.

Giger, U., Pagani, P., and Trepp, C. (1971). "The Low-Temperature Plants for the Big European Bubble Chamber 'BEBC'." *Prog. Refrig. Sci. Technol.* (Proc XIII Int. Cong. of Ref., Vol. 1, Washington). AVI Publishing Co., Inc., Westport, Connecticut, pp. 71–81.

Giger, U., Pagani, P., and Trepp, C (1974). "Refrigerating Systems for the Big CERN Bubble Chamber." *Sulzer Tech. Rev.* **56**(2), 54–60.

Giger, U., Quack, H., and Senn, A. (1974). "Combined Helium Refrigerator and Liquefier with Gas Bearing Turbines." *Proc. 5th Int. Cryog. Eng. Conf.*, Kyoto, Japan. IPC Sci. and Technol. Press, Sussex, England, 380–383.

Giger, U., Trepp, C., and Pagani, P. (1974). "Refrigeration Plants for the Large European Bubble Chamber of CERN." *Klima Kaelte Ing.* **2**(9), 369–374.

Glassford, A. P. M. (1962). "An Oil-Free Compressor, Based on the Stirling Cycle." M.Sc. Thesis, Dept. of Mech. Eng., MIT, U.S.A.

Glimstedt, L. (1964). "Termodynamiska Berakningar for Kylprocesser Bid Laga Temperaturer." ("Thermodynamic Calculations for Low-Temperature Refrigeration"). Report No. ZG-O-3-809. SAAB, AB, Goteberg, Sweden.

Gogolina, T. B., and Fomin, A. N. (1974). "Utilization of Low-Temperature Refrigerating Plants in the USSR Industry." *Kholod. Tekh.* No. 1, 19–23.

Goldschvartz, J. M., Vander Merwe, V. P., and Kollen, A. (1975). "Reaching a Temperature of 0.64 K in an He^4 Cryostat with a New Geometry." *Cryogenics* **15**, 153.

Goldschvartz, J. M., and Van der Merwe, W. P. (1976). "Temperatures of 0.5 K Obtained by Pumping He^4." *Cryogenics* **16**, 615.

Goldwater, B., and Morrow, R. B. (1977). "Demonstration of a Free-Piston Stirling Linear Alternator Power Conversion System." Proc. 12th I.E.C.E.C., Paper No. 779249, pp. 1488–1495, Washington, D.C.

Goodstein, D. L., McCormick, W. D., and Dash, J. G. (1966). "Sintered Copper Sponges for Use at Low Temperatures." *Cryogenics* **6**, 167.

Gorlin, G. B. (1974). "The Vacuum Sealing of Optical Windows for Operation at Low Temperature." *Cryogenics* **14**, 407.

Gorshkov, A. M. (1972). "Effect of Hydraulic Resistance on Cold Productive Capacity of Gas Refrigerating Machine." *Deep Cold and Conditioning* (Ed. G.I. Voronin), Trudy MVTU, No. 149, Moscow, USSR, pp. 100–105.

Gotoh, M. (1976). "A Simple Helium-3 Refrigerator for Producing Temperatures down to 0.4 K." *Tamagawa Daigaku Kogakubu Kiyo* **11**, 35–42.

Grachev, A. G. (1975). "Calculation of the Work of Cooling and Freezing of Cryogenic Liquids by Vacuum Evaporation." *Izv. Vyssh. Uchebn. Zaved., Energ.* **18**(1), 89–93.

Grachev, A. G., and Voroshilov, B. S. (1969). "Determination of Losses of Energy from Regenerative Heat Transfer in the Cylinder of a Piston Expansion Cooler." *J. Eng. Phys.* (*USSR*) **17**(6), 1573–1576.

Grassman, P. (1967). "Ungewohnliche Kalteprozesse im Bereich Sehr Tiefer Temperaturen." ("Unusual Refrigeration Cycles for Very Low Temperatures"). *Kaltetechnik* **19**(6), 158–161.

Grigorenko, N. M., Sauchenko, V. I., and Prusman, Yu. O. (1975). "Results of Test of a Heat-Using Cryogenic Machine. *Chem. Pet. Eng.* (*USSR*) **11**(9–10), 808–810. (Transl. of *Khim. Neft. Mashinostr.* No. 9, 17–18).

Gromyshev, A. V., Lutsenko, E. L., and Sinitskii, V. V. (1971). "A Temperature Control Device for Heating at Constant Rate." *Cryogenics* **11**, 409.

Grushevskii, V. M., and Pshennikov, F. M. (1972). "Experience in Finishing Small-Size Gas Refrigerators." *Chem. Pet. Eng.* (*USSR*) **8**(7–8), 671–673.

Gunn, R. D., Chueh, P. L., and Prausnitz, J. M. (1966). "Inversion Temperatures and Pressures for Cryogenic Gases and Their Mixtures." *Cryogenics* **6**, 324.

Gylling, R. G. (1971). "Construction and Operation of a Nuclear Refrigeration Cryostat." Report No. APS-CH-81. Pub. by the Finnish Academy of Tech. Sciences, Helsinki, NTIS PB-205 950. Acta Polytechnica Scandinavica, Stockholm, Sweden.

Haarhuis, G. J. (1967). "New Type Helium Liquefier." *Prog. Refrig. Sci. Technol.* (Proc. 12 Int. Cong. of Ref., Madrid, Paper No. 136, paper 1.36), pp. 121–128.

Haarhuis, G. J. (1974). "Stirling Cryogenerators and Cryopumping." *Le Vide* **29**(171–172), 351.

Haarhuis, G. J. (1978). "The MC 80-A Magnetically Driven Stirling Refrigerator." *Proc. 7th Int. Cryog. Eng. Conf., London*, IPC Business Press, Guildford, U.K.

Hahnemann, H. (1948). "Approximate Calculation of Thermal Ratios in Heat Exchangers

Including Heat Conduction in the Direction of Flow." N.G.T.E. Mem. 36, National Gas Turbine Establishment, Pyestock, U.K.

Haisma, J., and Roozendaal, K. (1967). "Investigation of the Behaviour of an Expansion-Ejector in the Low-Temperature Region beyond the λ-Transition of Helium." *Prog. Refrig. Sci. Technol.* (Proc. XII Int. Cong. of Refrig., Madrid, Vol. 1. paper 1.37), pp. 111–120.

Hall, H. E., Ford, P. J., and Thompson, K. (1966). "Helium-3 Dilution Refrigerator." *Cryogenics* **6**, 80.

Halley, J. A. (1958). "The Robinson-Type Air Engine." *J. Stephenson Eng. Soc. Kings Coll. Newcastle* **2**(2), 49.

Hammel, E. F., and Rogers, J. D. (1970). "Cryogenics and Nuclear Physics, Part I," *Cryogenics* **10**, 5.

Hammel, E. F., Rogers, J. D., and Hassenzahl, W. V. (1970). "Cryogenics and Nuclear Physics, Part II." *Cryogenics* **10**, 186.

Harkless, L. B., (1973). "Reliability Test Results on V-M Coolers." Proc. Closed-Cycle Cryogenic Cooler Conf. USAF Academy, Colorado, October, 18 pp. (AFFDL-TR-73-149, Vol. 1, AD No. 918234).

Harley, R. T., Gastafson, J. C., and Walker, C. T. (1970). "A Simple Magnetic Thermometer for Use below 1K." *Cryogenics* **10**, 510.

Harness, G. C., and Newmann, P. E. L. (1972). "A Theoretical Solution of the Shuttle Heat Transfer Problem." *Proc. Fourth Int. Cryog. Eng. Conf.* IPC Sci. and Technol. Press, Guildford, England, pp. 97–100.

Harper, D. B., and Rohsenow, W. M. (1953). "Effect of Rotary Regenerator Performance on Gas-Turbine Plant Performance." *Trans. Am. Soc. Mech. Eng.* **75**, 759–765.

Harris, R. E., and Breckenridge, R. W. Jr., (1973). "Rotary-Reciprocating Refrigerator Performance Spectrum." Proc. Closed-Cycle Cryogenic Cooler Conf. USAF Academy, Colorado, 22 pp. (AFFDL-TR-73-149, Vol. 1, AD No. 918234).

Harris, W. S., Rios, P. A., and Smith, J. L., Jr. (1971). "The Design of Thermal Regenerators for Stirling-Type Refrigerators." *Adv. Cryog. Eng.* **16**, 312–323.

Harrowell, R. V. (1972). "Preliminary Studies of Superconducting Alternators." *Cryogenics* **12**, 109.

Harvell, J. T., and Hogan, W. H. (1965). "Kalteanlagen mit Geschlossenem Kreislauf." (Refrigerating Systems with Closed Cycle"). *Kaltetechnik*, **17**(6), 180–185.

Haseler, L. E., Scurlock, R. G., and Thornton, G. K. (1976). "Thermodynamic Considerations for the Refrigeration of Rotating Superconducting Machinery." *Cryogenics* **16**(6), 337–342.

Haseler, L. E., Scurlock, R. G., and Thornton, G. K. (1976). "Copper Conduction Precooling of Helium Refrigerant in Rotating Superconducting Machines." *Cryogenics* **16**(6), 337–342.

Hassenzahl, W. V. (1975). "Will Superconducting Magnetic Energy Storage Be Used On Electric Utility Systems?" Applied Superconductivity Conf., Illinois, *IEE Trans. Mag.* **MAG-11**(3), 482–488.

Hausen, H. (1929). "Über die Theorie des Wärmeaustausches in Regeneratoren." *Z. Angew. Math. Mech.* **9**, 173–200. (On the Theory of Heat Exchange in Regenerators." R.A.E. Library Translation, No. 126).

Hausen, H. (1931). "Naherungsverfahren zur Berechnung des Wärmeaustausches in Regeneratoren." *Z. Angew. Math. Mech.* **11**, 105–114. (An Approximate Method of Dimensioning Regenerative Heat-Exchangers." R.A.E. Library Translation, No. 98).

Hausen, H. (1942). "Vervollstandigte Verrechnung des Wärmeaustauches in Regeneratoren." *A. ver. dt. ing. Beiheft Verfahrenstechnik. No.* 2, 31. (M.A.P. Reports and Translations, No. 312, 1946).

Herbert, I. R., and Campbell, S. J. (1976). "An Efficient Continuous Flow Helium Cooling Unit for Mössbauer Experiments." *Cryogenics* **16**(12), 717–719.

Herschel, J. (1850). "Making Ice." *The Athenaeum* (January 5), 22.

Hersh, D. J., and Abrardo, J. M. (1977). "Air Separation and Plant Design." *Cryogenics* **17**, 383.

Higa, W. H. (1965). "A Practical Philips Cycle for Low-Temperature Refrigeration." *Cryog. Technol.* **1**(5), 203–208.

Higa, W. H. (1969). "Cryogenic Applications in Space Communications." *Appl. Cryog. Technol.* **1**, 248–272.

Higa, W. H., and Wiebe, E. (1967). "A Simplified Approach to Heat Exchanger Construction for Cryogenic Refrigerators." *Cryog. Technol.* **3**(2), 47–48.

Higa, W., and Wiebe, E. (1977). "One Million Hours at 4.5 Kelvin." App. of Closed-Cycle Cryocooler to Small Superconducting Devices, Proc. of Conf. NBS Boulder, October, pp. 99–108.

Hill, R. W., Martin, D. L., and Osborne, D. W. (1968). "Calorimetry below 20 K." Chapter VII. Pub. In *Experimental Thermodynamics*, Vol. 1. AD-681 277. National Research Council of Canada, Ottawa.

Hoettinger, G. C., Skinner, R. P., and Trentham, R. A. (1961). "Design Considerations for Cryogenic Liquid Refill Systems for Cooling Infrared Detection Cells." *Adv. Cryog. Eng.* **6**, 354–362.

Hoffman, T. E. (1963). "Reliable Continuous Closed Circuit 4°K Refrigeration for a Maser Application." *Adv. Cryog. Eng.* **8**, 213–220.

Hofmann, A., Köfler, H., and Schapper, C. (1977). "Thermodynamics of a Self-Pumping Cooling Cycle for Superconducting Generator Applications." *Cryogenics* **17**, 429.

Hofmann, A., and Komarek, P. (1976). "Report on the Cryogenic Engineering Conference 1975 and the International Refrigeration Congress at Moscow." *Klima Kaelte Ing.* **4**(4), 157–162.

Hogan, W. H. (1967). "Refrigeration Means and Design Considerations for Cooling Masers and Parametric Amplifiers." *Low Temperature Refrigeration for Microwave Systems* (Proc. of Conf. on Low Noise Microwave Amplifiers and Low Temperature Refrig., Frankfurt am Main, Germany) (Ed. W. H. Hogan, *et al.*). Boston Technical Publishers Inc., Cambridge, Massachusetts, pp. 13–14.

Hogan, W. H. (1968). "Reliability of Cryogenic Refrigeration Systems. Proc. Second Int. Cryog. Eng. Conf. Iliffe Sci. and Technol. Pubs., Guildford, England, pp. 27–31.

Hogan, W. H. (1976). "Reliability Aspects of Cryogenic Refrigeration." *Adv. Cryog. Eng.* **21**, 187–189.

Hogan, W. H., and Stuart, R. W. (1963). "Design Considerations for Cryogenic Refrigerators." ASME Paper No. 63-WA-292.

Hogan, W. H., and Stuart, R. W. (1964). "Cryogenic Refrigerator." *Mech. Eng.* **86**, No. 10.

Holman, J. P. (1976). *Heat Transfer*. 3rd Ed. McGraw-Hill, New York.

Hood, C. B., Vogelhuber, W. W., and Barnes, C. B. (1964). "Helium Refrigerators for Operation in the 10–30 K Range." *Adv. Cryog. Eng.* **9**, 496–506.

Horikawa, N. (1975). "Helium 3 Refrigerator of Polarized Proton Target." *Cryog. Eng. (Tokyo)* **10**(1), 2–12.

Horn, S. B., Acord, T. T., Raimondi, P. K., and Walters, B. T. (1975). "Miniature Cryogenic Coolers." *Adv. Cryog. Eng.* **21**, 428–434.

Horn, S. B., Cowan, K. C., and Berry, R. L. (1973). "Low Production Cost VM Coolers." Proc. Closed-Cycle Cryogenic Cooler Conf. USAF Academy, Colorado, October, 10 pp. (AFFDL-TR-73-149, Vol. 1. AD No. 918234).

Horn, S. B., and Lumpkin, M. E. (1973). "Theoretical Analysis of Pneumatically Driven Split-Cycle Cryogenic Refrigerators." *Adv. Cryog. Eng.* **19**, 221–230.

Horn, S. B., Lumpkin, M. E., and Walters, B. T. (1973). "Pneumatically Driven Split-Cycle Cryogenic Refrigerator." *Adv. Cryog. Eng.* **19**, 216–220.

Horn, S. B., Lumpkin, M. E., Walters, B. T., and Acord, T. T. (1973). "Miniature Cryogenic Cooler for TOW Night Sight." Proc. Closed-Cycle Cryogenic Cooler Conf. USAF Academy, Colorado, 18 pp., October (AFFDL-TR-73-149, Vol. 1, AD No. 918234).

Hougen, J. O., and Piret, E. L. (1951). "Effective Thermal Conductivity of Granular Solids through Which Gases Are Flowing." *Chem. Eng. Prog.* **47**, 295–303.

Howard, C. P. (1963). "Heat-Transfer and Flow-Friction Characteristics of Skewed-Passage and Glass-Ceramic Heat-Transfer Surfaces." A.S.M.E. Paper No. 63-WA-115.

Howard, C. P. (1964). "The Single-Blow Problem Including the Effects of Longitudinal Conduction." A.S.M.E. Paper No. 64-GTP-11.

Hrycak, P. (1963). "Thermodynamic Analysis of a New Gas Refrigeration Cycle." *Cryogenics*, **3**, 23–26.

Hrycak, P., and Levy, M. J. (1966). "Pulse-Tube Refrigerator Analysis." Paper No. 66-WA/PID/3. A.S.M.E. Winter Annual Meeting, New York, 5 pp.

Hrycak, P., and Levy, M. J. (1968). "Thermodynamics of Pulse-Tube Refrigeration." *Proc. First Int. Cryog. Eng. Conf.*, Heywood-Temple Ind. Pubs., London, pp. 231–233.

Hudspeth, W. S. (1977). "Whats New in Cool(ing)?" *Electro-opt. Syst. Des.* **9**(5), 32–37.

Hunik, R., Konter, J. A., and Huiskamp, W. J. (1978). "Nuclear Demagnetization Experiments with the Aid of Precooling by PRCU (6)." *Physics at Ultralow Temps.*, Proc. Int. Symp., Hakone Japan Phys. Soc. of Japan, Tokyo (September), pp. 287–290.

Hunt, R. (1967). "Cryogenic Refrigerators for Cooling Aircraft Infrared Systems." *Cryog. Eng. News* **2**(6), 24–30.

Ijzer, J. A. L. (1964). "Gas Refrigerator with Stirling Cycle." *Chem. Ing. Tech.* **36**(5), 562.

Iliffe, C. E. (1948). "Thermal Analysis of the Contra-flow Regenerative Heat Exchanger." *Proc. Instn. Mech. Eng.* **159**, 363–372.

Imes, J. L., Neiheisel, G. L., and Pratt, W. P., Jr. (1975). "Magnetic Susceptibility Measurements on Some New Cerium Compounds for Adiabatic Demagnetization." *J. Low Temp. Phys.* **21**(1–2), 1–19.

Itoh, J., and Sano, N. (1978). "Nuclear Adiabatic Demagnetization of Two Proton Systems in Singlet Ground State Magnetic Salt." *Phys. at Ultralow Temps.*, Proc. Int. Symp., Hakone, Japan. Phys. Soc. of Japan, Tokyo (September), pp. 263–273.

Jacobs, R. B. (1962). "The Efficiency of an Ideal Refrigerator." *Adv. Cryog. Eng.* **7**, 567–571.

Jakob, M. (1957). *Heat Transfer.* Vol. II. Wiley and Sons, New York.

Jensen, J. E. (1960). "Performance of an Air Expansion Engine." *Adv. Cryog. Eng.* **1**, 105–110.

Jensen, J. E. (1975). "Advanced Refrigeration System for the Brookhaven Superconducting Cable Project." Adv. Refrig. Sys. Symp., Moscow, USSR. Brookhaven National Lab., Upton, New York, Report No. BNL-20443 (September), 42 pp.

Johnson, J. C., (1973). "Cryogenic Cooling Technology—An Air Force Perspective." Proc. Closed Cycle Cryogenic Cooler Conf. USAF Academy, Colorado, October, 7 pp. (AFFDL-TR-73-149, Vol. 1, AD No. 918234).

Johnson, J. E. (1952). "Regenerator Heat Exchangers for Gas Turbines." U.K. Aero. Res. Council, Tech. Rep., R and M, No. 2630, U.K.

Johnson, R. P., Bennett, A., Emigh, S. G., Griffith, W. R., Hoble, J. D., Penrome, R. E., and White, M. A. (1977). "Stirling-Hydraulic Artificial Heat Power Source." Proc. 12th I.E.C.E.C., Paper No. 779016, pp. 104–111, Washington, D.C.

Johnson, R. W., Collins, S. C., and Smith, J. L. (1971). "An Hydraulically Operated Two-Phase Helium Expansion Engine." *Adv. Cryog. Eng.* **16**, 171–177.

Jones, E. A., and Van der Sluijs, J. C. A. (1972). "On Indium Seals in Low Temperature Devices." *Cryogenics* **12**, 135.

Jones, M. C., Giarratano, P. J., McConnell, P. M., and Arp, V. (1973). "Refrigeration with Forced Flow of Helium." Proc. Cryogenic Cooler Conf., USAF Academy, Colorado, (October), pp. 441–462, NTIS COM-74-50509/0.

Jones, M. C., and Peterson, R. G. (1975). "A Study of Flow Stability in Helium Cooling Systems." *J. Heat Transfer* **97**(4), 521–527.

Joy, P. (1972). "Optimum Cryogenic Heat Pipe Design." *Adv. Cryog. Eng.* **17**, 438–448.

Jurriens, R. G., Pennings, H. N., and Satoh, T. (1978). "The (3)He–(4)He Dilution Refrigerator through Which (4)He is Circulated and Cooling by Compression of the (3)He in the Mixing chamber of this Refrigerator." *Phys. at Ultralow Temps.*, Proc. Int. Symp., Hakone, Japan. Phys. Soc. of Japan, Tokyo (September), pp. 226–241.

Kadi, F. J., and Longsworth, R. C. (1977). "Optimization of Helium Refrigerators for Superconducting Power Transmission Lines in Terms of Cost and Reliability." *J. Eng. Ind.* **99**(3), 551–557.

Kapitza, P. (1934). "The Liquefaction of Helium by an Adiabatic Method." *Proc. R. Soc., London Ser. A* **147**, 189.

Kapitza, P. L., and Danilov, I. B. (1961). "Expansion Engine for Liquefaction of Helium." *Zh. Tekhn. Giz.* **31**. [Reprinted in *Collected Papers of P.L. Kapitza*, Vol. 2 1938–1964, Pergamon Press Ltd., Oxford (1965), pp. 816–27.

Kapitza, P. L., and Danilov, I. B. (1968). "Cascade Helium Liquefiers with Piston-Type Engines." *Proc. First Int. Cryo. Eng. Conf.*, Heywood-Temple Ind. Publs. Ltd., London, pp. 228–231.

Karavansky, I. I., and Meltser, L. Z. (1959). "Thermodynamic Investigation of the Working Cycle of the Philips Machine." Proc. 10th Int. Cong. Refrig., Copenhagen, Vol. 3, 29, 209 Pergamon Press, Oxford.

Kasamatsu, K., Majima, K., Moriya, J., *et al.* (1971). "An Experiment and Consideration on the Reliability of the Cryogenic Refrigerator." *Cryog. Eng. (Tokyo)* **6**(5), 132–139.

Katheder, H. (1974). "Operating Experience Gained with an He-1 Cycle in a 300-Watt Refrigerator Plant." Arbeitskreis Festkoerperphysik of the Deutsche Physikalische Gesellschaft, Proc. Spring Mtg., Freudenstadt, Germany (April), B1–B7.

Katheder, H., Lehmann, W., and Spath, F. (1974). "Test and Operation of Helium Low-Temperature Devices for Applied Superconductivity." Inst. für Experimentelle Kernphysik, Karlsruhe, West Germany Report No. KFK-EXT.-3/74-9 (April), 37 pp.

Kayen, C. F. (1960). "Heat Exchanger Performance Prediction by Electrical Analog." *Adv. Cryog. Eng.* **2**, 282–294.

Kays, W. M., and London, A. L. (1958). *Compact Heat Exchangers*, 2nd Edition. McGraw-Hill, New York.

Keesom, W. H. (1933). *Commun. Phys. Lab. Univ. Leiden, suppl.* 76a.

Keilin, V. E., Klimenko, E. Ju., Kovalev, I. A., and Samoilov, B. N. (1970). "Force-Cooled Superconducting Systems." *Cryogenics* **10**, 224.

Kenoldt, W. (1965). "Selected Examples of European Cryogenic Practice." *Adv. Cryog. Eng.* **10**, 392–402.

Kessler, G. (1970). "Joule–Thomson Hydrogen Refrigerator Target." *Adv. Cryog. Eng.* **15**, 443–446.

Khalil, A., and McIntosh, G. (1977). "Thermodynamic Optimization Study of the Helium Multi-Engine Claude Refrigeration Cycle." Paper FC-1. *Adv. Cryog. Eng.* **23**, 431–437.

Khan, M. I. (1962). "The Application of Computer Techniques to the General Analysis of the Stirling Cycle." M.Sc. thesis, Univ. of Calgary.

Kim, J. C., and Qvale, E. B. (1971). "Analytical and Experimental Studies of Compact Wire-Screen Heat Exchangers." *Adv. Cryog. Eng.* **16**, 302–311.

Kim, J. C., Qvale, E. B., and Helmer, W. A. (1971). "Apparatus for Studies of Regenerators and Heat Exchangers for Pulse-Tube, Vuilleumier and Stirling-Type Refrigerators." *Prog. Refrig. Sci. and Technol.* (Proc. XIII Int. Cong. of Ref. Vol. 1, Washington). AVI Publishing Co., Inc., Westport, Connecticut, pp. 113–117.

Kirk, A. (1874). "On the Mechanical Production of Cold." *Proc. Inst. Civil Eng.* **37**, 244–315.

Kirk, W. P., and Adams, E. D. (1974). "He^4 Film Flow Suppressor for a Dilution Refrigeration Still and He^3 Purification." *Cryogenics* **14**, 147–149.

Kitami, T., Akino, Y., and Okuno, N. (1975). "Liquid Hydrogen Target with a Helium Refrigerator." Institute for Nuclear Study, Japan, Report No. INS-TH-94. Available INIS 28 pp.

Klann, J. L. (1973). "Analysis of a Combined Refrigerator–Generator Space Power System." Report No. NASA-TM-X-71433, E-7683. Presented at Cryog. Cooler Conf., Colorado, October NTISN73-31992/3. NASA, Cleveland, Ohio.

Kläy, H. R. (1975). "Reciprocating Compressors with Labyrinth Pistons for Helium." *Cryogenics* **15**, 569.

Klemens, P. G. (1963). "Applications for Superconductivity." *Prog. Refrig. Sci. Technol.* (Proc. XI Int. Cong. of Ref., Munich. Vol. 1. Paper No. 11, pp. 9–13), Pergamon Press, Oxford.

Knapton, J. D. (1963). "Bellows Operated Cryostat." *Cryogenics* **3**, 175.

Kneuer, R., and Turnwald, E. (1968). "Measurements and Experience with Valveless Piston Expansion Engines." *Proc. Second Int. Cryog. Eng. Conf.*, Iliffe Sci. and Technol. Pubs., Guildford, England, pp. 12-15.

Koh, J. C. Y., and Stevens, R. L. (1975). "Enhancement of Cooling Effectiveness by Porous Materials in Coolant Passage." *J. Heat Transfer* **97**(2), 309-311.

Köhler, J. W. L. (1960). "The Gas Refrigerating Machine and Its Position in Cryogenic Technique." *Prog. Cryog.* **2**, 43-67.

Köhler, J. W. L. (1960). "Principles of Gas Refrigerating Machines." *Adv. Cryog. Eng.* **2**.

Köhler, J. W. L. (1960). "Refrigeration below −100°C." *Adv. Cryog. Eng.* **5**, 518-525.

Köhler, J. W. L. (1964). "La Production du Froid aux Basses Temperatures a l'Aide des Machines Frigorifiques a Gaz Froid. Journees des Basses et Tres Basses Temperatures." ("Production of Cold at Low Temperatures with Cold Gas Refrigeration Machines. Meeting on Low and Very Low Temperatures") Paris (June 1-2), 1964. Paper No. MF 277-0.

Köhler, J. W. L. (1964). "La Production du Froid Aux Basses Temperatures a l'Aide des Machines Frigorifiques a Gaz Froid." Phillips (N.V.) Glofilampenfabrieken, Eindhoven, Netherlands. ("Refrigeration at Low Temperature by Means of Cold Gas Refrigerating Machines"). *Rev. Gen. Froid* **55**(9), 1119-1125.

Köhler, J. W. L. (1965). "The Stirling Refrigeration Cycle." *Sci. Am.* **212**(4), 119-127.

Köhler, J. W. L. (1968). "The Application of the Stirling Cycle at Cryogenic Temperatures." *Proc. First Int., Cryog. Eng. Conf.*, Heywood-Temple Ind. Pubs. Ltd., London, pp. 9-12.

Köhler, J. W. L. (1969). "Computation of the Temperature Field of Regenerators with Temperature-Dependent Parameters." *Proc. Second Int. Cryog. Eng. Conf.*, Iliffe Sci. and Technol. Press, Guildford, England, pp. 44-46.

Köhler, J. W. L., and Jonkers, C. O. (1954). "Construction of a Gas Refrigerating Machine. *Philips Tech. Rev.* **16**(5), 105-115.

Köhler, J. W. L., and Jonkers, C. O. (1954). "Fundamentals of the Gas Refrigerating Machine." *Philips Tech. Rev.* **16**(3), 69-78.

Köhler, J. W. L., Stevens, P. F., de Jonge, A. K., and Beuzekom, D. C. (1975). "Computation of Regenerators Used in Regenerative Refrigerators." *Cryogenics* **15**, 521-531.

Köhler, J. W. L., and Van der Ster, J. (1960). "A Small Liquid Nitrogen Plant Using a Gas Refrigerating Machine." *Adv. Cryog. Eng.* **2**, 351-356.

Kolm, H. H. (1975). "The Future of Superconducting Technology." *Cryogenics* **15**, 63.

Konter, J. A., Hunik, R., and Huiskamp, W. J. (1977). "Nuclear Demagnetization Experiments on Copper." *Cryogenics* **17**(3), 145-154.

Korrovits, V. Kh., Liidya, G. G., and Mikhkelsoo, V. T. (1974). "Thermostating Crystals at Temperatures below 1 K by Using the Electrocaloric Effect." *Cryogenics* **14**(1), 44-45.

Kovachev, V. T., and Georgiev, Y. K. (1976). "On the Thermodynamic Efficiency of Cooling with a Two-Stage Stirling–Phillips Cryogenerator." *Kokl. Bolg. Akad. Nauk* **29**(1), 41.

Kraus, J. (1977). "New Condensation Stage for a He^3-He^4 Dilution Refrigerator." *Cryogenics* **17**, 173-175.

Kroeger, P. G. (1967). "Plated Tube Heat Exchanger: Analytical Investigation of a New Surface Concept." *Adv. Cryog. Eng.* **12**, 340-351.

Kroeger, P. G. (1967). "Performance Deterioration in High Effectiveness Heat Exchangers Due to Axial Heat Conduction Effects." *Adv. Cryog. Eng.* **12**, 363-371.

Kropschot, R. H. (1973). "Helium Heat Transfer." (Final Rept.) Pub. in *Proceedings of Application of Superconducting Cable in Elec. Eng. and High Energy Phys.*, Titisee, Germany, NTIS COM-74-40382/2. *National Bureau of Standards*, Washington, D.C.

Kun, L. C., and Ranov, T. (1965). "Efficiency of Low-Temperature Expansion Machines." *Adv. Cryog. Eng.* **10**, 433-440.

Kurti, N. (1967). "Low Temperatures in the Generation and Transmission of Electric Power." *Prog. Refrig. Sci. Technol.* (Proc. XII Cong. of Refrig, Vol. 1., Madrid), pp. 1-13.

Kurti, N. (1973). "Low-Temperature Terminology." *Prog. Refrig. Sci. Technol.* (Proc. XIII Int. Cong. of Ref., Vol. 1, Washington). AVI Publishing Co., Inc., Westport, Connecticut, pp. 593-597.

Kuznetsov, B. G., and Arkharov, A. M. (1970). "Method of Determination of Continuously Variable Temperature in Working Cavities of Ideal Prototypes of Gas Refrigerating Machines Operating on the Stirling Cycle." *Deep Cold and Conditioning* (Ed. G. I. Voronin) Trudy MVTU, No. 138, Moscow (USSR), pp. 28–38.

Lambertson, T. J. (1958). "Performance Factors of a Periodic-Flow Heat-Exchanger." *Trans. ASME* **80**, 586–592.

Land, M. L. (1960). "Expansion Turbines and Engines for Low-Temperature Processing." *Adv. Cryog. Eng.* **5**, 250–260.

Larin, M. P. (1976). "Production of Temperature over the 29–63 K Range Using Solid Nitrogen and Its Application." *Instrum. Exp. Tech.* (Engl. Transl.) **19**(6), Pt. 2, (Nov–Dec) 1824–1827 (Transl. of *Prob. Tekh. Eksp.* No. 6, 208–211).

Lavrenchenko, G. K., and Trotsenko, A. V. (1976). "Study of Optimum Parameters of the Linde Refrigeration Cycle." *Kholod. Tekh. Tekhnol.* No. 23, 54–8.

Lawless, W. N. (1973). "Dielectric Cooling Technology: 15–4.2 K." Proc. Closed-Cycle Cryogenic Cooler Conf. USAF Academy, Colorado, October, 24 pp. (AFFDL-TR 73–149, Vol. 1, AD No. 918234).

Lax, B. (1963). "High Magnetic Field Research." *J. Appl. Phys.* **33**(3), Suppl. (Mar.), 1025–1029. (National Magnet Lab., Massachusetts).

Lechner, R. A., and Ackermann, R. A. (1973). "Concentric Pulse Tube Analysis and Design." *Adv. Cryog. Eng.* **18**, 467–475.

Lee, K. (1976). "The Stirling Cycle with Adiabatic Compression and Expansion." M.Sc. Thesis, University of Calgary.

Leger, P., Thomas, P., and Zermati, C. (1971). "Design of Valveless-Type Piston Expanders with Variable Clearance for a 1.5 W, 4.5 K Helium Refrigerator." *Proc. Third Int., Cryog. Eng. Conf.*, Illife Sci. and Technol. Pubs., Guildford, England, pp. 259–263.

Lehmann, W. (1974). "Operating Experiences with the First Cryostat of the Superconducting Linear Accelerator in the 1.8 K, 300-W Refrigerator." Arbeitskreis Festkoerperphysik of the Deutsche Physikalische Gesellschaft, Proc. Spring Mtg., Freudenstadt, Germany (April), pp. Cl–C12.

Lehrfeld, D., and Pitcher, G. (1973). "An Oil-Lubricated, Triple Expansion VM Cooler for Long Duration Space Missions." Proc. Closed-Cycle Cryogenic Cooler Conf. USAF Academy, Colorado, October, 22 pp. (AFFDL-Tr 73-149, Vol. 1, AD No. 918234).

Lenfestey, A. G. (1961). "Low-Temperature Heat Exchangers." *Prog. in Cryogenics* (Ed. K. Mendelssohn), Vol. 3, pp. 23–48, Heywood and Co., London, England.

Lenfestey, A. G. (1968). "Compact Heat Exchanger Assemblies for Gas Separation Plants." *Proc. 2nd Int. Cryog. Eng. Conf.*, pp. 47–49, Iliffe Sci. and Technol. Pubs. Ltd., Guildford, U.K.

Lerner, E., and Daunt, J. G. (1975). "A He³ Cryostat-Liquefier with Mechanical Precooling." *Cryogenics* **15**, 548.

Levi, E. (1971). "The Stirling Cycle and Its Various Applications to Cryogenerators." *Termotecnica* **25**(7), 362–370.

Leyarowski, E. J., and Kowachev, V. T. (1969). "Three-Flow, Low-Temperature Heat Exchanger." *Cryogenics* **9**, 337.

Lindale, E. (1978). "Stirling Cycle Refrigerators for Gamma Ray Detector." Report No. PL-42-Cr78-0713, Johns Hopkins Univ., Applied Physics Lab., Laurel, Maryland.

Lins, R. C., and Elkon, M. A. (1975). "Design and Fabrication of Compact High-Effectiveness Cryogenic Heat Exchangers Using Wire Mesh Surfaces." *Adv. Cryog. Eng.* **20**, 283–299.

Locke, G. L. (1950). "Heat-Transfer and Flow-Friction Characteristics of Porous Solids." T.R. No. 10, Dept. of Mech. Eng., Stanford University.

Long, H. M. (1963). "Refrigeration at Cryogenic Temperatures." Paper No. 63-WA-287. ASME Winter Ann. Meet., Philadelphia, 23 pp.

Long, H. M., and Simon, F. E. (1954). *Appl. Sci. Res.* **4**, 237.

Longsworth, R. C. (1966). "An Analytical and Experimental Investigation of Pulse Tube Refrigeration." Syracuse Univ. Ph.D. Thesis, 196 pp.

Longsworth, R. C. (1967). "An Experimental Investigation of Pulse Tube Refrigeration Heat Pumping Rates." *Adv. Cryog. Eng.* **12**, 608–618.

Longsworth, R. C. (1970). "Modified Solvay Cycle Refrigerator." *Cryogenics Ind. Gases* **5**(6), 33–35.

Longsworth, R. C. (1971). "A Modified Solvay Cycle Cryogenic Refrigerator." *Adv. Cryog. Eng.* **16**, 195–204.

Longsworth, R. C. (1971). "Thermal Efficiency of the Displacer Annular Clearance Space as a Regenerator." *Prog. in Refrig. Sci. and Technol.* (Proc. XIII Int. Cong. of Refrig., Washington) AVI Publishing Co., Inc., Westport, Connecticut, 129–136.

Longsworth, R. C. (1973). "Small Split System Closed-Cycle Cryogenic Cooler," Conf. USAF Academy, Colorado, October, 10 pp. (AFFDL-TR-73-149, Vol. 1, AD No. 918234).

Longsworth, R. C. (1974). "Application of a 20 K Refrigerator to Reducing Boil-Off Rate of a Liquid-Helium Dewar." *Proc. 5th Int. Cryog. Eng. Conf.*, Kyoto, Japan. IPC Sci. and Technol. Press, Sussex, England, pp. 390–394.

Longsworth, R. C. (1977). "Performance of a Cryopump Cooled by a Small Closed-Cycle 10 K Refrigerator" Paper KC-1. Cryog. Eng. Conf./Int. Cryog. Mats, Conf., Colorado, (August), 22 pp.

Lounasmaa, O. V. (1974). *Experimental Principles and Methods below 1 K*. Academic Press, New York, 325 pp.

Lounasmaa, O. V. (1978). "Nuclear Refrigeration of Liquid (3) He." *Phys. at Ultralow Temps.*, Proc. Int. Symp., Phys. Soc. of Japan, Tokyo, pp. 246–262, September.

Lyapin, V. K., Prusman, Yu. O., and Bakhnev, V. G. (1975). "Effect of Efficiency of the End Heat Exchanger on the Start-up Period of a Helium Cooler. *Chem. Pet. Eng.* (*USSR*) **11**(9–10), 803–805 (Transl. of *Khim. Neft. Mashinostr.* No. 9, 15–6.)

Madden, H. H., Fizzino, A. S., and Bohm, H. V. (1966). "Increased Liquefaction Rate on Early Model ADL-Collins Helium Liquefier." *Rev. Sci. Instrum.* **37**, 1409–1910.

Madocks, F. E. (1968). "Application of Turbomachinery to Small-Capacity Closed-Cycle Cryogenic Systems." *Adv. Cryog. Eng.* **13**, 463–473.

Maddox, R. N., and Erbar, J. H. (1976). "Some Aspects of Expander Processing." Gas Proc. Assoc., Proc. 55th Annual Conv., Oklahoma, pp. 99–102.

Maissin, J., and Le Diouron, R. (1976). "Liquid Nitrogen Refrigeration Units for Temperature Control in the Chemical Pharmaceutic and Food Industry." *Int. Cryog. Eng. Conf., Proc. 6th*, Grenoble, France. IPC Sci. and Technol. Press, Guildford, England, pp. 72–73.

Maki, E. R., and de Hart, A. O. (1971). "A New Look at Swash-Plate Drive Mechanisms." *S.A.E. Trans.* **80**, Paper No. 710829.

Mälek, Z., Pust, L., and Ryska, A. (1977). "Porous Heat Exchangers for Continuous Flow Helium Cryostats." *Cryogenics* **17**, 543.

Malkov, M. P., and Danilov, I. B. (1968). "Cryogenic Equipment of the Institute for Physical Problems." *Proc. First Int. Cryog. Conf. Tokyo*, pp. 225–228, Heywood-Temple Ind. Publns., London.

Malkov, M. P., and Sytchev, V. V. (1975). "Soviet Research in Cryogenic Technology and Applied Superconductivity." *Cryogenics* **15**, 65.

Mann, D. B., Bjorkland, W. R., Macinko, J., and Hiza, M. J. (1960). "Design, Construction and Performance of a Laboratory Size Helium Liquefier." *Adv. Crog. Eng.* **5**, 346–353.

Marler, J. M., and Gelezunas, V. L. (1973). "Operational Characteristics of a High Purity Germanium Photon Spectrometer Cooled by a Closed-Cycle Cryogenic Refrigerator." *Trans. IEEE Nucl. Sci.* **NS2**(1), 522–527.

Mart, J. W., Steel, A. J., and Clarke, M. E. (1976). "An Automatic Helium Purifier for Integration with Small Laboratory Liquefiers." *Int. Cryog. Eng. Conf., Proc. 6th*, Grenoble, France. IPC Sci. and Technol. Press, Guildford, England, pp. 69–71.

Martynovsky, V. S., Aleckseev, V. P., and Bondarenko, L. F. (1971). "Thermodynamic Analysis of Low-Temperature Utilization Cycles." *Prog. Refrig. Sci. Technol.* (Proc. XIII Int. Cong. of Refrig. Vol. 1 Washington). AVI Publishing Co., Inc., Westport, Connecticut pp. 171–177.

Matsubara, Y. (1969). "Recent Refrigeration Cycle." *Cryog. Eng. (Tokyo)* **4**(1), 8–19.

Matsubara, Y., Ishizaki, Y., and Oshima, K. (1968). "New Type Expansion Engine for Refrigeration." *Proc. First Int. Cryo. Eng. Conf.*, Heywood Temple Ind. Pubs., London, pp. 210–212.

McInroy, J. (1967). "Miniature Joule–Thomson Coolers." *Prog. Refrig. Sci. Technol.* (Proc. XII Int. Cong. of Refrig., Vol. 1, Madrid). pp. 59–67.

McMahon, H. O. (1960). "Recent Developments in Gas Cryogenics." *Cryogenics* 1(2), 65–70.

McMahon, H. O., and Gifford, W. E. (1960). "A Closed-Cycle Helium Refrigerator for Maintaining Very Low Temperature." *Bull. Inst. Int. Froid Annexe* 1960–1, 15–27.

McMahon, H. O., and Gifford, W. E. (1960). "A New Low-Temperature Gas Expansion Cycle, Parts I and II." *Adv. Cryog. Eng.* 5, 354–372.

Meek, R. M. G. (1961). "The Measurement of Heat-Transfer Coefficients in Packed Beds by the Cyclic Method." Int. Heat-Trans. Conf. (A.S.M.E.), Boulder, Colorado, pp. 770–780.

Meier, R. H., and Currie, R. B. (1968). "A 4.0 K Single Engine Cycle Helium Refrigerator." *Adv. Cryog. Eng.* 13, 441–449.

Meijer, H. C. (1972). "Low-Temperature-Operating Cooling Machine." *Chem. Tech. (Amsterdam)* 27(18), 489–491).

Meijer, R. J. (1959). "The Philips Hot-Gas Engine with Rhombic Drive Mechanism." *Philips Tech. Rev.* 20(9), 245–276.

Meijer, R. J. (1959). "The Philips Thermal Engine." *Philips Res. Rep. Suppl.* No. 1, Philips Research Labs, Eindhoven.

Meisler, J. (1960). "Ultra Low Temperature Production and Control in Environmental Testing and Application Facilities." *Adv. Cryog. Eng.* 4, 160–173.

Meisler, J. (1960b). "Applications and Economics of the Norelco Gas Liquefier for Recovery of Flash Gases." *Adv. Cryog. Eng.* 4, 454–463.

Meulenberg, R. E., and Abell, T. W. D. (1969). "Marine Applications of Stirling Cycle Refrigerators." *Inst. Mar. Engrs. Trans., U.K.* 61(7), 225–248. Reprinted *J. Refrig. (London)* 12(8), 243–253.

Mijnheer, A. (1972). "Experiments on a Two-Stage Stirling Cryogenerator with Unbalanced Regenerators." *Proc. Fourth Int. Cryog. Eng. Conf.*, IPC Sci. and Technol. Press, Guildford, England, pp. 83–86.

Modest, M. F. and Tien, C. L. (1974). "Thermal Analysis of Cyclic Cryogenic Regenerators." *Int. J. Heat Mass Transfer* 17(1), 37–49.

Moiseev, V. A., Pshisukha, A. M., and Zvyagin, A. I. (1970). "A Cryostat for Measuring EPR at Temperatures between 2 and 300 K." *Cryogenics* 10, 332.

Mole, C. J., Litz, D. C., and Feranchak, R. A. (1974). "Cryogenic Aspects of Superconducting Electrical Machines for Ship Propulsion." Amer. Soc. of Mech. Engs. Annual Mtg., New York, Paper No. 74-WA/PID-9.

Montgomery, D. B. (1966). "Magnets for Fields above 100 KG," INTERMAG Conf. (Stuttgart, Germany, April). *IEEE Trans. Mag.* **MAG-2**(3), 154–158.

Morain, W. A., and Holmes, J. W. (1963). "An Analysis of the Performance of Large Reciprocating Expansion Engines with the Aid of a Computer and Laboratory Prototype." *Adv. Cryog. Eng.* 8, 228–235.

Morales, J. R., Romero, J. L., and Brandan, M. E. (1974). "A Cooling System for Thin Tritiated Titanium Targets." *Nucl. Instrum. Methods* 119(1), 91–92.

Morgan, G., Aggus, J., and Bamberger, J. (1975). "Superconductive Magnet System for the AGS High Energy Unseparated Beam." *IEE Trans. Nucl. Sci.* **NS-22**(3), 1164–1167.

Morgan, G. H., and Jensen, J. E. (1977). "Counter-Flow Cooling of a Transmission Line by Supercritical Helium. *Cryogenics* 17(5), 259–267.

Morihara, H., and Terbot, J. W. (1975). "Liquid-Helium Refrigerator for Testing Superconduction System." Final Report Elec. Power Res. Inst., California, Report No. EPRI No. 7839, (August), 40 pp.

Moriya, J. (1972). "The Reliability of the Cryogenic System." *Cryog. Eng. (Tokyo)* 7(4), 181–186.

Moriya, J., Kasamatsu, K., and Okubo, I. (1975). "Kirk Cycle Cryogenic Miniature Refrigerator." *Cryog. Eng. (Tokyo)* 10(2), 55–64.

Moulenberg, R. E. (1966). "Refrigeration–Stirling Cycle Reliquefaction Plant for Small Gas Carriers." *Shipbld. Shipping Rec.* **107**(1), 77–79.

Mullen, L. O., and Hiza, M. J. (1966). "Experimental Apparatus and Procedures for Evaluating Parameters Affecting the Pumping Efficiency of a Cryogenically Cooled Plane." *J. Vac. Sci. Technol.* **4**, 219–229. *CFSTI PB*-176 589.

Muller, H. J. (1964). "Improvements in Non-Lubricated Compressor Design." *Linde Rep. Sci. Technol.* **6**, 3–8.

Muller, H. J. (1971). "Advances in Non-Lubricated Compressor Design." *Linde Rep. Sci. Technol.* **17**, 3–11.

Murray, J. A., Martin, B. W., Bayley, F. J., and Rapley, C. W. (1961). "Performance of Thermal Regenerators under Sinusoidal Flow Conditions." Int. Heat-Trans. Conf., A.S.M.E., Boulder, Colorado, pp. 781–796.

Nadolnikov, A. G., Fastovskii, V. G., and Petrovskii, Yu. V. (1965). "A Miniature Refrigerating Machine." *Cryogenics* **5**(6), 342–343.

Naer, V. A., Khirich, I. Ya., and Belozerova, L. A. (1975). "Thermoelectric Cells for Low-Temperature Cooling Devices." *Kholod. Tekh.* No. 11, 39–41.

Nagano, H. (1974). "Design and Performance of a 30-l h^{-1} Helium Liquefier/Refrigerator with Plastic Piston Expansion Engines." *Cryogenics*, **14**, 654.

Nagano, H., Kurichara, A., Kikuchi, K., and Kanazawa, Y. (1971). "A Helium Liquefier with Gas Purification Circuit." *Proc. Third Int. Cryog. Eng. Conf.*, Iliffe Sci. and Technol. Pubs., Guildford, England, pp. 271–273.

Nagano, H., Watanabe, N., and Suzuki, M. (1971). "A Miniature Expansion Turbine for a Cryogenic Refrigerator." *Prog. Refrig. Sci. Technol.* (Proc. XIII Int. Cong. of Refrig. Vol. 1, Washington). AVI Publishing Co., Inc., Westport, Connecticut, pp. 165–170.

Nast, T. C., Bell, G. A., and Wedel, R. K. (1975). "Orbital Performance of a Solid Cryogenic Cooling System for a Gamma Ray Detector." *Adv. Cryog. Eng.* **21**, 435–442.

Neelen, G. T. M., Ortegren, L. G. H., Kuhlmann, P., and Zacharias, F. (1971). "Stirling Engines in Traction Applications." C.I.M.A.C., **A26**, 9th Int. Congress on Combustion Engines, Stockholm, Sweden.

Neeper, D. A. (1967). "Rapid Start-Up of a He-3/He-4 Dilution Regrigerator." *Cryogenics* **7**, 307.

Newhouse, V. L., and Atherton, D. L. (1969). "New Materials and More Applications for Superconductivity." *Cryogenics* **9**, 80.

Nicholds, K. E. (1963). "Compact Cryogenic Refrigerators for Special Purposes." *New Scientist* **18**(334), 98–99.

Nicholds, K. E. (1966). "Expansion Engines." *Mod. Refrig.* **69**(821), 681–684.

Nicholds, K. E. (1968). "Low-Temperature Devices for Laboratory Operation." *Proc. 2nd Int. Cryog. Eng. Conf.*, Brighton Iliffe Sci. and Technol. Publ., Guildford, U.K., pp. 65–66.

Nicholds, K. E. (1970). (a) "Performance of Self-Regulating Joule–Thomson Minicoolers." (pp. 277–282); (b) "Miniature Cryogenic Cooling Systems for an Upper Atmosphere Infrared Research Programme." (pp. 283–286). *Proc. 3rd Int. Cryog. Eng. Conf.*, Berlin, Iliffe Sci. and Technol. Publ., Guildford, U.K.

Niinikoski, T. O. (1971). Construction of Sintered Copper Heat Exchangers. *Cryogenics* **11**, 232.

Niinikoski, T. O. (1976). "Dilution Refrigeration. New Concepts." *Proc. 6th Int. Cryog. Eng. Conf.*, Grenoble, France. IPC Sci. and Technol. Press, Guildford, England, pp. 102–111.

Norton, M. T., (1963). "Miniature Helium Turbo-Expander for Cryogenic Refrigeration Systems." *Prog. Refrig. Sci. Technol.* (Proc. XI Int. Cong. of Refrig., Munich. Vol. 1. Paper 1–9, Pergamon Press, Oxford, pp. 131–135).

Novotelnov, V. N. (1974). "Thermodynamic Analysis of the Non-Calculable Range of a Cryogenic Expander." *Izv. Vyssh. Uchebn. Zaved., Energ.* **17**(10), 86–90.

Nusselt, W. (1927). "Die Theorie des Winderhitzers." *Z. Ver. Dt. Ing.* **71**, 85.

Nusselt, W. (1928). "Der Beharrungszustand im Winderhitzer." *Z. Ver. dt. Ing.* **72**, 1052.

Oberly, C. E., and Barthelemy, R. R. (1967). "Low-Temperature Magnets for MHD Power Generation." *Prog. Refrig. Sci. Technol.* (Proc. XII Int. Cong. of Refrig, Madrid. Vol. 1. Paper 1.27), pp. 129–139.

Oda, K. (1974). "Refrigeration at Extremely Low Temperatures (Below 1° K)." *Reito* **49**(557), 239–244.

Ohtsubo, A., Satoh, T., Terui, G., Inoue, M., and Kanda, E. (1970). "A Dilution Refrigerator for Nuclear Building Designed to Minimize Eddy Current Heating." Annexe 2, *Cryophysics and Cryoengineering*. IIR. Paris. pp. 63–66.

Okhrem, V. G., and Samoilovich, A. G. (1977). "Physical Bases for Operation of Round–Cylindrical Galvanothermomagnetic Refrigerating Element." *Ukr. Fiz. Zh.* (*Russ. Ed*) **22**(1), 39–44.

O'Neill, P. S., Gottzmann, C. F., and Terbot, J. W. (1972). "Novel Heat Exchanger Increases Cascade Cycle Efficiency for Natural Gas Liquefaction." *Adv. Cryog. Eng.* **17**, 420–437.

Oshima, K., Ishizaki, Y., and Matsubara, Y. (1967). "An Electrical Indicator for Expansion Engines at Low Temperature." *Adv. Cryog. Eng.* **12**, 602–607.

Oshima, K., Matsubara, Y., and Kubo, T. (1968). "Experimental Investigation of Pulse Tube Refrigerator." *Cryog. Eng.* (*Tokyo*) **3**(2), 72–7.

Otop, H., and Sujak, B. (1974). "A Gifford-Type 10 K Miniature Refrigerator with Liquid Nitrogen Precooling Stage." *Acta Phys. Pol. A* **45**(3), 485–488.

Parkinson, D. H. (1959). "Some Problems in the Design of Helium Liquefiers Based on the Joule–Thomson Effect." *Prog. Refrig. Sci. Technol* (Proc. X Int. Cong. in Ref., Copenhagen. Vol. 1. Paper 1-a-10, Pergamon Press, Oxford, pp. 53–57).

Parkinson, D. H. (1963). "Infrared Spectroscopy and Low Temperatures." *Cryogenics* **3**, 1.

Parkinson, D. H. (1965). "Cryogenic Electronic Developments." *Adv. Cryog. Eng.* **10**, 411–418.

Parkinson, D. H. (1967). "Miniature Refrigeration Systems—A Review." *Prog. Refrig. Sci. Technol.* (Proc. 12 Int. Cong. of Refrig., Madrid), Paper No. 1.28, pp. 69–77.

Parulekar, B. B., Bijlani, C. A., Narayankhedkar, K. G., and Khadilkar, J. S. (1971). "Gifford–McMahon Type Cryorefrigerator." *Prog. Refrig. Sci. Technol.* (Proc. XIII Int. Cong. of Refrig., Washington), AVI Publishing Co., Inc., Westport, Connecticut, pp. 47–54.

Parulekar, B. B., and Narayankhedkar, K. G. (1972). "Small-Capacity Valveless Piston Expansion Engine Type Cryorefrigerator." *Proc. Fourth Int. Cryog. Eng. Conf.* IPC Sci. and Technol. Press, Guildford, England, pp. 90–92.

Parulekar, B. B., and Narayankhedkar, K. G. (1976). "Performance of the Valveless Low-Temperature Heat Pump." *Proc. 6th Int. Cryog. Eng. Conf.*, Grenoble, France. IPC Sci. and Technol. Press, Guildford, England, pp. 77–79.

Pasotti, G., and Spadoni, M. (1976). "A New Approach to the Cooling of Superconducting Magnets for Fusion Reactors." *Proc. 6th Int. Cryog. Eng. Conf.*, Grenoble, France. IPC Sci. and Technol. Press, Guildford, England, pp. 325–326.

Pastuhov, A. (1960). "Helium Refrigeration." *Adv. Cryog. Eng.* **5**, 41–43.

Pavlov, V. N., and Eremenko, V. V. (1968). "A Cryostat for Magneto-Optical Studies." *Cryogenics* **8**, 170.

Pavlov, V. N., Neganov, B. S., Konicek, J., and Ota, J. (1978). "A Combined (3)He-(4)He Dilution Refrigerator." *Cryogenics* **18**(2), 115–119.

Pennings, N. H., Taconis, K. W., and De Bruyn Ouboter, R. (1974). "An Improved Version of the He^3–He^4 Refrigerator through Which He^4 is Circulated." *Cryogenics* **14**, 53–454.

Pennings, N. H., Taconis, K. W., and De Bruyn Ouboter, R. (1975). "The Leiden Dilution Refrigerator." *Proc. European Phys. Soc. Topical Conf.*, Haifa, Israel. John Wiley and Sons, New York, pp. 397–399.

Pennings, N. H., Taconis, K. W., and De Bruyn Ouboter, R. (1976). "The Leiden Dilution Refrigerator." *Physica* (*Utrecht*) *B* + *C* **81**(1), 101–106.

Pennings, N. H., Taconis, K. W., and De Bruyn Ouboter, R. (1976). "The Leiden Dilution Refrigerator II." *Physica* (*Utrecht*) *B* + *C* **84**(1), 102–109.

Penrod, E. B. (1963). "Concepts of Thermoelectric Refrigeration." *Prog. Refrig. Sci. Technol.* (Proc. XI Int. Cong. of Ref. Munich. Vol. 1. Paper No. 8), Pergamon Press, Oxford, pp. 15–22.

Permyakov, U. V., and Kovriga, N. N. (1972). "A Device for Controlling the Operation of a Metal Helium Cryostat." *Cryogenics* **12**, 137.

Peshkov, V. P. (1970). "An He^3 Dilution Cryostat for Operation in the Millidegree Region." *Cryogenics* **10**, 250.

Petersen, V. F. (1973). "Facing the Cryogenic Unit Interface Problems." Proc. Closed Cycle Cryogenic Cooler Conf. USAF Academy, Colorado, October, 10 pp.

Pippard, A. B. (1965). "Continuous Refrigeration of Current Leads." *Cryogenics* **5**, 81.

Pitcher, G. K., and Dupre, F. K. (1970). "Miniature Vuilleumier Cycle Refrigerator." *Adv. Cryog. Eng.* **15**, 447–451.

Plechac, L. (1974). "Cooling of Large Masses down to the Temperature of Liquid Helium I." *Elektrotech. Obz.* **63**(11), 666–670.

Podolskii, A. G. (1974). "Calculation of Thermo-Gas-Dynamic Parameters of a Cryogenic Refrigerator." *Izv. Akad. Nauk SSR. Energ. Transp.* (**Nov–Dec**), 164–169.

Polturak, E., and Rosenbaum, R. (1978). "Properties of the Double-Mixing Chamber Dilution Refrigerator." *Proc. Int. Symp.*, Hakone, Japan. Phys. Soc. of Japan, Tokyo, pp. 274–276.

Powell, R. L., and Clark, A. F. (1977). "Definitions of Terms for Practical Superconductors: 1. Fundamental States and Flux Phenomena." *Cryogenics* **17**, 697.

Prast, G. (1963). "A Philips Gas Refrigerating Machine for 20°K." *Cryogenics* **3**, 156–160.

Prast, G. (1965). "A Gas Refrigerating Machine for Temperatures down to 20°K and Lower." *Philips Tech. Rev.* **26**(1), 1–11.

Prast, G. (1965). "A Modified Philips-Stirling Cycle for Very Low Temperatures." *Adv. Cryog. Eng.* **10**, Pt. 2, 40–45.

Prast, G. (1967). "Closed-Cycle Cooling Systems for Microwave Equipment." *Low Temperature Refrigeration for Microwave Systems* (Proc. Conf. on Low Noise Microwave Amplifiers and Low Temperature Refrig., Frankfurt am Main, West Germany) (Eds. W. H. Hogan *et al*). Boston Technical Pubs., Inc., Cambridge, Massachusetts., pp. 247–251.

Prast, G. (1968). "A 3.5 K Refrigerator Based on the Three-Space Stirling Refrigerator." *Proc. Second Int. Cryog. Eng. Conf.*, Iliffe Sci. and Technol. Pubs., Guildford, England, pp. 19–22.

Prast, G. (1969). "European State of the Art in Cryogenics." Proc. of 1968 Summer Study on Superconducting Devices and Accelerators. Brookhaven National Lab., Upton, New York, pp. 205–228.

Prast, G. (1970). "Choice of Cryogenic Cycles." *Proc. Third Int. Cryog. Eng. Conf.*, Iliffe Sci. and Technol. Pubs., Guildford, England., pp. 21–24.

Prast, G., and Haarhuis, G. J. (1964). "The Philips-Gas Refrigerating System for Very Low Temperatures." *Kaltetechnik* **16**(8), 232–235.

Prast, G., Hargreaves, C. M., Mijnheer, A., and Van Mal, H. H. (1971). "A New Small Refrigerating Device for Temperatures in the Region of 20 K." *Prog. Refrig. Sci. Technol.* (Proc. XIII Int. Cong. of Refrig. Washington), AVI Publishing Co., Inc., Westport, Connecticut, Vol. 1. pp. 37–41.

Pratt, W. P., Jr., Rosenblum, S. S., Steyert, W. A., and Barclay, J. A. (1977). "A Continuous Demagnetization Refrigerator Operating near 2 K and a Study of Magnetic Refrigerants." *Cryogenics* **17**, 689–693.

Pronko, V. G., Amamchyan, R. G., Guilman, I. I., and Raygorodsky, A. I. (1976). "Some Problems of Using Adsorbents as a Matrix Material for Low-Temperature Regenerators of Cryogenic Refrigerators." *Proc. 6th Int. Cryog. Eng. Conf.*, Grenoble, France. IPC Sci. and Technol. Press, Guildford, England, pp. 86–88.

Pronko, V. G., Onosovskii, E. V., and Usanov, V. V. (1975). "Manufacture of High-Efficiency Heat Exchangers for Cryogenic Systems." *Chem. Pet. Eng.* **11**(9–10), 791–793 (Transl. of *Khim. Neft. Mashinostr.* No. 9, 8–10).

Quack, H. (1974). "Contribution to the Theory and Implementation of a Magnetic Cooling Cycle with New Thermal Switches." Eidgenoessische Technische Hochschule, Zurich, Switzerland, Ph.D. thesis, 136 pp.

Qvale, E. B. (1963). "An Analytical Model of Stirling-Type Engines." D.Sc. Thesis, M.I.T.

Qvale, E. B., and Smith, J. L. Jr. (1968). "A Mathematical Model for Steady Operation of Stirling-Type Engines." *J. Eng. Power A*, No. 1, 45–50.

Qvale, E. B., and Smith, J. L., Jr. (1969). "An Approximate Solution for the Thermal Performance of a Stirling-Engine Regenerator." *J. Eng. Power A*, No. 2, 109–112.

Qvale, E. B., and Smith, J. L. Jr. (1969). "A Simple Correlation for the Heat Transfer Characteristics of a Family of Matrices Subjected to Complex Flow Conditions." *Cryogenics* **9**, 62.

Raab, B., Schock, A., and King, W. C. (1975). "Nuclear Heat Source for Cryogenic Refrigerators in Space." Proc. 10th Annual Intersociety Energy Conv. and Eng. Conf., Del. Inst. of Elec. and Electronics Engs. Inc., New York, pp. 894–900.

Rabinowitz, M. (1977). "Cryogenic Power Generation." *Cryogenics* **17**, 319.

Radcenco, V. (1969). "Theoretical Study of Thermo-Gas-Dynamic Processes in Expansion Engines." *Rev. Roum. Tech. Ser. Electrotech. Energet.* **14**(1), 111–128.

Radcenco, V. (1974). "Establishing the Expression of the Exergetic Efficiency of Piston Expanders and Turboexpanders Used in the Low Temperature Technique." *Rev. Chim. (Bucharest)* **25**(10), 816–821.

Radcenco, V. (1975). "Thermodynamics of Piston-Type Gas Expanders." *Rev. Chim. (Bucharest)* **26**(3), 231–236.

Radebaugh, R., and Siegwarth, J. D. (1970). "Numerical Analysis of Continuous and Discrete Heat Exchangers for Dilution Refrigerators." Report No. NBS-R-629. *Conf. on Cryophysics and Cryoengineering*, Tokyo, Japan (September), NTIS COM-71-00819. National Bureau of Standards, Boulder, Colorado.

Radebaugh, R., and Siegwarth, J. D. (1970). "Numerical Analysis of Continuous and Discrete Heat Exchangers for Dilution Refrigerators." Annexe 2, *Cryophysics and Cryoengineering*. IIR, Paris, pp. 57–62.

Radziwill, W. (1969). "A Highly Efficient Small Brushless D.C. Motor." *Philips Tech. Rev.* **30**(1), 7–12.

Rallis, C. J., Urieli, I., and Berchowitz, D. M. (1977). "A New Mathematical Model for Stirling Cycle Machines." Proc. 12th I.E.C.E.C. Paper No. 779254, pp. 1522–1527, Washington, D.C., August 28–September 2.

Rapley, C. (1960). "Heat Transfer in Thermal Regenerators." M.Sc. thesis, Durham University.

Rechowicz, M. (1975). *Electric Power at Low Temperatures*. Clarendon Press, Oxford, England, 152 pp.

Reed, R. P. (1971). "Materials at Low Temperatures." *Cryogenics* **11**, 347.

Reitlinger, J. (1876). "Über Kreisprozesse zwischen zwei Isothermen." *Z. Ost. Ing. Arch. Ver.*

Renard, M. (1974). "Cryogenics and Electrotechnics. *Rev. Gen. Electr.* **83**(10), 727–728.

Renyer, B. L. (1973). "High-Capacity, Long-Life Vuilleumier-Cycle Refrigerator." Proc. Closed-Cycle Cryogenic Cooler Conf. USAF Academy, Colorado, October, 16 pp. (AFFDL-TR 73-149, Vol. 1, AD No. 918234).

Rettori, C., Kim, H. M., and Davidov, D. (1974). "He³ Coldfinger Cryostat EPR Experiments." *Cryogenics* **14**, 285.

Richter, R., and Mahefkey, E. T. (1973). "The Applicability of a Solar Collector Thermal Power System to a Vuilleumier Cooler." Proc. Closed-Cycle Cryogenic Cooler Conf. USAF Academy, Colorado, October (AFFDL-TR 73-149, Vol. 1, AD No. 918234).

Rietdijk, J. A. (1965). "A New Positive Seal for Pistons and Axially Moving Rods." *Adv. Cryog. Eng.* **10**, 464–467.

Rietdijk, J. A. (1966). "The Expansion-Ejector, A New Device for Liquefaction and Refrigeration at 4°K and Lower." Paper No. 111-3. Comm. I Mtg., Boulder, Colorado, 6 pp.

Rietdijk, J. A., Van Beukering, H. C. J., van der Aa, H. H. M., and Meijer, R. J. (1965). "A Positive Rod or Piston Seal for Large Pressure Differences." *Philips Tech. Rev.* **26**, 287–296.

Rietdijk, J. A. (1966). "The Expansion-Ejector—A New Device for Liquefaction and Refrigeration at 4 K and Lower." *Liquid Helium Technology, Bull. Int. Inst. Refrig.* Annexe 1966–65, p. 241.

Rios, P. A., Qvale, E. B., and Smith, J. L., Jr. (1969). "An Analysis of the Stirling-Cycle Refrigerator." *Adv. Cryog. Eng.* **14**, 332–342.

Rios, P. A., and Smith, J. L. Jr. (1968). "The Effect of Variable Specific Heat of the Matrix on the Performance of Thermal Regenerators." *Adv. Cryog. Eng.* **13**, 566–573.

Rios, P. A., and Smith, J. L., Jr. (1969). "An Analytical and Experimental Evaluation of the Pressure-Drop Losses in the Stirling-Cycle." Paper No. 69-WA/Ener-8. ASME Winter Ann. Meeting, Los Angeles, California.

Rizzuto, C., Vaccarone, R., and Vivaldi, F. (1975). "Construction and Performance of Type He^3/He^4 Dilution Refrigerators." *Termotecnic* **29**(12), 666–670.

Robbins, R. F., Weitzel, D. H., and Herring, R. N. (1962). "The Application and Behavior of Elastomers at Cryogenic Temperatures." *Adv. Cryog. Eng.* **7**, 343–352.

Robinson, G. Y. (1965). "Large Capacity Helium Liquefier." *Adv. Cryog. Eng.* **10**, 22–26.

Robinson, G. Y. (1968). "Large Scale Helium Refrigeration System for the Range 2–20 K." *Proc. First Int. Cryo. Eng. Conf.*, Heywood-Temple Ind. Pubs., London, pp. 219–220.

Robinson, G. Y. (1969). "Production of Liquid Helium and Helium Refrigeration 1.8° to 20° K." *Cryogenic Technology* (Ed. H. Weinstock) Boston Technical Publishers, Inc., Boston, pp. 43–50.

Rogers, G. F. C., and Mayhew, Y. R. (1967). *Engineering Thermodynamics Work and Heat Transfer.* 2nd Ed. S.I. Units, Longmans Group Ltd., London.

Rose, K. (1969). "Superconducting Materials for Devices Spanning the Infrared–Microwave Gap." *Cryogenics* **9**, 227.

Rosenblum, S. S., Sheinberg, H., and Steyert, W. A. (1976). "Continuous Refrigeration at 10 mK Using Adiabatic Demagnetization." *Cryogenics* **16**, 245–246.

Roubeau, P. (1976). "A Dilution Refrigerator for Neutron Experiments." *Proc. 6th Int. Cryo. Eng. Conf.*, Grenoble, France. IPC Sci. and Tech. Press, Guildford, England, pp. 99–101.

Roubeau, P., and Varozuaux, E. (1970). "Copper Foil Heat Exchangers for Dilution Refrigerators." *Cryogenics* **10**, 255.

Rousar, D., and Millek, F. (1975). "Cooling with Supercritical Oxygen." AIAA/SAE 11th Prop. Conf., Paper No. 75–1248, (September), 6 pp.

Ruelle, G. (1975). "Main Problems Encountered in the Study of Cryogenic Generators." *Cryogenics* **15**, 69.

Ruhemann, M. (1961). "Low-Temperature Refrigeration." *Cryogenics* **1**, 193.

Ruhemann, M., and Manley, E. (1964). "Low-Temperature Engineering." *Chartered Mech. Engr.* **11**(5), 254–271.

Rule, T. T., and Qvale, E. B. (1969). "Steady-State Operation of the Idealized Vuilleumier Refrigerator." *Adv. Cryog. Eng.* **14**, 343–352.

Runge, E. (1973). "Thermodynamic Processes and Apparatus of Low-Temperature Technology." *Klima Kalte Ing.* **1**(10), 25–32.

Sato, T. (1968). "Some Experiments on the Reciprocating Expander for Large Helium Liquefiers." *Proc. First Int. Cryog. Eng. Conf.* Heywood-Temple Ind. Pubs., London, pp. 217–218.

Sato, T. (1975). "Superleak-Operated Refrigerators." *Cryog. Eng. (Tokyo)* **10**(6), 205–213.

Sato, T. (1978). "The New-Type Pomeranchuk Cooling Machine." *Phys. at Ultralow Temps.* Proc. Int. Symp., Hakone, Japan. Phys. Soc. of Japan, Tokyo, pp. 242–245.

Sato, T., and Takada, J. (1967). "Methods of Generating Low Temperatures." *Kagaku Kogaku* **31**(8), 730–735.

Saunders, O., and Ford, H. (1940). "Heat Transfer in the Flow of Gas through a Bed of Solid Particles." *J. Iron Steel Inst.*, No. 1, 291.

Saunders, O. A., and Smoleniec, S. (1948). "Heat Regenerators." Proc. 7th Int. Congress Appl. Mech., Vol. 3, pp. 91–105.

Schaffers, T. W. (1960). "Design and Application of a Gas Liquefier." *Adv. Cryog. Eng.* **3**, 92–98.

Schalwijk, W. E. (1959). "A Simplified Regenerator Theory." *J. Eng. Power* **A81**, 142–150.

Schmid, C. (1974). "Gas Bearing Turboexpanders for Cryogenic Plants." *Proc. 6th Int. Gas Bearing Symp.*, Southampton, Eng. British Hydromechanics Res. Assoc., Cranfield, England, B1-1–B1-8.

Schmidt, G. (1861). "Theorie der Geschlossenen Calorischen Maschine Von Laubroy und Schwartzkopff in Berlin." *Z. Ver. Oster. Ing.*, 79.

Schmidt, G. (1871). "Theorie der Lehmannschen Calorischen Maschine." *Z. Ver. dt. Ing.* **15**, No. 1.

Schock, A. (1978). "Nodal Analysis of Stirling Cycle Devices." (a) Draft Report DOE; (b) Proc. 13th I.E.C.E.C., San Diego, California.

Schubert, R. (1971). "The Influence of a Gas Atmosphere and Moisture on Sliding Wear in PTFE Compositions." *Linde Rep. Sci. Technol.* **17**, 12–20.

Schulte, C. A., Fowle, A. A., Huechling, T. P., and Kronauer, R. E. (1965). "A Cryogenic Refrigerator for Long-Life Applications in Satellites." *Adv. Cryog. Eng.* **10**, 477–485.

Schultz, B. H. (1951). "Regenerators with Longitudinal Heat Conduction. General Discussion on Heat Transfer." *A.S.M.E. J. Mech. E.*

Schultz, B. H. (1953). "Approximate Formulae in the Theory of Thermal Regenerators." *Appl. Sci. Res.* A **3**, 165–173.

Schumann, T. E. W. (1929). "Heat Transfer to a Liquid Flowing through a Porous Prism." *J. Franklin Inst.* **208**, 405–416.

Scott, R. B. (1966). *Cryogenic Engineering*. Van Nostrand Co. Inc., New Jersey.

Scott, L. E. (1963). "Ideal Yield of a Simon Liquefier." *Cryogenics* **3**, 111.

Severijns, A. P., Staas, F. A., and Cense, W. A. (1978). "An Improved He(3)-He(4) Mixing Chamber for Single-Cycle Experiments." *Cryogenics* **18**(2), 87–89.

Shelpuk, B., Crouthamel, M. S., and Cygnarowicz, T. A. (1970). "Icicle-Integrated Cryogenic Isotope Cooling Engine System." Paper No. 70-HT/SPT-30. ASME Space Technology and Heat Transfer Conference, Los Angeles, California, 8 pp.

Sherman, A. (1971). "Mathematical Analysis of a Vueilleumier Refrigerator." Paper No. 71-WA/HT-33. ASME 92nd Annual Meeting, Washington, D.C. (November 28–December 2, 1971), 15 pp.

Sherman, A. (1973). "Selected Vuilleumier Refrigerator Performance Characteristics." *Adv. Cryog. Eng.* **18**, 352–359.

Sherman, A. (1978). "Cryogenic Cooling for Spacecraft Sensors, Instruments, and Experiments." *Astronaut. Aeronaut.*, **16**(11), 39–47.

Sherman, A., Gasser, M., Goldowsky, M., Benson, G., and McCormick, J. (1979). "Progress on the Development of a 3–5 Year Lifetime Stirling Cycle Refrigerator for Space," *Adv. Cryog. Eng.* **25**, 791–800.

Sherman, A. L., and Gershman, R. (1971). "Design of a Cryogenic Heat Exchanger." *Prog. Refrig. Sci. Technol.* (Proc. XIII Int. Cong. of Ref., Vol. 1, Washington). AVI Publishing Co. Inc., Westport, Connecticut, pp. 107–112.

Shigi, T. (1972). "Refrigeration Cycle." *Cryogen. Eng. (Tokyo)* **7** (4), 157–164.

Shimotomai, M., Omar, A. M., and Robinson, J. E. (1977). "A Closed-Cycle Helium Cooled Cryostat." *Cryogenics* **17**, 47.

Shmalko, K. Ya., Musatova, L. G., Salenko, V. P. *et al.* (1970). "Numerical Theoretical Analysis of a Single-Step Regeneration Cycle at 80, 60, 40, 20 K." *Sb. Rab. Stud. Nauch. Obshchest., Leningrad. Takhnol. Inst. Kholod. Prom.* **1**, 42–50.

Shnide, I. M. (1969). "Thermodynamic Characteristics of Refrigerating Machines Using the Pulse Tube." *Bull. IIR Annexe*, 59–67.

Shnide, I. M. (1971). "Regenerative Heat Exchange in Gas-Refrigeration Machines." *Prog. Refrig. Sci. Technol.* (Proc. XIII. Int. Cong. of Refrig. Vol. 1., Washington). AVI Publishing Co., Inc., Westport, Connecticut, pp. 179–187.

Shvets, A. D. (1966). "Production of Temperatures below 1 K by Pumping on Liquid Helium-4." *Cryogenics* **6**, 270.

Shvets, A. D. (1967). "Apparatus for Obtaining Temperatures from 4.2 to 0.3 K. *Cryogenics* **7**, 294.

Siegel, R., and Perlmutter, M. (1961). "Two-Dimensional Pulsating Laminar Flow in a Duct with a Constant Wall Temperature." Int. Heat Trans. Conf. (ASME), Boulder, Colorado, pp. 517–535.

Siegwarth, J. D., and Radebaugh, R. (1971). "Analysis of Heat Exchangers for Dilution Refrigerators." (Final Rept.) *Rev. Sci. Instrum.* **42**, 111–119. NTIS COM-72-50311. National Bureau of Standards, Boulder, Colorado.

Siemens, C. W. (1882). *Proc. Inst. Civil Eng.* **68**, 179–186.

Sims, W. S. (1973). "Army Requirements for Cryogenic Cooling of Infrared Detectors." Proc. Closed-Cycle Cryogenic Cooler Conf. USAF Academy, Colorado, October, 10 pp. (AFFDL-TR-73-149, Vol. 1, AD No. 918234).

Smith, H. J. (1966). "Analysis of a Vibrating Heat Engine Using Pactolus." *Simulation* **6**(1), 63–8.

Smith, H. J. (1969). "Vibrating Heat Engine." *Cryog. Ind. Gases* **4**(6), 72–75.

Smith, J. L. (1965). "Some Aspects of the Selection of Regenerators." *Cryogenics* **5**(6), 305–314.

Smith, J. L. (1967). "A Metal Bellows Expansion Engine." *Adv. Cryog. Eng.* **12**, 595–601.

Smith, J. L., Jr., and Keim, T. A. (1974). "Applications of Superconductivity of AC Rotating Machines." *NATO Adv. Study Inst., Lectures*, Entreves, Italy, Plenum Press, New York, pp. 279–345.

Smith, S. C., and Anderson, A. C. (1971). "A Simple Josephson Junction Galvanometer." *Cryogenics* **11**, 53.

Sochacka, Z. (1971). "Remarks on the Cold Regenerators in the Stirling Cryogenic Refrigerators." *Chlodnictwo* **6**(4), 7–9.

Solente, P. (1970). "Cooling of Superconducting Magnets, Principles and Uses." Panel on Superconductors, Paris, France, March NTIS CONF-700341-6. Air Liquide, Sassenage, France.

Solente, P., and Marinet, D. (1968). "Refrigerateur Helium 7w à 4.5 K Application à La Refrigeration Par Convection Gazeuse de Bubinages Supraconducteurs." ("A 7-W, 4.5-K Helium Refrigerator Cooling of Superconducting Coils by Refrigeration and Gaseous Convection"). *Proc. Second Int. Cryog. Eng. Conf.*, Iliffe Sci. and Technol. Pubs., Guildford, England., pp. 23–26.

Spath, F. (1974). "Testing and Operation of He-II Low Temperature Plants." Arbeitskreis Festkoerperphysik of the Deutsche Physikalische Gesellschaft, Proc., Freudenstadt, Germany, A1–A14.

Spies, A. (1963). "Helium Refrigerator for the Production of Cold at Temperatures down to 2.4 K." *Prog. Refrig. Sci. Technol.* (Proc. XI Int. Cong. of Refrig., Munich. Vol. 1, Paper No. 1–21. pp. 123–129), Pergamon Press, Oxford.

Springford, M., and Stockton, J. R. (1969). "Simple Cryopumped Cryostat for Use with a Superconducting Magnet." *Cryogenics* **9**, 390.

Staas, F. A. (1976). "Continuous Cooling in the Millikelvin Range." *Philips Tech. Rev.* **36**(4), 104.

Staas, F. A., Severijns, A. P., and Van Der Waerden, H. C. M. (1975). "A Dilution Refrigerator with Superfluid Injection." *Phys. Lett. A* **53**(4), 327–328.

Staas, F. A., Weiss, K., and Severijns, A. P. (1974). "Surface Efficiency of Various Types of Heat Exchangers in an He^3-He^4 Dilution Refrigerator." *Cryogenics* **14**(5), 253–263.

Starr, C. (1941). "The Design of Hydrogen Liquefiers." *Rev. Sci. Instrum.* **12**, 193–198.

Steel, A. J., Bruzzi, S., and Clarke, M. E. (1976). "A 300 W 1.8 K Refrigerator and Distribution System for the CERN Superconducting RF Particle Separator." *Proc. 6th Int. Cryog. Eng. Conf.*, Grenoble, France. IPC Sci. and Technol. Press, Guildford, England, pp. 58–61.

Stephens, S. (1970). "A Self-Regulating Miniature Joule–Thomson Refrigerator." *Appl. Cryog. Technol.* **3**.

Steyart, W. A. (1978). "Magnetic Refrigerators for Use at Room Temperatures and Below." Report No. LA-UR 78-1764, Los Alamos Scientific Laboratory, New Mexico.

Stoddart, D. (1960), "Generalized Thermodynamic Analysis of Stirling Engines." B.Sc. thesis, Durham University.

Storace, A. (1971). "A Miniature, Vibration-Free, Rhombic Drive Stirling Cycle Cooler." *Adv. Cryog. Eng.* **16**, 185–194.

Strobridge, T. R. (1968). "Refrigeration Techniques." Report No. R-540. Pub. in *Proceedings of the Helium Appl. Symp.*, Wash., D.C., October. Helium Centennial Symp., Atlantic City, New Jersey, September, CFSTI PB-188 214. NBS, Boulder, Colorado.

Strobridge, T. R. (1969). "Refrigeration for Superconducting and Cryogenic Systems." Report No. R-552. *IEEE Trans. Nucl. Sci.* **NS-16**(3) 8; PB-189070. NBS, Boulder, Colorado.

Strobridge, T. R. (1974). "Cryogenic Refrigerators— An Updated Survey." NBS Tech. Note 655, 12 pp., June.

Strobridge, T. R. (1975). "Multipurpose Refrigerator for a Superconducting Cable Test Facility." NBS, Colorado. Final Report Electrical Power Research Institute, Palo Alto, California, Report No. EPRI 282, Research Project 282, 52 pp.

Strobridge, T. R., and Chelton, D. B. (1967). "Size and Power Requirements of 4.2 K Refrigerators." *Adv. Cryog. Eng.* **12**, 576–584 CFSTI AD-642 444. NBS, Boulder Colorado.

Strobridge, T. R., and Mann, D. B. (1965). "A Pulsed Refrigeration System for Cryogenic Magnet Application." *Adv. Cryog. Eng.* **10**, 54–61.

Strobridge, T. R., and Voth, R. O. (1977). "Refrigeration Technology for Superconductors." *IEEE Trans. Nucl. Sci.* **NS-24**(3), 1222–1226.

Stuart, R. W., Cohen, B. M., and Hartwig, W. H. (1970). "Operation and Application of a Three-Stage Closed-Cycle Regenerative Refrigerator in the 6.5°K Region." *Adv. Cryog. Eng.* **15**, 428–435.

Stuart, R. W., and Hogan, W. H. (1965). "A Small Helium Liquefier." *Adv. Cryog. Eng.* **10**, 62–68.

Stuart, R. W., and Hogan, W. H. (1971). "A Balanced Pressure Engine." *Prog. Refrig. Sci. Technol.* (Proc. XIII Int. Cong. of Refrig., Washington). AVI Publishing Co., Inc., Westport, Connecticut, pp. 119–128.

Stuart, R. W., Hogan, W. H., and Rogers, A. D. (1967). "Performance of a 4 K Refrigerator." *Adv. Cryog. Eng.* **12**, 564–575.

Sujak, B., and Otop, H. (1974). "The Liquefaction of Helium in the First Microvolume Experimental Liquefier Made in Poland." *Acta Phys. Pol. A* **45**(3), 493–496.

Sujak, B., Otop, H., and Gruszcynski, A. (1974). "A Small Condensation Liquefier for the Gases of the A Group Based on a 60 K Minirefrigerator." *Acta Phys. Pol. A* **45**(3), 489–491.

Sujak, B., Otop, H., and Pochaba, A. (1973). "A 60 K Minirefrigerator of the Gifford Type." *Acta Phys. Pol. A* **44**(2), 351–358.

Suslov, A. D. (1972). "Basic Features of Working Process of a Machine with Fixed Heat-Transfer Apparatus." *Deep Cold and Conditioning* (Ed. G. I. Voronin), Trudy MVTU, No. 149, Moscow, USSR, pp. 81–91.

Suslov, A. D., and Ruban, A. N. (1970). "Pulsation Machines." *Deep Cold Conditioning* (Ed. G. I. Voronin), Trudy MVTU, No. 138, Moscow, USSR, pp. 21–27.

Svetlov, Yu. V., Krasnoselskii, V. Ya., and Sarmatova, E. M. (1974). "Design of Efficient Coiled-Tube Heat Exchangers of Cryogenic Installations." *Chem. Pet. Eng. (USSR)* **10**(5–6), 415–419. (Transl. of *Khim. Neft. Mashinostr.* No. 5, 16–19).

Swearingen, J. S. (1947). "Expansion Turbines for Low Temperature Processing." *Trans. Am. Inst. Chem. Eng.* **43**, 83–90.

Swearingen, J. S. (1974). "Turboexpanders and Expansion Processes for Industrial Gases." *Cryotech 73—British Cryog. Council, Proc. Conf.*, Brighton, England, IPC Sci. and Technol. Press, Guildford, England, pp. 36–42.

Sweet, R. C. (1973). "Metal Enclosure for Cryocooled Multi-Element Infrared Detectors." Proc. Closed Cryogenic Cooler Conf. USAF Academy, Colorado, 8 pp. (AFFDL-TR-73-149, Vol. 1, AD No. 918234).

Swift, D. A. (1967). "Prospects for the Superconducting AC Power Cable." *Prog. Refrig. Sci. Technol.* (Proc. XII Int. Cong. of Refrig. Madrid, Vol. 1. Paper No. 1.46), pp. 173–185.

Taconis, K. W. (1968). "Refrigeration Systems and Cooling Devices, Including Solid State Cooling." *Prog. Refrig. Sci. Technol.* (Proc. XII Int. Cong. of Refrig., Madrid, Vol. 1, Paper No. 1.08), pp. 109–110.

Taconis, K. W., Das, P., and De Bryn Ouboter, R. (1967). "A Small Refrigerator below 0.5°K." Proc. XII Int. Cong. of Refrig., Madrid, Spain, Paper No. 1.08, 2 pp.

Tanner, T. K. (1970). "Cryogenic Research Spin-off Leads to the Development of a New, More Economical Type of Refrigerated System that can be Used with Transports." *Ind. Res.* **12**, (1), 1–2.

Terbot, J. W. (1976). "A New Helium Refrigerator for Superconducting Cable Systems." *Adv. Cryog. Eng.* **21**, 190–196.

Testard, O. A., Leny, J., Leszczyszyn, J., Jehanno, C., Cheron, C., Rothenflug, R., and Griffith, M. (1974). "A Cryogenic Device for a Low-Energy X-Ray Space Detector." *Cryogenics* **14**, 509.

Thullen, P., Stecher, R. W., Jr., and Bejan, A. (1975). "Flow Instabilities in Gas-Cooled Cryogenic Current Leads." Applied Superconductivity Conf., Illinois, *IEE Trans. Magn.* **MAG-11**(2), 573–575.

Timmerhaus, K. D., and Schoenhals, R. J. (1974). "Design and Selection of Cryogenic Heat Exchangers." *Adv. Cryog. Eng.*, **19**, 445–462.

Tipler, W. (1947). "A Simple Theory of the Heat Regenerator." Tech. Report No. ICT/14, Shell Petroleum Co. Ltd.

Timpler, W. (1948). "An Electrical Analogue to the Heat Regenerator." *Proc. Int. Cong. Appl. Mech.* **3**, 196–210.

Tomasz, S. (1975). "Liquefaction of Helium." *Gospod. Paliwami Energ.* **23**(5), 15–8.

Tsukamoto, O. (1974). "Stationary Stability of Hollow Superconductors." *Electr. Eng. Jpn.* **94**(4), No. 4).

Turnwald, E. (1973). "A New Piston Expansion Machine for Helium Refrigeration." *Linde Ber. Tech. Wiss.*, No. 34, 25–28.

Turnwald, E. (1973). "A New Reciprocating Expansion Machine for Helium-Refrigeration Plants." *Linde Rep. Sci. Technol.*, No. 19, 25–28.

Turnwald, E. (1974). "A New Reciprocating Expansion Machine for Helium-Refrigeration Plants." *Kerntechnik* **16**(12), 517–522.

Tutton, R. C., Knight, H. A., Halford, P., Malyn, T. H., and Tantum, D. H. (1970). "Cryogenics Safety Manual—A Guide to Good Practice." *Cryogenics* **10**, 367.

Tyler, J. S., and Potter, J. A. (1961). "Infrared Detector Refrigerators." *Adv. Cryog. Eng.* **6**, 363–371.

Urieli, I. (1977). "A Computer Simulation of Stirling Cycle Machines." Ph.D. Thesis, Univ. of Witwatersrand, Johannesburg, South Africa.

Urieli, I. (1979). "A Review of Stirling Cycle Machine Analysis." paper No. 799236, Proc. 14th I.E.C.E.C., Boston, Massachusetts, pp. 1086–1090.

Urwin, R. J., Dean, R. H., Scurlock, R. G., and Lund, F. P. (1971). "Simple Servo-Controlled Gas-Flow Cryostat for any Temperature between 4.2 and 300 K." *Cryogenics* **11**, 225.

Van Beelen, H., and Hartoog, A. (1977). "On the Use of Sintered Materials for the Exchange of Heat with a Flow of Pure Helium II." *Cryogenics* **17**, 435.

Vander Arend, P. C., Stoy, S. T., and Richied, D. (1975). "Cooling of a System of Superconducting Magnets by Means of Pumped Subcooled Liquid Helium." Applied Superconductivity Conf., *IEE Trans. Magn.* **MAG-11**(2), 565–568.

van der Laan, G. M. J., and Roosendaal, K. (1961). "A Snow Separator for Liquid–Air Installations." *Philips Tech. Rev.* **23**(2), 48–54.

Van der Ster (1960). "The Production of Liquid Nitrogen from Atmospheric Air Using a Gas Refrigerating Machine." Delft Technische Hochschule Thesis.

Vasil'ev, D. I. (1968). "Simple System for Obtaining Temperatures down to 0.55 K by Pumping Helium-3." *Cryogenics* **8**, 252.

Vasilev, L. M., Dmitrevskii, Yu. P., and Melnik, Yu. M. (1974). "Two-Meter Liquid-Hydrogen Target without Liquid-Nitrogen Cooling for Muon Experiments." *Instrum. Exp. Tech.* **17**(6), Pt. 1, 1578–1580. (Transl. of *Prob. Tekh. Eksp.* No. 6, 30–32).

Vasilyev, A. A., Khrychikov, E. Ye., Naumov, A. V., and Sosedov, I. B. (1968). "Optical Oscillator with an Injection Gas Diode and a Closed Cycle Cooling System." *Radio. Eng. Electron. Phys.* **13**(7), 1169–1170. (Transl. of *Radiotekhn. Elektron.* **13**, 1339–1340).

Vasishta, V. (1969). "Heat-Transfer and Flow-Friction Characteristics of Compact Matrix Surfaces for Stirling-Cycle Regenerators." M.Sc. thesis, University of Calgary.

Veksler, M. D., Gustov, V. F., and Klimenko, A. P. (1974). "Algorithmization of the

Calculation of Vapor Condensation in Regenerators of Air-Fractionating Units." *Algorit-miz Rascheta Protsessov I Apparatov Khim. Prioz-V, Tekhol. Pererabotki I Transp. Nefti I Gaza Na EUM* No. 8, 61–67.

Verbeek, H. J. (1969). "An Industrial Gas Refrigerating Machine for the Temperature Range from Room Temperature down to 20°K and Lower." Proc. IIR-British Cryogenics Council Conference on Low Temperatures and Electric Power, Royal Society, London, pp. 91–97.

Verkin, B. I., and Zhitomirskii, I. S. (1974). "Determination of the Flow Rate of the Refrigerant for Thermostatic Control of the Windings of Cryoelectromachines. *Izv. Akad. Nauk SSR, Energ. Transp.* No. 5 (Sept–Oct) 102–108.

Vidwans, V. V., and Ratnum, B. A. (1978). "A Portable Liquid-Helium Laboratory Cryostat." *Cryogenics* **15**, 423.

Vogelhuber, W. W., and Parish, H. C. (1968). "Compact LNG System Using Large Stirling Cycle Cold Gas Refrigerator." Proc. of the First Int. Conf. On LNG. Inst. of Gas Tech., Chicago, Illinois, 28–36.

Vonk, G. (1968). "A New Type of Compact Heat Exchanger with a High Thermal Efficiency." *Adv. Cryog. Eng.* **13**, 582–589.

Von Minnigerode, G. (1974). "Importance of Helium-3 Isotopes for Cooling below 1 K." *Umsch. Wiss. Tech.* **74**(20), 648–649.

Voth, R. O., Norton, M. T., and Wilson, W. A. (1966). "A Cold-Moderator Refrigerator Incorporating a High-Speed Turbine Expander." *Adv. Cryog. Eng.* **11**, 126–138. CFSTI-PB-173 781. National Bureau of Standards, Boulder, Colorado.

Voth, R. O., and Petropoulos, S. K. (1973). "Cryogenic Refrigerators for Shipboard Forward Looking Infrared Applications." Proc. Closed-Cycle Cryogenic Cooler Conf. USAF Academy, Colorado. (AFFDL-TR-73-149, Vol. 1, AD No. 918234) pp.27–33. NTIS COM-74-50663/5.

Wadsworth, J. (1961). "An Experimental Investigation of the Local Packing and Heat-Transfer Processes in Packed Beds of Homogeneous Spheres." Int. Heat-Trans. Conf. (ASME), Boulder, Colorado, pp. 760–769.

Walker, G. (1961). "The Operational Cycle of the Stirling Engine with Particular Reference to the Function of the Regenerator." *J. Mech. Eng. Sci.* **3**, No. 4.

Walker, G. (1962). "An Optimization of the Principal Design Parameters of Stirling-Cycle Machines." *J. Mech. Eng. Sci.* **4**(3), 226–240.

Walker, G. (1963). "Density and Frequency Effects on the Pressure Drop across the Regenerator of a Stirling-Cycle Machine." *Engineer, London* **216**, 1063–1065.

Walker, G. (1963). "Machining Internal Fins in Components for Heat-Exchangers." *Machinery, London* **101**, No. 2590.

Walker, G. (1963). "Regeneration in Stirling Engines." *Engineer, London* **216**, No. 5631.

Walker, G. (1965). "Some Aspects of the Design of Reversed Stirling-Cycle Machines." Paper No. 231, ASHRAE Ann. Summ. Mtg., Portland, Oregon.

Walker, G. (1969). "Dynamical Effects of the Rhombic Drive for Miniature Cooling Engines." *Adv. Cryog. Eng.* **14**, 370–377.

Walker, G. (1972). "Stirling Engines for Isotope Power Systems." Proc. 2nd Int. Conf. on Power from Radioisotopes, Madrid.

Walker, G. (1972). "Stirling Engines–The Second Coming." *Chart. Mech. Eng.* **19**(4), 54–57.

Walker, G. (1973). *Stirling Cycle Machines.* Oxford University Press, Oxford, 156 pp.

Walker, G. (1974). "Stirling Cycle Cooling Engine with Two-Phase, Two-Component Working Fluid." *Cryogenics* **14**(2), 459–462.

Walker, G. (1980). *Stirling Engines.* Oxford University Press, Oxford.

Walker, G., and Burn, K. (1976). "Exploratory Study of the Rainbow Variant Stirling Engine." Proc. 11th I.E.C.E.C., Lake Tahoe, Nevada.

Walker, G., and Khan, M. I. (1965). "Theoretical Performance of Stirling Cycle Engines." Paper No. 949A, SAE Int. Auto. Eng. Congr., Detroit, Michigan, 7 pp.

Walker, G., and Vasishta, V. (1971). "Heat-Transfer and Flow-Friction Characteristics of Dense Mesh Wire Screen Stirling Cycle Regenerators." *Adv. Cryog. Eng.* **16**, 324–332.

Walker, G., and Wan, W. K. (1972). "Heat-Transfer and Fluid-Friction Characteristics of

Dense-Mesh Wire Screen Regenerator Matrices at Cryogenic Temperatures." Proc. Fourth Int. Cryog. Eng. Conf., IPC Sci. and Technol. Press, Guildford, England.

Wan, W. K. (1971). "The Heat-Transfer and Friction-Flow characteristics of Dense-Mesh Wire-Screen Regenerator Matrices." M.Sc. thesis, University of Calgary.

Ward, D. E. (1961). "Some Aspects of the Design and Operation of Low Temperature Regenerators." *Adv. Cryog. Eng.* **6**, 525–536.

Weber, H. W., Westphal, G. P., and Goblirsh, R. (1976). "Installation of a Dilution Refrigerator at a Polarized Neutron Facility." *Nucl. Instrum. Methods* **134**(1), 55–60.

Weimer, R. F., and Hartzog, D. G. (1973). "Effects of Maldistribution on the Performance of Multistream, Multipassage Heat Exchangers." *Adv. Cryog. Eng.* **18**, 52–64.

Weinhold, J. (1964). "Einsatz der Gaskaltemaschine als Universelle Tieftemperatur-Quelle." ("Installation of a Gas Refrigerator for Universal Low Temperature Production"). *Chem. Ing. Tech.* **36**(5), 562–563.

Weinhold, J. (1967). "Cryogeneratoren fur Tiefe und Tiefste Temperaturen." ("Cryogenerators for Low and Very Low Temperatures"). *Kalte* (*Hamburg*) **20**(7), 313–315. 318–320.

Weinstock, H. (1968). "A Review of Cryogenic History and Principles of Refrigeration." *Appl. Cryog. Technol.* **I** 13–44.

Weiss, K. (1975). "Cooling near Extremely Low Temperatures." *Natuurkd, Voordi.* **53**, 103.24.

West, C. (1971). "The Fluidyne Heat Engine." Report No. AERE-R-6776, Atomic Energy Research Establishment, Harwell, Berks, England.

West, F. B., and Taylor, A. T. (1952). "The Effect of Pulsations on Heat Transfer—Turbulent Flow of Water Inside Tubes." *Chem. Eng. Prog.* **48**(1), 39–43.

Wheatley, J. C. (1975). "Helium Three." *Proc. 14th Int. Conf.*, Otaniemi, Finland (August). American Elsevier Pub. Co., New York, pp. 6–51.

Wheatley, J. C., Allen, P. C., Knight, W. R., and Paulson, D. N. (1980). "Principles of Liquids Working in Heat Engines." (In press). Dept. of Physics, Univ. of California at San Diego.

White, R. (1973). "Program for Predicting V–M Cooler Off-Design Performance." Proc. Closed-Cycle Cryogenic Cooler Conf. USAF Academy, Colorado, October, 26 pp. (AFFDL-TR-73-149, Vol. 1. AD No. 918234).

Wiedemann, W., Probst, C., and Kraus, J. (1970). "A He^3–He^4 Dilution Refrigerator Using Novel Types of Heat Exchangers." Annexe 2, *Cryophysics and Cryoengineering.* IIR. Paris, pp. 73–78.

Wigley, D. A. (1971). *Mechanical Properties of Materials at Low Temperatures.* Int. Cryogenics Monograph Series, Plenum Press, New York.

Wigley, D. A. (1978). *Properties of Materials at Low Temperature.* Engineering Design Guides, No. 27, Oxford University Press, Oxford, England.

Wilkes, W. R. (1970). "A Recirculating Helium-3 Refrigerator not Requiring a Pumped Helium-4 Bath." Annexe 2, *Cryophysics and Cryoengineering*, IIR. Paris, pp. 67–72.

Wilkes, W. R. (1972). "A 200 mW Recirculating Helium^3 Refrigerator." *Cryogenics* **12**, 180.

Wilks, J. (1963). "Technological Applications of Very Low Temperatures." *Prog. Refrig. Sci. Technol.* (Proc. XI Int. Cong. of Refrig. Munich. Vol. 1, paper 3, pp. 3–7), Pergamon Press, Oxford.

Williams, M. (1970). "Practical Hints on the Construction of Magneto-Optic Apparatus for Cryogenic Applications." *Cryogenics* **10**, 394.

Williamson, J. M. (1959). "The Effectiveness of the Periodic-Flow Heat-Exchanger." English Electric Report, No. W/M/4B.

Wilson, D. G., and D'Arbeloff, B. J. (1965). "The Performance of Refrigeration Cycles below 100 R." *Adv. Cryog. Eng.* **11**, 160–170.

Winters, A. R., and Snow, W. A. (1966). "Capacity and Economic Performance of a Large 5 K Helium Refrigerator." *Adv. Cryog. Eng.* **11**, 116–125.

Wittner, C. E. (1966). "Design of a Closed Cycle Helium Temperature Refrigerator." *Adv. Cryog. Eng.* **11**, 107–115.

Wittig, M., Forth, H. J., and Hofmann, A. (1974). "Development and Design of an He3/He4-

Dilution Refrigerator." Bundesministerium Fuer Forschung and Technologie, Bonn, Germany. Report No. BMFT-FB-T-74-18 (July), 42 pp.

Wolf, A. A., Davis, J. R., and Nisenoff, M. (1974). "Superconducting Extremely Low Frequency (Elf) Magnetic Field Sensors for Submarine Communications." *IEE Trans. Commun.* **COM-22**(4), 549–554.

Wood, G. H. (1971). "Phenolics for Cryogenic Apparatus." *Cryogenics* **11**, 234.

Wood, H. J., and Morgan, N. E. (1961). "Comparative Rating of Positive-Displacement Engines and Turbines for Cryogenic Power Systems." Prog. in Astros. and Rock. Proc. of ARS Symp., Santa Monica, Calif., Vol. **3**, 565–592.

Yagi, S., Kunii, D., and Wakao, N. (1961). "Radially Effective Thermal Conductivities in Packed Beds." Int. Heat-Trans. Conf. (ASME), Boulder, Colorado, pp. 742–749.

Yamaguchi, M. (1970). "Chart for Temperature Change in Joule–Thomson Expansion of Helium." *Cryogenics* **10**, 72.

Yan, S. S. (1975). "Refrigeration by Dilution." *Wuli.* **4**(2), 111–114.

Yanai, M. (1968). "Pulse-Tube Refrigerator." *Cryog. Eng. (Tokyo)* **3**, (5), 201–207.

Yanai, M. (1972). "Small Cryogenic Refrigerator." *Cryog. Eng. (Tokyo)* **7**(4), 173–180.

Yaqub, M. (1960). "Cooling by Adiabatic Magnetization of Superconductions." *Cryogenics* **1**, 101.

Yates, B. (1968). "Symposium on Low Temperature Devices." *Cryogenics* **8**, 349.

Yates, B., and Hoare, F. E. (1968). "Small Scale Hydrogen Liquefaction." *Cryogenics* **2**, 84.

Yendall, E. F. (1960). "A Novel Refrigerating Machine." *Adv. Cryog. Eng.* **2**, 188–196.

Yonemitsu, H. (1969). "Helium Refrigerator for Superconducting Magnet." *Cryog. Eng. (Tokyo)* **4**(1), 42–50.

Yonemitsu, H., Maeda, J., and Ohkawa, M. (1963). "A Simple Cooling System with a Cryogenic Pump." *Prog. Refrig. Sci. Technol.* (Proc. XI Int. Cong. of Refrig. Munich. Vol. 1, Paper 1.25, pp. 121–2), Pergamon Press, Oxford.

Zeitz, K., and Meier, R. (1967). "The Evolution of Helium Refrigerators at Air Products and Chemicals, Inc." *Low-Temperature Refrigeration for Microwave Systems.* Proc. of Conf. on Low-Noise Microwave Amplifiers and Low-Temperature Refrigeration Means, Frankfurt am Main, West Germany (Eds. W. H. Hogan *et al.*). Boston Technical Pubs., Cambridge, Massachusetts, pp. 219–245.

Zeitz, K., and Woolfenden, B. K. (1963). "A Closed-Cycle Helium Refrigerator for 2.5 K." *Adv. Cryog. Eng.* **8**, 206–212.

Zel'dovich, A. G., and Pilipenko, Y. K. (1961). "A Hydrogen Liquefier with an Output of 50 l per Hour of Liquid Hydrogen." *Cryogenics* **2**, 101.

Zel'dovich, A. G., and Pilipenko, Y. K. (1969). "Improvement and Boosting of 50 l per Hour Hydrogen Liquefier." *Cryogenics* **5**, 45.

Zimmerman, F. J., and Longsworth, R. C. (1971). "Shuttle Heat Transfer." *Adv. Cryog. Eng.* **16**, 342–351.

Zimmerman, J. E. (1972). "Josephson Effect Devices and Low-Frequency Field Sensing." *Cryogenics* **12**, 19.

Zimmerman, J. E., Bohm, H. V., and McNutt, J. D. (1962). "A Magnetic Refrigerator Employing Superconducting Solenoids." *Cryogenics* **2**, 153.

Zimmerman, J. E., and Radebaugh, R. (1977). "Operation of a SQUID in a Very Low-Power Cryocooler." Appl. of Closed-Cycle Cryocoolers to Small Superconducting Devices, Proc. of Conf. NBS, Boulder, October, pp. 59–66.

Zimmerman, J. E., Radebaugh, R., and Siegwarth, J. D. (1977). "Possible Cryocoolers for SQUID Magnetometers." *Proc. Int. Conf., Berlin*, Walter De Gruyter and Co., New York, pp. 287–296.

GOVERNMENT REPORTS

Ackerman, R. A. (1970). "Investigation of Gifford–McMahon Cycle and Pulse Tube Cryogenic Refrigerators." Report No. Ecom-3245, Army Electronic Command, Fort Monmouth, New Jersey.

Ackermann, R. A. (1970). "The Design of a Closed-Cycle Refrigerated Test Facility for High-Q Superconducting Tunable Circuit Studies." Report No. Ecom-3327, Army Electronic Command, Fort Monmouth, New Jersey.

Ackermann, R. A. (1971). "Vuilleumier Refrigerator Analysis. Low-Temperature Refrigerator Design Optimization." (Final Report). Report No. 1001, NTIS AD-724 770. Cryomech. Inc., Jamesville, New York.

Adams, D. W., and Weiland, W. F., Jr. (1969). "Heat-Transfer Characteristics of a Water-to-Cryogenic-Hydrogen Heat Exchanger." Report No. NASA-TN-D-5429, CFSTI N69-36701. NASA, Cleveland, Ohio.

AiResearch Manufacturing Co. (1972). "Vuilleumier Program Engineering Notebook." Contract NAS 5-21096, Goddard Space Flight Center, Greenbelt, Maryland, August.

AiResearch Manufacturing Co. (1972). "Fractional Watt Vuilleumier Cryogenic Refrigerator Program Engineering Notebook," Vol. 1, Thermal analysis. Contract NAS 5-21715, Goddard Space Flight Center, Greenbelt, Maryland, May.

Alfeev, V. N., and Ekimov, V. D. (1967). "Problems of Designing Cooled Mixers for the Superhigh-Frequency Range." Report No. FTD-HT-67-375, CFSTI AD-660 729. WPAFB, Dayton, Ohio.

Alfimenko, V. P., et al. (1975). "(3)He–(4)He Dilution Refrigerator Used in the Slow Neutron Beam." Joint Inst. for Nuclear Research, Dubna, U.S.S.R., Report No. JINR-P8-9168, 20 pp.

Althouse, E. L. (1968). "The Production of Temperatures below 1°K by the Adiabatic Demagnetization of 50% Ce (PO3)3-Ba(PO3)2 Class." Report No. NRL-6711, CFSTI AD-673 525. Naval Research Lab., Washington, D.C.

Ammann, H. H., and Morihara, H. (1968). "Miniature Turbo-Expander for a Closed Cycle Helium Refrigerator." AFFDL-TR-68-58.

Anon. (1962). "New Reciprocating IR Cooling Technique." Report No. SSD-TOR-62-178-AF 04 (695)-144. A. D. Little, Inc., Cambridge, Massachusetts, 81 pp.

Anon. (1964). "The Design of a Cryostat for the Air Force Nuclear Engineering Test Facility." (Master's Thesis). Report No. AFIT-GNE/Phys/653, CFSTI AD-603 609. WPAFB, Dayton, Ohio.

Anon. (1964). "Miniature Stirling Cycle Cooler." Report No. AL TR-65-15. AF 33 657 7889. Malaker Labs. Inc., High Bridge, New Jersey, 42 pp.

Anon. (1965). "A Bibliography of References for the Thermophysical Properties of Helium-4. Hydrogen, Deuterium, Hydrogen Deuteride, Neon, Argon, Nitrogen, Oxygen, Carbon Dioxide, Methane, Ethane, Krypton, and Refrigerants 13, 14 and 23." Report No. NBS-8808, CFSTI AD-467 519. National Bureau of Standards, Boulder, Colo.

Anon. (1965). "Design, Fabrication, Modification and Testing of Cryogenic-Solid Cooler," Aerojet Model K6 Final Engineering Report. Report No. NASA-CR-75768, CFSTI N66-28764. Aerojet-General Corp., California.

Anon. (1965). "Cryogenic Technology Research at MSFC." Report No. NASA-TM-X-53515, CFSTI N66-37993. NASA, Huntsville, Alabama.

Anon. (1968). "Cryogenic Storage Systems Design, Fabrication and Evaluation." Volume 1 Final Report. Report No. NASA-CR-92118, CFSTI N68-23659. Bendix Corp., Davenport, Iowa.

Anon. (1968). "Cryogenic Storage Systems Design, Fabrication and Evaluation." Volume 2—Appendices. 1. "Summary of Cryogenic Storage System Evaluation." 2. "Ardeform Pressure Vessel Program Final Report." Report No. NASA-CR-92119, CFSTI N68-23792. Bendix Corp., Davenport, Iowa.

Anon. (1970). "ICICLE Feasibility Study." Report No. NASA-CR-109880. RCA Tech. Labs./Goddard Space Flight Center, Greenbelt, Maryland, 300 pp.

Anon. (1971). "Design and Development of a Prototype Static Cryogenic Heat Transfer System." (Final Report). Report No. NASA-CR-121939, DTM-50-50, NTIS N71-36355. Dynatherm Corp., Cockeysville, Maryland.

Anon. (1972). "Final Engineering Report on the Design and Development of Two Miniature Cryogenic Refrigerators." Cont. No. DAAK02-73-C-0495, NTIS AD-784 437/8. Army Elect. Command, Fort Belvoir, Virginia.

Anon. (1973). "Reliability/Design Handbook. Thermal Applications." Volume III. Report No. NAVELEX-0967-437-7030, NTIS AD-A009 015/9ST. Naval Elec. Systems Command, Washington, D.C.

Anon. (1973). "Shuttle Cryogenic Supply System Optimization Study." Volume 4: "Cryogenic Cooling in Environmental Control Systems." (Final Report). Report No. NASA-CR-133952, LMSC-A991396-Vol-4, NTIS N73-26895/5. Lockheed Missiles and Space Co., California.

Anon. (1974). "Electrocaloric Refrigeration for Superconductors." (Semi-Annual Tech. Rept.). Contract No. ARPA Order 2535, NTIS AD-787 660/0SL. National Bureau of Standards, Boulder, Colorado.

Anon. (1974). "Energy and Superconductivity." Report No. B-1153, NTIS COM-74-10713/7. National Bureau of Standards, Boulder, Colorado.

Anon. (1974). "Physical Sciences: Thermodynamics, Cryogenics, and Vacuum Technology: A Compilation." Report No. NASA-SP-5973(01), NTIS N74-34173/6SL. NASA, Washington, D.C.

Anon. (1975). "Helium Research in Support of Superconducting Power Transmission." (First Quarterly Report). Contract No. AT(49-1)-3719, NTIS TID-26808, National Bureau of Standards, Boulder, Colorado.

Anon. (1976). "Design and Analysis of a Cryogenic Variable Conductance Axial Grooved Heat Pipe." Report No. NASA-CR-137882, BK008-1009. NTIS N76-25525/6ST. B&K Engineering Inc., Towson, Maryland.

Antal, J. J. (1971). "Automatic Start-Up and Running of an ADL-Collins Helium Refrigerator." (Tech. Rept.). Report No. AMMRC-TR-71-6, NTIS AD-725 517. Army Materials and Mechanics Research Center, Watertown, Massachusetts.

Arend, P. C., Chelton, D. B., and Mann, D. B. (1964). "Satellite Refrigeration Study. Part II Technical Analysis." Report No. NASA-CR-81134, NBS-8444, CFSTI N67-15256, National Bureau of Standards, Boulder, Colorado.

Arkharov, A. M., Butkevich, K. S., Golovintsov, A. G., *et al.* (1964a). "Liquefaction of Hydrogen and Helium, Obtaining Ultralow Temperatures." *Tekhnika Nizkikh Temperatur.* Chapter 4 [Transl. by Foreign Technology Div., Wright–Patterson AFB, Dayton, Ohio, Transl. No. FTD-MT-65-167 (January 1967), 77 pp.].

Arkharov, A. M., Butkevich, K. S., Golovintsov, A. G., *et al.* (1964b). *Cryogenic Engineering,* Vol. 2. *Tekhnika Nizkikh Temperatur* [Transl. by Foreign Technology Div., Wright–Patterson AFB, Dayton, Ohio, Transl. No. FTD-MT-24-216-68. (November 1968), 748 pp.].

Arkharov, A. M., Butkevich, K. S., Golovintsov, A. G., Kulakov, V. M., and Marfenina, I. V. (1968). *Cryogenic Engineering,* Vol. 1 (selected chapters). Report No. FTD-MT-24-216-68-Vol-1, CFSTI AD-685 948. WPAFB, Dayton, Ohio.

Arkharov, A. M., Butkevich, K. S., Golovintsov, A. C., Kulakov, V. M., and Marfenina, I. V. (1968). *Cryogenic Engineering,* Vol. 2 (selected chapters). Report No. FTD-MT-24-216-68-Vol-2, CFSTI AD-685 977. WPAFB, Dayton, Ohio.

Arnett, R. W., and Muhlenhaupt, R. C. (1970). "Study of Oxygen and Nitrogen Generating Systems." Report No. 9757, National Bureau of Standards, Boulder, Colorado.

Arp, V. D. (1973). "Refrigeration of Superconducting Rotating Machinery." (Interim Rept.). Report No. NBSIR-73-331, NTIS AD-774 892/4, National Bureau of Standards, Boulder, Colorado.

Arp, V. D. (1973). "Refrigeration of Superconducting Rotating Machinery." (Interim Rept.) Report No. NBSIR-73-331, NTIS COM-74-10238/5, National Bureau of Standards, Boulder, Colorado.

Arp, V. D., Daney, D. E., Frederick, N. V., and Jones, M. C. (1975). "Helium Research in Support of Superconducting Power Transmission." (Annual Rept.) Report No. NBSIR-75-823, NTIS PB-246 658/9SLT, National Bureau of Standards, Boulder, Colorado.

Balas, C., Jr. (1973). "Design and Fabrication of a Rhombic Drive Stirling Cycle Cryogenic Refrigerator." (Final Rept.) Contract No. DAAK02-72-C-0224, NTIS AD-A015 197/7ST. Army Mobility Equipment Research and Development Center, Fort Belvoir, Virginia.

Balas, C., Jr. (1975). "Design and Fabrication of a Rhombic Drive Stirling Cycle Cryogenic Refrigerator." Supplement to Final Report, Contract No. DAAK02-72-C-0224, DDC AD A015 198, (April) 8 pp.

Beam, J. E. (1976). "Evaluation of Eutectic Fluoride Thermal Energy Storage Unit Compatibility, Part II. Test Procedures and Post-Test Evaluation Results." (Interim Tech. Rept.) Report No. AFAPL-TR-75-92-Pt-2, NTIS Pt. 1 AD-A022 060, Pt. 11 AD-A042 234/5ST. WPAFB, Dayton, Ohio.

Beam, J. E., and Mahetkey, T. (1976). "Demonstration Testing of a Vuilleumier Cryocooler with an Integral Heat Pipe/Thermal Energy Storage Unit." Report No. AFAPL-TR-77-10, NTIS AD-A042 786/4ST. WPAFB, Dayton, Ohio.

Ben-Israel, D., and Skala, L. (1964). "Cryogenic Liquid Transfer Techniques." Report No. AL-TOR64 126, CFSTI AD-606 496. Materials Research Lab., Orangeburg, New York.

Bennet, M. D. (1975). "Variable Inductive Reactance Temperature Controller." AFFDL-TR-75-7. Kinergetics Inc., Tarzana, California.

Bennett, M. O., and Lins, R. C. (1976). "Bonded Wire Screen Compact Heat Exchanger." Report No. CP-2028F, NTIS AD-AC33 632/1ST. Kinergetics Inc., Tarzana, California.

Berry, R. L. (1974). "Ultraminiature Vuilleumier Refrigeration System." Report No. AFFDL TR-74-16, DDC AD 917 677L. (March). 133 pp. Hughes Aircraft Co., Culver City, California.

Brand, F. S., Jacobs, H., LoCasio, C., Novick, G., and Schick, D. (1964). "Properties of Cooled, Uncoated Ruby Laser Oscillators." AD-611 661. Army Elec. Labs., Fort Monmouth, New Jersey.

Brandt, R. B. (1971). "Superconducting Technology in Japan." (Trip Rept.) Report No. ONR-28, NTIS AD-727 094. Office of Naval Research, Virginia.

Brawe, D. E., Coe, H. H., and Scibbe, H. W. (1970). "Cooling Requirements of Ball Bearings Lubricated by Glass-Fiber-Filled Polytetrafluorethylene Retainers in Cold Hydrogen Gas." Report No. NASA-TN-D-5607, E-5222, CFSTI N70-19124. NASA, Cleveland, Ohio.

Breckenridge, R. W., Jr. (1967). "Development of a Miniature, Reciprocating Cryogenic Refrigerator for Space Applications." (Final Rept.) Report No. AFFDL-TR-67-78, NTIS AD-817 952/5ST. Arthur D. Little, Inc., Cambridge, Massachusetts.

Breckenridge, R. W. (1968). "Exploratory Development of a 1 Watt 3.6°K Reciprocating Refrigerator for Space Application." AFFDL-TR-68-59. AD 844 314, (April), A. D. Little Inc., Cambridge, Massachusetts.

Breckenridge, R. W., Jr., Heuchling, T. P., and Moore, R. W., Jr. (1971). "Rotary-Reciprocating Cryogenic Refrigeration Systems Studies." Part I. "Analysis." (Final Rept.) Report No. AFFDL-TR-115-Pt-1, NTIS AD-888 464/L. A. D. Little Inc., Cambridge, Massachusetts.

Breckenridge, R. W., Jr., Hidden, W. P., Kinckley, R. B., et al. (1976). "Preliminary Cryo Cooler Design for Satellite Sensors." Report No. AFFDL TR-76-5, Vol. I and II, Cont. No. F33615-75-C-3156 (April) 132 pp. WPAFB, Dayton, Ohio.

Breckenridge, R. W., Jr., and Moore, R. W., Jr. (1975). "A Superfluid Helium System for an LST IR Experiment." (Final Rept.) Report No. NASA-CR-144688, ADL-C-76779, NTIS N76-13311/5ST. A. D. Little, Inc., Cambridge, Massachusetts.

Breckenridge, R. W., Jr., et al. (1972). "Development of a Rotary-Reciprocating Cryogenic Refrigerator for Space Applications." Technical Report AFFDL-TR-72-88. AD 908 360L. A. D. Little, Inc., Cambridge, Massachusetts.

Breckenridge, R. W., Jr. et al. (1974). "Design Study of a Rotary Reciprocating Thermal Compressor." AFFDL-TR-74-127. AD B00 2185L. A. D. Little, Cambridge, Massachusetts.

Breckenridge, R. W., Jr. et al. (1975). "Development of a Rotary Reciprocating Refrigerator for Space Applications." AFFDL-TR-75-77. A. D. Little, Cambridge, Massachusetts.

Brennan, P. J., and Kroliczek, E. J. (1975). "Erts-C (Landsat 3) Cryogenic Heat Pipe Experiment Definition." (Final Rept.) Report No. NASA–CR–143797, BK005-1009, NTIS N75-23882/4ST. B&K Engineering Inc., Towson, Maryland.

Brewer, D. F. (1976). "Experiments on the Formation and Surface States of Adsorbed Helium

Films." (Final Tech. Rept.) Report No. ARDG(E)-R/D-2092, NTIS AD-A031 258/7ST. Sussex Univ., Brighton, England.

Brewer, G. D., and Morris, R. E. (1975). "Study of Active Cooling for Supersonic Transports." Report No. NASA-CR-132573, NTIS N75 17336. Lockheed-California Co., Burbank, California.

Brodyanskii, V. M. (1973). "Prospects for the Development of Cryogenic Technology." Report No. FTD-MT-24-241-73, NTIS AD-762 315. WPAFB, Dayton, Ohio.

Brown, D. P. (1976). "Isabelle Forced-Circulation Cooling System—Proposed Method of Producing and Distributing Helium Refrigerant for 4.5 K Superconducting Magnets." Report No. BNL-50514 (April) 18 pp. Brookhaven National Lab., Upton, New York.

Browning, C. W., Miller, W. S., and Potter, V. L. (1972). "75°K Vuilleumier Refrigerator—Final Report for Task IV Model Fabrication." Report 72-8687, prepared for Goddard Space Flight Center, Greenbelt, Maryland.

Browning, C. W., and Potter, V. L. (1972). "75°K Vuilleumier Cryogenic Refrigerator Final Report for Task II Analytical and Test Program." Report 72-8497, prepared for Goddard Space Flight Center, Greenbelt, Maryland.

Brubaker, D. C. (1975). "Investigation of Maintenance Problems Associated with the F/FB-111 AN/AAR-34 Cryogenic Converter." AFFDL-TR-75-113. In-House, WPAFB, Dayton, Ontario.

Buist, R. J. (1974). "Feasibility Study for a High-Power, Low-Temperature Thermoelectric Cooler." (Final Tech. Rept.) Contract No. N60921-73-C-0296, DDC AD8002 643L. Naval Surface Weapons Center, Silver Spring, Maryland.

Buist, R., Fenton, J., Lichniak, G., and Norton, P. (1976). "Low-Temperature Thermoelectric Cooler for 145K Detector Array Package." (Final Tech. Rept.) Contract No. DAAK02-71-C-0009, DDC AD B008 934L. Army Electronics Command, Fort Belvoir, Virginia.

Caren, R. P., and Coston, R. M. (1968). "Design and Construction of an Engineering Model Solid Cryogen Refrigerator for Infrared Detector Cooling at 50 Deg. K." Report No. NASA-CD-988, CFSTI N68-13115. Lockheed Missiles and Space Co., California.

Chellis, F. F., and Stewart, R. W. (1972). "Multiple Cold Finger Refrigeration System for Cooling of Infrared Detectors." AFFDL-Tr-72-84. AD 903 739L. Cryogenic Technology Inc., Latham, Massachusetts.

Church, J. F., and Peirce, R. M. (1973). "Flight Testing of a Cryogenically Cooled Hygrometer." Report No. AFCRL-TR-73-0292, AFCRL-IP-188, NTIS AD-764 718/3. Air Force Cambridge Research Labs., Massachusetts.

Clarke, J., Dick, G. J., Langenberg, D. N., Little, W. A., and Mercereau, J. E. (1971). "Device Applications of Cryogenics." Part I. "Superconducting Electronics." (Final Rept.) Report No. RAI-100. NTIS AD-729 697. Research Advisory Inst., Newport Beach, California.

Clarke, R. G., et al. (1967). "Research on Materials Essential to Cryocooler Technology." AFML-TR-67-229.

Coggins, J. L., Breckenridge, R. W., Jr., Perry, S. H., and Moore, R. W., Jr. (1971). "Rotary-Reciprocating Cryogenic Refrigeration System Studies." Part II. "Computer Program." (Final Tech. Rept.) Report No. AFFDL-TR-115-Pt-2. NTIS AD-891 602/5ST. Arthur D. Little, Inc., Cambridge, Massachusetts.

Cohen, B., and Daniels, A. (1973). "Design and Development of a Free-Displacer/Thermal Compressor Cryogenic Refrigeration System." (Final Rept.) Contract No. DAAK02-71-C-0421, NTIS AD-920 028/8ST. Philips Labs., Briarcliff Manor, New York.

Coles, W. D., and Lawrence, J. C. (1965). "Design, Construction, and Performance of Cryogenically Cooled and Superconducting Electromagnets." Report No. NASA-TM-X-52121, CFSTI N66-29352. NASA, Cleveland, Ohio.

Colyer, D. B. (1972). "Design and Development of Cryogenic Turborefrigerator Systems." AFFDL-TR-72-35. AD 890-372L, General Electric Co., Schenectady, New York.

Colyer, D. B., Gessner, R. L., and Fleming, R. B. (1969). "Liquid-Helium, Closed-Cycle Cryogenic Refrigerator." AFFDL-TR-69-26. AD 872 659, November, WPAFB, Dayton, Ohio.

Colyer, D. B., and Oney, W. R. (1970). "High-Speed Cryogenic Alternator Development,"

Phase 1 Final Report. Contract DAAK02-68-C0320, U.S. Army Mobility Equipment Research and Development Center, Ft. Belvoir, Virginia.

Colyer, D. B., Schoch, K. R., Oney, W. R., and Terbush, R. K. (1972). "High-Speed Motor-Driven Helium Compressor." (Final Tech. Rept.) Report No. SRD-72-055. NTIS AD-773 665/5. General Electric Co., Schenectady, New York.

Colyer, D. B., et al. (1971). "Design and Development of Cryogenic Turborefrigerator Systems." Vol. 1, Report No. AFFDL-TR-71-117. AD 891-2552. General Electric Co., Schenectady, New York.

Colyer, D. B., et al. (1971). "Design and Development of Cryogenic Turborefrigerator Systems." Vol. 2, Report No. AFFDL-TR-71-117. AD 891-256L. General Electric Co., Schenectady, New York.

Colyer, D. B., et al. (1972). "High-Speed Motor-Driven Helium Compressor." Final Technical Report, Contract DAAK01-69-C-0171, U.S. Army Mobility Equipment Research and Development Center, Ft. Belvoir, Virginia. (General Electric Co., Schenectady, New York).

Colyer, D. B., et al. (1972). "Design and Development of Cryogenic Turborefrigerator Systems." Vol. I, Report No. AFFDL-TR-72-154. AD 911-062. General Electric Co., Schenectady, New York.

Colyer, D. B., et al. (1973). "Design and Development of Cryogenic Turborefrigerator Systems." Vol. II. AFFDL-TR-72-154. AD 911 063. General Electric Company, Schenectady, New York.

Colyer, D. B., et al. (1973). "Design and Development of Cryogenic Turborefrigerator Systems." Vol. III, Report No. AFFDL-TR-72-154. AD 912 949L. General Electric Company, Schenectady, New York.

Colyer, D. B., et al. (1974). "Design and Development of Cryogenic Turbo-refrigerator Systems." AFFDL-TR-74-93. AD 922 382. General Electric Co., Schenectady, New York.

Copeland, R. J., and Oren, J. A. (1974). "Cooling Systems for Satellite Remote Sensing Instrumentation." Report No. REPT-2-53002/4R-3182. NASA Contractor Report No. NASA-CR-132517, Contract No. NASI-10900 (September) 128 pp. LTV Aerospace Corp., Dallas, Texas.

Copeland, R. J., and Oren, J. A. (1974). "Cooling Systems for Satellite Remote Sensing Instrumentation." Report No. NASA-CR-132517, Rept-2-53002/4R-3182, NTIS N75-182 83/2ST. LTV Aerospace Corp., Dallas, Texas.

Cowans, K. W. (1972). "5 K Vuilleumier Cryogenic Refrigerator." AFFDL-TR-7210. AD 902 552. Sub-Marine Systems Inc., Tarzana, California.

Cowans, K. W. (1972). "Development of a Small Lightweight 77 K Vuilleumier Cryogenic Battery-Operated Hot Gas Cooler." Army Electronics Command, Fort Belvoir, Virginia. Night Vision Lab., Final Report DAAK02-70-C-0436.

Cox, J. E. (1974). "Potential Cooling Methods for an ELF Squid." (Final Rept.) Report No. NRL-MR-2899, NTIS AD A002 727. Naval Research Lab., Washington, D.C.

Cox, J. E. (1975). "Evaluation of a 4.5 K Closed-Cycle Refrigerator." (Final Rept.) Report No. NRL-MR-3088, NTIS AD-A015 308/0ST. Naval Research Lab., Washington, D.C.

Cox, J. E., and Edelsack, E. A. (1971). "Proceedings of the Workshop on Naval Applications of Superconductivity," held at Naval Ship Research and Development Laboratory, Panama City, Florida November 4–6, 1970. Report No. NRL-7302, NTIS AD-727 573. Naval Research Lab., Washington, D.C.

Curwen, P. W., White, H. V., and Gray, S. (1974). "Design and Test of a Helium Gas Spring." (Final Rept.) Report No. MTI-74TR28, NTIS AD-787 142/9SL. Mech. Tech. Inc., Latham, New York.

Daggerhart, J. A., and Smetana, F. V. (1969). "An Experimental Study of the Cryoentrainment Pump Progress Report." Report No. NASA-CR-66873, CFSTI N70-12948. Dept. of Mech. and Aerospace Eng., North Carolina State Univ., Raleigh, North Carolina.

Damiano, R., and Kliphuis, J. (1964). "Development of Integrated X-band Parametric Amplifier/Cryogenic Refrigerator System." Report No. 1, DA36-039-AMC-03728(s). TRG Inc., Melville, New York.

Daney, D. E. (1974). "Refrigeration for an 8 K to 14 K Superconducting Transmission Line." Report No. NBSIR 74-375 (Oct) 58 pp. National Bureau of Standards, Boulder, Colorado.

Daney, D. E., McConnell, P. M., and Strobridge, T. R. (1972). "Low-Temperature Nitrogen Ejector Performance." (Final Rept.) V18, 478-485, NTIS COM-73-50712/1, National Bureau of Standards, Washington, D.C.

Dannan, J. H. (1974). "Superconducting Propulsion System and Ship Interface Study," Vol. 1, "DC–DC System." (Final Rept.) Report No. 74-9972-1-BK-3, DDC AD 918 651L. Contract No. N00024-73-C-5487. Naval Ship Systems Command, Washington, D.C.

Dannan, J. H. (1974). "Superconducting Propulsion System and Ship Interface Study," Vol. 2, "AC–DC System." (Final Rept.) Report No. 74-9972-2, DDC AD 918 652L. Contract No. N00024-73-C-5487. Naval Ship Systems Command, Washington, D. C.

Daunt, J. G. (1970). "Cryogenic Refrigerators." Proc. Workshop on Naval App. of Superconductivity. Naval Ship Research and Development Lab., Panama City, Florida, pp. 40–63.

Daunt, J. G. (1971). "Experimental Investigations of Desorption Cooling Methods at Cryogenic Temperatures," Final Report, May 1968–June 1970. Report No. NASA-CR-119184, NTIS N71-30291. Stevens Institute of Technology, Hoboken, New Jersey.

Daunt, J. G. (1975). "Thermoelectricity and Refrigeration at Low Temperatures." (Final Report) Contract No. N00014-67-A-0202-0027, NTIS AD-A013 606/9ST. Stevens Institute of Technology, Hoboken, New Jersey.

Daunt, J. G., and Goree, W. S. (1969). "Miniature Cryogenic Refrigerators." (Tech. Rept.) Report No. TR-1(ONR), CFSTI AD-697 972. Stevens Institute of Technology, Hoboken, New Jersey.

Daunt, J. G., and Goree, W. S. (1969). "Minature Cryogenic Refrigerators." (Final Rept.) Contract No. N00014-67-C-0393. Stanford Research Institute, Menlo Park, California.

Daunt, J. G., and Rossi, R. A. (1963). "Investigation of Gas Liquefiers for Space Operation." Report No. ASD-TDR-63-775. Malaker Labs., Inc., High Bridge, New Jersey.

Dean, J. W., and Mann, D. B. (1965). "The Joule–Thomson Process in Cryogenic Refrigeration Systems." Report No. NBS-TN-227, CFSTI PB-184 473, National Bureau of Standards, Boulder, Colorado.

Dehne, A. G. (1975). "Improved Ultraminiature Split Vuilleumier Refrigeration System." (Final Rept.) Report No. AFFDL-Tr-75-118, DDC AD 8010 949L. Contract No. F33615-73-C-3003. Hughes Aircraft Co., Culver City, California.

Desjardins, L. F., and Hooper, J. (1973). "Analysis and Test of a Breadboard Cryogenic Hydrogen/Freon Heat Exchanger." (Tech. Rept.) Report No. NASA-CR-128987, SVHSER-6180, NTIS N73-26959/9. Hamilton Standard, Windsor Locks, Connecticut.

Dinger, R. J., Davis, J. R., and Nisenoff, M. (1976). "Long-Hold-Time Liquid-Helium Dewar for Cooling of a Squid ELF Antenna" (Final Memorandum Rept.) Report No. NRL-MR-3256, NTIS AD-A023 467/4ST. Naval Research Lab., Washington, D.C.

Doering, R. D. (1968). "Vuilleumier Cycle Cryogenic Refrigeration Development." AFFDL-TR-68-68. AD 841 543 (April), WPAFB, Dayton, Ohio.

Donabedian, M. (1972). "Survey of Cryogenic Cooling Techniques." Report No. SAMSO-TR-73-34, NTIS AD-755 780. Aerospace Corp., El Segundo, California.

Doody, R. D. (1970). "77 Degrees K Vuilleumier Cycle Cryogenic Refrigeration System for Ground Applications." Report No. HAC-P-70-343. Army Electronics Command, Night Vision Lab., Fort Belvoir, Virginia.

Doody, R. D. (1971). "Two-Stage Vuilleumier Cycle Cryogenic Refrigeration System for Advanced Forward Looking Infrared (ARLIR) Applications." (Final Rept.) Report No. AFFDL-TR-71-17, NTIS AD-886 822/6ST. Hughes Aircraft Co., Culver City, California.

Doody, R. D. (1975). "Long-Life, High-Capacity Vuilleumier Refrigerator for Space Applications." (Final Rept.) Report No. AFFDL-TR-75-108. AD B007 180L, Hughes Aircraft Co., Culver City, California.

Dybwad, J. P., and Logan, L. M. (1972). "Liquid-Helium-Cooled Photon Detectors: Cause, Prevention and Remedy of Contamination." Report No. AFCRL-TR-73-0189, NTIS AD-758 529. Air Force Cambridge Research Labs., Massachusetts.

Earley, R., and Tabakoff, W. (1969). "An Analytical and Experimental Study of a Cryogenic Evaporator for Space Application." (Tech. Rept.) Report No. THEMIS-AE-69-2, CFSTI AD-696 444. Aerospace Engineering, Ohio.

Ekern, W. F., Glassford, A. P., Jensen, H. L., and Nast, T. C. (1971). "Investigation of External Refrigeration Systems for Long Term Cryogenic Storage Final Report." Report No. NASA-CR-114920, LMSC-A981632, NTIS N71-20279. Lockheed Missiles and Space Co., Sunnyvale, California.

Fastovskii, V. G., Petrovskii, Yu. V., and Rouinskii, A. E. (1973). "Infrared Radiation Receivers." Report No. FSTC-HT-23-1370-73, NTIS AD/A-001 202/1SL. Army Foreign Sci. and Technol. Center, Charlottesville, Virginia.

Fleming, R. B. (1968). "Regenerators in Cryogenic Refrigerators." AFFDL-TR-68-143. AD 844 687 (September) WPAFB, Dayton, Ohio.

Fleming, R. B. (1970). "Fabrication and Evaluation of Advanced High-Effectiveness Cryogenic Heat Exchangers." Final Technical Report No. AFFDL-TR-70-7. AD 872 503, WPAFB, Dayton, Ohio.

Fleming, R. B., Colyer, D. B., Terbush, R. K., and Gamble, B. B. (1975). "Development of a Miniature Gas-Bearing Cryogenic Turbo Refrigerator." (Final Rept.) Report No. SRD-75-097, NTIS AD-A018 047/1ST. Army Mobility Equip. Res. and Devel. Center, Fort Belvoir, Virginia.

Fleming, R. B., Terbush, R. K., and Colyer, D. B. (1972). "Development of a Single-Stage Cryogenic Turborefrigerator." Phase II Final Report, Contract DAAK02-68-C-0320, U.S. Army Mobility Equipment Research and Development Center, Fort Belvoir, Virginia.

Forsyth, E. B., and Gibbs, R. J. (1976). "Brookhaven Superconducting Cable Test Facility." Rept. No. CONF-760829-27, NTIS BNL-217780. Brookhaven National Lab., Upton, New York.

Fowle, A. A., Heuchling, T. P., and Schulte, C. A. (1965). "Reciprocating Cooling Techniques for Space Based Infrared Sensors." Report No. AFFDL-TR-65-14. A. D. Little, Inc., Cambridge, Massachusetts.

Fradkin, J., Okwit, S., Siegel, K., and Smith, J. G. (1965). "Closed-Loop Cryogenic System for Advanced 2200-MC Traveling-Wave Maser Amplifier." Report No. AFAL-TR-65-41, CFSTI AD-460 561. Airborne Instruments Lab., Deer Park, New York.

Friedly, J. C., Kroeger, P. G., and Manganaro, J. L. (1967). "Stability Investigation of Thermally Induced Flow Oscillations in Cryogenic Heat Exchangers." Final Report. Report No. NASA-CR-61745, CFSTI N68-23231. General Electric Co., Schenectady, New York.

Fulton, J. T., and Kaplan, I. (1975). "High Average Power Diode Laser Illuminator" (Final Rept.) Contract No. F33615-73-C-1045, NTIS AD-A011 692/1ST. Grumman Aerospace Corp., Bethpage, New York.

Galzin, F. (1971). "Vacuum Radiation Test of Thermal Mock-Up of Infrared Detector Cooling System." Report No. NASA-TT-F-13833, Rept. 70.CT.TA.TP(METEOSAT-0-492, NTIS N72-14464. NASA, Washington, D.C.

Geyer, R. (1975). "Design, Assembly and Installation of Electronic Control Circuits for VM Cooler Drive Motors." AFFDL-TR-75-104. AD B00 8179. Philips Labs, Briarcliff Manor, New York.

Gibbons, R. M., and Kuebler, G. P. (1968). "Research on Materials Essential to Cryocooler Technology. Thermophysical and Transport Properties of Argon, Neon, Nitrogen and Helium-4." (Final Rept.) Report No. AFML-TR-68-370, NTIS AD-852 010/8ST. Air Prods. and Chems. Inc., Allentown, Pennsylvania.

Gifford, W. E. (1962). "Design and Construction of $4.2°$ K Maser Refrigerator System." Final Report C-63774, NAS7-100, A. D. Little, Inc., Cambridge, Massachusetts.

Gifford, W. E. (1966). "Basic Investigation of Cryogenic Refrigeration Methods." AFFDL-TR-66-26, WPAFB, Dayton, Ohio.

Gifford, W. E. (1966). "Study of Novel Refrigeration Methods," Final Report. Report No. NASA-CR-80513, CFSTI N67-13167. Jet Propulsion Lab., Pasadena, California.

Gifford, W. E. (1968). "Basic Investigation of Cryogenic Refrigeration Methods." AFFDL-TR-68-61, AD 839 424 (June), WPAFB, Dayton, Ohio.

Gifford, W. E. (1970). "Low-Temperature Refrigerator Design Optimization." AFFDL-TR-70-158, AD 882 760. (September), WPAFB, Dayton, Ohio.

Glode, J. B., Riha, F. J. III, Gainey, R. T., Aske, H. D., and Doody, R. D. (1971). "Vuilleumier-Cycle Cryogenic Refrigeration Systems for Infrared Scanner Applications." (Final Rept.) Report No. AFFDL-TP-71-18, NTIS AD-886 823. Hughes Aircraft Co., Culver City, California.

Goff, J. F. (1975). "The Low-Temperature Thermoelectric Cooler for Navy Use." (Interim Rept.) Report No. NSWC/WOL/TR-75-17, NTIS AD-A008 145/5ST. Naval Surface Weapons Center, Silver Spring, Maryland.

Goree, W. S., and Daunt, J. G. (1969). "Miniature Cryogenic Refrigerators." (Final Rept.) NTIS AD-860 866/3ST. Stanford Research Inst., Menlo Park, California.

Goree, W. S., Hesterman, V. W., and Chilton, F. (1970). "Theoretical and Experimental Study of Noise in Superconductors." (Final Rept.) Contract No. N00014-67-C-0393, NTIS AD-871 770/4ST. Stanford Research Inst., Menlo Park, California.

Gothe, K. H., Froehlich, H. J., and Martini, L. (1968). "On the Construction of a Ruby Maser for the X-Band with Liquid Hydrogen as Coolant." Report No. FTD-HT-23-1589-67, NTIS AD-849 480/9ST. WPAFB, Dayton, Ohio.

Gray, P. F. (1967). "Experimental Testing of Size 204 Ball and Roller Bearings for Use in Miniature Cryogenic Compressors." AFFDL-TR-66-225, WPAFB, Dayton, Ohio.

Groll, M., Pittman, R. B., and Eninger, J. E. (1975). "Parametric Performance of Circumferentially Grooved Heat Pipes with Homogeneous and Graded-Porosity Slab Wicks at Cryogenic Temperatures." Report No. NASA-TM-X-73095, A-6405, NTIS N76-17325/1ST. NASA, Ames Res. Center, Moffett Field, California.

Gunn, R. D., Cheuh, P. L., and Prausnitz, J. M. (1966). "Inversion Temperatures and Pressures for Cryogenic Gases and Their Mixtures." Report No. R-415, CFSTI PB-174 687, National Bureau of Standards, Boulder, Colorado.

Gylling, R. G. (1971). "Construction and Operation of a Nuclear Refrigeration Cryostat." Report No. APS-CH-81. Pub. by the Finnish Academy of Tech. Sciences, Helsinki, NTIS PB-205 950. Acta Polytechnica Scandinavica, Stockholm, Sweden.

Hahnemann, H. (1948). "Approximate Calculation of Thermal Ratios in Heat-Exchangers Including Heat Conduction in the Direction of Flow." N.G.T.E. Me. 36.

Hardgrove, W. F., and John, J. E. (1967). "The Attainment of Clean Vacuum with a One-Watt Refrigerator." Report No. NASA-TN-D-4006, CFSTI N67-30121. NASA, Greenbelt, Maryland.

Harkee, J. (1964). "Study and Fabrication of a Solid–Refrigeration Superinsulation Cooling System." Quarterly Progress Report NTIS AD-454 029/0ST. Aerojet-General Corp., Azusa, California.

Harkless, L. B. (1974). "Demonstration of Advanced Cryogenic Cooler–Infrared Detector Assembly." Report No. AFFDL TR-74-15, DDC AD 917 941L. Honeywell Radiation Center.

Harris, R. E. (1966). "77 Degrees Kelvin Rotary-Reciprocating Cryogenic Cooler Program." Res. Tech. Briefs, 4, No. 2. AFFDL Wright–Patterson AFB, Dayton, Ohio, 26 pp.

Harris, R. E. (1973). "77 Kelvin Rotary-Reciprocating Cryogenic Cooler Evaluation Program." AFFDL-TR-73-39, AD 912 949L. In-House, Wright–Patterson AFB, Dayton, Ohio.

Hartmann, R. T., and Kercheval, J. E. (1968). "Design and Development of a 30 K Closed-Cycle Refrigerator." Report No. 67-2567 AFAL TR 67-328. Garrett Corp., Los Angeles, California.

Haskin, W. L. (1967). "Cryogenic Heat Pipe." (Final Tech. Rept.). Report No. AFFDL-TR-66-228, CFSTI AD-657 025. WPAFB, Dayton, Ohio.

Haskin, W., and White, R. (1977). "Initial Testing of High-Capacity VM Cryo-Coolers." (Final Rept.) Report No. AFFDL-TR-76-160, DDC ADB 019 990L. WPAFB, Dayton, Ohio.

Hatch, B. D., Fleming, R. B., Jones, D. W., and Minnich, S. H. (1972). "Cryogenic Systems and Superconductive Power." (Semi-Annual Tech. Rept.) Report No. SRD-73-022, NTIS AD-758 431. General Electric Co., Schenectady, New York.

Hatch, B. D., Fleming, R. B., Jones, D. W., and Kerr, D. L. (1973). "Cryogenic Systems and Superconductive Power." (Semi-Annual Tech. Rept.) Report No. SRD-73-139, NTIS AD-775 228/0. General Electric Co., Schenectady, New York.

Hatch, B. D., and Kerr, D. L. (1974). "Superconducting Propulsion System." (Final Rept.). Report No. SRD-74-019, DDC AD 917 198L. General Electric Co., Schenectady, New York.

Hatch, D. B. *et al.* (1972). "Cryogenic Systems and Superconductive Power." First Semiannual Technical Report. Contract DAHC-15-72-C-0235, Department of Defense Advanced Research Projects Agency, Washington, D.C.

Haumesser, R., Lockwood, D., McNall, R., and Welsh, J. P. (1975). "Development of Lightweight Transformers for Airborne High Power Supplies." Vol. I. (Final Tech. Rept.) Report No. AFAPL-TR-75-15-Vol-1, NTIS AD-A018 545/4ST. Thermal Technology Lab. Inc., Buffalo, New York.

Hein, R. A., Gubser, D. U., and Takken, E. H. (1970). "Proceedings of the 1970 Ultralow Temperature Symposium," held at Naval Research Laboratory, Washington, D.C., April 23–24, 1970. Report No. NRL-7133, CFSTI AD-712 061. Naval Research Lab., Washington, D.C.

Hicks, D. C. (1974). "Application of Superconducting Electrical Machinery to the Propulsion Systems of Commercial Vessels." (Kings Point Scholars Series Rept.) NTIS COM-75-10137/8ST. Merchant Marine Academy, Kings Point, New York.

Hirschy, D. C., and Napolitano, M. C. (1973). "Design, Fabricate and Test a Pre-Prototype Cryogenically Cooled Interferometer Spectrometer." (Final Rept.) Report No. AFCRL-TR-73-0599, NTIS AD-777 890/5. Idealab Inc., Franklin, Massachusetts.

Hudson, R. P. (1972). "Principles and Application of Magnetic Cooling." Final Report, Publ. in Series in Low Temperature Physics, V2, 1–230. NTIS COM-73-50541. National Bureau of Standards, Washington, D.C.

Hughes, R. C., Sweet, R. C., and Bronnes, R. L. (1969). "Study and Investigation of Infrared Detector Envelope-Enclosures." Air Force Avionics Laboratory Technical Report, AFAL-TR-69-85.

Jacobs, J. A. H. (1962). "Survey of Closed-Cycle Helium-Temperature Refrigerators." Report No. RADC-TDR-62-319, NTIS AD-283 101/4ST, Martin Marietta Corp., Baltimore, Maryland.

Janocko, M. A., and Jones, C. K. (1972). "Superconducting Quantum Devices for Use at Liquid Hydrogen Temperatures." (Final Rept.) Report No. 72-9J2-SUCON-P1, NTIS AD-902 717/8ST. Westinghouse Research Labs., Pittsburgh, Pennsylvania.

Jensen, H. L., Nast, T. C., and Glassford, A. P. (1970). "Investigation of External Refrigeration Systems for Long-Term Cryogenic Storage." (System Review Rept.) Report No. NASA-CR-115191, LMSC-A903162, NTIS N71-37105. Lockheed Missiles and Space Co., Sunnyvale, California.

Jensen, H. L., Nast, T. C., Glassford, A. P., and Vernon, R. M. (1971). "Handbook of External Refrigeration Systems for Long Term Cryogenic Storage." Report No. NASA-CR-115190, LMSC-A984158, NTIS N71-37561. Lockheed Missiles and Space Co., Sunnyvale, California.

Jensen, H. L., Nast, T. C., Glassford, H. P., Vernon, R. M., and Ekern, W. F. (1971). "Investigation of External Refrigeration Systems for Long-Term Cryogenic Storage." (Summary Rept.) Report No. NASA-CR-115192, LMSC-A984159, NTIS N71-37104. Lockheed Missiles and Space Co., Sunnyvale, California.

Johnson, J. C., ed. (1973). "Proc. Closed Cycle Cryogenic Cooler Technology and Applications Conf.," U.S. Air Force Academy, Boulder, Colorado, AFFDL-TR-73-149, (Vol. I), December AD No. 918234.

Johnson, J. D., and Bonner, J. C. (1975). "Quantitative Studies of Magnetic Cooling on a Magnetic System Which Obeys the Third Law." Report No. CONF-751209-10, NTIS LA-UR-75-2199. Los Alamos Scientific Lab., New Mexico.

Johnson, J. E. (1952). "Regenerator Heat-Exchangers for Gas Turbines." Aero Research Council Tech. Rep., R&M No. 2630.

Johnson, V. J., and Stewart, R. B. (1960). "A Compendium of the Properties of Materials at Low Temperature." (WADD Tech. Rep. 60-56).

Jones, M. C., and Johnson, W. W. (1976). "Heat Transfer and Flow of Helium in Channels—Practical Limits for Applications in Superconductivity." (Tech. Note). Report No. NBS-TN-675, NTIS PB-250 725/9ST. National Bureau of Standards, Boulder, Colorado.

Kadi, F. J., and Longsworth, R. C. (1975). "Assessment and Study of Existing Concepts and Methods of Cryogenic Refrigeration for Superconducting Transmission Cables." Progress Report, 1 May 1975–30 June 1975. NTIS COO-2552-2. Air Products and Chemicals Inc., Allentown, Pennsylvania.

Kadi, F. J., and Longsworth, R. C. (1975). "Assessment and Study of Existing Concepts and Methods of Cryogenic Refrigeration for Superconducting Transmission Cables." Progress Report, 1 September 1975–31 October 1975. Contract No. E(11-1)-2552, NTIS COO-2552-4. Air Products and Chemicals Inc., Allentown, Pennsylvania.

Kadi, F. J., and Longsworth, R. C. (1975). "Assessment and Study of Existing Concepts and Methods of Cryogenic Refrigeration for Superconducting Transmission Cables." Air Products and Chemicals Inc., Allentown, Pennsylvania, Progress Report, Chicago Opers. Office (ERDA), Illinois, Report No. COO-1198-1137, Contract No. E(11-1)-2552 (June), 33 pp.

Kadi, F. J., and Longsworth, R. C. (1976). "Assessment and Study of Existing Concepts and Methods of Cryogenic Refrigeration for Superconducting Transmission Cables." Chicago Opers. Office (ERDA), Illinois, Final Report No. COO-2552-6 (February), 223 pp.

Kadi, F. J., and Longsworth, R. C. (1976). "Assessment and Study of Existing Concepts and Methods of Cryogenic Refrigeration for Superconducting Transmission Cables." (Final Report) NTIS COO-02552-7. Air Products and Chemicals Inc., Allentown, Pennsylvania.

Katheder, H., Lehmann, W., and Spath, F. (1974). "Test and Operation of Helium Low-Temperature Devices for Applied Superconductivity." Institute für Experimentelle Kernphysik, Karlsruhe, West Germany, Report No. KFK-EXT-3/74-9 (April), 37 pp.

Kazovskii, E. Ya. (1968). "Prospects of Application Superconductors in Electrical Machines and the Problem of Boosting Superconducting Systems." Report No. FTD-MT-24-212-68, CFSTI AD-681 601. WPAFB, Dayton, Ohio.

Kercheval, J. E. (1974). "Low Production Cost Vuilleumier Cycle Cryogenic Refrigeration System." (Final Rept.) Report No. HAC-P74-235, DDC AD B008 201L, Contract No. DAAK02-72-C-0277. Army Electronics Command, Fort Belvoir, Virginia.

Kilgore, R. A., and Adcock, J. B. (1977). "Specific Cooling Capacity of Liquid Nitrogen." Report No. NASA-TM-X-74015, NTIS N77-21261/1ST. Langley Research Center, Langley Station, Virginia.

Kirtley, J. L., Jr., and Smith, J. L., Jr. (1977). "Superconducting Electric Machines for Ship Propulsion." (Final Tech. Rept.) NTIS AD-A038 825/6ST. M.I.T., Dept. of Mech. Engineering, Cambridge, Massachusetts.

Kitchens, T. A. (1976). "Observations on European Low-Temperature Physics Research: An Annotated Directory of Low-Temperature Physics in British Universities and Some Comments on the Low Temperature Physics Programs in Europe." (Tech. Rept.) Report No. ONRL-R-13-76, NTIS DODXA. Office of Naval Research, London, England.

Klann, J. L. (1973). "Analysis of a Combined Refrigerator–Generator Space Power System." Report No. NASA-TM-X-71433, E-7683. Presented at Cryog. Cooler Conf., Colorado (October), NTIS N73-31992/3. NASA, Cleveland, Ohio.

Kovac, L., and Balla, J. (1974). "Combined Helium-3 Cryostats and Helium-3, Helium-4 Dilution Refrigerators." Univerzida Pavla Jozefa Safarika, Kosice, Czechoslovakin, Report No. KFK1-74-60. Available INIS, 15 pp.

Kroebig, H. L. (1975). "Plasma Deposited Rider Rings for Hot Displacer." Report No. Pat-APPL-556 988, NTIS AD-D000 686/6ST. D.A.F., Washington, D.C.

Kroebig, H. L. (1977). "Fusible Heat Sink for a Cryogenic Refrigerator." Report No. Pat-APPL-758 879, NTIS AD-D003 515/4ST. D.A.F., Washington, D.C.

Kropschot, R. H., Birmingham, B. W., and Mann, D. B. (1968). "Technology of Liquid Helium." CFSTI NBS MONO-111, National Bureau of Standards, Boulder, Colorado.

Kuebler, G. P., Weimer, R. F., and Clark, R. G. (1967). "Research on Materials Essential to Cryocooler Technology." Report No. 6, AF33(615)-2191. Air Products and Chemicals, Inc., Allentown, Pennsylvania, 39 pp.

Kulakov, V. M. (1967). "Experience in Design of a Low-Temperature Helium Turbine Refrigerator." Report No. FTD-HT-67-227. Trans. of Khimicheskoe i Neftyanoe Mashinostroenie (USSR) N6, 8–10. CFSTI AD-670 065. WPAFB, Dayton, Ohio.

Lechner, R. A. (1971). "Investigation of Regenerators and Pulse Tube Cryogenic Coolers." Report No. ECOM-3409, NTIS AD-725 514. Army Electronics Command, Fort Monmouth, New Jersey.

Lehrfeld, D., and Boser, O. (1974). "Absorption–Desorption Compressor for Spaceborne/Airborne Cryogenic Refrigerators." (Final Rept.) Report No. AFFDL-TR-74-21, Philips Labs., Briarcliff Manor, New York.

Leiby, C. C., Jr., and Ryan, T. G. (1973). "Thermophysical Properties of Thermal Energy Storage Materials—Aluminum." Report No. AFCRL-PSRP-554, NTIS AD-914 187/0ST. Air Force Research Labs., Cambridge, Massachusetts.

Leo, B. (1970). "Designers Handbook for Spaceborne Two-Stage Vuilleumier Cryogenic Refrigerators." Report No. AFFDL-TR-70-54. AD 872 995. 105 pp. (June), WPAFB, Dayton, Ohio.

Leo, B. (1971). "Vuilleumier Cycle Cryogenic Refrigeration System Technology Report." AFFDL-TR-71-85, AD 888992L, WPAFB, Dayton, Ohio.

Lerma, G. (1975). "Insulation for Piping." Report No. PAT-APPL-643 895, NASA-CASE-MSC-19523-1, NTIS N76-16245/2ST. NASA, Houston, Texas.

Liang, S. T. W., and Martin, L. F. (1971). "Superconducting Machinery for Ship Propulsion Systems." (Summary Rept.) Report No. NSRDC-3787, NTIS AD-893 605/6ST. Naval Ship Research and Development Center, Bethesda, Maryland.

Linnet, C., Purdy, V., Chang, Y. W., and Frederking, T. H. K. (1970). "Unsaturated Helium Cooling Limits." (Summary of Liquid Helium-4 Boundary and Heat Transport Studies). (Final Rept.) Report No. UCLA-ENG-70-91, NTIS AD-717 568. Univ. Calif. School of Engineering and Applied Science, Los Angeles, California.

Lins, R. C. (1974). "Manufacturing Methods and Technology for Closed-Cycle Cryogenic Coolers, III." Quarterly Progress Report No. 3136-7Q3, DDC AD B000 664L. Contract No. DAAB05-74-C-2523. Army Electronics Command, Fort Monmouth, New Jersey.

Linsteadt, G. F. (1965). "Evaluation of a Closed-Cycle Refrigerator." Report No. NOTS-TP-3809, CFSTI AD-616 119. Naval Ordinance Test Station, China Lake, California.

Longsworth, R. C. (1974). "Split-Stirling Cycle Cryogenic Refrigerator." (Final Tech. Rept.) DDC AD 920 029L. Contract No. DAAK02-72-C-0316. Army Electronics Command, Fort Belvoir, Virginia. NTIS AD-920 029/6ST. Air Products and Chemicals Inc., Allentown, Pennsylvania.

Lucek, R., Damaz, G., and Daniels, A. (1967). "Adaptation of Rolling-Type Seal Diaphragms to Miniature Stirling Cycle Refrigerators." Report No. TR-216, AFFDL-TR-67-96. Philips Labs., Briarcliff Manor, New York, 48 pp.

Magee, F. N., and Doering, R. D. (1968). "Vuilleumier Cycle Cryogenic Refrigerator Development." Report No. AFFDL-TR-68-67. Hughes Aircraft Co., Culver City, California, Research and Development Div., 256 pp.

Magee, P. R., and Datring, R. (1969). "Vuilleumier Cycle Cryogenic Refrigerator Development." Technical Report, TR 68-69. U.S. Air Force Flight Dynamics Lab., WPAFB, Dayton, Ohio.

Mann, D. B., Norton, M. T., and Strobridge, T. R. (1964). "Cryogenic Magnet Refrigeration." Report No. 8239, National Bureau of Standards, Washington, D.C. 129 pp.

Marshak, H., and Dove, R. B. (1970). "A Right Angle 3He Cryostat Incorporating a High-Field Superconducting Solenoid." (Tech. Note) NTIS COM-71-00128. National Bureau of Standards, Washington, D.C.

Martini, W. (1978). "Design Manual for Stirling Engines." NASA CR 135382, DOE/NASA Contractor Report, NASA, Greenbelt, Maryland.

McCarthy, J. S., Whitney, R., Healy, D. C., Parks, D., and Fisher, T. R. (1969). "A Polarized ^{59}Co Target Using a ^{3}He-^{4}He Dilution Refrigerator." Report No. HEPL-619, CFSTI AD-703 241. High-Energy Physics Lab., Stanford Univ., California.

McCarty, R. D. (1972). "Thermophysical Properties of Helium-4 from 2 to 1500 K with Pressures to 1000 Atmospheres." NBS Tech. Note No. 631, US Government Printing Office, Washington, D.C.

McConnell, P. M. (1973). "Liquid Helium Pumps." (Interim Rept.) Report No. AFAPL-TR-73-72, NTIS COM-73-11954/7. National Bureau of Standards, Boulder, Colorado.

McCullough, J. E. (1967). "Exploratory Development of Gas Springs as Applied to Rotary-Reciprocating Cryogenic Coolers." AFFDL-TR-67-62, WPAFB, Dayton, Ohio.

McLaughlan, P. B. (1967). "Performance Evaluation of a Liquid-Shrouded Cryogenic Storage System." Report No. NASA-TM-X-64392, MSC-EP-R-67-41, CFSTI N70-34535. NASA, Houston, Texas.

McPherson, D. J. (1974). "Trap MATS Canopy Modification Acceptance Test Report and Appendix A Through M." Report No. G8484.16.23. NTIS AD-A011 940/4ST. E-Systems Inc., Greenville, Texas.

McPherson, D. J. (1974). "Trap MATS Canopy Modification Acceptance Test Report. Appendix N. Flight Acceptance Test, Drum Resonance." Report No. G8484.16.23.-App. N, NTIS AD-A011 941/2ST. E-Systems Inc., Greenville, Texas.

McWilliams, D., Doherty, P. (1974). "Operators Manual Supplement. Model 1400 System for Office of Naval Research USN Ship Research and Development Center in Accord with CTI Specification A3542561." (Final Rept.) Report No. CTI-A3542561, DDC AD 776 248. Contract No. N00014-73-00468. Cryogenic Technology Inc., Waltham, Massachusetts.

Mebus, E. A., and Fitti, N. S., Jr. (1969). "An Evaluation of Small Closed-Cycle Cryogenic Refrigerators as Cooling Devices for Infrared Detectors." (Phase Rept.) Report No. NADC-AE-6843, NTIS AD-852 931/5ST. Naval Air Devel. Center, Johnsville, Pennsylvania.

Miller, W. S., and Potter, V. L. (1973). "Fractional-Watt Vuilleumier Refrigerator—Final Report of Task I Preliminary Design." Report 72-8846, prepared for Goddard Space Flight Center, Greenbelt, Maryland.

Miller, W. S., and Potter, V. L. (1974). "Fractional-Watt Vuilleumier Cryogenic Refrigerator—Final Report for Task II Analytical and Test Programs." Contract NAS 5-21715, Goddard Space Flight Center, Greenbelt, Maryland.

Moore, R. W. (1970). "An Investigation of Thermal Regenerators for the 4 to 20°K Temperature Range." AFFDL-TR-70-64. AD 876 605. A. D. Little, Inc., Cambridge, Massachusetts.

Moore, R. W., Jr. (1971). "Development and Testing of Heat Exchangers for a 1 Watt, 3.6°K Rotary-Free-Piston Refrigerator." Technical Report AFFDL-TR-71-27. AD 885 288. A. D. Little, Inc., Cambridge, Massachusetts.

Moore, R. W., Jr. (1974). "Investigation of Scroll Fluid Machinery for Cryogenic Helium Refrigerators." (Final Rept.) Report No. ADL-C-75976-F, NTIS AD-787056/1SL. A. D. Little, Inc., Cambridge, Massachusetts.

Moore, R. W., Jr. (1975). "Investigation of Scroll Fluid Machinery for Cryogenic Helium Refrigerators." (Final Rept.) Report No. ADL-C-75976, NTIS AD-A014 070/7ST. A. D. Little, Inc., Cambridge, Massachusetts.

Morgan, N. E. (1971). "Analysis and Preliminary Design of Airborne Air Liquefiers." (Final Tech. Rept.) Report No. AFFDL-TR-71-171. AD 892 500/0ST, Hughes Aircraft Co., Culver City, California.

Morgan, N. E. (1975). "Preliminary Design of a Cryo-Cooler for Satellite Sensors (CCSS)," Vol. I. (Final Rept.) Report No. AFFDL-TR-75-154. AD B011 745L, Hughes Aircraft, Culver City, California.

Morgan, N. E. (1975). "Preliminary Design of a Cryo-Cooler for Satellite Sensors (CCSS),"

Vol. II. (Final Rept.) Report No. AFFDL-TR-75-154. AD C006 6534, Hughes Aircraft Co., Culver City, California.

Muhlenhaupt, R. C., and Strobridge, T. R. (1967). "The Single-Engine Claude Cycle as a 4.2 K Refrigerator." CFSTI NBS-TN-354. National Bureau of Standards, Boulder, Colorado.

Muhlenhaupt, R. C., and Strobridge, T. R. (1968). "An Analysis of the Brayton Cycle as a Cryogenic Refrigerator." CFSTI NBS-TN-366. National Bureau of Standards, Boulder, Colorado.

Murray, D. O. (1976). "Preliminary Design of a Cryogenic Radiator System." AFFDL-TR-76-136. AD B017 810L. Lockheed Missiles and Space Co.

Norman, R. H. (1976). "Joule–Thomson Expander and Heat Exchanger." (Final Tech. Rept.) Report No. NASA-CP-151978, NTIS N77-22424/4ST. AiResearch Mfg. Co., Torrance, California.

Norman, R. H. (1976). "Preliminary Design Study of Astronomical Detector Cooling System." Report No. NASA-CR-151979, NTIS N77-24024/0ST. AiResearch Mfg. Co., Los Angeles, California.

Ohara, T., Moore, K. G., and Gordon, M. N. (1976). "Vuilleumier Cooler Wear Rate Test Program." AFFDL-TR-76-135. AD B019 525L. Hughes Aircraft Co., Culver City, California.

Oney, W. R., and Colyer, D. B. (1970). "High-Speed Cryogenic Alternator Development. Phase I Final Report Prepared for the U.S. Army Mobility Equipment Research and Development Center." Contract No. DAAK02-68-C-0320, General Electric Company, Schenectady, New York.

Peterson, R. E., and Ramsey, J. W. (1973). "Solar Absorber Coating Study." Report No. AFML-TR-73-80, NTIS AD-760 577. Honeywell Inc., Minnesota.

Pitcher, G. K. (1970). "Design and Development of the Laboratory Model Vuilleumier Cycle Cryogenic Refrigerator." Report No. 243 DAAK02-69-C-0364. Army Electronics Command, Fort Belvoir, Virginia.

Pitcher, G. K. (1971). "Development of Spacecraft Vuilleumier Cryogenic Refrigerators—Part I. Thermodynamic Design and Study of Life-Limiting Components." U.S. Air Force Tech. Report AFFDL-TR-71-147. AD 891 461L, Philips Labs, Briarcliff Manor, New York.

Pitcher, G. K. (1972). "Development of Spacecraft Vuilleumier Cryogenic Refrigerators. Part II. Development and Fabrication." Report No. AFFDL-TR-71-147-Pt. 2 Air Force Flight Dynamics Lab., Wright–Patterson AFB, Ohio, 66 pp. AD 962 944L, Philips Labs, Briarcliff Manor, New York.

Pitcher, G. K. (1973). "Mechanical Life of Space Cryocoolers. Closed-Cycle Cryogenic Cooler Technology and Applications." AFFDL-TR-73-149, Vol. 1, pp. 211–224, WPAFB, Dayton, Ohio.

Pitcher, G. K. (1975). "Spacecraft Vuilleumier Cryogenic Refrigerator Development." (Final Rept.) Report No. AFFDL-TR-114. AD B007 675L, Philips Labs, Briarcliff Manor, New York.

Pitcher, G. K. (1977). "Development of a Flight Design for an Oil-Lubricated VM Cryocooler." AFFDL-TR-77-95, WPAFB, Philips Laboratories, Briarcliff Manor, New York.

Plaks, N. (1966). "Study and Fabrication of Superinsulated Solid-Cryogen Cooler Model K5." (Final Rept.) Report No. ECOM-00171-F, NTIS AD-487 618/1ST. Aerojet-General Corp., Azusa, California.

Puckett, L. J., Teague, M. W., and McCoy, D. G. (1970). "Refrigerating Vapor Bath." (Memorandum Rept.) Report No. BRL-MR-2065, NTIS AD-714 198. Ballistic Res. Labs., Maryland.

Pytlik, W. E., and Barnoski, J. J. (1974). "Modified AN/AAS-18 Infrared Detecting Set Cryogenic Cooler." Report No. TR-MMER/RM-74-130, DDC AD 919 848L. Ogden Air Logistics Center, Hill AFB, Utah.

Quadrini, J., and Kosson, R. (1974). "Design, Fabrication, and Testing of a Cryogenic Thermal Diode." (Interim Research Rept.) Report No. NASA-CR-137616, NTIS N75-21568/1ST. Grumman Aerospace Corp., Bethpage, New York.

Raab, B., Schock, A., King, W. G., Kline, T., and Russo, F. A. (1975). "Nuclear Heat Sources for Cryogenic Refrigerator Applications." Report No. ERDA-SNS-3063-7, NTIS FSECNSG-217-75/50. Fairchild Space and Electronics Co., Germantown, Maryland.

Radebaugh, R. (1967). "Thermodynamic Properties of He3–He4 Solutions with Applications to the He^3–He^4 Dilution Refrigerator." (Tech. Note) Report No. NBS-TN-362, NTIS COM-74-10482/9. National Bureau of Standards, Boulder, Colorado.

Radebaugh, R. (1977). "Refrigeration Fundamentals: A View Toward New Refrigeration Systems. Applications of Closed-Cycle Cryocoolers to Small Superconducting Applications." NBS Spec. Pubn. 508, Supt. of Documents, U.S. Govt. Printing Office, Washington, D.C.

Radebaugh, R., Lawless, W. N., and Siegwarth, J. D. (1974). "Electrocaloric Refrigeration for Superconductors." (Semi-Annual Tech. Rept.) NTIS AD-A008 852/6ST. National Bureau of Standards, Boulder, Colorado.

Radebaugh, R., and Siegwarth, J. D. (1970). "Numerical Analysis of Continuous and Discrete Heat Exchangers for Dilution Refrigerators." (Final Rept.) NTIS COM-72-50639. National Bureau of Standards, Boulder, Colorado.

Radebaugh, R., and Siegwarth, J. D. (1971). "Dilution Refrigerator Technology." Report No. NBS-R-657, NTIS COM-72-10217, National Bureau of Standards, Boulder, Colorado.

Radebaugh, R., Siegwarth, J. D., Lawless, W. N., and Morrow, A. (1975). "Electrocaloric Refrigeration for Superconductors." (Final Rept.) Report No. NBSIR-76-847, NTIS AD-A037 413/2ST. National Bureau of Standards, Boulder, Colorado.

Radebaugh, R., Siegwarth, J. D., Lawless, W. N., and Morrow, A. J. (1977). "Electrocaloric Refrigeration for Superconductors." National Bureau of Standards, Internal Report No. NBSIR 76-847, 194 pp.

Ramsey, J. W., Petersen, C. B., and Schmidt, R. N. (1970). "Solar Heat Source Feasibility Study." (Final Tech. Rept.) Report No. AFML-TR-70-294, NTIS AD-879 939/7ST. Honeywell Inc., Minneapolis, Minnesota.

Reed, W. E. (1975). "Cryogenic Refrigeration." Vol. 1, 1964–1972. (A Bibliography with Abstracts) NTIS/PS-75/825/0ST. National Technical Information Service, Springfield, Virginia.

Reed, W. E. (1977). "Cryogenic Refrigeration." Vol. 2, 1973–1977. (A Bibliography with Abstracts) NTIS/PS-77/1158. National Technical Information Service, Springfield, Virginia.

Reed, W. E. (1978). "Cryogenic Refrigeration." Vol. 3, Nov. 1977–Nov. 1978. (A Bibliography with Abstracts) NTIS/PS-78-1262/IPNM. National Technical Information Service, Springfield, Virginia.

Rhia, F. J., III (1971). "Development of Long-Life, High-Capacity Vuilleumier Refrigeration System for Space Applications." Part 1, AFFDL-TR-71-92, AD 887399L, WPAFB, Dayton, Ohio.

Rhia, F. J., III (1971). "Vuilleumier Cycle Cryogenic Refrigeration System for Missile Guidance Applications." AFFDL-TR-31, AD 886 824L. Hughes Aircraft Co., Los Angeles, California.

Rhia, F. J., III. (1971). "Development of Long-Life, High-Capacity Vuilleumier Refrigeration System For Space Applications, Part II, Description of Preliminary Refrigerator Design and Results of Thermal Analyses and Life Studies." AFFDL-TR-71-92, Part II, AD 890601L, WPAFB, Dayton, Ohio.

Rhia, F. J., III. (1971). "Development of Long-Life, High-Capacity Vuilleumier Refrigeration System For Space Applications, Part III, Refrigerator Design and Thermal Analyses." AFFDL-TR-71-92, Part III, AD 901235L, WPAFB, Dayton, Ohio.

Rhia, F. J., III. (1971). "Development of Long-Life, High-Capacity Vuilleumier Refrigeration System For Space Applications," Part IV, "Design and Mechanical Analyses of Refrigerator and Interface Unit." AFFDL-TR-71-92, Part IV, AD 920 255, WPAFB, Dayton, Ohio.

Richter, R. (1976). "Thermal Energy Storage Demonstration Unit for Vuilleumier Cryogenic

Cooler." (Interim Rept.) Report No. AFAPL-TP-76-110, NTIS AD-A040 895/5ST. Xerox Corp./Electro Optical Sys., Pasadena, California.

Rogers, A. D., and O'Neil, J. A. (1970). "Closed-Cycle Cryogenic Coolers." Tech. Rept. No. ECOM-0417-F. Army Electronics Command, Fort Monmouth, New Jersey, 37 pp.

Roubeau, P., and Vandevyver, M. (1968). "Variable Temperature Cryostat—1.3 to 300 Deg. K." Report No. CEA-P-3463. CFSTI N68-31245. Commissariat A L Energie Atomique, Saclay, France.

Russo, S. C. (1976). "Study of a Vuilleumier Cycle Cryogenic Refrigerator for Detector Cooling on the Limb Scanning Infrared Radiometer." Report No. NASA-CR-145078, NTIS N77-11211/8ST.

Schmidt, W. F. (1971). "Fabrication and Instrumentation of an Experimental Passive Radiative Infrared Detector Cooler for Spacecraft Applications." AFFDL-TR-71-125. AD 893-616L. Philco-Ford Corp.

Schwettman, H. A., Turneaure, J. P., Fairbank, W. M., Smith, T. I., and McAshan, M. S. (1967). "Low-Temperature Aspects of a Cryogenic Accelerator." Report No. HEPL-503, CFSTI AD-651 410. High-Energy Phys. Lab., Stanford Univ., California.

Sheinberg, H., and Steyert, W. A. (1969). "Fabrication of Porous Copper Heat Exchanger Media for ^{3}He–^{4}He Dilution Refrigerators." CFSTI LA-4259. Los Alamos Sci. Lab., New Mexico.

Shelpuk, B., and Crouthamel, M. (1972). "VM Thermal Actuation and Burner Investigation." USAFCOM Report, Ft. Belvoir, Virginia, pp. 13–14.

Shelpuk, B., Crouthamel, M. S., Amith, A., and Yim, M. (1968). "Low-Temperature Solid-State Cooling Technology." AFFDL-TR-68-128, AD 841 559, August, WPAFB, Dayton, Ohio.

Sherman, A. (1971). "Mathematical Analysis of a Vuilleumier Refrigerator." Report No. NASA-TM-X-65534, X-763-71-125, NTIS N71-25812. NASA, Goddard Space Flight Center, Greenbelt, Maryland.

Sickles, J. E. (1966). "Experimental Investigation of Pulse Tube Refrigerator." Report No. AFFDL-TR-66-168. System Research Labs., Inc., Dayton, Ohio, 98 pp.

Siegwarth, J. D., and Radebaugh, R. (1971). "Analysis of Heat Exchangers for Dilution Refrigerators." Report No. NBS-R-647, NTIS COM-72-10223. National Bureau of Standards, Boulder, Colorado.

Siegwarth, J. D., and Radebaugh, R. (1972). "The Design of Optimum Heat Exchangers for Dilution Refrigerators." (Final Rept.) Report No. NBS-R-682, NTIS COM-72-50817. National Bureau of Standards, Boulder, Colorado.

Smith, R. V. (1963). "Choking Two-Phase Flow Literature Summary and Idealized Design Solutions for Hydrogen, Nitrogen, Oxygen and Refrigerants 12 and 11." (Tech. Note) Report No. NBS-TN-179. CFSTI PB-190 610. National Bureau of Standards, Boulder, Colorado.

Spencer, R. H. (1975). "Development of a Temperature Controller for a Vuilleumier Cycle Power Cylinder." AFFDL-75-99, WPAFB, Dayton, Ohio.

Spencer, S. E. (1977). "Double-Acting Dynamic Seal." Report No. PAT-APPL-776 039, NTIS AD-D003 863/8ST. D.A.F., Washington, D.C.

Starr, E. F., Jr. (1968). "A Hydraulically Driven Ericsson Cycle Refrigerator." Report No. AFFDL TR-67-191. Systems Research Labs., Inc., Dayton, Ohio, 25 pp.

Stevenson, R., and Marston, P. (1966). "A Cryogenic Magnet System for Quasi-Continuous Operation." (Tech. Rept.) Report No. TR-8. CFSTI AD-641 919. Eaton Elec. Res Lab., Montreal, Canada.

Stochly, C. A., and Nolan, E. R. (1964). "Current Status and Future Trends of Cryogenic Coolers for Electronic Applications." Report No. ECOM-2524, CFSTI AD-610 015. Army Elec. Labs., Fort Monmouth, New Jersey.

Strobridge, T. R. (1968). "Review of the Cryogenics Session—Second Week of the Brookhaven Summer Study on Superconducting Devices and Accelerators." Report No. BNL-50155, CFSTI PB-184221. National Bureau of Standards, Boulder, Colorado.

Strobridge, T. R. (1974). "Cryogenic Refrigerators—An Updated Survey." (Tech. Note)

Report No. NBS-TN-655, NTIS COM-74-50542/1. National Bureau of Standards, Boulder, Colorado.

Strobridge, T. R. (1975). "Feasibility Study of a Multipurpose Refrigerator for a Superconducting Cable Test Facility." (Final Rept.) NTIS PB-244 630/0ST. National Bureau of Standards, Boulder, Colorado.

Strobridge, T. R., and Chelton, D. B. (1966). "Size and Power Requirements of 4.2 K Refrigerators." CFSTI AD-643 327. National Bureau of Standards, Boulder, Colorado.

Strobridge, T. R., Muhlenhaupt, R. C., and Mann, D. B. (1965). "Analysis of a Pulsed Refrigeration System." Report No. 8817. National Bureau of Standards, Boulder, Colorado.

Stump, F. B. (1972). "AN/AAS-18 Infrared Detecting Set Cryogenic Cooler." (Reliability Test Rept.) Report No. TR-MMER/RM-72-107, NTIS AD-903 404/2ST. Hill AFB, Utah.

Sukhia, S. P., Coletta, G. C., and Pellow, H. C. (1967). "Design Considerations for Capillary Heat Pipes at Cryogenic Temperatures." NTIS DRNL-MIT-28. M.I.T., Oak Ridge, Tennessee.

Texas Instrument Equipment Group—Anon. (1974). "Final Engineering Report on the Design and Development of Two Miniature Cryogenic Refrigerators." Contract No. DAAK02-73-0495, Night Vision Lab., Ft. Belvoir, Virginia, NTIS Report No. AD784436.

Timberlake, A. B., and Shilliday, T. S. (1965). "Thermoelectric and Thermomagnetic Cooling, A Literature Survey." Contract No. DA-01-021-AMC-11706(Z), CFSTI AD-478 668. Battelle Memorial Inst., Columbus, Ohio.

Trimmer, D. S. (1974). "Design, Development and Testing of a Cryogenic Temperature Heat Pipe for the Icicle System." (Final Tech. Rept.) Report No. NASA-CR-143710, DTM-73-15, NTIS N75-21559/0ST. Dynatherm Corp., Cockeysville, Maryland.

Triplett, M. J. (1973). "Transient Optimization of a Brayton Cycle Refrigerator." (Final Rept.) Report No. AEDC-TR-73-193, NTIS AD-771 645/9. Arnold Air Force Station, Tennessee.

Uhl, A. J., Jr., and Stidhaun, F. R. (1963). "Air Force Cryogenic Cooling." Technical Memorandum. ASRMD-TM-63-12.

Vance, R. W. (1967). "Predictions for Future of Cryogenic Applications." Report No. TR-0158(3710-01)-1, CFSTI AD-664 560. Aerospace Corp., El Segundo, California.

Vance, R. W. (1967). "Space Applications of Cryogenic Technology." Report No. TR-1001 2710-01-1, SSD-TR-67-65. Aerospace Corp., El Segundo, California, 19 pp.

Vance, R. W. (1974). "Cryogenic Coolers for Space Systems." (Final Rept.) Report No. TR-0074(4127-01)-1, NTIS AD-785 083/7. Aerospace Corp., El Segundo, California.

Vennell, R. R. (1965). "Feasibility Study on EOD Applications for Liquid Nitrogen." (Tech. Memo.) Report No. TM-1667, CFSTI AD-618 484. Picatinny Arsenal, Dover, New Jersey.

Voth, R. O. (1974). "Cryogenic Refrigerators for Shipboard Forward Looking Infrared Applications." (Final Rept.) Report No. NBSIR-74-372, NTIS AD/A-006 037/6ST. National Bureau of Standards, Boulder, Colorado.

Weitzel, D. H., Robbins, R. F., and Ludtke, P. R. (1965). "Elastomeric Seals and Materials at Cryogenic Temperatures." Report No. ML-TDR-64-50, Pt. II, U.S. Air Force, Wright-Patterson Air Force Base, Dayton, Ohio.

White, R. (1976). "Computer Program for Optimizing Three-Stage Vuilleumier Cycle Cryogenic Refrigerators." (Final Rept.) Report No. AFFDL-TR-76-27. DDC ADB 011 379L. In-house, WPAFB, Dayton, Ohio.

White, R. (1976). "Vuilleumier Cycle Cryogenic Refrigeration." (Final Rept.) Report No. AFFDL-TR-76-17, NTIS AD A027 055/3ST. In-house, WPAFB, Dayton, Ohio.

White, R., and Haskins, W. (1976). "Initial Testing of High Capacity VM Cryocoolers." AFFDL-TR-76-160. AD B019 990. In-house, WPAFB, Dayton, Ohio.

Wiebelt, J. A. (1971). "Analytical Study to Determine the Feasibility of a Radiantly Cooled Cryostat for Spacecraft." AFFDL-TR-69-88. AD 702 940. (February), WPAFB, Dayton, Ohio.

Wise, J. (1975). "Integral Heater Thermal Energy Storage Device." Report No. PAT-APPL-602 039, NTIS AD-D001 447/2ST.

Wittig, M. (1976). "Low-Temperature Heat Exchangers." (Final Rept.) Report No. BMFT-FB-T-76-07, NTIS N77-11350/4ST. Leybold-Heraeus G.M.B.H., Cologne, West Germany.

Wolf, S. A., Nisenhoff, M., and Davis, J. R. (1974). "Superconducting ELF Magnetic Field Sensors for Submarine Communications." Report No. NRL 7720 (May) 14 pp. Naval Research Lab., Washington, D.C.

Woodard, R. S., Welch, P. H., and Jansson, R. M. (1978). "Manufacturing Methods and Technology for the Establishment of Production Techniques for a Split-Cycle Stirling Cryogenic Cooler." Report No. 15.181. 3rd Quarterly Progress Report April–June, 1978, Contract No. DAAB07-77-C-0631. U.S. Army Elect. Res. and Dev. Command, Fort Monmouth, New Jersey.

Wyatt, C. L. (1974). "Rocketborne Near Infrared Radiometer System." Model NR-1 (Final Rept.) Report No. AFCRL-TR-74-0059, NTIS AD-778 105/7. Utah State Univ., Logan Electro-Dynamics Lab.

Yoshikawa, D. K. (1970). "75°K Miniature Vuilleumier Cryogenic Engine—Final Report for Task 1—Preliminary Design." Report 70-6854, prepared for Goddard Space Flight Center, Greenbelt, Maryland.

Zenner, G. H. (1963). "Cycles and Equipment for Producing Low Temperatures." Report No. TDR-169(3711-01)TN-1. Aerospace Corp., Physical Res Lab., El Segundo, California.

Zheleznov, N., Gulovanov, Y., Pokrouskiy, A., and Knovalov, B. (1975). "Its-K Infrared Cryogenic Telescope-Spectrometer Aboard Salyut-4." Report No. NASA-TT-F-16355, NTIS N75-23936/8ST. Joint Publications Research Service, Arlington, Virginia.

Zierman, G. A. (1969). "Feasibility Study and Development Design of a Passive Radiative Cooler for Infrared Detectors." AFFDL-TR-69-22. AD 871 914. (December), WPAFB, Dayton, Ohio.

Zimmerman, J. E., and Flynn, T. M. (1978). "Applications of Closed-Cycle Cryocoolers to Small Superconducting Devices." NBS Spec. Publ. 508. National Bureau of Standards/U.S. Department of Commerce. Washington, D.C.

RELEVANT CONFERENCES

Anon. (1974). "Meeting on Technology Arising from High-Energy Physics." European Organization for Nuclear Research, Geneva, Report No. CERN 74-9, Vol. 2 (June), pp. 153–202.

Kitchens, T. A., Jr., Klein, B. M., and Gubser, D. U. (1975). "The International Conference on Low-Temperature Physics (14th)." Helsinki University of Technology, Otaniemi, Finland, 14–20 August, 1975. (Conference Rept.) Report No. ONRL-C-22-75, NTIS AD-A022 709/0ST. Office of Naval Research, London, England.

PATENTS

Acord, T. T. (1976). "Solenoid-Controlled Cold Head for a Cryogenic Cooler." U.S. Patent 3,991,586.

Acord, T. T. (1976). "Modified Rotary Compressor Yielding Sinusoidal Pressure Wave Outputs." U.S. Patent 3,941,522.

Admiral, P. S. (1977). "Device for Liquefying Gases." U.S. Patent 4,055,961.

Andres, K., Bucher, E., and Wernick, J. H. (1975). "AUIN(2) as a Cryogenic Thermometer or Refrigerant." U.S. Patent 3,865,557.

Andres, K., and Schmidt, P. H. (1977). "PRNI(5) As A Cryogenic Refrigerant." U.S. Patent 4,028,905.

Anon. (1975). "Helium-3–Helium-4 Dilution Refrigerator and a Method for Starting the Same." British Patent 1,382,376.

Aoki, I., and Kitsukawa, Y. (1974). "Absorption-Multicomponent Cascade Refrigeration for Multi-Level Cooling of Gas Mixtures." U.S. Patent 3,817,046.

Appleton, A. D., and Tinlin, F. (1974). "Superconducting Dynamo-Electric Machines." U.S. Patent 3,835,663.

Asztalos, S., Kneuer, R., Stephan, A., and Glatthaar, R. (1975). "Method of and Apparatus for the Cooling of an Object." U.S. Patent 3,882,687.

Baldwin, D. M. (1974). "Probe for Cryomagnetic Resonance." U.S. Patent 3,805,883.

Bamberg, W. H. (1977). "Lost-Motion Refrigeration Drive System." U.S. Patent 4,036,027.

Bamberg, W. H. (1978). "Lost-Motion Refrigeration Drive System." U.S. Patent 4,078,389.

Bamberg, W. H., and O'Neil, J. A. (1974). "Temperature-Staged Cryogenic Apparatus of Stepped Configuration with Adjustable Piston Stroke." U.S. Patent 3,802,211.

Banike, R. A. (1974). "Refrigerating Apparatus." U.S. Patent 3,818,719.

Barger, J. P. (1977). "Cryosurgical Probe Tip." U.S. Patent 4,029,102.

Barkey, J., Zuckerman, H., and Pundak, N. (1975). "Freezing Apparatus Particularly Useful for Freezing Spermatozoa." U.S. Patent 3,893,308.

Baumgardner, A. R., Johnston, R. P., Martini, W. R., and White, M. A. (1969). "Stirling Cycle Machine with Self-Oscillating Regenerator." U.S. Patent 3,484,616.

Becker, R. (1963). "Process and Apparatus for Refrigeration by Work-Producing Expansion." U.S. Patent 3,091,941.

Berry, R. L. (1968). "Low Temperature Refrigerating Arrangement." U.S. Patent 3,365,896.

Berry, R. L. (1975). "Reduction in Cooldown Time for Cryogenic Refrigerator." U.S. Patent 3,913,339.

Berry, R. L., and Cowans, K. W. (1967). "Cryogenic Refrigerator." U.S. Patent 3,315,490.

Berry, R. L., and Dehne, A. G. (1977). "Vuilleumier Refrigerator with Separate Pneumatically Operated Cold Displacer." U.S. Patent 4,024,727.

Brooks, F. P. (1965). "Low Temperature Refrigerator." U.S. Patent 3,222,877.

Brown, G. V. (1978). "Magnetic Heat Pumping." U.S. Patent 4,069,028.

Browning, C. W. (1977). "Hydraulically Actuated Split Stirling Cycle Refrigerator." U.S. Patent 4,019,335.

Bruno, R. P., and Naugler, A. W. (1974). "Cooling Apparatus for Infrared Detectors." U.S. Patent 3,836,779.

Buller, J. S. (1976). "Quick Cooling Cryostat with Valve Utilizing Simon Cooling and Joule–Thomson Expansion." U.S. Patent 3,952,543.

Burnier, P. H., and Bricout, D. D. (1974). "Method and Apparatus for Supercooling of Electrical Devices." U.S. Patent 3,795,116.

Bush, V. (1938). "Apparatus for Transferring Heat." U.S. Patent 2,127,286.

Bush, V. (1939). "Apparatus for Compressing Gases." U.S. Patent 2,157,229.

Byrd, P. N. (1967). "Thermal Transfer Arrangement for Cryogenic Device Cooling and Method of Operation." U.S. Patent 3,302,429.

Campbell, D. N. (1974). "Cryogenic Cooling Apparatus." U.S. Patent 3,818,720.

Campbell, D. N. (1976). "Joule–Thomson Liquefier Utilizing the Leidenfrost Principle." U.S. Patent 3,990,265.

Caren, R. P., and Coston, R. M. (1969). "Dual Solid Cryogens for Spacecraft Refrigeration." U.S. Patent 3,545,226.

Chamberlain, W. C., and Sneller, J. A., Sr. (1978). "Cryogenic Freezer." U.S. Patent 4,078,394.

Chellis, F. F. (1965). "Pneumatically-Operated Refrigerator with Self-Regulating Valve." U.S. Patent 3,188,821.

Chellis, F. F. (1967). "Closed Cycle Cryogenic Refrigerator." U.S. Patent 3,333,433.

Chellis, F. F. (1967). "Fluid Actuated Cryogenic Refrigerator." U.S. Patent 3,321,926.

Chellis, F. F. (1971). "Rotary-Valved Cryogenic Apparatus." U.S. Patent 3,625,015.

Chellis, F. F. (1978). "Refrigeration System With Magnetic Linkage." U.S. Patent 4,118,943.

Chellis, F. F., and Hogan, W. H. (1965). "Cryogenic Refrigeration Apparatus Operating on an Expansible Fluid and Embodying a Regenerator." U.S. Patent 3,218,815.

Chellis, F. F., and O'Neil, J. A. (1971). "Cryogenic Heat Station and Apparatus Incorporating Same." U.S. Patent 3,600,903.

Chovet, P., Rollin, C., Galasso, H., and Prost, R. (1974). "Method of Regulation of the Frigorific Power of a Joule–Thomson Refrigerator and a Refrigerator Utilizing Said Method." U.S. Patent 3,827,252.

Clark, A. C. (1973). "Magnetic Refrigeration." U.S. Patent 3,841,107.

Claudet, G., Koubeau, P., and Verdier, J. (1976). "Method for the Production of Superfluid Helium under Pressure at Very Low Temperature and an Apparatus for Carrying out Said Method." U.S. Patent 3,992,893.

Collins, S. C. (1964). "Expansion Engine." U.S. Patent 3,131,547.

Collins, S. C. (1968). "Method and Apparatus for Continuously Supplying Refrigeration below 4.2°K." U.S. Patent 3,415,077.

Collins, S. C. (1969). "Expansion Engine for Cryogenic Refrigerators and Liquefiers and Apparatus Embodying the Same." U.S. Patent 3,438,220.

Collins, S. C. (1975). "Apparatus for Liquefying a Cryogen by Isentropic Expansion." U.S. Patent 3,864,926.

Cowans, K. W. (1967). "Self-Contained Cryogenic Refrigerator." U.S. Patent 3,334,491.

Cowans, K. W. (1968). "Heat Powered Engine." U.S. Patent 3,379,026.

Cowans, K. W. (1969). "Cryogenic Refrigerator Adapted to Miniaturization." U.S. Patent 3,423,948.

Cowans, K. W. (1969). "Cryogenic Refrigerator Arrangement." U.S. Patent 3,469,409.

Cowans, K. W., and Glode, J. B. (1966). "Refrigerant Regeneration and Purification as Applied to Cryogenic Closed-Cycle Systems." U.S. Patent 3,282,064.

Cramer, R. L., and Mientus, J. A. (1977). "Self-Regulating Cryostat." U.S. Patent 4,002,039.

Crandell, W. H., Lisenbee, W. F., and Nelson, K. E. (1976). "Cryosurgical Probe." U.S. Patent 3,951,152.

Crawford, J. W., and White, R. (1972). "Solid Cryogen Heat Transfer Apparatus." U.S. Patent 3,745,785.

Cytryn, E. P. (1974). "Cryogenic Absorption Cycles." U.S. Patent 3,854,301.

Damsz, G. J. (1965). "Cooldown Time of Installation Incorporating Stirling Cycle Refrigerator." U.S. Patent 3,220,200.

Daniels, A. (1974). "Pulse Tube Refrigerator." U.S. Patent 3,817,044.

Daniels, A. (1975). "Vuilleumier Refrigerator." U.S. Patent 3,862,546.

Darredeau, B., Cappiello, P., and Le Bihan, H. (1976). "Method and Apparatus for Cooling a Gas." U.S. Patent 3,945,214.

Daunt, J. G. (1965). "Remotely-Located Cold Head for Stirling Cycle Engine." U.S. Patent 3,188,822.

Daunt, J. G. (1968). "Cyclic Desorption Refrigerator." U.S. Patent 3,397,549.

Daunt, J. G. (1968). "Regenerator Matrix Systems for Low-Temperature Engines." U.S. Patent 3,397,738.

De Baufre, W. D., and Filipi, T. A. (1941). "Expansion Engine." U.S. Patent 2,239,883.

Dehne, A. G. (1970). "Hydraulically Driven Cryogenic Refrigerator." U.S. Patent 3,530,681.

Dehne, A. G. (1972). "Cryogenic Refrigerator Cycle." U.S. Patent 3,640,082.

Dehne, A. G. (1974). "Double Acting Expander Engine and Cryostat." U.S. Patent 3,788,088.

Dehne, A. G. (1974). "Decontamination Method and Apparatus for Cryogenic Refrigerators." U.S. Patent 3,793,846.

Dehne, A. G. (1974). "Double-Acting Expander Engine and Cryostat." U.S. Patent 3,834,172.

De Lange, L. (1956). "Cooler Construction in a Hot-Gas Engine." U.S. Patent 2,764,879.

Dix, R. M., Tobias, S. F., and Whicker, S. L. (1977). "Modular Magnetically-Coupled Drive for a Cryogenic Refrigerator." U.S. Patent 4,044,567.

Doll, R., and Eder, F. X. (1966). "Piston-Type Cryogenic Apparatus." U.S. Patent 3,285,142.

Doody, R. D. (1975). "Tubular Regenerator for a Cryogenic Refrigerator." U.S. Patent 3,933,000.

Doody, R. D. (1975). "Tubular Regenerator for a Cryogenic Refrigerator." Report No. PAT-APPL-547 664, NTIS AD-D000 675/9ST.

Doody, R. D. (1975). "Cold Cylinder Assembly for Cryogenic Refrigerator." Report No. PAT-APPL-561 767, NTIS AD-D000 780/7ST.

Doody, R. D. (1975). "Cold Cylinder Assembly for Cryogenic Refrigerator." U.S. Patent 3,969,907.

Doody, R. D. (1976). "Tubular Regenerator for a Cryogenic Refrigerator." U.S. Patent No. 3,933,000.

Doody, R. D. (1976). "Cold Cylinder Assembly for Cryogenic Refrigerator." U.S. Patent 3,969,907.

Dros, A. A. (1963). "Multi-Stage Refrigerating Arrangement." U.S. Patent 3,091,092.

Dros, A. A. (1963). "Gas Refrigerator." U.S. Patent 3,101,597.

Dros, A. A. (1967). "Cold-Gas Refrigerator." U.S. Patent 3,323,314.

Durenec, P. (1978). "Balanced Compressor." U.S. Patent 4,092,829.

Durenec, P. (1978). "Split-Phase Cooler with Expansion Piston Machine Enhancer." U.S. Patent 4,092,833.

Gifford, W. E., and McMahon, H. O. (1962). "Pneumatic Expansion Method and Apparatus." U.S. Patent 3,045,436.

Gifford, W. E. (1964). "Entropy Balancing Method of Refrigeration and Apparatus Therefor." U.S. Patent 3,138,004.

Gifford, W. E. (1964). "Gas Balancing Refrigeration Method." U.S. Patent 3,119,237.

Gifford, W. E. (1967). "Method and Apparatus for Refrigeration Utilizing Stirling Cycle Type of Operation." U.S. Patent 3,312,072.

Gifford, W. E. (1974). "Gas Separation and Purification Utilizing Time Sequenced Flow through a Pair of Regenerators." U.S. Patent 3,851,493.

Gifford, W. E. (1977). "Air Distillation Apparatus Comprising Regenerator Means for Producing Oxygen." U.S. Patent 4,017,284.

Guiller, J. (1978). "Apparatus for Freezing Small Drops of Liquid." U.S. Patent 4,073,158.

Hanson, C. M. (1977). "Vuilleumier Cycle Thermal Compressor Air Conditioner System." U.S. Patent 4,060,996.

Hanson, C. M. (1978). "Automotive Accessory Engine." U.S. Patent 4,070,860.

Hanson, C. M. (1978). "Dual-Displacer Two-Stage Split Cycle Cooler." U.S. Patent 4,090,859.

Herrington, R. E., and Taylor, C. O. (1977). "Adjustable Joule–Thomson Cryogenic Cooler with Downstream Thermal Compensation." U.S. Patent 4,028,907.

Heuchling, R. P., and Fowle, A. A. (1965). "Cryogenic Refrigerator Operations on the Stirling Cycle." U.S. Patent 3,220,201.

Higa, W. H. (1969). "Refrigeration Apparatus." U.S. Patent 3,421,331.

Hoffman, T., Hogan, W. H., and Stuart, R. W. (1964). "Refrigeration Apparatus." U.S. Patent 3,148,512.

Hogan, W. H. (1963). "Refrigeration Apparatus and Method." U.S. Patent 3,115,015.

Hogan, W. H. (1963). "Refrigeration Method and Apparatus." U.S. Patent 3,115,016.

Hogan, W. H. (1964). "Closed-Cycle Cryogenic Refrigerator and Apparatus Embodying Same." U.S. Patent 3,151,446.

Hogan, W. H. (1965). "Fluid Expansion Refrigeration Apparatus." U.S. Patent 3,188,820.

Hogan, W. H. (1965). "Refrigeration Method and Apparatus." U.S. Patent 3,188,819.

Hood, C. B., Jr. (1968). "Cryogenic Refrigerator." U.S. Patent 3,371,498.

Horn, S. B. (1973). "Pneumatic Stirling Cycle Cooler with Non-Contaminating Compressor." U.S. Patent 3,765,187.

Horn, S. B. (1976). "Low-Void-Volume Regenerator for Vuilleumier Cryogenic Cooler." U.S. Patent 3,853,437.

Horn, S. B., and Walters, B. T. (1973). "Split-Cycle Cryogenic Cooler with Rotary Compressor." U.S. Patent 3,853,437.

Jakob, F. (1968). "Process for Heat Exchange and Cleansing of Gases in Periodically Reversible Regenerators." U.S. Patent 3,375,672.

Jonkers, C. O., and De Lange, L. (1956). "Freezer and Cooler Chamber Construction of a Cold-Gas Refrigerator." U.S. Patent 2,770,109.

Jonkers, C. O., and Köhler, J. W. L. (1959). "Gas Liquefying Apparatus." U.S. Patent 2,872,789.

Jonkers, C. O., and Köhler, J. W. L. (1959). "System Comprising a Cold-Gas Refrigerator and a Heat Exchanger." U.S. Patent 2,897,655.

Kataoka, H., Takayama, Y., and Iwata, Y. (1974). "Cooling Method for Transmission Cables." U.S. Patent 3,800,062.

Kelly, D. A. (1970). "Rotary Stirling Cycle Refrigerating System." U.S. Patent 3,527,269.

Kirk, A. C. (1862). "Refrigerating Apparatus." British Patent 1,218.

Klee, D. J., and Howells, R. A. (1975). "Cryogenic Freezer with Variable Speed Gas Control System." U.S. Patent 3,892,104.

Köhler, J. W. L. (1956). "Refrigerator Gas Liquefication Device." U.S. Patent 2,734,354.

Köhler, J. W. L. (1956). "Apparatus for Liquefying Air." U.S. Patent 2,764,877.

Köhler, J. W. L. (1956). "Refrigerator Gas Liquefier." U.S. Patent 2,745,262.

Köhler, J. W. L. (1957). "Hot-Gas Reciprocating Engine for Refrigerating." U.S. Patent 2,784,570.

Köhler, J. W. L. (1957). "Method of Separating Gas Mixtures Into Fractions of Different Volatility." U.S. Patent 2,808,709.

Köhler, J. W. L. (1958). "Cold-Gas Refrigerating Machine and Method." U.S. Patent 2,856,756.

Köhler, J. W. L. (1959). "Gas Fractionating System Including a Vapor Lift Pump." U.S. Patent 2,889,686.

Köhler, J. W. L. (1968). "Device for Producing Cold with Cold Loss Prevention Means." U.S. Patent 3,383,862.

Köhler, J. W. L., and Bloem, A. T. (1957). "Cold-Gas Refrigerator." U.S. Patent 2,781,647.

Köhler, J. W. L., De Lange, L., and Schalkwijk, W. F. (1957). "Regenerator Construction of a Cold-Gas Refrigerator." U.S. Patent 2,775,875.

Köhler, J. W. L., Dros, A. A., Geuns, J. R., and Prast, G. (1967). "Device for Producing Cold at Low Temperatures and Cold-Gas Refrigerator Particularly Suitable for Use in Such a Device." U.S. Patent 3,327,486.

Köhler, J. W. L., Dros, A. A., and Ijer, J. A. L. (1962). "Apparatus for Condensing Shipboard Cargoes of Vaporizable Liquid." U.S. Patent 3,033,003.

Köhler, J. W. L., and Fokker, H. (1959). "Cold-Gas Refrigerating Apparatus." U.S. Patent 2,907,175.

Köhler, J. W. L., Rinia, H., and Piet, J. G. (1962). "Thermal Regenerator." U.S. Patent 3,045,982.

Köhler, J. W. L., and Schalkwijk, W. F. (1956). "Cold-Gas Refrigerator." U.S. Patent 2,750,765.

Korsgren, T. Y., Sr. (1965). "Heat Engine." U.S. Patent 3,183,662.

Kroebig, H. L. (1975). "Plasma Deposited Rider Rings for Hot Displacer." U.S. Patent 3,981,155.

Kroebig, H. L. (1978). "Fusible Heat Sink for a Cryogenic Refrigerator." U.S. Patent 4,079,595.

Kugler, S., and Clarke, M. E. (1975). "Superconducting Device." British Patent 1,395,707.

La Fleur, J. K. (1965). "Power Refrigeration System." U.S. Patent 3,194,026.

Lagodmos, G. P. (1974). "Thermal Coupling Device for Cryogenic Refrigeration." U.S. Patent 3,807,188.

Leger, P. A. M. (1959). "Refrigeration Liquefaction Device." U.S. Patent 3,487,424.

Leo, B. S. (1973). "Magnetically Driven Cryogenic Vuilleumier Refrigerator." U.S. Patent 3,774,405.

Leo, B. S. (1975). "Free-Piston Cryogenic Refrigerator with Phase Angle Control." U.S. Patent 3,877,239.

Leo, B. S. (1975). "Vuilleumier Refrigerator Hot Cylinder Burner Head." U.S. Patent 3,892,102.

Longsworth, R. C. (1975). "Vibration-Free Refrigeration Transfer." U.S. Patent 3,894,403.

Maeda, H., and Matsuda, T. (1975). "Refrigerating Apparatus." U.S. Patent 3,889,488.

Malaker, S. F. (1973). "Transportable Refrigeration System." U.S. Patent 3,719,051.

Malaker, S. F., and Daunt, J. G. (1963). "Miniature Cryogenic Engine." U.S. Patent 3,074,244.

Malaker, S. F., and Daunt, J. G. (1964). "Cascade Cryogenic System." U.S. Patent 3,118,285.

Malaker, S. F., and Daunt, J. G. (1964). "Closed-Cycle Cryogenic System." U.S. Patent 3,128,605.

Malaker, S. F., and Daunt, J. G. (1964). "Multi-Stage Cryogenic Engine." U.S. Patent 3,147,600.

Malaker, S. F., and Daunt, J. G. (1965). "Compensating-Pressure Piston and Cylinders for Gas Compressors and Expanders." U.S. Patent 3,204,864.

Markum, A. D. (1976). "Cryostat Control." U.S. Patent 3,993,003.

Martin, P. S., and Moody, B. M. (1974). "Refrigeration Apparatus." U.S. Patent 3,802,212.

McMahon, H. O., and Gifford, W. E. (1959). "Fluid Expansion Refrigeration Method and Apparatus." U.S. Patent 2,906,101.

Melchior, A. (1962). "Regeneration System for Gas Fractionating Devices and the Like." U.S. Patent 3,056,269.

Mijnheer, A., and De Jonge, A. K. (1977). "Refrigerator." U.S. Patent 4,019,336.

Miller, C. G., and Stephens, J. B. (1976). "Cryostat System for Temperatures on the Order of 2 K or Less." U.S. Patent 3,983,714.

Morain, W. A., Ostborg, J. L., and White, D. (1966). "Cryogenic Expansion Engine." U.S. Patent 3,274,781.

Muenger, J. P., and Alexander, O. L. (1970). "Self-Cleaning Regenerators for Cryogenic Systems." U.S. Patent 3,490,245.

Mulder, J. (1976). "Cold-Gas Refrigerator." U.S. Patent 3,991,585.

Nesbitt, L. B. (1966). "Low-Temperature Thermal Regenerator." U.S. Patent 3,262,277.

Parel, J. M. (1974). "Cryogenic Apparatus." U.S. Patent 3,823,575.

Payne, W. H., and Henry, R. E. (1968). "Expansion Engine." U.S. Patent 3,394,633.

Pomerroy, B. D., and Kosky, P. G. (1977). "Flow Control Device for Superconductive Rotor Refrigerant." U.S. Patent 4,048,529.

Post, A. H., Jr., and Streeter, M. H. (1965). "Cryogenic Transport Tube Incorporating Liquefaction Apparatus." Canadian Patent 718,618.

Prast, G. (1968). "Arrangement for Producing Cold at Very Low Temperatures." U.S. Patent 3,372,554.

Prast, G. (1968). "Fluid Cooling Employing Plural Cold Producing Machines." U.S. Patent 3,400,544.

Promish, D. I. (1963). "Photodetecting Apparatus Having Cryogenic Cooling and Flushing Means." U.S. Patent 3,258,602.

Prost, R., and Broche, J. (1976). "Method and Apparatus for Generating Low Temperatures." U.S. Patent 3,986,338.

Quack, H. (1977). "Refrigerating Plant Using Helium as a Refrigerant." U.S. Patent 4,048,814.

Raimondi, P. K. (1975). "Variable Pneumatic Volume for Cryogenic Coolers." U.S. Patent 3,906,739.

Rietdijk, J. A. (1969). "Device for Producing Cold and/or Liquefying Gases." U.S. Patent 3,427,817.

Rietdijk, J. A. (1969). "Apparatus and Ejector for Producing Cold." U.S. Patent 3,434,298.

Rietdijk, J. A. (1969). "Apparatus and Ejector for Producing Cold." U.S. Patent 3,442,093.

Rietdijk, J. A. (1969). "Cold Producing Systems." U.S. Patent 3,447,339.

Rietdijk, J. A. (1969). "Cryogenic Apparatus for Producing Cold." U.S. Patent 3,456,456.

Rietdijk, J. A. (1969). "Systems for Producing Cold and Ejectors in Such Systems." U.S. Patent 3,464,230.

Rinia, H., Dros, A. A., Van Weenen, F. L., et al. (1958). "Cold-Gas Refrigerator Control System." U.S. Patent 2,824,430.

Rinia, H., and Meijer, R. J. (1963). "Cold-Gas Refrigerator." U.S. Patent 3,101,596.

Robinson, R. L., and Le Nguyen, Van. (1966). "Cryogenic Closed-Cycle Power System." U.S. Patent 3,232,050.

Roozendaal, K., and Hellingman, R. (1958). "Refrigerating Device Comprising a Gas Refrigerator." U.S. Patent 2,836,964.

Roubeau, P. (1975). "Method for Reducing the Consumption of a Cryostat and a Device for Carrying out Said Method." U.S. Patent 3,892,106.

Sarsten, J. A. (1977). "Fractional Condensation of an NG Feed with Two Independent Refrigeration Cycles." U.S. Patent 4,057,972.

Schalkwijk, W. F., De Lange, L., and Koopmans, A. (1957). "Regenerator Construction of a Cold-Gas Refrigerator." U.S. Patent 2,775,876.

Scheibel, E. G. (1977). "Cascade Refrigeration System." U.S. Patent 4,028,078.

Schmidt, F. (1976). "Cooling Apparatus for an Electric Cable." U.S. Patent 3,946,141.

Schulze, R. R., and Ladd, F. R. (1978). "Cryogenic Probe." U.S. Patent 4,074,717.

Severijns, A. P., and Staas, F. A. (1976). "Refrigeration Method and Apparatus by Converting (4)He to a Superfluid." U.S. Patent 3,978,682.

Severijns, A. P., and Aarts, F. H. E. (1977). "(3)He–(4)He Dilution Refrigerating Machine." U.S. Patent 4,047,394.

Shaievitz, S. (1966). "Apparatus for Liquefying Helium." U.S. Patent 3,233,418.

Simpson, A. U. (1970). "Gas Engine-Refrigerator." U.S. Patent 3,523,427.

Solin, S. A., and Doehler, J. (1973). "Recirculating Liquid-Nitrogen-Coolant System for Solid-State Lasers." U.S. Patent 3,851,274.

Solvay, E. (1887). German Patent No. 39,280.

Spies, A., Stephan, A., and Sellmaier, A. (1974). "Closed Refrigerant Cycle for the Liquefaction of Low-Boiling Gases." U.S. Patent 3,828,564.

Staas, F. A., and Severijns, A. P. (1974). "Device for Transporting Heat from a Lower to a Higher Temperature Level." U.S. Patent 3,835,662.

Steyert, W. A., Jr., and Rosenblum, S. S. (1977). "Continuous, Noncyclic Magnetic Refrigerator and Method." U.S. Patent 4,033,734.

Stirling, R. (1817). "Improvements for Diminishing the Consumption of Fuel and in Particular, An Engine Capable of Being Applied to the Moving of Machinery on a Principle Entirely New." British Patent No. 4081.

Strong, W. K. (1978). "Cryogenic Freezer," U.S. Patent 4,077,226.

Swearingen, J. S. (1975). "Process and Apparatus for Low-Temperature Refrigeration." U.S. Patent 3,889,485.

Taconis, K. W. (1951). "Process of, and Apparatus for Heat Pumping." U.S. Patent 2,567,454.

Tam, W. A. (1967). "Refrigerating Machine." U.S. Patent 3,358,459.

Tanner, A. B., Senftle, F. E., and Moxham, R. M. (1971). "Meeting Cryogen Cooling for Radiation Logging Probe." U.S. Patent 3,702,932.

Taylor, C. E., Hunt, A. L., and Omohundro, J. E. (1964). "Process and Device for Cryogenic Adsorption Pumping." U.S. Patent 3,144,200.

Tyree, L., Jr. (1974). "System for Cooling Material Using CO(2) Snow." U.S. Patent 3,815,377,

Vander Arend, P. C. (1974). "Cryogenic Helium Refrigeration System." U.S. Patent 3,850,004.

Vander Arend, P. C., and Fowler, W. B. (1977). "Superconducting Magnet Cooling System." U.S. Patent 4,048,437.

Van Geuns, J. R. (1967). "Device for Producing Cold at Low Temperatures and Compression Devices Suitable for Use in Said Devices." U.S. Patent 3,318,101.

Van Geuns, J. R., and Prast, G. (1965). "Method of Absorbing Thermal Energy at Low Temperatures and Apparatus for Carrying out Such Methods." U.S. Patent 3,214,924.

Van Heeckeren, W. J. (1949). "Hot-Air Engine Actuated Refrigeration Apparatus." U.S. Patent 2,484,392.

Von Bredon, H., and Vogelhuber, W. W. (1971). "Cryogenic Expansion Engine." U.S. Patent 3,574,998.

Vuilleumier, R. (1918). "Method and Apparatus for Inducing Heat Changes." U.S. Patent 1,275,507.

Webster, R. J., and Garrett, M. E. (1969). "Pulse Tube Refrigeration Process." U.S. Patent 3,431,746.

Whicker, S. L., and Taylor, C. O. (1975). "Cryogenic Cooler Off-Axis Drive Mechanism for an Infrared Receiver." U.S. Patent 3,889,119.

Wiebe, E. R. (1971). "Improved Helium Refrigerator and Method for Decontaminating the Refrigerator." U.S. Patent APPL-SN-112999, NTIS N72-11567.

Wiebe, E. R. (1975). "Helium Refrigerator." U.S. Patent 3,914,950.

Wiebe, E. R. (1976). "Multistation Refrigeration System." Report No. PAT-APPI-712 981, NTIS N77-15219/7ST.

Wiebe, E. R. (1978). "Multistation Refrigeration System." U.S. Patent 4,077,231.

Zeist, K., Geist, J. M., and Lashmet, P. K. (1965). "Methods for Producing Low-Temperature Refrigeration." U.S. Patent 3,199,304.

Zimmerman, J. E. (1979). "Cryogenic Refrigeration System." U.S. Patent 4,143,520.

Glossary of Terms for Cryocoolers and List of Organizations

Adiabatic Compression and Expansion: Thermodynamic process of volume, pressure, temperature change, and also adiabatic process change that occurs without heat transfer to or from the system.

Aftercooler: Water- or air-cooled heat exchanger used to cool compressed fluid leaving a compressor.

Axial Compressor: A type of fluid compressor with a rotor carrying blades arranged radially on a drum or disks; corresponding blades are arranged on the stator. The fluid flows through the compressor in the axial direction increasing in pressure and density during the compression process.

Beale Free-Piston Stirling Engine: A type of Stirling engine in which the piston and displacer move entirely under the action of fluidic forces. There are no connecting mechanisms between the piston and displacer. The load is direct coupled to the piston.

Brayton Cycle: See Joule–Brayton cycle.

Bucket Brigade Loss: Finkelstein's term for shuttle heat transfer.

Centrifugal Compressor: A type of fluid compressor with a high-speed rotating impeller which accelerates fluid to a high centrifugal velocity, the energy of which is subsequently transformed to pressure energy in a volute casing.

Claude, Georges: French scientist who conceived the combination of expansion engine and Joule–Thomson valve for gas liquefaction.

Claude Cycle: An idealized thermodynamic cycle in which a fraction of the high-pressure fluid is expanded in an expansion engine and the remainder in a Joule–Thomson valve. The cold, low-pressure fluid from the engine cools the remaining high-pressure fluid passing to the JT valve. All the fluid is compressed adiabatically at ambient temperature and is cooled (at high pressure) prior to expansion in contraflow recuperative heat exchangers by the low-pressure return stream.

Clearance: The amount by which a cylinder is greater in diameter than a piston or a bearing than the shaft rotating in it.

Clearance Space: (a) The minimum volume of the compression and expansion spaces of Stirling or Vuilleumier engines. (b) The small volume in the cylinder above the piston at the end of compression (in a compressor) or at the start of admission in an expander.

Coefficient of Performance (COP): The ratio of heat transferred to input work. For refrigeration the COP = heat (refrigeration effect)/work supplied. For a heat pump the COP = heat rejected/work supplied (i.e., the inverse of thermal efficiency).

Coldfinger: The long, thin cylinder of a cryocooler containing a displacer, or regenerative displacer. Refrigeration is generated at the end of the coldfinger. Also cold sting.

Collins, Samuel: American cryogenic engineer of the 20th century best known for development of helium liquefiers working on the Claude cycle.

Collins Cryostat: A helium liquefier working on the Claude cycle.

Compound Working Fluid: The working fluid of a Stirling engine that consists of two or more components and which may exist as a liquid gas, vapor, or dissociated elements.

Compression Space: The variable volume of the working space in a Stirling engine where the working fluid is principally concentrated when the total system volume is decreased, the pressure rises, and heat is rejected to the cooling medium. In a prime mover, the compression space is cooler than the expansion space. In a refrigerator or heat pump, the compression space is warmer than the expansion space.

Compressor: A machine used to elevate the pressure of the fluid; may be a reciprocating, rotary, or screw compressor.

Constant-Enthalpy Process: Thermodynamic compression or expansion at constant enthalpy—e.g., a Joule–Thomson expansion.

Constant-Entropy Process: Thermodynamic compression or expansion process that occurs reversibly with no transfer of heat and hence no change in entropy.

Constant-Pressure Process: Thermodynamic heating or cooling process that occurs at constant pressure. This may or may not be regenerative.

Constant-Temperature Process: Thermodynamic heating or cooling process that occurs at constant temperature. This may or may not be regenerative.

Constant-Volume Process: Thermodynamic heating or cooling process that occurs at constant volume. This may or may not be regenerative.

Cooler: The heat exchanger provided to facilitate the transfer of thermal energy *from* the working fluid to the cooling medium, water, air, or some other fluid.

Crank Drive: One form of kinematic drive consisting of a crank and connecting rod used to convert reciprocating to rotary motion and to convey power between pistons and drive shaft.

Cryocooler: Any device, system, or ensemble capable of generating refrigeration at cryogenic temperatures, i.e., less than 120 K.

Cryogenerator: A cryocooler capable of achieving refrigeration at cryogenic temperatures (less than 120 K).

Dead Volume Ratio: That part of the total working space not included in the variable volumes of the expansion and compression spaces, expressed in terms of the variable volume of the expansion space.

Direct Heating: A system in which the hot products of combustion pass directly over the heater tubes in which the working fluid flows, so that heat is transferred directly from the combustion products to the heater tube walls and hence to the working fluid.

Discontinuous Piston Motion: The nonsinusoidal motion of the piston and displacers required to achieve the necessary volume variations of the idealized thermodynamic cycles.

Displacer: A lightweight structural reciprocating element in a Stirling engine characterized by a large temperature difference but a negligible pressure difference across the upper and lower transverse faces.

Double-Acting Engines: A family of Stirling engines having a single reciprocating element per thermodynamic system. There is a minimum number of two cylinders but no maximum number.

Dual-Pressure Cycle: A thermodynamic cycle with two or more stages of expansion in engines or JT valves. Many variations are possible involving several stages of expansion and intermediate pressure separation of saturated liquid and vapor.

Duplex Stirling Engine: Two Stirling engines arranged so that one, operating as a prime mover, receives heat at a high temperature and produces work to drive the second Stirling engine, acting as a cooling engine, refrigerator, or heat pump.

Ericsson Cycle: An idealized thermodynamic cycle consisting of isothermal compression and expansion processes at different temperatures bounded by constant-pressure regenerative processes.

Exhaust Gas Heat Exchanger: See Regenerative Cycle.

Expansion Space: The variable volume of the working space in a Stirling engine where the working fluid is principally concentrated when the total system volume is increased, the pressure falls, and heat is absorbed. In a prime mover, the expansion space is hotter than the compression space. In a refrigerator or heat pump the expansion space is cooler than the compression space.

Finkelstein Adiabatic Cycle: An idealized thermodynamic cycle for Stirling engines with no heat transfer in the compression and expansion spaces and infinite rates of heat transfer in the heat exchangers.

Free-Displacer Engines: A form of Ericsson regenerative engine (Bush type) where the displacer moves under the action of fluidic forces. Used principally as a pressure generator or pump.

Freezer: The heat exchanger provided in a refrigerator or heat pump to facilitate the transfer of heat to the working fluid from an external low-temperature source.

Gifford–McMahon Engine: A regenerative expansion engine to generate refrigeration at cryogenic temperatures. Valves regulate flow of compressed gas to and expanded gas from the expansion cylinder.

Harmonic Piston Motion: The near sinusoidal motion of the pistons and displacers used in practical Stirling engines.

Heater: The heat exchanger provided in a prime mover to facilitate the transfer of thermal energy from an external source to the working fluid.

Heat Pipe: A device used in an indirect heating system in which an intermediate fluid is used to transfer heat from an external energy source to the working fluid. Usually the intermediate fluid a liquid metal, i.e., solium) is evaporated at the thermal inlet and condenses at the thermal outlet. Large rates of heat transfer can be effected with minimal temperatures differences.

Heat Pump: A machine driven from external power supply absorbing heat at ambient temperature and rejecting the heat at some higher temperature.

Heylandt Crown: The addition of an extension to a reciprocating piston to remove the hot or cold fluid from the region where the piston rings and seals operate.

Hybrid Free-Displacer–Crank-Controlled Piston Engine: A form of Stirling engine where the reciprocating piston has kinematic coupling to a rotating shaft but the displacer is oscillated under the action of fluidic forces.

Indirect Heating: A system in which thermal energy from an outside source heats an intermediate fluid (i.e., sodium) which conveys the energy to the heater tubes and hence to the working fluid (see Heat Pipe).

Intercooler: Water- or air-cooled heat exchanger used to cool compressed fluid between stages in a multistage compressor.

Intermediate (Capacity) Cryocooler: Cryocooler having a refrigerating capacity less than 25 W at 1 K, 100 W at 4 K, 1 kW at 20 K, or 15 kW at 80 K.

Isentropic Process: Thermodynamic process of volume, pressure, and temperature change that takes place at constant entropy.

Isobaric Process: See Constant-Pressure Process.

Isometric Process: See Constant-Volume Process.

Isothermal Compression and Expansion: The process of volume and pressure change that occurs without change in the temperature of the system.

Isothermal Process: See Constant-Temperature Process.

Joule, J.P.: English scientist of the 19th century best known for understanding that energy can be transformed from one type to another, but not created or destroyed.

Joule–Brayton Cycle: An idealized thermodynamic cycle comprising adiabatic compression and expansion separated by constant-pressure heating and cooling processes.

Joule Cycle: See Joule–Brayton cycle.

Joule–Thomson Expansion: Expansion of fluid to a low pressure constricted so that it occurs slowly with much frictional dissipation of energy. An irreversible thermodynamic process occurring at constant enthalpy. Much used for the final stage expansion in gas liquefaction.

JT Valve: A valve designed to accomplish isenthalpic Joule–Thomson expansion.

Kapitza, Peter: Russian cryogenic scientist of the 20th century. First used gas-lubricated piston in Claude cycle helium liquefier and discovered thermal boundary resistance to liquid helium.

Kinematic Drive: A system of cranks, connecting rods, levers or swashplates used to regulate and control the reciprocating motion of pistons or displacers and to convey power between the pistons and drive shafts.

Large (Capacity) Cryocooler: Cryocooler having a refrigerating capacity exceeding 25 W at 1 K, 100 W at 4 K, 1 kW at 20 K, or 15 kW at 80 K.

Linde, Karl von: German scientist of the 19th century best known for first liquefaction of air in quantity.

Linde Cycle: See Linde–Hampson cycle.

Linde–Hampson Cycle: A cryocooler process for gas liquefaction with isentropic compression and isenthalpic (JT) expansion separated by constant pressure heating and cooling in recuperative contraflow heat exchangers.

Metallurgical Limit: The maximum temperature of operation for the materials used in the hot spaces of the engine.

Microminiature (Capacity) Cryocooler: Very small cryocooler having a refrigerating capacity less than $\frac{1}{4}$ W at 20 K or 1 W at 80 K.

Miniature (Capacity) Cryocooler: Small cryocooler having a refrigerating capacity less than $\frac{1}{2}$ W at 4 K, 2 W at 20 K, or 8 W at 80 K.

Multifuel Capacity: The ability of an engine to operate on various fuels or energy sources.

Oil-Flooding: The process of adding oil to a compressor to prevent over-heating and to assist cooling.

Phase Angle: The angle by which volume variations in the expansion space lead those in the compression space.

Piston: A heavy structural reciprocating element of a Stirling engine characterized by a large pressure difference but a negligible temperature difference across the upper and lower transverse faces.

Porosity: The total volume of void volume expressed as a fraction of the volume envelope of the porous solid (frequently expressed also as a percentage).

Postle, Davy: Nineteenth century Australian inventor of the Postle refrigerator.

Postle Engine: A form of regenerative free-displacer refrigerator with self-acting valves regulating admission and exhaust of fluid in the expansion cylinder (invented about 1873).

Precooled Cycle: The process of adding supplementary refrigeration from an external source to assist the cooling of compressed fluid *en route* to an expansion engine or JT valve. Used in multistage gas liquefiers operating on the Linde–Hampson or Claude cycles.

Pressure Drop, Pressure Loss: The difference in pressure that arises when fluid flows through a duct or heat exchanger because of aerodynamic friction effects.

Pressure Excursion: The range of variation of the cylical pressure change of the working fluid in the cylinder.

Pressure Ratio: The ratio of the maximum and minimum pressures of the working fluid.

Prime Mover: A Stirling engine used to produce mechanical work from heat supplied at high temperatures.

Rallis Cycle: An idealized thermodynamic cycle with regenerative processes that occur partly at constant volume and partly at constant pressure. The process of compression and expansion may occur isothermally or adiabatically.

Reciprocating Machine: Compressor or expander with reciprocating pistons operating in cyclinders.

Recuperator (Recuperative Heat Exchanger): A form of heat exchanger (tube and shell, or finned tube) with separate channels for the hot and cold fluids. Usually the flow is continuous and constant in the channels.

Refrigerating Capacity: The rate of refrigeration generated by a cryocooler measured in watts.

Refrigeration Load: The extra refrigeration required following the addition of a detector element and its associated leads to the coldfinger of a cryocooler.

Refrigerator Temperature: The temperatures at which the refrigeration generated by a cryocooler is available.

Regenerative Annulus: A narrow annular gap between the displacer and cylinder through which the working fluid passes *en route* from the expansion or compression spaces. There is a temperature difference along the length of the annulus and as the gas passes through, a measure of regenerative heat exchange is accomplished.

Regenerative Cycle: A thermodynamic cycle in which some attempt is made to utilize the heat in the fluid being rejected from the cycle at low temperatures to heat the incoming fluid and so reduce the amount of "new" heat required and hence improve the efficiency of the cycle. The regenerative action may take place periodically as in the Stirling engine or continuously as in the Brayton cycle gas turbine. In the latter case, the heat transfer unit which accomplishes the regenerative action may be either a regenerative or a recuperative heat exchanger. Great care must be exercised to avoid confusion when discussing exhaust gas heat exchangers for regenerative thermodynamic cycles.

Regenerative Matrix: A porous volume of finely divided material (usually metallic) contained in the working space between the compression and expansion spaces. It acts as a reservoir of thermal energy.

Regenerator (Regenerative Heat Exchanger): A form of heat exchanger consisting of a porous solid mass with a single set of flow passages through which pass periodic, alternate flows of hot and cold fluids.

Regulation: The process of temperature or power control used to regulate the output of a Stirling engine.

Reitlinger Cycle: Generalized thermodynamic ideal cycle with isothermal compression and expansion processes at different temperatures bounded by regenerative processes of any nature.

Rhombic Drive: A special kinematic drive for Stirling engines which regulates the motion of the piston and displacer in single-acting-type engines. It is possible to achieve perfect dynamic balance while operating the reciprocating elements at the required phase difference. There are no side forces on the cylinder walls.

Roll-Sock Seal: A rolling diaphragm seal developed by Philips for containing the working fluid in the working space.

Rotary Machine: Compressor or expander with no reciprocating parts.

Schmidt Cycle: An idealized thermodynamic cycle for Stirling engines with sinusoidal volume variation of the isothermal compression and expansion spaces at different temperatures.

Screw Compressor: A form of fluid compressor with two long contrarotating rotors with meshing lobes.

Shuttle Heat Transfer: Heat transfer similar in effect to conduction heat transfer arising from the displacer reciprocating in a cylinder with the result that surfaces at different temperature levels are put close together and heat flow facilitated.

Siemens, Karl Wilhelm: An inventor extraordinaire, born German and became naturalized English, Sir Charles William Siemens. Credited with conception of the contraflow heat exchanger, the multiple-cylinder Stirling engine with adjacent cylinders coupled, and much else.

Siemens Cycle: An idealized thermodynamic cycle with isothermal compression, isentropic expansion separated by constant-pressure cooling and heating processes in contraflow recuperative heat exchangers.

Siemens Engine: An arrangement of three or more cylinders for a Stirling engine in which the cylinders are interconnected so that only one reciprocating element is required per Stirling system.

Single-Acting Engine: A family of Stirling engines with two reciprocating elements per thermodynamic system.

Small (Capacity) Cryocooler: Cryocooler having a refrigerating capacity less than 1 W at 1 K, 10 W at 4 K, 100 W at 20 K, 0.8 kW at 80 K.

Space Power System: An energy conversion device used to provide power for spacecraft.

Sting: See Coldfinger.

Stirling Cycle: An idealized thermodynamic cycle consisting of isothermal compression and expansion processes at different temperatures bounded by constant-volume regenerative processes.

Swash-Plate Drive: A system used in double-acting Siemens-type Stirling engines for regulating the motion of the displacer pistons and transmitting power to the drive shaft. The pistons are connected to an inclined disk on a rotating shaft which causes the pistons to reciprocate as the disk rotates.

Swept Volume Ratio: The volume variation in the compression space expressed in terms of the volume variation in the expansion space.

Temperature Ratio: The ratio of the temperatures of the working fluid in the compression and expansion space.

Thermal Efficiency: The fraction of total heat supplied that is converted to useful work.

Thomson, William (later Lord Kelvin): renowned English scientist of the 19th century.

Total Working Space: See Working Space.

Turbocompressor: Rotary compressor of the centrifugal or axial flow variety.

Turboexpander: Rotary expander of the centrifugal or axial flow variety.

Two-Phase Two-Component Working Fluid: See Compound Working Fluid.

VM Cooler: A Vuilleumier engine.

Void Volume: The total volume of the void spaces in the working space of a Stirling engine including the porous volume of the regenerator and the associated heat exchangers and connecting ducts or ports.

Volume Compression Ratio: The ratio of the maximum and minimum volumes of the total working space.

Vuilleumier, Rudolph: Inventor of the engine which bears his name. An American citizen, resident in New York at the time his patent was granted in 1918.

Vuilleumier Engine (also VM engine): A regenerative cryocooler sometimes described as a Stirling engine with a thermal rather than mechanical compressor. Pressure perturbations are generated in a large hot cylinder by the motion of a displacer. Refrigeration is generated in a small cold cylinder connected to the hot cylinder and utilizing the pressure perturbations.

Wobble-Plate Drive: See Swash-Plate Drive.

Work Done: The work done by or on the working fluid during a change in volume.

Working Fluid: The gas, liquid, or vapor which experiences periodic compression and expansion at different temperatures in the working space of a Stirling engine.

Working Space: The ensemble of variable volumes and constant volumes comprising the Stirling engine system, including an expansion space, a compression space, void volumes of the regenerator, heater, cooler, and the volumes of clearance spaces and connecting ducts or ports.

Appendix II

Organizations Having Substantial Interest in Cryocoolers and Cryocooler Manufacturing

ORGANIZATIONS

AFFDL: Air Force Flight Dynamics Laboratory at WPAFB.
ERDA: Energy Research and Development Administration (now the Department of Energy), Washington, D.C.
GSFC: NASA Goddard Space Flight Center, Greenbelt, Maryland.
JPL: Jet Propulsion Laboratory, California Institute of Technology, Pasadena, California.
NASA: National Aeronautics and Space Administration.
NEL: Naval Engineering Laboratory, Washington, D.C.
NVL: U.S. Army Night Vision and Electro-Optics Laboratory, Fort Belvoir, Virginia.
NWRL: Naval Weapons Research Laboratory, China Lake, California.
RRE: Royal Radar Establishment, Malvern, Worcester, England.
SAMSO: U.S. Air Force Space and Missile Systems Command, Los Angeles, California.
WPAFB: U.S. Air Force, Wright–Patterson Air Force Base, Dayton, Ohio.

MANUFACTURERS

ADL: Arthur D. Little Incorporated, Cambridge, Massachusetts.
AiResearch Manufacturing: Torrance, California.
CTI: Cryogenic Technology Incorporated, Waltham, Massachusetts.
Cryomech: Jamestown, New York.
ERG: Energy Research and Generation, San Francisco, California.

HAC: Hughes Aircraft Company, Culver City, California.

MM: Martin Marietta Corporation, Orlando, Florida.

Malakar Labs: now defunct, High Bridge, New Jersey.

Magnavox: Philips Company in the United States.

Hymatic Engineering Ltd: Redditch, Worcester.

NAP: North American Philips, Briarcliff Manor, New York.

PRL: Philips Research Laboratories, Briarcliff Manor, New York; also Eindhoven, Holland.

SBRC: Santa Barbara Research Center, Santa Barbara, California.

TI: Texas Instruments, Dallas, Texas.

Appendix III

Guide to the Cryogenic Engineering Literature

INTRODUCTION

The literature of cryogenic engineering is surprisingly diverse and much is concerned with cryocoolers. It can be divided into two principal groups: (a) government reports and (b) open literature.

It is important to have access to both, for many government reports are never summarized in the open literature. Furthermore, many of the reports, although unclassified, are restricted in circulation to agencies of government and their contractors. Still others are classified at various levels of restriction.

GOVERNMENT REPORTS

Government reports on cryocoolers are, in the main, those of contractors to the government agencies, principally, NASA, the Army, the Navy, the Air Force, and the Department of Energy. These are concerned mostly with small or miniature cooling systems for infrared thermal sensing and other electronic applications. The main contractors include: Hughes Aircraft Co., Texas Instruments Inc., Martin Marietta Co., Cryogenic Technology Inc., Philips Laboratories, R.C.A., AiResearch Manufacturing Corp.

United States government reports can sometimes be obtained simply by requesting a copy from the contractors or the department of government concerned. This only works when the request is made soon after the report is published and spare copies are likely to be on hand. One must, therefore, know the report is due, but one then is probably already on the distribution list and receives the report anyway. An unofficial approach is always worth a try, particularly where a report is required urgently, but the chance of a favorable response declines exponentially with elapsed time following publication.

The official sources of United States government reports are

(a) National Technical Information Service (NTIS)
    U.S. Department of Commerce
    Springfield, Virginia 22161
(b) Defense Documentation Center (DDC)
    Defense Logistics Agency
    Cameron Station
    Alexandria, Virginia

NTIS handles the distribution of copies of all U.S. government unclassified, unrestricted reports in all fields. DDC supplies copies of reports on defense and security related matters at all levels of security classification to approved requesters. Only approved "DDC Users" will likely have success in requesting material from DDC. One becomes a "DDC User" by working for U.S. Government agency or government contractor on defense related matters.

NTISearches

In addition to simply filling orders for copies of reports, NTIS provides another important service. They will generate bibliographies on specified topics of reports that they have on file. The NTIS data base includes over 800,000 document/data records covering U.S. government sponsored research from 1964. Several hundred bibliographies have already been assembled and are listed in the NTISearch Subject Index, obtainable on request from NTIS.

In the subject field of our interest here, cryocoolers, three NTISearches already exist: Reed 1975, 1977, and 1978. All are by William Reed and entitled *Cryogenic Refrigeration—A Bibliography with Abstracts.* Volume 1 covers the period 1964 to 1972. Volume 2 covers the period 1973 to 1977. Volume 3 covers the period 1977 to 1978.

Superintendent of Documents (SupDocs)

Another U.S. government agency responsible for the dissemination of government information is the Superintendent of Documents. This is the sales organization of the U.S. Government Printing Office (GPO). Various reports from government research laboratories are published by the GPO and distributed through SupDocs. They do not handle government contractor reports nor is there any obligation to maintain stocks of particular items, so that frequently one's requests are met with an "out-of-print" response. NTIS handles the same material as SupDocs plus of course the government contractor reports.

The Cryogenic Data Center

Another U.S. government source that was important to those interested in cryogenic information was the Cryogenic Data Center at the laboratories of the National Bureau of Standards, U.S. Department of Commerce, Boulder, Colorado. Over the past 20 years the Cryogenic Data Center had amassed the world's best collection of cryogenic literature. The material was all included in a computerized data base system organized to facilitate subject, author or chronologic searching. This system has been recently turned over to Prof. H. Weinstock, Illinois Institute of Technology, Chicago, Illinois, for continuance.

Conference Proceedings

From time to time agencies of the U.S. government organize conferences dealing with cryocoolers on matters relating thereto.

The most recent open meeting (October 1977) was held at the National Bureau of Standards, sponsored jointly by the Bureau and the Office of Naval Research, Arlington, Virginia. The proceedings (Zimmerman, 1978) of the conference, edited by James Zimmerman and Thomas Flynn, both of the Boulder laboratories, have been published as NBS Special Publication 508, available from the superintendent of Documents, and NTIS (or simply by writing to the editor at NBS, Boulder). These proceedings are required reading for anyone interested in the contemporary status of cryogenic cooling systems for small superconducting or other low-capacity electronic applications.

Infra-Red Information Symposia (IRIS) are held annually (about May/June) at various centers in the United States. The symposia are a joint service-classified meeting organized by the Office of Naval Research devoted to military applications of infrared radiation. The proceedings of the symposia are edited by the Infrared Information and Analysis Center, Environmental Research Institute of Michigan, P.O. Box 8618, Ann Arbor, Michigan. The proceedings of the 1977 San Francisco IRIS meeting (Zissis, 1978) were published as *Proc. IRIS*, Volume 22, NAV SO-P 2315, February 1978, a volume of 810 pages.

Various other government-sponsored symposia have been held from time to time. Although not specific to cryocoolers the proceedings do contain significant contributions of interest. The Office of Naval Research organized a workshop on naval applications of superconductivity in November 1970 at Panama City, Florida. The proceedings (Cox and Edelsack, 1971) contain an authoritative review of cryogenic refrigerators by John Daunt, Professor of Physics, at the Stevens Institute of Technology, Hoboken, New Jersey.

Foreign Government Sources

The above review of government reports and information sources refers entirely to U.S. government activities. This comes about because of my close proximity to that country and the wonderfully refreshing American characteristic of frank, open disclosure. I know there must be corresponding efforts in cooler developments taking place in other parts of the world for both military and civil applications but am unaware of the details. I shall, therefore, be grateful to readers if they would send me any information or otherwise draw my attention to these deficiencies that I may seek to rectify in a subsequent edition (should the Editors believe it worthwhile!)

OPEN LITERATURE SOURCES

Advances in Cryogenic Engineering

The principal body of cryogenic engineering literature is exceedingly well organized. The jewel in the crown is the 25 or so volumes entitled *Advances in Cryogenic Engineering*, edited by Klaus Timmerhaus, Associate Dean of Engineering at the University of Colorado, Boulder, Colorado, and, more recently, by Ronald Fast of the Fermi National Accelerator Laboratory, Batavia, Illinois. These volumes are well presented, uniformly bound books containing the proceedings of the annual (now biannual) Cryogenic Engineering Conference extending back to the first conference in 1954. Volume 20 contains a complete subject/author index for the series up to that time. It is the first reference to be consulted when approaching a new topic in cryogenic engineering.

Cryogenics

The other important reference collection is the international journal *Cryogenics* published in Great Britain. This started in 1960 under the guidance of Professor K. Mendelssohn of the University of Oxford Clarendon Laboratory. *Cryogenics* publishes a broad range of papers on research topics, general topical and book reviews, reports of conferences, and notices of interest to the cryogenics community.

International Cryogenic Engineering Conference

The proceedings of the biannual International Cryogenic Engineering Conferences are sufficient in number (6 volumes in 1979) to be an important reference source, particularly for material emanating from sources outside the United States. The conference schedule now appears to have settled down to a harmonious relationship with the U.S. based Cryogenic Engineering Conference. Both are now biannual events, one occurring in years alternate to the other.

Applications of Cryogenic Technology

Proceedings of the conferences of the Cryogenic Society of America are published as *Applications of Cryogenic Technology*, volumes 1 through 7 (in 1979). The books are edited by Robert W. Vance of the Aerospace Corporation in Los Angeles. They are very well presented, and the papers contained therein are of exceptional interest for they tend to be lengthy reviews written by experts in their field.

International Institute of Refrigeration

The International Institute of Refrigeration (IIR), located in Paris, is the oldest organization devoted to refrigeration technology. Presently the main meetings of the Institute are held every four years and are called the International Congress of Refrigeration. The proceedings are published (in three or four volumes) under the title of *Progress in Refrigeration Science and Technology*.

The IIR is organized in ten interest groups, called Commissions, as follows:

1. Cryophysics and Cryoengineering
2. Heat and Mass Transfer
3. Refrigerating Machinery
4. Refrigeration of Perishable Produce
5. Cold-Storage Facilities
6. Air Conditioning
7. Refrigerated Land Transport
8. Refrigerated Sea Transport
9. Applications of Refrigeration to Chemical, Civil, and Industrial Engineering
10. Cryobiology and Freeze Drying

For the XIII Congress, the Proceedings were organized in four volumes.

Volume 1 dealt primarily with low-temperature applications and contained the papers of Commissions 1 and 9. Volume 2 consisted of the papers of Commissions 2 and 3. Volume 3 covered the papers of Commissions 4, 5, and 10. Volume 4 contained the papers of Commissions 6, 7, and 8.

In addition to the regular quadrennial Congress meetings, the various Commissions meet either singly or in combinations having mutual interests to consider specific topics or a range of topics. For example, one meeting of Commission 1 in 1970 addressed itself to: measurement of temperature, He^3/He^4 refrigeration, heat transfer to liquid gases, superconductivity, and experimental techniques. In London in 1969, one meeting of Commission 1 was concerned with liquid natural gas; simultaneously, but at another separate meeting in London, the topic was low temperature and electric power. The proceedings of all these meetings are published by the IIR as a supplement or annex of the *Bulletin*, a regular bimonthly publication of about 250 pages including review papers, original studies, information on current research in refrigeration, information on forthcoming meetings of interest, and the program abstracts of papers of the next Congress.

In addition to all these, the IIR has many other important publications including three *Bibliographic Guides to Refrigeration* for the periods 1953–1960, 1961–1964, and 1965–1968, an international dictionary of refrigeration, and many different charts and tables of thermodynamic properties.

The veritable plethora of activities, interests, and publications with many variations due to the input of local characteristics makes recovery of publications somewhat more difficult. To obtain a list of publications or further information, application may be due to:

Institut International du Froid
(International Institute of Refrigeration)
177 boulevard Malesherbes
F 75017 Paris
France

Low-Temperature Physics

The cryogenics field has been a playground of the physicist for a hundred years, four or five times as long as it has been a topic of substantial engineering interest. Therefore, one should not overlook the physics sections of the library. Many choice items of great interest to cryogenic engineers will be found there. For example, Daunt has given an excellent and very comprehensive treatment, 136 pages, a book in itself, in the *Handbuch der Physik*, Vol. XIV on "The Production of Low Temperatures down to Hydrogen Temperature."

Then there are the proceedings of the Conference of Low Temperature Physics. Meetings are held at three-year intervals in different parts of the world, with the proceedings of the conference produced by the local organizing group. The proceedings and sources known to the author are

LT 14 (1975) (Otaniemi, Finland) Ed. M. Krusius and M. Vuorio, American Elsevier Publishing Co., New York.

LT 13 (1972) (Boulder, Colorado) Ed. K. Timmerhaus, W. H. Sullivan, and E. F. Hammel, Plenum Press, New York.

LT 12 (1971) (Kyoto, Japan) Ed. E. Danda, Academic Press of Japan, Tokyo.

LT 11 (1968) (St. Andrews, Scotland) Ed. J. F. Allen, D. M. Finkyson, and D. M. McCall, University of St. Andrews, Printing Dept.

LT 10 (1967) (Moscow, USSR) Ed. unknown, USSR Academy of Sciences.

LT 9 (1964) (Columbus, Ohio) Ed. J. G. Daunt, D. O. Edwards, and Y. M. Milford, Plenum Press, New York.

LT 8 (1962) (London, England) Ed. R. D. Davies, Butterworths Press Ltd., London.

LT 7 (1960) (Toronto, Canada) Ed. G. M. Graham and A. S. Hollis-Hallett, University of Toronto Press.

Books, Monographs and Course Notes

In addition to the above, many excellent books too numerous to mention on cryogenics and experimental techniques may be found on the shelves of a good technical library. One volume of exceptional interest to the cryogenics engineer is the Russian work translated and published by Pergamon Press, entitled *Plant and Machinery for the Separation of Air by Low-Temperature Methods*, Ed. I. P. Usyukin.

Occasionally the proceedings of summer courses or workshops may be published in book form. One example of great interest is *The Science and Technology of Superconductivity*, Ed. W. D. Gregory, W. N. Mathews, and E. A. Edelsack, published by Plenum Press, New York, 1973, based on a summer course held in August 1971 at Georgetown University, Washington, D.C.

The proceedings of the 1968 summer study on superconducting devices and accelerators was published by the Brookhaven National Laboratory Associated Universities Inc. as BNL 50155 (C-55) under contract with the United States Atomic Energy Commission.

Again the proceedings of a Cryogenic Workshop, March 1972, were published by the NASA George C. Marshall Space Flight Center. This is an excellent summary of contemporary cryogenic matters with particular reference to spacecraft.

House Journals

Two house journals have, over the years, contributed very substantially to the cryogenic engineering literature, particularly with regard to cryogenic cooling systems.

One of these two journals is the *Philips Technical Review*. Since the first publication, by J. W. L. Köhler in 1954, of two articles about the new Philips Stirling cooling engines, there have been many other papers dealing with subsequent developments and improvements. The *Philips Technical Review* is an excellent journal devoted to research topics and new product development in the Philips groups of companies. Subscriptions or single copies of the *Review* may be obtained on application to

N.V. UITGEVERSMAATSCHAPPIJ CENTREX
(Centrex Publishing Co.)
N.W. Emmasingel 9
P.O. Box 76
Eindhoven, Netherlands

Another journal of exceptional relevance and interest to the field of cryocoolers is the *Linde Reports in Science and Technology*. This monthly publication contains six to ten articles describing recent Linde developments and plant construction in the field of cryogenic gas processing and chemical engineering. An index to the articles or copies of the reports may be obtained on application to

Pressestelle der Linde Aktiengesellschaft
Wiesbaden, Hildastrasse 2-10
West Germany.

Name Index

A. D. Little Company, [1] 10, 12, 14, 237-240
AEG-Telefunken A. G., [1] 14
AiResearch Manufacturing Company, [1]
    16, 199-202; [2] 387
Air Products and Chemical Incorporated
    (APCI), [1] 15, 77, 238, 240, 293
American Motors, [1] 111

Baumann Institute of the Moscow High
    Technological School, [2] 262
Beale, William, [1] 52, 166
British Oxygen Company, [1] 8, 19
Bush, Vannevar, [1] 188

Carnot, Sadi, [1] 39
Chellis, Fred, [1] 171
Claude, Georges, [1] 8, 9, 322, 323; [2] 375
Collins, Samuel, [1] 9, 10, 326-329; [2] 376
Cowans, Ken, [1] 14
Crummett, Charles, [1] 9
Crummett, Orin, [1] 9
Cryogenic Data Center, U.S., [2] 389
Cryogenic Society of America, [2] 391
Cryogenic Technology Incorporated (CTI),
    [1] 10, 12, 14, 15, 100, 103, 238; [2] 387
Cryomech Incorporated, [1] 14, 238

Daniels, A., [1] 101
Daunt, John, [1] 14
Davis, Harvey, [1] 9
Defense Documentation Center (DDC),
    U.S., [2] 388
du Pre, F. K., [1] 96, 101

Energy Research and Generation
    Incorporated (ERG), [1] 20

Fairchild Space and Electronics Company,
    [1] 150
Finkelstein, Theodore, [1] 109, 131, 142, 147
Flight Dynamics Laboratories, [1] 13
Ford Motor Company, [1] 111
Franchot, Charles Louis, [1] 109

General Motors, [1] 111
Gifford, William, [1] 14, 237-263
Gorrie, John, [1] 8

Hampson, W, [1] 8
Harwell Atomic Energy Establishment, [1]
    108
Herschel, John, [1] 6, 95
Heylandt, D., [1] 8
Higa, Walter, [1] 16, 171
Horn, Stuart, [1] 171
Hughes Aircraft Company, [1] 14, 103,
    191-193; [2] 387
Hughes Santa Barbara Research Center, [1]
    14
Hymatic Engineering Limited, [1] 19,
    288-290

International Institute of Refrigeration, [2]
    391-392

Subject Index